Law and Agroecology

Massimo Monteduro • Pierangelo Buongiorno •
Saverio Di Benedetto • Alessandro Isoni
Editors

Law and Agroecology

A Transdisciplinary Dialogue

Springer

Editors
Massimo Monteduro
Dipartimento di Scienze Giuridiche
Università del Salento
Lecce
Italy

Pierangelo Buongiorno
Dipartimento di Scienze Giuridiche
Università del Salento
Lecce
Italy

Institute for the History of Law
University of Münster
Münster
Germany

Saverio Di Benedetto
Dipartimento di Scienze Giuridiche
Università del Salento
Lecce
Italy

Alessandro Isoni
Dipartimento di Storia
Società e Studi sull'Uomo
Università del Salento
Lecce
Italy

ISBN 978-3-662-46616-2 ISBN 978-3-662-46617-9 (eBook)
DOI 10.1007/978-3-662-46617-9

Library of Congress Control Number: 2015939279

Springer Heidelberg New York Dordrecht London

Printed on acid-free paper

Springer-Verlag GmbH Berlin Heidelberg is part of Springer Science+Business Media (www.springer.com)

Preface

This collective volume is the first outcome of an experiment in transdisciplinary scientific research started in 2012 with the creation at the University of Salento (Italy) of a group of young researchers called LAIR (an acronym for *Law and Agroecology – Ius et Rus*), and continued in 2013 with the organization of an International Conference in Lecce entitled *Agroecology and Law: A Transdisciplinary Dialogue*.

The research was motivated by a growing awareness of profound changes in the socioeconomic paradigm that have taken place in agriculture. Agriculture has evolved from the monofunctional perspective, referring exclusively to the production of goods for private use (raw materials to be used for food or industrial purposes) and to the remuneration of producers for those goods, towards a multifunctional vision. It is recognized that agriculture provides fundamental ecosystem services, inspired by the principle of sustainable development and conforming to the rule of environmental cross-compliance.

This process of transformation has been accompanied by the emergence of a vibrant and expanding field of international research, namely agroecology.

Agroecology has progressively integrated the points of view of various disciplines: agronomy, ecology, environmental sciences, geography, sociology, anthropology, history, economics, ethics, and political science. Agroecology has evolved through overcoming the traditional frontiers between "natural" and "social" sciences and examining the concept of agroecosystem viewed as a socio-ecological system.

Law, however, has remained separate and very far from the debate within agroecology.

This volume proposes to explore, for the first time in a direct and broad-spectrum way, the relationship between law and agroecology. These two branches of knowledge that hitherto have not really communicated with each other are now called upon to become reciprocally acquainted, giving rise to a process of coevolution.

On the one hand, agroecology is called upon to integrate within itself the point of view of law. This means studying the complexities of agri-food systems also in the light of normative and institutional variables, with the lens of categories such as rights, duties, powers, responsibilities, and procedural safeguards. On the other hand, law is called upon to review its own "internal geometries," confronting them with the agroecological paradigm. In this sense, it must address the necessity of overcoming the divisive approach that so far has kept separate, on the disciplinary level, agricultural law and environmental law and, more generally, has disarticulated the legal regulation of closely linked matters, such as agriculture, environment, landscape, and food.

This volume intends to be the first moment of an open exchange between juridical and nonjuridical systems of thought with regard to agroecology. At the same time, it deals with the experiences of different countries, in order to start up a fertile dialogue destined to continue into the future.

On the level of the academic training of the authors, the approach based on transdisciplinarity explains why in this volume are included, besides legal scholars, also scholars of ecology, landscape ecology, agronomy, food governance, chemistry, engineering, history of agroecosystems and political institutions, rural sociology, and ethics. Among the legal scholars are representatives of various fields: from Roman law to international and comparative law; from constitutional, public, and administrative law to private and agricultural law; from environmental and landscape law to consumer law. The authors come from nine different countries (Italy, Denmark, France, Greece, Hungary, the Netherlands, Spain, the United Kingdom, and the United States of America).

On a structural level, the volume is composed of three parts. The first part addresses the methodological issues entailed in linking agroecology to law. The second part aims to identify some concrete challenges that agroecology presents to law, highlighting the correspondence between multifunctionality of agriculture and multidimensionality of the relationships between land, agriculture, and the environment on legal and scientific levels. The third and final part focuses on sustainable rural development and on rural civilization as paradigms in the new agroecological approach.

The volume is woven round a key concept that is ancient and modern at the same time, namely the concept of *rus*. *Rus* marks the plurality and the interdependence of different complex systems based jointly on the land as a central point of reference. "Rural" is more than "agricultural." If agriculture is traditionally understood as an activity aimed at exploiting the land for the production of material goods for use, consumption, and private exchange, rurality marks the collocation of agriculture into a wider sphere, which is not only economic but also social and cultural; not only material but also ideal, relational, historic, and symbolic; not only private but also public.

There is a need for a transdisciplinary approach in order to integrate agroecological paradigm in legal regulation: but this does not require a super-law that hierarchically purports to incorporate and replace the existing legal fields. It requires constructing a trans-law that progressively attempts to coordinate interlegalities between different legal fields by respecting their autonomy but emphasizing their common historical roots in *rus*.

Lecce, Italy — Massimo Monteduro
Lecce, Italy / Münster, Germany — Pierangelo Buongiorno
Lecce, Italy — Saverio Di Benedetto
22 December 2014 — Alessandro Isoni

Contents

About the Editors

Massimo Monteduro—Associate Professor of Administrative Law at the Department of Legal Sciences, University of Salento. Formerly Full Researcher and Assistant Professor of Administrative Law and Environmental Administrative Law at the Department of Legal Sciences, University of Salento. In 2000, he graduated in Law, final mark 110/110 *summa cum laude*, at the LUISS Guido Carli (Free International University of Social Studies) in Rome. In 2005, he received the Ph.D. degree with distinction in "Economics and Market Law" by the Advanced School ISUFI (Institute of Advanced Interdisciplinary Studies) of the University of Salento. In 2006 and 2007, he was Visiting Professor at the University of Granada, Department of Administrative Law. Since 2006, he has lectured in courses for the University of Salento in Administrative Law and Environmental Administrative Law. In 2009, he carried out teaching activities at the National School of Public Administration (SSPA) in Rome. In 2013, he was Visiting Professor at the University of Santiago de Compostela, Law School. On 24/12/2013, he got the National Scientific Qualification as Associate Professor in Administrative Law. He is a member of the International Research Group GI-1137, *Derecho de las Administraciones y Entidades del Sector Público* (DAESP), established at the Department of Public Law and Theory of State of the University of Santiago de Compostela (DAESP includes professors and researchers from Spain, Portugal, Italy, Austria, Chile, and Colombia). He is a founding member of the Italian Association for Sustainability Science (IASS) and a member of the Italian Association of Urbanism and Town-Planning Law (AIDU). He is a cofounder of the Group LAIR "Law and Agroecology – Ius et Rus" established at the University of Salento. He has been invited as speaker in national and international conferences. He is the author of several national and international studies on administrative agencies, theory and practice of public organizations, administrative procedure, public contracts and utilities, environmental law, town planning law, innovative aspects of administrative law, interpretation of administrative acts. Among his recent publications are: Monteduro M (2012) Un nudo problemático en la transposición de la Directiva 2006/123/CE en Italia: la nueva institución de la indicación certificada de comienzo de actividad. In: Nogueira López A (ed) La termita Bolkestein. Mercado único vs. Derechos ciudadanos. Thomson Reuters Civitas – Editorial Aranzadi, Cizur Menor, Navarra, pp 385–452; Monteduro M (2012) Provvedimento amministrativo e interpretazione autentica. Vol I: Questioni presupposte di teoria del provvedimento. Cedam – Wolters Kluwer Italia, pp 1–338; Monteduro M (2013) Environmental Law and Agroecology. Transdisciplinary Approach to Public Ecosystem Services as a New Challenge for Environmental Legal Doctrine. European Energy and Environmental Law Review 22(1): 2–11.

Pierangelo Buongiorno—Researcher of Roman Law and Ancient Law and Assistant Professor of History and Institutions of Roman Law at the Department of Legal Sciences, University of Salento; Nachwuchsgruppenleiter at the Institute for the History of Law, University of Münster. He graduated in Classical Philology, final mark 110/110 *summa cum laude*, at the University of Lecce. He also graduated in Law, final mark 110/110 *summa cum laude*, at the University of Bari. Since September 2004, he has cooperated to the teaching and research activities of the Roman Law Chairs at the University of Lecce. He was a Ph.D. student in Ancient History at the University of Bari (2004–2007). During and after his doctoral studies, he spent several times at the Ruprecht-Karls-University of Heidelberg (SS 2005, 2006, 2007, 2009, 2011 and WS 2006, 2014). He obtained a research fellowship of the Annemarie-und-Elise-Jacobi-Stiftung, which he spent at the Kommission für Alte Geschichte des Deutschen Archäologischen Instituts of München (2008, January–February). He spent a research fellowship of the University of Bari at the Law Faculty of the Johannes-Gutenberg University of Mainz (2008, April–December). He was 'Wissenschaftliche Hilfskraft' of the Roman and Civil Law Chair of the same University (2008, April–July). He was Research Fellow (*Assegnista di Ricerca*) at the University of Salento with a research about 'The Normative Production in the I Century A.D.: the Relations between the Princeps and the Senate' (2009–2011). He spent a research fellowship at the IUSS (Pavia) for the 8th CEDANT (Collegio di Diritto Romano) about 'Leges publicae. La legge nell'esperienza giuridica romana,' under the direction of Prof. Jean-Louis Ferrary and Prof. Dario Mantovani (2010, January and September). He won the VIII International Romanistic Prize 'Gérard Boulvert' for his first book (2010). He spent a research fellowship in Princeton University (2013, April) and DAI-AEK München (2013, May). He was also Visiting Professor in Heidelberg (SS 2013 and 2014) and in Münster (WS 2014). On October 2013, he won the Excellence Prize of the University of Lecce as the best researcher for the Juridical Area. On 5/02/2014, he got the National Scientific Qualification as Associate Professor in Roman Law and Ancient Law. On August 2014, he received the 'Sofja Kovalevskaja Award' tendered by the Alexander von Humboldt Foundation. He is a member of the editorial board of the 'Quaderni Lupiensi di Storia e Diritto' and of some scientific associations. He has been an invited speaker in national and international conferences. He is a cofounder of the Group LAIR "Law and Agroecology – Ius et Rus" established at the University of Salento. Among his recent publications are: Buongiorno P (2010) *Senatus consulta Claudianis temporibus facta*. ESI, Napoli, p XII-540; Buongiorno P (2010) CIL X 1401 e il senatoconsulto Osidiano. IVRA

58: 234–251; Buongiorno P, Lohsse S (eds) (2013) *Fontes Iuris*: Atti del VI Jahrestreffen Junger Romanistinnen und Romanisten [Lecce, 30–31 marzo 2012]. ESI, Napoli.

Saverio Di Benedetto—Full Researcher of International Law and Assistant Professor of International Economic Law at the Department of Legal Sciences, University of Salento. He graduated in Law in 1998, final mark 110/110 *summa cum laude*, at the University of Pisa. In 2000, he obtained the master's degree (L.L.M.) in "Public and International Management and Trade Law" at the Advanced School ISUFI (Institute of Advanced Interdisciplinary Studies) of the University of Salento. In 2005, he received the Ph.D. degree with distinction in "Economics and Market Law" by the Advanced School ISUFI. In 2006, he was Visiting Professor of "International and Armed Conflicts Law" at the Italian Naval Academy (Accademia Navale Italiana) of Livorno. Since 2006, he has lectured in courses for the University of Salento in International Law and International Economic and Trade Law. In 2008, he was invited as participant to the Centre for Studies and Research in International Law of the Hague Academy of International Law; his research work was included in the Law Book of the Hague Academy on the Implementation of International Environmental Law. On 18/12/2014, he got the National Scientific Qualification as Associate Professor in International Law. He is a founding member of the Italian Association for Sustainability Science (IASS) and of the Association "The Group of Lecce" for Global Governance (www.thegroupoflecce.org). He is a cofounder of the Group LAIR "Law and Agroecology – Ius et Rus" established at the University of Salento. He was invited as speaker in several national and international conferences, including the Biennal Conference of the European Society of International Law held in 2008 at the University of Heidelberg. He is the author of several articles and essays published in national and international journals and volumes, concerning several subjects of international law, such as international investment law, international trade law, global governance, international environmental law. He is also a coeditor of collected volumes on different themes of international and EU law. Among his recent publications are: Di Benedetto S (2011) Le rôle des tribunaux CIRDI au regard de la mise en œuvre de la protection de l'environnement. In: Maljean Dubois S, Rajamani L (eds) La mise en œuvre du droit de l'environnement. Hague Academy of International Law – Martinus Nijhoff, The Hague, pp 537–578; Di Benedetto S (2013) Sustainable Development or Strong Growth? Perspectives in G8 and G20 Declarations on Global Economic Governance in Time of Crisis. In: Bossone B, Cafaro S, Di Benedetto S, Malaguti MC (eds) Legitimacy and Effectiveness in Global Economic Governance. Cambridge Scholars, Newcastle Upon Tyne, pp 343–368; Di Benedetto S (2013) International Investment Law and the Environment. Edward Elgar, Cheltenham, p XVI – 250.

Alessandro Isoni—Associate Professor of History of Political Institutions at the Department of History, Society and Human Studies, University of Salento. Formerly Assistant Professor of Constitutional History and Full Researcher of History of Political Institutions at the Department of History, Society and Human Studies, University of Salento. In 1999, he graduated in Political Sciences, final mark 110/110 *summa cum laude*, at the University of Sassari. In 2000, he received a master's degree with distinction in "International, Supranational and National Law of Market and Public Policies" by the Advanced School ISUFI (Institute of Advanced Interdisciplinary Studies) of the University of Salento. In 2005, he received the Ph.D. degree with distinction in "Economics and Market Law" by the Advanced School ISUFI (Institute of Advanced Interdisciplinary Studies) of the University of Salento. Since 2005, he has lectured in courses for the University of Salento in Administrative Law, History of the Italian Administration, and Public Law. On 24/12/2013, he got the National Scientific Qualification as Associate Professor in History of Political Institutions. He is a member of the editorial board of some reviews, like *Eunomia* and *Historia Magistra*. He has been an invited speaker in national and international conferences. He is a cofounder of the Group LAIR "Law and Agroecology – Ius et Rus" established at the University of Salento. He has been an author of several articles and studies about the political and historical causes of the ECSC High Authority, together with other studies concerning the European integration institutional features and, in addition, other studies devoted to the parliamentary select committees in Italy under the Albertine Statute and the Republican Constitution. Among his recent publications are: Isoni A (2012) Farewell to the European Community. The Lisbon Treaty and the Conceptual Shifts of a *Sui Generis* Public Law Experience. Eunomia 1(1):85–98; Isoni A (2012) Federalismo e sussidiarietà nel processo di integrazione europea. In: Tedoldi L (ed) Alla ricerca della statualità. Un confronto storico-politico su Stato, federalismo e democrazia in Italia e in Europa. QuiEdit, Verona, pp 181–200; Isoni A (2014) "*Fonder la République*". The French National Convention and the Revolutionary Government (1793–94). Parliaments, Estates and Representation 34 (1):40–54.

Contributors

Mariagrazia Alabrese Institute of Law, Politics and Development (Dirpolis), Sant'Anna School of Advanced Studies, Pisa, Italy

Hugo F. Alrøe Department of Agroecology, Aarhus University, Aarhus, Denmark

Mariacristina Antonucci Faculty of Law, Department of Legal Sciences, University of Salento, Lecce, Italy

Roberta Aretano Landscape Ecology Laboratory, Department of Biological and Environmental Sciences and Technologies (DISTEBA), University of Salento, Lecce, Italy

Marco Brocca Department of Legal Sciences, University of Salento, Lecce, Italy

Giuseppina Buia Faculty of Law, Department of Legal Sciences, University of Salento, Lecce, Italy

Fabio Caporali Department of Agriculture, Forest, Nature and Energy, University of Tuscia, Viterbo, Italy

Idelberto Castorini Landscape Ecology Laboratory, Department of Biological and Environmental Sciences and Technologies (DISTEBA), University of Salento, Lecce, Italy

Eloisa Cristiani Institute of Law, Politics and Development (Dirpolis), Sant'Anna School of Advanced Studies, Pisa, Italy

Antonella De Marco Landscape Ecology Laboratory, Department of Biological and Environmental Sciences and Technologies (DISTEBA), University of Salento, Lecce, Italy

Achille de Nitto Department of Legal Sciences, University of Salento, Lecce, Italy

Antonello Denuzzo Faculty of Educational, Political and Social Studies, University of Salento, Lecce, Italy

Elizabeth Dooley Ecologic Institute, Berlin, Germany

Manuel González de Molina Navarro Department of History, Geography and Philosophy, University Pablo de Olavide, Sevilla, Spain

Stefano Grego Department of Agriculture, Forestry, Nature and Energy, University of Tuscia, Viterbo, Italy

Carole Hermon Institute of Legal Studies for Urban Planning, Construction, and the Environment (IEJUC), Toulouse 1 Capitole University, Toulouse, France

Otto Hospes Department of Social Sciences, Public Administration and Policy Group, Wageningen University, Wageningen, The Netherlands

Brian Jack School of Law, Queen's University of Belfast, Belfast, UK

Emmanuel G. Koukios National Technical University of Athens, Bioresource Technology Unit, Athens, Greece

Massimo Monteleone STAR*AgroEnergy Group, Department of Agri-Environmental Sciences, Chemistry and Plant Protection, University of Foggia, Foggia, Italy

Vincenzo Naso Department of Mechanical and Aeronautical Engineering, University "La Sapienza" of Rome, Rome, Italy

Egon B. Noe Department of Agroecology, Aarhus University, Aarhus, Denmark

Maria Rita Pasimeni Landscape Ecology Laboratory, Department of Biological and Environmental Sciences and Technologies (DISTEBA), University of Salento, Lecce, Italy

Irene Petrosillo Landscape Ecology Laboratory, Department of Biological and Environmental Sciences and Technologies (DISTEBA), University of Salento, Lecce, Italy

Maurizia Pierri Department of Legal Sciences, University of Salento, Lecce, Italy

Teodoro Semeraro Landscape Ecology Laboratory, Department of Biological and Environmental Sciences and Technologies (DISTEBA), University of Salento, Lecce, Italy

János Ede Szilágyi Department of Agricultural and Labour Law, University of Miskolc, Miskolc, Hungary

Sara Tommasi Department of Legal Sciences, University of Salento, Lecce, Italy

Michele Troisi Department of Legal Sciences, University of Salento, Lecce, Italy

Scilla Vernile Department of Law, Bocconi University of Milan, Milan, Italy

Nicola Zaccarelli Landscape Ecology Laboratory, Department of Biological and Environmental Sciences and Technologies (DISTEBA), University of Salento, Lecce, Italy

Giovanni Zurlini Landscape Ecology Laboratory, Department of Biological and Environmental Sciences and Technologies (DISTEBA), University of Salento, Lecce, Italy

Part I
Law and Agroecology: Crossing the Boundaries Between Natural, Social and Legal Sciences

History and Development of Agroecology and Theory of Agroecosystems

F. Caporali

Abstract This work provides a knowledge contribution in order to understand agroecology as evolution of both a scientific discipline and a philosophical paradigm for promoting sustainability in agriculture. The peculiar character of agroecology as an applied, transdisciplinary science based on the systems paradigm is explored in its theoretical and practical foundations. The agroecosystem concept is regarded as an epistemological tool for creating an ontology or representation of agriculture based on a systems view. Hierarchy, emergence, communication and control are shown as agroecosystem properties. Integration is viewed as an ontological link operating in the construction of agriculture as a human activity system. Integration is regarded as an organisational capability for connecting different hierarchical levels, which is critical for achieving the goal of agriculture sustainability. Development of sustainability indicators is considered a crucial step of enquiry for providing elements of assessment, evaluation and anticipation of solutions for both farm design and management, and land use policy. Sustainability in agriculture will depend on the capacity of harmonic integration between contrasting trade-offs in the search of a balance among human nutrition, ecological integrity and economic development. Convergence of agroecological principles between global and local levels of planning will be a crucial component for success towards sustainable agriculture.

Keywords Agroecology • Agroecosystem • Integration • Sustainable agriculture

1 Introduction

Starting from a simple and neutral definition, Agroecology is the science of ecology applied to agriculture. Looking at its methods, contents, aims and meaning, Agroecology is today a transdisciplinary field of enquiry that is capable of changing our common vision of both agriculture and society.[1] Assuming the postmodern

[1] Caporali (2010), pp. 1–71.

F. Caporali (✉)
Department of Agriculture, Forest, Nature and Energy, University of Tuscia, Viterbo, Italy
e-mail: caporali@unitus.it, http://www.unitus.it/dipartimenti/dpv/Caporali.htm

M. Monteduro et al. (eds.), *Law and Agroecology*,
DOI 10.1007/978-3-662-46617-9_1

terminology of Michel Foucault,[2] Agroecology is indeed a new "episteme", i.e. an underlying code for understanding and making agriculture, that governs its language, its logic, its schemas of perception, its values and its techniques.[3] For making an episteme operational, language, biology and economy must co-evolve and penetrate society.[4] As suggested by Chandrasoma and Lee,[5] technology is another essential component for making an episteme operational in society because knowledge is essentially a culture-oriented entity firmly anchored in a system of technology at a given period of time.

As an applied science, Agroecology focuses both on theoretical principles and on their practical applications in order to inform design and implement solutions to real problems. Agroecology makes collective meaning and sense-making possible for legitimising and justifying managerial actions and decision-making in agriculture and society. It is possible to outline the contribution of Agroecology, regarded as the ontological link between ecological theory and practice of agriculture, within the following sequence of terms concerning the dialectics of an evolving agriculture:

Ecology (science) → Agriculture (practice, what is) → Agroecology (applied science) → Sustainable Agriculture (potential practice, what should be) → Organic Agriculture (niche practice, a response to what should be).

This sequence unveils how the science of ecology, once applied to agriculture (agroecology), has generated a set of values or paradigm for an ideal model of agriculture (sustainable agriculture) that is currently being implemented (organic farming) in compliance with a practical platform (standards) defined and supported by law.[6] On this base, Gliessman[7] was right in defining "Agroecology as the application of ecological concepts and principles to the design and management of sustainable agroecosystems", as well as Francis[8] in labelling "Organic Farming" as the "Ecological System".

Since its very beginning, the science of Agroecology has established its own peculiar epistemological tool, an input/output model for representing and analysing agriculture reality, that has been defined as "agroecosystem".[9] It provides a basic methodology of enquiry in agriculture at any level of hierarchical organisation, from the field, the farm, the landscape to the national and international levels of organisation. Within a hierarchy of agroecosystems, the farm level of organisation has been regarded as a meaningful one in that it represents the crossroads between farmers' interests and society's expectations in terms of agricultural performance.[10]

---

[2] Foucault (1966).

[3] O'Leary and Chia (2007), pp. 392–406.

[4] Foucault (1966).

[5] Chandrasoma and Lee (2012).

[6] Caporali (2004).

[7] Gliessman (1998).

[8] Francis (2009).

[9] Caporali (2006), pp. 415–429.

[10] Caporali et al. (1989), pp. 579–595; Caporali (2010).

As a consequence of its epistemological tool, i.e. the concept of agroecosystem, Agroecology is said to be an "integrative" science, i.e. a science that looks for relationships in order to promote more understanding of the agricultural reality and its context. According to its systemic character, Francis et al.[11] define Agroecology as the *integrative* study of the ecology of the entire food system, encompassing biophysical, economic and social dimensions. "When agroecology is defined as an integrative area of study of the ecology of the food system, it is essential for us to examine the production process as well as the local and the landscape ecological impacts, the broad economics of farms and communities, and other social dimensions of agriculture including involvement of families and organizations in the political process".[12]

In order to understand the role that the science of Agroecology has played in developing a new meaning of agriculture, it is useful to read the under reported definitions of both Agriculture and Agroecology provided by the authoritative IAASTD[13]—a panel of international experts for International Assessment of Agriculture Knowledge, Science and Technology—in the glossary of its Global Report "Agriculture at a Crossroads":

Agriculture. A linked, dynamic social-ecological system based on the extraction of biological products and services from an ecosystem, innovated and managed by people [. . .] It encompasses all stages of production, processing, distribution, marketing, retail, consumption and waste disposal.

Agroecology. The science of applying ecological concepts and principles to the design and management of sustainable agroecosystems. It includes the study of the ecological processes in farming systems and processes such as: nutrient cycling, carbon cycling/sequestration, water cycling, food chains within and between trophic groups (microbes to top predators), lifecycles, herbivore/predator/prey/host interactions, pollination, etc. Agroecological functions are generally maximized when there is high species diversity/perennial forest-like habitats.

> Within the "intellectual landscape of agroecology",[14] agriculture is re-conceptualised according to an ecosystem paradigm and reframed as a coupled system (an "agroecosystem") involving a wide range of social and natural processes.

We can say that Agroecology starts rooting in society when a field or a farm is viewed first as an ecosystem. The relevance of Agroecology emerges both at local and global levels. A review from a global-change perspective[15] argues that a key challenge within agroecology is "to coherently understand manifold relationships

---

[11] Francis (2004), pp. 21–43.

[12] Francis (2004).

[13] International Assessment of Agricultural Knowledge, Science and Technology for Development (IAASTD) (2006).

[14] Tomich et al. (2011), pp. 193–222.

[15] Tomich et al. (2011).

and feedbacks across two broad cross-disciplinary boundaries", that of connecting agricultural sciences and environmental sciences and that of connecting agricultural sciences and social sciences. Mitigating environmental impacts of agriculture while dramatically increasing global food production and improving livelihoods is a global challenge that requires an "agroecological nexus", i.e. the convergence of social and environmental forces shaping agricultural and land-use decisions.

Today, society as a whole seems to be both reactive and proactive to agroecological pressure, as shown by recent innovations at academic, political and practical levels.

At academic level, new curricula in both Agroecology and Organic Farming have been implemented worldwide.[16] At political level, the Rio UN Conference of 1992 on "Sustainable Development" yielded the most important international document or platform for decision-making processes at institutional level (Agenda 21), where many chapters—7, 8, 10, 14, 15, 16, 28, 31, 32—have been devoted to directly or indirectly promoting sustainable development in agriculture in accordance with agroecological principles.[17] At practical level, IFOAM (International Federation of Organic Agriculture Movements) has been successfully established since 1972 as a network of farmers and other stakeholders in the agriculture chain in order to promote the continuing development of organic farming worldwide. According to the current IFOAM's mission statement, the principles inspiring organic farming are those of *health*, *ecology*, *equity* and *care*, i.e. ethical principles supported by the science of ecology.

The aim of this contribution is to illustrate the many patterns of the *integrative character of Agroecology* and its potential for bettering both theory and practice of agriculture.

2 Epistemic and Ontology of Agroecology

The most relevant epistemic character of ecology is its methodology based on the systems paradigm, which is "the integrative principle" by definition. Indeed, a system is a functional unit, made up of parts or components, which are interdependent while operating in a broader context. The relationships between each system and its context of existence are hierarchically determined, according to the so-called *levels of organisations* that operate both up (*external*) and down (*internal*) its level of existence. In ecology, the elemental functional unity of study and representation of reality is defined as *ecosystem*. Its level of organisation concerns the relationships between the community of living beings and the environment they live in. Four properties characterise the systems paradigm: *hierarchy*,

[16] Francis et al. (2003), pp. 99–118; Caporali (2004); Altieri and Toledo (2011), pp. 587–612.

[17] El-Hage Scialabba and Hattam (2002) and Caporali (2006).

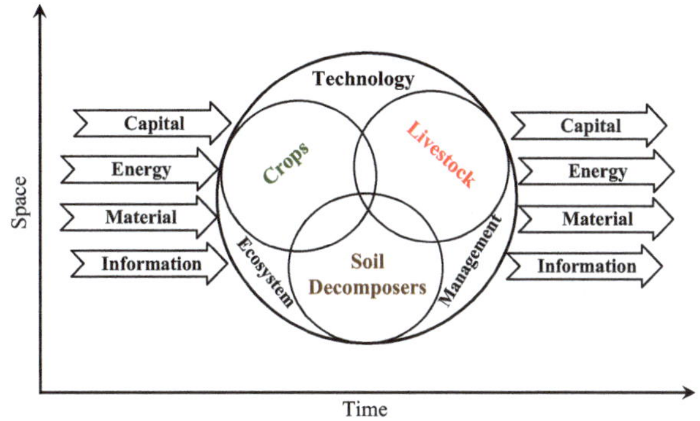

Fig. 1 Agroecosystem as an input/output model of agricultural reality (from Caporali 2010)

emergence, *communication* and *control*.[18] When applied to agriculture, the systems paradigm generates a representation model of agriculture or *agroecosystem*, which is both an epistemological and a scientific tool for enquiry at any spatial and temporal scale. In analogy with the ecosystem concept, the agroecosystem is both a real ecosystem modified and used for agricultural purposes and the model that represents it. In this sense, both ecology and agroecology exhibit a method that reflects contents, whereby ontology and epistemology coincide.[19] The general model for representing structure and functioning of an agroecosystem is based on an input/output functional scheme (Fig. 1), where socioeconomic and biophysical components are fully integrated in a process of continuing production and consumption that happens at any spatial–temporal scale. The agroecosystem concept is like a lens for focusing on rural reality at different levels of resolution. The systems paradigm is completely reflected in agroecological research and teaching.

2.1 Meaningful Historical Comparisons

According to Spedding,[20] who was a pioneer of systems thinking in agriculture with his book "The Biology of Agricultural Systems", the first step in understanding an agricultural system is to picture it in the mind and the second step is to describe it. The result of the whole process (or conceptualisation) is to establish a general picture (or concept) that is basic to all or to a large group of systems and that includes the following elements, as a minimum:

a) the *purpose* for which the system is being carried out;
b) the *boundary* that defines what is inside or outside the system;
c) the *context* or the external environment in which the system operates;

[18] Checkland (1993).
[19] Caporali (2006).
[20] Spedding (1975).

d) the main *components* that are involved to form the system;
e) the *relationships* between components;
f) the *resources* or internal components within the system that are used in its functioning;
g) the *inputs* or external resources that are used by the system;
h) the *outputs* (main, desired products or performances and by-products, useful but incidental).

The above list is itself a picture that can be improved in the layout of a diagram (or model) where items are related to each other to show how energy and information flow and matter circulates between inorganic and organic agroecosystem components in a general framework of purpose. The whole agroecosystem is usually represented with a network of functional components as boxes and processes of energy–matter–information transfer as arrows. Interesting is to note that Spedding's major focus is on both purpose and item relationships, especially the output of product per unit of resource, whose measure (assessment) expresses the inherent capacity of the system to transform a resource into a useful product (system efficiency). The meaning of using a systems approach, as described in the procedure followed by Spedding, is that of realising a whole process of understanding, which is articulated in four steps: *conceptualizing*, *modelling*, *assessing, predicting*. In introducing the first number of the new journal *Agro-Ecosystems*, Harper[21] had argued that there was a need for a strong predictive science and that the strength of the science of agroecosystems had to be proved by its ability to predict complex interactions of ecosystems. All that is about constructing a package of information for improving action. Later in chapter 8 (Agricultural Ecosystems) of his mentioned book, Spedding[22] explains that the term ecosystems was deliberately employed to emphasise the ecological approach to the whole agricultural systems and not merely to their internal structure and functions, since the intention was to discuss the interactions between agricultural ecosystems and the environments within which they operate.

If Ecology and Agroecology share both the same epistemology, it is meaningful to search the roots of the systems view as they appear in the evolution of human thought. As to Ecology, Caporali[23] has shown how the ecosystem concept is an epistemological tool that satisfies the four causes of knowledge mentioned in Aristotle's *Physics* as *efficient, formal*, *material* and *final* causes. Those philosophical foundations are further re-enforced by the Whitehead's process philosophy that is already well established in the early twentieth century. Instead, scientific legitimisation comes from cross-disciplinary fields such as thermodynamics, biogeochemistry and evolutionary biology that are already well developed in the second half of the nineteenth century. Lotka's book "Elements of physical

[21] Harper (1974), pp. 1–6.
[22] Spedding (1975).
[23] Caporali (2006).

biology"[24]—although the term ecology is never mentioned in it and the term ecosystem had still to be invented at that time—can be regarded as the first book in Ecology organised according to an ecosystem approach.

As to Agroecology, the first books of Latin tradition on agriculture already show their own way of systems thinking. For example, in Cato's *De Agri Cultura* (On Agriculture),[25] which dates back to the second century BC, the following suggestions are formulated in order to establish guidelines for evaluating a farm as a good one:

a) good environment (external context, notice how the neighbours keep up their places);
b) good climate (not subject to storm);
c) good soil (naturally strong);
d) good topography (possibly, it should lie at the foot of a mountain and face south);
e) good supply of water;
f) good supply of labourers;
g) good access (near it there should be a flourishing town or a sea or a navigable stream or a good and much-travelled road);
h) good equipment (oil presses and wine vats);
i) good buildings (well-built barn and storage rooms);
j) good size (a hundred *jugera*, i.e. about 30 ha, of land);
k) good biodiversity of both cropland and woodland (vineyard, watered garden, osier bed, olive yard, meadow, grain land, woodlot for both timber and fruits for livestock, tree plantation);
l) good care for livestock (*Boves maxima diligentia curatos habeto*);
m) good care for fields and crops based on organic management (litter composting, farmyard manuring, legume green manuring, crop rotation, intercropping, hedgerow planting).

All these recommendations are made under the general assumption that "it is from the farming class that the bravest men and the sturdiest soldiers come".

We can hardly contest that these points are still today relevant for a good modern farming carried out on a family-based organisation, even if labour as human slaves has been replaced by machine-driven labour. Indeed, agricultural tradition has kept these general recommendations valid for most countries in the world because their success has been assured by the experiential learning cumulated along continuing generations of farmers. Agroecology in general recognises the value of tradition in agriculture, establishes the scientific reasons of traditional agriculture's best practices and recognises their values as foundation for a sustainable agriculture.

---

[24] Lotka (1925).

[25] Cato Marcus Porcius, *De agricultura*. Varro, Marcus Terentius *Rerum rusticarum libri tres*. English edition: Cato and Varro On Agriculture (1934) (trans: Hooper WD, Ash HB). Harvard University Press (Loeb Classical Library 283), Cambridge, MA.

The scientific revolution has invested all human activity, including agriculture. With the introduction of agricultural sciences at academic level, which happened for the first time (1844) at Pisa University in the Grand Duchy of Tuscany, a three-year curriculum of study in "Agriculture and Animal Husbandry" was established with a specific focus on farm design and management. For the curriculum developers, Cosimo Ridolfi and Pietro Cuppari, the overall principle of the curriculum organisation was that of considering a farm as a "living body", made up of interacting parts to be organised harmonically under physical, biological, technological and economical constraints.[26] Experiential learning to be carried out in a pilot farm was regarded as an essential part of the curriculum. Many publications were developed for students with the aim to deepen specific knowledge of the farm components and processes, but the general rationale of considering the performance of the whole farm as the purpose of agriculture teaching and research was never neglected. About a century later, Alfonso Draghetti,[27] a leader researcher at the Experimental Station of the Italian Ministry of Agriculture in Modena, gave a new impetus to research focusing on a farm as a "living body" and published a seminal book with a meaningful self-explaining title: "Farm Physiology Principles". In that book, a farm is viewed as a functional unity (living body) where all parts (organs) are in connections through a "physiology" (organisation), provided by farmer's design and management, that allows "circulation" and "re-cycling" of materials in a synergistic framework between complementary components, according to their functional roles of "organs". Soil fertility maintenance is the main "physiological" objective for ensuring long-term productivity or "agroecosystem health", while crop rotations and mixed farming with farmyard manure are the main "organs" for supplying organic matter to soil.

These references can be considered as the most prominent contribution of Italian scholars for agriculture systems thinking development, and these scholars can be fairly recognised as proto-agroecologists in the evolving framework of Agroecology.

For a today's agroecologist, there is no doubt that agriculture is a complex business where decision-making processes are taken at different scales of spatial and temporal organisation. As Spedding[28] points out, the farming systems actually found in any locality are the results of past as well as present decisions by individuals, communities or governments and their agencies. They cumulate the sequence of considerations (see Fig. 2) about the biophysical and socioeconomic environment that determine the feasibility, profitability, practicability and preferences that are involved in the choice of systems. What is common in both any place and time is the process of agroecosystem organisation that makes agriculture operational in a framework of changing expectations.

[26] Cuppari (1862).

[27] Draghetti (1948).

[28] Spedding (1975).

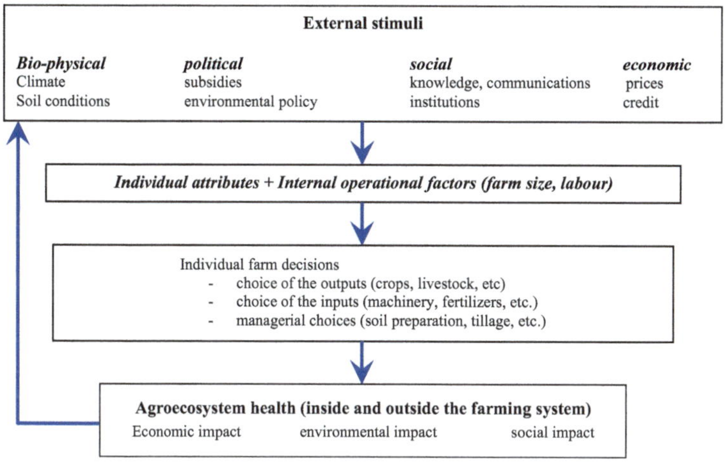

Fig. 2 Recursive factors affecting the choice of farming systems (modified from Smit et al. 1996)

2.2 *Hierarchy*

Hierarchy in agriculture is to be meant as a spatial–temporal continuum or an open, interconnected sequence or stratification of multi-layered agroecosystems, which are isolated only for a necessity of study and management (Table 1).

The functional characteristics behind this hierarchical representation model of agriculture is the *openness* of each level in the sense that each level is at the same time the context of the next lower level and the component of the next upper level. Therefore, integration between different levels is an ontological necessity for the existence and the functioning of both each level and the whole system.

A meaningful case for describing hierarchical integration is focusing on soil as a component of all levels in the hierarchy of agroecosystems. Soil in itself can be considered as an ecosystem,[29] if viewed as the combination of the organism-population complex with the complex of physical factors that make up the environment, plus the organic and inorganic constituents that are basic sources of food materials and energy. In a given spatial–temporal context of interest, a field, a farm or a watershed, soil is a sub-system or component[30] of a broader ecosystem and its field of influence can be detected according to the level of organisation under enquiry. At the field level, where detecting soil–plant–atmosphere relationships is relevant for crop nutrition, it is important, for example, to investigate the biochemical chain in which organic material is broken down and its constituents released for reutilisation. However, nutrient release depends on the quantity and quality of crop residues left by the ongoing cropping system and therefore by the choice made by

[29] Auerbach (1958), pp. 522–529.

[30] Witkamp (1971), pp. 85–110.

Table 1 Levels of agroecosystem hierarchy and their main fields of interest (modified from Caporali 2006)

Agroecosystem hierarchical levels	Main fields of study and management
Field system	Soil–plant–atmosphere relationships
Cropping system	Relationships between crops in a spatial–temporal scale
Farming system	Relationships between crops, livestock and management
Landscape system	Land use, biodiversity conservation, soil conservation practices and aesthetics
Regional system	Rural–urban integration; socioeconomic development
National and international system	Globalisation aspects (market, economy, policy and environment)

the farmer in terms of crop rotation, intercropping or mono-cropping. Cropping systems, in their turn, are components of the broader farming systems, where supply of organic matter to the crops and the fields depends on the availability of organic manure from livestock or composting. A mosaic of farming systems constitute the structure and functioning of the entire ecosystems at the landscape level, as a consequence of land use, biodiversity conservation and land conservation practices. Diffusion or restriction of agriculture land use on regional, national and international levels is a consequence of political and economic choices that depend mostly on the ongoing process of globalisation, which is already a driving force in channelling the urban–rural destination of land use also at the local level.

For agroecosystem design, planning and implementation according to an ecological perspective, hierarchical organisation of levels of enquiry should be better meant as *hierarchical integration* of levels of enquiry because there is the need to harmonise all levels of hierarchy for the common goal and good that sustainability of agriculture is. To Ikerd,[31] the explicit purpose of integrating the science of ecology and agriculture is to enhance the sustainability of agriculture. He mentions that the concept of agroecology was first used by scientists in the 1970s in questioning the ecological, social and economic sustainability of Green Revolution era agroecosystems. A 1982 symposium at the annual meetings of the Ecological Association of America was the first effort to bring scientists from the various disciplines associated with agroecology together in a national academic forum.[32] "Agroecology provided a logical, conceptual framework for integrating the disciplines of ecology, economics, and sociology for the purpose of enhancing research and education related to sustainability of agriculture".[33]

Indeed, integration is a kind of organisation that is purposefully oriented. The ultimate purpose of component integration is the functioning of the whole system. The final aim of ecosystems is to maintain themselves through reproduction of their

[31] Ikerd (2009), pp. 41–52.

[32] Lowrance et al. (1984).

[33] Ikerd (2009).

biological individual components, whose species are well integrated in their environment through adaptation. Ecosystems are self-organising and self-sustainable parts of biosphere, and biosphere is the whole self-organising and self-sustainable planetary ecosystem (until they all will receive solar energy input). The ecosystem representation of reality unveils how between ecology and sustainability there is an ontological link.[34] Sustainability is the ultimate property of ecological systems, stemming from their inherent capacity of self-organisation, resilience and adaptability. Ecology and sustainability are conceptually interconnected, and we cannot speak of ecology without meaning sustainability as well as speak of sustainability without meaning ecology. The hierarchical order in agroecology opens the view beyond the field and the farm, embraces the landscape and reaches up to the great balance between contrasting land use at the biosphere level.

2.3 Emergence, Communication and Control

Scaling up hierarchical levels of reality, new properties emerge that are manifestations of innovation, communication and control. Emergence is an effect that stems from the interaction of components (communication) and can be useful or harmful for conferring more or less *coherence* among components within a hierarchical level and/or more or less *correspondence* between different levels (control).

Agriculture as a whole can be regarded as an "emergence" (a novelty or innovation) nested in the pristine ecosystems that human beings have modified and controlled for agricultural purposes. An emergence is a systemic property that occurs under an organisation that can facilitate (or depress) the integration among the composing parts. The extent of components' integration depends on their ecological complementary, i.e. synergy of functional roles. Crop seed, bare soil and water availability are the basic components of an agroecosystem; the latter *emerges* when the components can act together and not in isolation. Indeed, agriculture has been a revolutionary emergence of innovation driven by cultural information. In Table 2, the main emergent properties of agroecosystems (under conventional agriculture) are reported, in comparison with those of pristine ecosystems.

As Gliessman[35] notes, "an agroecosystem is created when human manipulation and alteration of an ecosystem take place for the purpose of establishing agricultural production". This introduces changes in key system level qualities or *emergent qualities* (see Table 2), which manifest themselves once all of the component parts of the system are organised. They concern structure and dynamics or metabolism of the whole system. Those same qualities can also serve as indicators of agroecosystem sustainability.[36]

---

[34] Caporali (2006).

[35] Gliessman (2004), pp. 61–80.

[36] Gliessman (2001).

Table 2 Main *agroecosystem properties emerging* from the use of pristine, mature ecosystems

Kind of ecosystem	State	Input	Output (yield)	Output (environmental impact)
Agroecosystem (under human control in conventional agriculture)	Disturbed, biologically simplified (human dependent) *Changes in energy flow, nutrient cycling, population regulation mechanisms, resilience*	From natural sources, from fossil fuel energy and cultural information (tillage, fertilisers, pesticides, irrigation, etc.)	*Maximum export of productivity*	*Erosion chemical pollution eutrophication biodiversity loss*
Pristine, mature ecosystem (without human control)	Undisturbed, biologically complex (*self-maintenance*)	From natural sources	*Minimum export of productivity*	*Minimum impact* (except for catastrophic events)

High productivity of conventional agriculture agroecosystems is a systemic property that emerges from an integration of main factors such as high-yielding crop varieties, chemical fertilisation, irrigation and chemical crop protection against weed and pests. Inputs in excess leave the cropping system and affect the adjacent ecosystem generating pollution and eutrophication. One of the best documented ecosystem "emergence" generated in agroecosystems of conventional industrialised agriculture has been the contamination of the entire food web from DDT spraying,[37] the so-called "biological magnification", due to the increasing concentration of DDT detected in the tissues of species along the trophic levels of food chains.[38]

Agroecology's purpose is to develop a scientific basis for an agriculture that is sustained by internal processes that reinforce synergistic effects due to more agroecosystem biodiversity (*biological buffering*) and avoid pollution. Like in many forms of traditional and organic agriculture,[39] productivity is maintained on a more stable level, without the ups and downs typical of conventional agriculture in a variable environment.

As conclusion, from the ecosystem analysis of Table 2 comes out that most of today's agricultural productivity (or food, fibre and fuel for man) occurs at the expense, of other ecological services, or useful functions for life maintenance, in terms of both support and regulation of environmental conditions, and human cultural and social development.[40] The challenge for Agroecology as a scientific discipline is to use ecological theory to study, design, manage and evaluate agricultural systems that are productive and also resource conserving.[41]

[37] Richardson (1977).

[38] Woodwell et al. (1967), pp. 821–824.

[39] Phelan (2009), pp. 97–135.

[40] Swinton et al. (2007), pp. 245–252.

[41] Altieri (1995), pp. 31–36.

Structural composition (number of crops) and configuration (their allocation in space/time) of cropping systems have, for instance, dramatic consequences on the agroecosystem performances, involving key processes at system level such as hydrological cycling; transport of soil and chemicals; soil, water and air qualities; environmental pollution; and weed and pests population dynamics.

Characteristics called emergent qualities, which exist at the ecosystem level, such as productivity, system stability and dynamic functioning, are of maximum importance for the design and management of sustainable agroecosystems. In ecological terms, more complex and well-diversified agroecosystems bear more integration of emergent qualities and benefit from synergistic effects that stem from the interactions between the same agroecosystem components. In structurally and functionally well-diversified agroecosystems, internal resources constitute the base for functioning in a sustainable fashion.

As a consequence of this assumption, Caporali[42] has proposed the following ten rules for design and management of sustainable agroecosystems at the farming system level:

1) to create diversity within the farm;
2) to integrate crop production and livestock husbandry;
3) to adopt soil conservation measures and minimum tillage practices;
4) to adopt crop rotations;
5) to adopt intercropping and cover cropping;
6) to use genotypes resistant to parasitic attacks;
7) to treat the soil with manure and composted organic matter;
8) to practise green manuring;
9) to foster the biological control of weeds and pests;
10) to plant and protect hedges.

Most of these points are best practices already established by tradition in agriculture and recommended in agronomy books. Others, such as 1), 9) and 10), derive from an integrated view of agroecosystem, where cropped and un-cropped area as field margins (hedges, strips, etc.) coexist and interact with exchanges of organisms, water and nutrients.[43]

In a complex human activity system like agriculture, where socioeconomic components interact with biophysical ones, regulations are good examples of emergence, communication and control at the same time.[44]

Being agriculture appropriately represented as a network of socioeconomic relationships nested upon a rural environment, a reorganisation of its system components is required when a new regulation is issued in agriculture. Changes should follow at any hierarchical level, i.e. field, farm, landscape and region. In essence, a new regulation in agriculture can be regarded as an "emergence" or

[42] Caporali (2004).

[43] Caporali (1991); Marshall (2006), pp. 365–404.

[44] Caporali (2004).

innovation in organisation due to a new flow of information. Through communication (extension service, training, etc.) and control (incentives, taxes, etc.), changes in individual and institutional behaviour can be promoted, as well as effects concerning the properties of the whole system and its components.

Organic farming regulations, recently approved by many countries worldwide, can be regarded as new "emergences" in both national and global agricultural systems. Certification procedures and produce labelling possess a peculiar character of communication and control purposefully credited to the civil society for reorienting preferences in favour of agriculture sustainability. This kind of "emergence" can be considered as an adaptation process for reconfirming agriculture as a human activity system operating for the common good.

3 Indicators of Sustainability as Tools for Integration in Agroecology Research

Agricultural sustainability is a complex property of agroecosystem organisation that emerges from the harmonic integration of the components, or sub-systems, under the constraints operating at different hierarchical scales (Table 3).

Agronomic constraints operate at the field and cropping system levels, where soil fertility maintenance depends on the appropriate integration of functional groups of crops over the course of multiple growing seasons. Microeconomic viability is dependent on the ability of the farmer to stay in business and must be performed at the farm level, being the farm the basic economic unit of management in agriculture. Macroeconomic sustainability depends on constraints established at the national or international scale, especially in terms of monetary and fiscal policies, that can even have a greater effect on farm than conventional microeconomic policies aimed at the farm level.[45]

Beyond providing marketable, private goods as food and fibre, agriculture can provide many public goods and services, like socioeconomic viability of rural areas, landscape maintenance, biodiversity preservation, soil and water conservation and soil carbon sequestration. Increasing scientific and political recognition of agriculture as a multipurpose activity in society has promoted the need to invest more intellectual and financial resources in research for monitoring and measuring sustainability conditions in agriculture, in order to appropriately inform decision-making processes for developing multifunctional agroecosystems.[46]

The necessity to stress the importance of the decision-making process in society is already well documented by chapter 8 of Agenda 21, "Integrating environment and development in decision-making", where it is stated that a) prevailing systems for decision-making in many countries tend to separate economic, social and

[45] Lowrance et al. (1986), pp. 169–173.

[46] Westra and Boody (2009), pp. 213–233.

Table 3 A hierarchical approach to agricultural sustainability (modified from Lowrance et al. 1986)

Main components of agricultural sustainability	Level of organisation	Major constraints	Agroecological, transdisciplinary field of enquiry
Yield stability	Field and cropping system	Agronomic	Agronomy/ecology
Economic viability	Farm	Microeconomic	Agronomy/ecology/economy
Environmental stewardship	Landscape (watershed)	Ecological	Agronomy/ecology/economy/policy/law
Local community building capacity	Large institutional bodies	Macroeconomic	Agronomy/ecology/economy/sociology/policy/law

environmental factors at the policy, planning and management levels; b) there is the necessity for a better integration among national and local governments, industries, sciences, environmental groups and the public in the process of developing effective approaches to environment and development; c) responsibility for bringing about changes lies with governments in partnership with the private sector and local authorities and in collaboration with national, regional and international organisations; d) the overall objective is to improve or restructure the decision-making process so that consideration of socioeconomic and environmental issues is fully integrated and a broader range of public participation is assured.

Among the activities for improving planning and management systems, data and information collection is crucial; therefore, it is recommended that "countries could develop systems for monitoring and evaluation of progress towards achieving sustainable development by adopting *indicators* that measure changes across economic, social and environmental dimensions". Two features of indicators, such as quantification of information and simplification of complex phenomena, are important in order to facilitate the communication process among users. Usually, indicators quantify the relationships between important system components and take the form of a ratio or a flow. In the decision-making process, they serve as tools for monitoring/assessing and facilitating judgement before action.

Their use has been largely promoted in agroecological research as a necessary instrument for understanding agroecosystem organisation and performances, facilitating judgements and suggesting solutions for improving sustainability in agriculture.[47]

In general, *indicators of agricultural sustainability* (IASs) provide meaningful information about structure/function/performance of an agroecosystem or its components (sub-systems) at any hierarchical scale of organisation. Usually, IASs are

[47] Caporali et al. (1989); Tellarini and Caporali (2000), pp. 111–123; Caporali et al. (2003), pp. 67–72; Organisation for Economic Cooperation and Development (OECD) (1997, 1999, 2001, 2013).

developed on the base of the general input/output model of agroecosystem analysis derived by the systems paradigm.[48] With the development and use of IASs, agroecological research is getting more and more integrated in the structure of civil society, improving its role of scientific service for public utility. IASs would be useful in giving a broad picture of development to try to better understand the relationships between policy, agricultural management and production, and the environment.[49]

The aim to develop IASs has both an epistemological and a practical meaning, as they represent, respectively, a) an efficient instrument of enquiry for studying agroecosystem structure, functioning and performance according to an input/output approach and b) a relevant knowledge base for both the designing of sustainable agroecosystems and decision-making processes.

By analysing a farm as a process that transforms inputs into outputs in terms of energy, money, material and information, it is possible to calculate indicators of efficiency, to express judgements about the agroecosystem functioning and to suggest solutions in order to optimise the process for improving its sustainability.

This methodology allows rationalising the on-farm choices since it is possible to focus on three modifiable factors: the type and amount of input, the type and amount of output and the organisation of the agroecosystem components.

For an assessment and a comparison of agroecological performances in different farming systems, it is essential to establish how the latter are composed (i.e., land use, kind of crops and livestock), how they are organised (i.e., prevalent use of solar energy or auxiliary energy) and how they transform resources. In order to describe these characteristics, indicators are used according to the necessity of making understandable the agroecosystem's complexity. Indicators can be distinguished in agroecosystem structural and functional indicators.[50] The former are related to the description of composition and organisation of the system components, i.e. concern the resources that are used in the production process such as land, crops, animals and materials. The latter concern the efficiency with which those resources are used, as expressed by output/input or input/output ratios. Both kinds of indicators can be calculated in terms of money, energy and nutrient content on all incoming and outgoing materials chosen by the farmers and therefore dependent on their choices. With this approach, and using energetic and monetary accounting criteria, a farm can be studied as both an economic (with sociocultural values) and a thermodynamic (with biophysical behaviour) unit. Money, i.e. a sign representing what is potentially available, is to be considered a means of great semantic content and therefore a powerful tool or package of information able to channel a choice.

IASs were intended to highlight those forms of organisation capable of better exploiting the system's native resources (solar radiation, soil organic matter, atmospheric nitrogen, etc.) rather than the imported, nonrenewable ones.

---

[48] Edwards et al. (1993), pp. 99–121; Tellarini and Caporali (2000).

[49] Brouwer and Crabtree (1998) and Ministry of Agriculture, Fisheries and Food (MAFF) (2005).

[50] Tellarini and Caporali (2000); Caporali et al. (2003); Di Felice et al. (2012), pp. 119–126.

As a follow-up to the Agenda 21, the OECD Council approved in 1991 a Recommendation on Environmental Indicators and Information to further develop sets of reliable, readable, measurable and policy-relevant environmental indicators. The indicators chosen cover the range of primary agriculture's impacts on the environment that are policy relevant and that are practical to measure.[51]

The conceptual model for the search of agri-environmental indicators (AEIs) was defined by OECD "Driving Forces-State-Response (DSR) framework".

Driving forces are those elements that cause changes in the state of environment and include natural environmental processes and factors, biophysical inputs and outputs at the farm level, economic and social driving forces. The concept of driving forces recognises that agricultural activities can both produce beneficial impacts to enhance environmental quality, such as increasing the water storage capacity, and have harmful impacts, such as pollution.

The state or condition of the environment in agriculture refers to changes in environmental conditions that may arise from various driving forces. The impact of agriculture on the environment can occur both on-farm (internal environment) and off-farm (external environment). The state of the environment includes a) state of the natural resources; b) composition, structure and functioning of the ecosystem; and c) state of human health and environmentally related welfare.

Responses refer to the reaction by groups in society and policymakers to the actual and perceived changes in the state of the environment in agriculture. They include farmer behaviour, consumer reactions, responses by the agri-food chain and government actions.

The DSR framework denotes much of its agroecological foundation. It can provide a flexible framework to improve understanding of the complexity of linkages and feedback between the causes and effects of agriculture's impact on the environment and the responses of the main stakeholders. While agriculture can affect the state of the environment, changes in environmental conditions can also have an impact on agricultural production activities. Therefore, analysis of the linkages and feedback between driving forces, state and responses is a key element in shedding light on the dynamic functioning of agriculture as a human activity system.

The choice of ISAs is an evolving process depending on societal pressures and political choices. Some environmental areas are gaining in importance (e.g., soil greenhouse gas sinks), while others are diminishing in the context where some control measure is already running. Due to the importance of understanding the linkages between policies, agricultural production and environmental quality, the interpretation of any one indicator may need to be complemented with other indicators and be seen within the overall context of the set or appropriate subset of indicators.[52] According to OECD,[53] the recent environmental performance of

[51] OECD (1999).

[52] OECD (2001).

[53] OECD (2013).

agriculture provides some encouraging signs that agriculture is capable of meeting future environmental challenges. Evidence for OECD countries from 1990 to 2010 shows that improvements have been made in nutrient, pesticides, energy and water management using less of this input per unit volume of output. Enhanced environmental performance has also flowed from the widespread adoption of environmentally beneficial practices by farmers, such as conservation tillage, improved manure storage, soil nutrient testing and drip irrigation. Identifying the extent to which the total payments from OECD taxpayers to agricultural producers has shaped the environmental performance of OECD agriculture is, however, complex and not fully understood.[54]

4 Integrating Research with Practice in Agriculture

In a paper devoted to define the role of agroecology for agriculture sustainability, Edwards et al.[55] were able to define a strategy for successful research involving the following steps:

1) description of the target agroecosystem, including its goals, boundaries, components, functioning, interactions among components and interactions across its boundaries;
2) detailed analysis of the agroecosystem to determine factors that limit or could contribute to the attainment of productive and social goals;
3) design of interventions and identification of actions to overcome the constraints;
4) on-farm experimental evaluation of interventions;
5) review of the effectiveness of newly designed systems;
6) redesigning, as necessary.

All this approach could be defined as *participatory research*, which involves not only researchers but also farmers in the whole process, from planning to implementation and evaluation. All steps should be conducted *on farms* by an interdisciplinary team of agricultural, social and ecological scientists and with full participation of farmers. Understanding the farmer's goals is crucial, as the role of the proposed interventions is to help the farmer attain these goals.

Phelan[56] suggests the following principles for aiming at an ecology-based agriculture and a *next green revolution*:

a) redesigning future farming systems with an understanding of how ecosystems are organised and function;

---

[54] OECD (2013).

[55] Edwards et al. (1993).

[56] Phelan (2009).

b) making management decisions that allow for the special behaviours and properties of complex systems;
c) developing a philosophy of research and experimental design that better reveals the relative importance of systems components and their interactions;
d) adopting a perspective of costs and benefits that is more inclusive in space and time to assess the true value of technologies.

Agriculture should be designed to cope with fluctuating environmental conditions with the main purpose of increasing the stability of agroecosystems since "success on this front will likely lead to progress towards other goals".[57] From the evidence accumulated from both natural and managed ecosystems, any ecological-based approach to designing agricultural systems should start with soil fertility maintenance. "The shift from a C-based system of fertility to an inorganic one did more to reduce the ecological stability of agricultural than any other practice [. . .] Understanding the operation of detritus food webs and designing nutrient management that is more consistent with the nutrient cycles of natural systems is the single most important step that can be taken to increase economic sustainability, environmental compatibility, and biological resilience of agricultural systems".[58]

The farm level is the more appropriate one in the hierarchical scale for doing research and making decisions in favour of agroecosystem sustainability. Indeed, the farm is the management unit of agriculture with a biological base, easily identifiable because of its boundaries, and that represents the meeting point between human interests and the natural environment.[59] The most advanced regulations for sustainable agriculture, like those concerning organic farming, provide a framework of legitimisation based on the agreement between the civil society and the farmer. That means that there is an explicit recognition of the farm as the crucial level of organisation of resources—both biophysical and socioeconomic—in the food system. Performances at the farm level are able to affect characters and sustainability of the next upper hierarchical levels of agricultural systems (e.g., landscape and region) as well as the next lower level or field level, where interactions between crop, animals and microorganisms affect soil fertility, which is the basis for agriculture sustainability.

Research based on IASs at the farm level is therefore of great importance to decision-making processes, especially when groups of farm of contrasting management are involved. Organic farming systems are being considered as a long-term benchmark for the evaluation of apparently environmentally benign agricultural production systems.[60] Therefore, the aim of some recent research includes the comparison of organic farming systems with conventional ones on the base of appropriate IASs.[61] Results of this kind of research are easily shown graphically,

[57] Phelan (2009).

[58] Phelan (2009).

[59] Caporali et al. (1989).

[60] OECD (1999).

[61] Reganold et al. (2001), pp. 926–930; Mader et al (2002), pp. 1694–1697; Caporali et al. (2003); Di Felice et al. (2012).

with the help of a so-called sustainability polygon or web, which simultaneously displays scores for different indicators and avoids having to aggregate across different scales.

4.1 The Challenge of Integrating Research with Practice in the Search for Sustainable Land Use

Enlarging the scale of agroecosystem analysis at the landscape level, competing land use emerges between agricultural and urban systems, which prospects new emergent systemic properties and challenges for human society, such as that of *soil consumption*. Recent trends about land use in Central Italy[62] reveal that while agroecosystems have encroached upon natural ecosystems, urban sprawl encroaches upon existing agroecosystems, eroding substantial cropland acreage, especially in the most fertile soils of flatlands (Fig. 3).

Ecological and social services from different ecosystems, such as those reported in Table 4 (modified from House[63]), are largely affected.

The challenge of sustainability for today's society as a whole is a coin with two faces. On the one hand, sustainable development includes the necessity for humanity to grow food through agriculture as well as to maintain natural environments for ecological services other than food.[64] The search for a balance between production and protection in land use is therefore a major challenge to future society at both local and global levels. On the other hand, demographic growth and concentration of human population on flat areas of costs and internal lands bring about a consumption process of former, fertile agricultural soils that creates unbalances of land use at both local and international levels, thus undermining the sustainability of society as a whole. If local soil for food nutrition is eroded by alternative land use, other agricultural areas in the world must provide food for local people. However, the enlarging ecological footprint for food nutrition of an increasing human population has its own limit in the physical boundaries of the planet. Therefore, a balance must be necessarily found among competing land uses. An emergent research community, increasingly referred to as land change science,[65] seeks to address the major components and advances in global environmental change, such as observation and monitoring; understanding the coupled systems—causes, impacts and consequences; modelling; and synthesis issues. In Europe, new regulations about landscape, such as the European Landscape Convention (2000) and the Water Management Directive (2000) with an emphasis on

[62] Caporali and Mancinelli (2013), pp. 29–36.

[63] House (2009), pp. 283–296.

[64] Daily (1997) and Millennium Ecosystem Assessment (MEA) (2005).

[65] Turner et al. (2007), pp. 20666–20671.

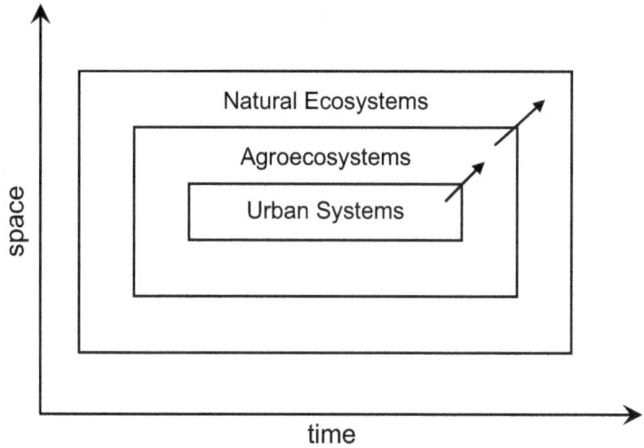

Fig. 3 Trends of encroachment between natural, agricultural and urban ecosystems

Table 4 Types of ecosystems and relative types of ecological and social services

Natural ecosystems	Agroecosystems	Urban systems
Life-support processes	Human-life-support services	Human-life-support services
Nutrient cycling	Food	Human habitation
Soil formation	Fibre	Industry
Air and water purification	Biofuels	Commerce
Flood control	Employment	Entertainment
Biodiversity	Agricultural biodiversity	Urban biodiversity
Supporting habitat	Rural community	Urban community

the watershed as an appropriate basis for institutional enquiry and intervention, can be considered as promising steps in the process of land use design and management according to an ecological paradigm. Enlarging the scale of research up to an appropriate watershed level allows for incorporation of interactions among agricultural and urban systems and of decisive political and socioeconomic influences that are external and yet extremely important to farm systems.[66] In such a case, the focus of research shifts from sustainable agriculture per se to sustainability in general and poses the practical question of how to articulate an efficient structuring of cooperation among ecological, social, economic and political aspects of land use design and management. It is obvious that in enlarging the scale of enquiry from the farm level to the landscape and regional levels, political and socioeconomic linkages are more determinant as driving forces of land use development than biophysical constraints. To stop soil consumption with an adequate legislation that allows for farmland preservation is today a new challenge for society as a whole.[67]

[66] Stinner (2009), pp. 13–20.

[67] Bunce (1998), pp. 233–247; Engelke and Vancutsem (2010).

In a recent book entitled "Agroecology in Action", Warner[68] claims that agroecology can be put into action effectively only when networks of farmers, scientists and other stakeholders learn together. Farmers, scientists and their organisations must work collaboratively to share knowledge. Warner describes successful agroecological initiatives in several USA federal states to show how mutual feedback between popular concern and environmental policy has stimulated creative solutions from scientists, extension agents and growers. This process of collaborative social learning has been identified as "agroecological partnership", i.e. a dialogical platform for producing knowledge needed for multifunctionality in working agroecosystems.[69] The concept of *multifunctional agriculture* (MFA) has been meant as a project for "sustainable land architecture", i.e. a joint production system of both agricultural commodities and a range of ecological services beneficially influencing biodiversity, pest and nutrient management, water quality and quantity, and amenity values.[70] The step of generating a fresh view of the relationships between agriculture and society, guided by the integration of multiple social goals, is crucial for the promotion of a more "civic agriculture",[71] i.e. "an agrifood system that is integrated into the social and economic development of a local community and that fosters participation in civic society".[72] The spread of such communities of ethical concern is viewed as a sign of change witnessing the demand for more capability in balancing economic development, social equity and environmental protection.

More accurate methodologies need to be developed in order to allow for bringing stakeholders, including researchers, to become aware of their role in transformational change. European research experiences on governance of complex environmental situations through social learning have produced a "heuristic" or diagnostic framework on how changes of understanding and practice can be brought about by the facilitation of the relationships among stakeholders, the ecological dynamic, institutions and policy. On the assumption that changing how people perceive ecosystems may change how people act, Steyaert and Jiggins[73] recommend an epistemology of action (knowledge that is produced during the process of acting) based on the following principles:

1. placing scientific knowledge effectively within the contexts where landscape scale impacts are desired, and in ways meaningful to others, while also acknowledging these stakeholders' own knowledge, understanding and experience;
2. learning as situated in practice;

[68] Warner (2007).

[69] Jordan and Warner (2010), pp. 60–66.

[70] Wilson (2007); Jordan et al. (2007), pp. 1570–1571; Renting et al. (2009), pp. S112–S123.

[71] Lyson (2004).

[72] Jordan and Warner (2010).

[73] Steyaert and Jiggins (2007), pp. 575–586.

3. acting professionally in interaction with others and developing the personal and organisational skills and capacities for this.

5 Integration of Agroecology into Academic Curricula

In the current situation, Agroecology can occupy different levels of academic organisation, the discipline, the degree course and the doctoral level. This multiple dimension denotes that agroecology is a scientific discipline but its transdisciplinary epistemology (systems paradigm) and its main purpose (agriculture sustainability) are so culturally relevant as to provide a framework within which a shared platform of contents for teaching and research at the academic level can occur.

At the discipline level, a detailed description of agroecological contents can be found in Altieri and Francis[74] and in Caporali.[75] At the curriculum level, integration is a more complex issue in that it has something to do with the harmonisation of relationships with both the other disciplines of the curriculum and the external context of agriculture as a human activity system. *Coherence* needs to be established with the other disciplines of the curriculum and *correspondence* with the driving forces of the external context.

Internal coherence is based on systems paradigm sharing. Its articulation in systemic knowledge fields (ecosystemic approach, input/output analysis) opens up the way towards a new cultural broader horizon that includes ethical and aesthetical insights and demands new research tools (indicators) for assessing the search for sustainability from information to action. Methodological tools inspired by the systems paradigm in terms of social linking can be helpful in bettering connections between a curriculum as a whole and its context (external tools) and among the curriculum components themselves (internal tools). The external context is made up by all kinds of information inputs coming from international, national and local levels. That external input forms a general framework of reference for internal decision-making. The internal context is created by the different feelings and prospects of those involved originally in the development of a curriculum, who also respond to other internal inputs, coming from the university, the departments and the personal attitudes of all people involved.

External methodological tools help introduce a broad concept of action-based learning. Integrating the expertise of farmers, business owners, government specialists and nonprofit groups can enrich the educational process by offering different perspectives and ways of knowing.[76] Case studies, interview and survey techniques, time-series measurements and activity calendars can be taught and applied to answer questions about integration within the whole agroecosystem

[74] Altieri and Francis (1992), pp. 89–93.

[75] Caporali (1991, 2004).

[76] Francis et al. (2001), pp. 89–95.

hierarchy (cropping systems–farming systems–regional systems–global systems). These approaches require several changes in attitude and organisation. Department members, administrators and others must invest time and money to establish research and learning in institutional networks. New sources of funding and revised systems of administering research funds will be required to promote this approach successfully (Stark Jr 1995).[77]

6 Concluding Remarks

Agroecology has emerged from agricultural tradition (or experiential learning), philosophical foundation (or systems thinking) and scientific innovation (or ecological approach). Its establishment as a scientific discipline has occurred since the definition of its epistemological tool, i.e. the concept of the agroecosystem. The agroecosystem is a flexible tool for representing, monitoring, valuating, designing and managing agriculture at any kind of organisation level according to an input/output model or a processual scheme derived from both philosophical and scientific roots. The emergent characteristic of Agroecology is that of a trans-discipline as it integrates other fields of knowledge into the concept of agroecosystem viewed as a socioecological system. Agroecology models agriculture as a human activity system, i.e. integrates biophysical and socioeconomic components into organised agroecosystems at different hierarchical levels. A sustainable integration is today required in agriculture at any hierarchical level, from the field to the regional and global levels, and the concept of integrated rural development has been created to revitalise together agriculture and economy while safeguarding the environment. Agroecology, both as a scientific discipline and a philosophical and organisational paradigm, has potential for constructing more integrated academic curricula as well as promoting more integrated, participatory research. New epistemological, ontological and methodological tools based on the systems paradigm, like that professed by Agroecology, are necessary in order to tackle the challenge of establishing a new culture and praxis for a sustainable development in agriculture and society. On the base of agroecological principles, indicators of agricultural sustainability have been developed worldwide for monitoring, assessing and planning at farm, local, regional, national and international levels. That process is still in progress with its revolutionary power of cognitive, technological and sociopolitical innovation.

With a research agenda that aims at creating nature-friendly and socially just agricultural systems, Agroecology is a science that introduces an ecological ethics in a world of business and competition, representing a turning point in favour of sustainable development, justice and peace.

[77] Stark Jr (1995), pp. 180–183.

References

Altieri MA (1995) Agroecology. In: Nierenberg WA (ed) Encyclopedia of environmental biology, vol 1. Academic, New York, pp 31–36

Altieri MA, Francis CA (1992) Incorporating agroecology into the conventional agricultural curriculum. Am J Altern Agric 7(1–2):89–93

Altieri MA, Toledo VM (2011) The agroecological revolution in Latin America: rescuing nature, ensuring food sovereignty and empowering peasants. J Peasant Stud 38(3):587–612

Auerbach SI (1958) The soil ecosystem and radioactive waste disposal to the ground. Ecology 39 (3):522–529

Brouwer F, Crabtree B (1998) Environmental indicators and agricultural policy. CABI Publishing, Wallingford

Bunce M (1998) Thirty years of farmland preservation in North America: discourses and ideologies of a movement. J Rural Stud 14(2):233–247

Caporali F (1991) Ecologia per l'Agricoltura. Utet-Libreria, Torino

Caporali F (2004) Agriculture and health. The challenge of organic farming. Editeam, Cento (Ferrara)

Caporali F (2006) Ecological agriculture: human and social context. In: Clini C, Musu I, Gullino ML (eds) Sustainable development and environment protection. Springer, Dordrecht, pp 415–429

Caporali F (2010) Agroecology as a transdisciplinary science for a sustainable agriculture. In: Lichtfouse E (ed) Biodiversity, biofuels, agroforestry and conservation agriculture. Springer, Dordrecht, pp 1–71

Caporali F, Mancinelli R (2013) La sostenibilità dell'uso del territorio nel Bacino del Tevere: la prospettiva agroecosistemica. In: Terzo Rapporto Annuale del Consorzio Tiberina. Per un futuro sostenibile della Tiberina: coesione territoriale e sviluppo endogeno fra tradizione e innovazione. Editrice Punto Uno, Terni, pp 29–36

Caporali F et al (1989) Concepts to sustain a change in farm performance evaluation. Agric Ecosyst Environ 27(1–4):579–595

Caporali F, Mancinelli R, Campiglia E (2003) Indicators of cropping system diversity in organic and conventional farms in Central Italy. Int J Agric Sustain 1(1):67–72

Cato Marcus Porcius, *De agricultura*. Varro Marcus Terentius, *Rerum rusticarum libri tres*. English edition: Cato and Varro On Agriculture (1934) (trans: Hooper WD, Ash HB). Harvard University Press (Loeb Classical Library 283), Cambridge, MA

Chandrasoma R, Lee JE (2012) Technocultural epistemes: a novel approach to manifestation of knowledge. Engineering & Technology Digital Library, Singapore

Checkland PB (1993) Systems thinking, systems practice. Wiley, Chichester

Cuppari P (1862) Saggio di ordinamento dell'azienda rurale. Cellini, Firenze

Daily GC (1997) Nature's services. Societal dependence on natural ecosystems. Island Press, Washington

Di Felice V, Mancinelli R, Proulx R, Campiglia E (2012) A multivariate analysis for evaluating the environmental and economical aspects of agroecosystem sustainability in Central Italy. J Environ Manage 98:119–126

Draghetti A (1948) Principi di fisiologia dell'azienda agraria. Istituto Edizioni Agricole, Bologna

Edwards CA, Grove TL, Harwood RR, Colfer CJP (1993) The role of agroecology and integrated farming systems in agricultural sustainability. Agric Ecosyst Environ 46:99–121

El-Hage Scialabba N, Hattam C (eds) (2002) Organic agriculture, environment and food security. FAO, Rome. http://www.fao.org/docrep/005/y4137e/y4137e00.htm. Accessed 28 Mar 2014

Engelke D, Vancutsem D (eds) (2010) Sustainable land use management in Europe. Providing strategies and tools for decision-makers. Certu, Lyon

Foucault M (1966) Le mots et les choses. Edition Gallimard, Paris

Francis CA (2004) Education in agroecology and integrated systems. J Crop Improv 11 (1–2):21–43

Francis CA (2009) Organic farming: the ecological system (Agronomy Monograph 54). American Society of Agronomy, Madison
Francis CA et al (2001) Challenges in designing ecological agriculture education: a Nordic perspective on change. Am J Altern Agric 16(2):89–95
Francis CA et al (2003) Agroecology and agroecosystems: the ecology of food systems. J Sustain Agric 22:99–118
Gliessman SR (1998) Agroecology: ecological processes in sustainable agriculture. CRC, Boca Raton
Gliessman SR (2001) Agroecological sustainability: towards practical strategies. CRC, Boca Raton
Gliessman SR (2004) Integrating agroecological processes into cropping systems research. J Crop Improv 2(1–2):61–80
Harper JL (1974) Agricultural ecosystems. Agro Ecosyst 1:1–6
House G (2009) Retrofitting suburban landscapes with sustainable agroecosystems. In: Bohlen PJ, House G (eds) Sustainable agroecosystem management. Integrating ecology, economics, and society. CRC, Boca Raton, pp 283–296
Ikerd JE (2009) Rethinking the first principles of agroecology. In: Bohlen PJ, House G (eds) Sustainable agroecosystem management. Integrating ecology, economics, and society. CRC, Boca Raton, pp 41–52
International Assessment of Agricultural Knowledge, Science and Technology for Development (IAASTD) (2006) Agriculture at a crossroads. Global report of the international assessment of agriculture knowledge, science and technology for development. Island Press, Washington
Jordan N, Warner KD (2010) Enhancing the multifunctionality of US agriculture. BioScience 60 (1):60–66
Jordan N et al (2007) Sustainable development of the agricultural bio-economy. Science 316:1570–1571
Lotka AJ (1925) Elements of physical biology. Williams & Wilkins Company, Baltimore
Lowrance R, Stinner BR, House GJ (1984) Agricultural ecosystems: unifying concepts. Wiley, New York
Lowrance R, Hendrix PF, Odum EP (1986) A hierarchical approach to sustainable agriculture. Am J Altern Agric 1(4):169–173
Lyson TA (2004) Civic agriculture: reconnecting farm, food, and community. University Press of New England, Lebanon
Mader P et al (2002) Soil fertility and biodiversity in organic farming. Science 296:1694–1697
Marshall EJP (2006) Agricultural landscapes: field margin habitats and their interaction with crop production. J Crop Improv 12(1–2):365–404
Millennium Ecosystem Assessment (MEA) (2005) Ecosystems and human well-being: biodiversity synthesis. World Resources Institute, Washington
Ministry of Agriculture, Fisheries and Food (MAFF) (2005) Towards sustainable agriculture. A pilot set of indicators. MAFF Publications, London. http://www.maff.gov.uk. Accessed 28 Mar 2014
O'Leary M, Chia R (2007) Epistemes and structures of sense-making in organisational life. J Manage Inq 16(4):392–406. doi:10.1177/1056492607310976
Organisation for Economic Cooperation and Development (OECD) (1997) Environmental indicators for agriculture: concepts and frameworks, vol 1. OECD, Paris
Organisation for Economic Cooperation and Development (OECD) (1999) Environmental indicators for agriculture: issues and design, vol 2. OECD, Paris
Organisation for Economic Cooperation and Development (OECD) (2001) Environmental indicators for agriculture: methods and results. OECD, Paris
Organisation for Economic Cooperation and Development (OECD) (2013) OECD compendium of agri-environmental indicators. OECD, Paris

Phelan PL (2009) Ecology-based agriculture and the next green revolution. In: Bohlen PJ, House G (eds) Sustainable agroecosystem management. Integrating ecology, economics, and society. CRC, Boca Raton, pp 97–135

Reganold JP et al (2001) Sustainability of three apple production systems. Nature 410:926–930. doi:10.1038/35073574

Renting H et al (2009) Exploring multifunctional agriculture. A review of conceptual approaches and prospects for an integrative transitional framework. J Environ Manage 90(Suppl 2):S112–S123

Richardson JL (1977) Dimensions of ecology. The Williams & Wilkins Company, Baltimore

Smit BE, McNabb D, Smithers JA (1996) Agricultural adaptation to climatic variation. Clim Change 33(1):7–29

Spedding CRW (1975) The biology of agricultural systems. Academic, London

Stark CR Jr (1995) Adopting multidisciplinary approaches to sustainable agriculture research: potentials and pitfall. Am J Altern Agric 10(4):180–183

Steyaert O, Jiggins J (2007) Governance of complex environmental situations through social learning: a synthesis of SLIM's lessons for research policy and practice. Environ Sci Policy 10(6):575–586

Stinner DH (2009) Evolution of agroecosystem management in the life of Benjamin R. Stinner. In: Bohlen PJ, House G (eds) Sustainable agroecosystem management. Integrating ecology, economics, and society. CRC, Boca Raton, pp 13–20

Swinton SM et al (2007) Ecosystem services and agriculture: cultivating agricultural ecosystems for diverse benefits. Ecol Econ 64:245–252

Tellarini V, Caporali F (2000) An input/output methodology to evaluate farms as sustainable agroecosystems. An application of indicators to farms in Central Italy. Agric Ecosyst Environ 77(1–2):111–123

Tomich TP, Brodt S, Ferris H, Galt R, Horwath WR, Kebreab E, Leveau JHJ, Liptzin D, Lubell M, Merel P, Michelmore R, Rosenstock T, Scow K, Six J, Williams N, Yang L (2011) Agroecology: a review from a global-change perspective. Annu Rev Environ Resour 36:193–222

Turner BL II, Lambin EF, Reenberg A (2007) The emergence of land change science for global environmental change and sustainability. Proc Natl Acad Sci U S A 104(52):20666–20671

Warner KD (2007) Agroecology in action. Extending alternative agriculture through social networks. MIT Press, Massachusetts

Westra J, Boody G (2009) Challenges and benefits of developing multifunctional agroecosystems. In: Bohlen PJ, House G (eds) Sustainable agroecosystem management. Integrating ecology, economics, and society. CRC, Boca Raton, pp 213–233

Wilson GA (2007) Multifunctional agriculture: a transition theory perspective. CABI Publishing, Wallingford

Witkamp M (1971) Soils as components of ecosystems. Annu Rev Ecol System 2:85–110

Woodwell GM, Wurster CF Jr, Isaacson PA (1967) DDT residues in an East Coast Estuary: a case of biological concentration of a persistent insecticide. Science 156(3776):821–824

Regulation of Agroecosystems: A Social Systems Analysis of Agroecology and Law

E.B. Noe and H.F. Alrøe

Abstract There are two main challenges for law and policy to foster a sustainable development of agri-food systems through regulation. The first challenge is that the regulation of a certain aspect often does not lead to the intended outcomes. Regulative measures can only perturb, disturb or irritate agroecosystems because they are self-organising and autopoietic systems. All regulation of autopoietic systems depends on self-regulation because it is the ability of the agroecosystem to observe the disturbance and its internal schema of logic that will define the reaction. The second challenge is that regulation of one aspect often leads to unforeseen and unwanted side effects regarding other aspects. These unintended effects call for more regulations to deal with them, leading to a paradoxical situation of an increasingly growing web of regulation and effects, a situation that is concretely reflected in the exponential growth rate of the amount of positive law on agriculture and environment. These challenges are amplified by the increasing complexity created by specialisations in science, law and farming practice, a complexity that cannot be dealt with by further specialisation. In this chapter, we argue, based on social systems theory, that there is a need for a second-order platform of agroecological regulation where different scientific and law perspectives can meet and communicate about sustainable development and regulation of agroecosystems. But it requires that each perspective acknowledges its own blind spots and acknowledges that the agroecosystem can be seen from many other perspectives.

Keywords Agri-food systems • Perspectivism • Polyocular framework • Second-order science • Sustainability

E.B. Noe (✉) • H.F. Alrøe
Department of Agroecology, Aarhus University, Aarhus, Denmark
e-mail: egon.noe@agro.au.dk, http://www.agrsci.au.dk; hugo.f.alroe@gmail.com, http://hugo.alroe.dk

M. Monteduro et al. (eds.), *Law and Agroecology*,
DOI 10.1007/978-3-662-46617-9_2

1 Introduction

In the last 30 years, there has been an increasing focus on agri-environmental policy and regulation to support a sustainable development of agri-food production systems, and the growth rate of positive law on agriculture and environmental regulation has been exponential.[1] The sustainability focus not only is on safe and healthy food production but also includes a range of other concerns like protection of environment and nature, biodiversity, landscape, rural livelihood and food sovereignty. Different approaches have been followed to deal with these issues, like multifunctionality, cross-compliance, voluntary measures, organic agriculture, etc. Despite these attempts, law and policy are increasingly being challenged in fostering sustainable development of agri-food production, and there is a need for a more holistic and integrative approach to deal with policy and regulation of the area of agro food systems and thereby a need for better interaction and trans-disciplinarity between law and agroecology.[2] Agroecology is globally seen as a concept that can host such a holistic effort,[3] especially among the Latin American scholars.[4] In this chapter, we will address the challenges and potentials of using the notion of agroecology as a platform for holistic agri-environmental policy and regulation from a social science perspective, or more specifically a social systems theoretical perspective.

2 The Problems of Regulation of Agroecosystems

As we see it, there are two main challenges in sustainable regulation of agri-food systems. The first challenge is that the intended regulation of a certain aspect often does not lead to the expected outcomes, e.g. pesticides tax does not have the expected effect on the use of pesticides, or the regulation of the use of manure does not have the intended effect on reduction of nutrient losses to the environment. There may be several possible explanations of this lack of effects. It could be due to the fact that the technical/biological understanding underlying the regulation is not good enough, e.g. that the effect of a certain action is over- or underestimated. Another explanation could be that the real systems do not resemble the models and understanding that underlies the calculation of the expected outcome (e.g., due to different cropping systems, different technologies or different farming practices). A further explanation could be that the farmers or the management of the agro-systems do not behave as expected by the technical and economic calculation, e.g. pesticide taxation does not lead to the estimated reduction in use of pesticides

[1] Monteduro (2013), pp. 2–11.

[2] Monteduro (2013).

[3] Dalgaard et al. (2003), pp. 39–51; Pretty (2008), pp. 447–65.

[4] Altieri (1995); Wezel et al. (2009), pp. 503–515.

because the farmers use other logics of calculation. This could be in the case of very large farms that use the rationale that when they are driving with the sprayer with fungicides anyway, they might just as well add an insecticide against aphids, to avoid the risk of being forced to drive again for an extra treatment. In this calculus, the concern for organising manpower and the capacity of equipment is much more important than the particular cost of a pesticide. Or this could be because the farmers hold other values than the partial optimisation of profit underlying the pesticide taxation, e.g. in the case of farming systems where clean fields and high-yielding crops have a value in itself, strongly linked to the professional identity of the farmer. Such values also change the calculus behind the decision-making completely. Prescriptive regulation, like the growth of green catch crops, is another example of possible discrepancy between expected and actual outcome because the effect of green catch crops is dependent not only on the establishment of catch crop (that the action is taken) but to a large extent on how the crop is established (how the action is taken).

These examples illustrate that it is not only a matter of understanding the technical and economic aspects of the farming systems but also an issue of conflicting values. For instance, in the case of a very dedicated arable farmer, he will by all means try to obtain high yields and clean and professionally looking fields. And the more legal measures of regulation that are put forward to reduce the intensity of arable production to protect the environment and nature quality, the more this farmer will move into opposition the legislation and search for ways of avoidance. This can be described as a dilemmatic situation of regulation.[5]

The second challenge is that regulation of one aspect often leads to unforeseen and unwanted side effects regarding other aspects. For example, the regulation of animal welfare in terms of outdoor pigs can result in increased loss of nutrients to the environment,[6] and support to biogas production can lead to reduction of organic farming due to the competition for land between growing corn for biogas and roughage for organic dairy farming.[7] Another example is that legal regulation of food safety in chicken production contributes to the closure of small poultry slaughter houses,[8] which amplifies industrialisation of chicken production, decrease in animal welfare and rural depopulation in terms of small holders. These unintended effects call for more regulations to deal with them and lead to the increasingly paradoxical situation[9] that the more regulation that is put forward to deal with the unintended side effects, the more likely they will produce new unintended side effects in an increasingly growing web of correlated effects and side effects. This mechanism is concretely reflected in the exponential growth rate of the amount of positive law on agriculture and environment.[10]

---

[5] Alrøe and Noe (2011), pp. 152–67.

[6] Eriksen et al. (2006), pp. 256–266.

[7] Schwarz et al. (2012), pp. 235–62.

[8] Antle (1996), pp. 1242–47.

[9] Alrøe and Noe (2011).

[10] Monteduro (2013).

In addition, this paradox of regulation is exacerbated by the fact that within the fields of science, law and policy there is an ongoing process of specialisation and differentiation into sub-branches focusing on different aspects, to deal with this increasing complexity.[11] And due to this increasing complexity and specialisation, there is literally no communication between the different bodies of policymaking and regulation.

The point is that how farming systems react on a certain legal regulation cannot be predicted or understood from an isolated economic, biologic, agronomic or social perspective but must be understood from an insight into the farming systems themselves. We thereby share the realisation that there is a need for new and integrative way to deal with the challenge of regulation of sustainable agri-food systems. We so also share the idea that the notion of agroecology can serve as a platform to establish and build such integrative approaches.[12] However, we do not see any possibility to turn the wheels back and establish a unitarian perspective on agroecology, and agroecological regulation, basically because the increase of complexity is irreversible and therefore also the need for specialisation and differentiation.[13] Instead, our argument is that we should see agroecology as a polyocular platform of second-order observations.

The aim of this chapter is to present our theoretical understanding of agroecology as a polyocular platform for second-order observation of the sustainability of agroecosystems based on a perspectivist theoretical framework[14] and our theoretical understanding of agroecosystems as autopoietic self-organising systems based on a social systems theoretical approach[15] and, furthermore, to discuss how these two insights can be used to develop an integrative platform for sustainable agroecological regulation. First, we will go into a deeper discussion of how to understand agroecology as a case of second-order observation.

3 Agroecology as a Platform for Second-Order Observation of Sustainable Agriculture

The notion of agroecology is widely used in the literature in the meaning of a study of interactions between soil, plants, animals and humans[16] or a study of technical, natural, social and human aspects.[17] It is strongly linked to the normative perspec-

---

[11] Noe and Alrøe (2015).

[12] Monteduro (2013).

[13] Luhmann (1984); Noe and Alrøe (2015).

[14] Alrøe and Noe (2014).

[15] Noe and Alrøe (2006), pp. 34–48.

[16] Dalgaard et al. (2003).

[17] Pretty (2008).

tives of sustainable farming based on ideals of how the interactions between these systems or aspects should be managed in a holistic and sustainable way, taking into account the environmental, agronomical, social and societal dimensions. Especially in the South American tradition, agroecology is seen also as a movement against large industrial farming systems, favouring small-scale family farming,[18] parallel to the partly organic farming discourse in the global north.[19]

Our way of understanding and using the notion of agroecology in this context is as an epistemological approach to observe agroecosystems based on the wider perspective of sustainability and sustainable development. We see agroecosystems as heterogeneous systems,[20] hybrids between technical, economical, biological and social systems, with the potential of many different kinds of observations from many different and relevant perspectives with regard to different values.

We share the common understanding that agroecology needs a multidisciplinary systems approach. But from our point of view, this also means that none of the perspectives involved in themselves can observe "agroecosystems" as such. In science, "agroecosystem" will always be defined by the "scientific eyes" observing. For example, a biological perspective observing an agroecosystem will focus on the biological processes and interactions going on, an economic perspective will define the system as a flow of money and transformations of assets, a sociological perspective will focus on human interactions and how the involved humans interact with nature, technology and economy. Our point is that, on the one hand, these different perspectives are needed to observe the system from an agroecological perspective, but, on the other hand, the perspectives are not simply puzzle bricks that add up to an agroecological understanding of the agroecosystem (see Fig. 1).

None of the individual perspectives can observe the agroecosystem as such; agroecology as a research perspective relies on these first-order perspectives and does not have its own first-order perspective to observe an agroecosystem as such. None of the disciplinary first-order perspectives involved offers a point of observation that can include the other observations, which means that no position has a privilege point of observing an agroecosystem.

To illustrate this, we can use the examples of "nature quality" and "rural livelihood". The research perspective of nature quality will focus on biodiversity and how different aspects and actions of the agroecosystem affect or will affect biodiversity. The description of the agroecosystem will mainly be a description of the prevalence of species in small biotopes and the degree of disturbance of these biotopes. The perspective of rural livelihood will focus on the living conditions of the people involved in the agroecosystem as well as living conditions in the communities surrounding the agroecosystem. Are there work opportunities, good working conditions, good conditions for family life, etc.? A description from this perspective will typically include the number of employees, salaries, work hours,

---

[18] Altieri (1987).

[19] Alrøe and Noe (2008), pp. 5–22.

[20] Noe and Alrøe (2006).

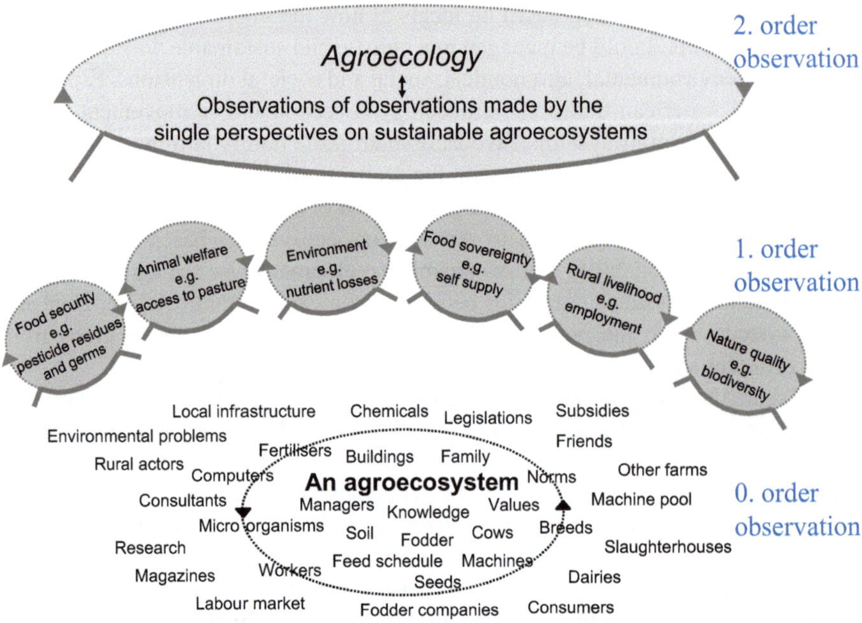

Fig. 1 Agroecology as a second-order science based on observation of first-order observations, with the agroecosystem as the shared object (0. order observation) (modified after Noe et al. 2008, pp. 1–15)

degree of self-supply and infrastructure. None of these descriptions can be integrated in the other perspectives. Biodiversity is only relevant for the rural livelihood perspective if it, e.g., increases the value of living there, increases the possibilities for rural tourism or increases the possibilities to collect food. And the employees are only relevant to the description of biodiversity if they disturb or contribute to maintaining biodiversity, e.g. by keeping sheep or cattle for nature conservation purposes. The two perspectives are so to speak blind to each other's perspectives. And none of them are able to see the potential synergy effects between the two aspects. The potential synergies can only be explored as a process of another logic than that belonging to the first-order observations. But this requires the involvement of both perspectives, the same way that binocular sight depends on the observations from two eyes to see depth. Depth is a mental construction that does not belong to either of the two eyes but cannot be made without synchronised observations of both eyes.[21]

Our claim is therefore that we need to understand agroecology as a second-order observation of the first-order observations of the single perspectives involved in what we call a multi-perspectival and polyocular approach.[22] We see it as a

[21] Alrøe and Noe (2014); Bateson (1979).

[22] Alrøe and Noe (2014).

multidisciplinary way to study agriculture and food production that focus on the sustainability of the agroecosystems studied, implicitly based on the values underlying the discourses of sustainability.

However, if no first-order perspectives have access to understanding and observing the agroecosystem as such (0.-order), how can we then operate with the ontological idea of an agroecosystem and thereby coordinate and synchronise so that the different perspectives observe the same object? To address this issue, we draw on the systems theoretical idea of autopoiesis, the idea that agroecosystems are self-organising systems. This is what makes it possible to observe the agroecosystem as an object from different angles, and which also makes agroecosystems able to observe themselves as organisation systems but not necessarily or likely as agroecosystems.

4 The Autopoietic Understanding of an Agroecosystem

Maturana and Varela define all living organisms as autopoietic systems, which means that they are self-creating and self-organising systems. Living organisms are operationally closed but open to material flows. Niklas Luhmann has adapted this understanding to encompass social systems, which he claims operate in communication.[23] We have further developed this theory to comprehend hybrids like agroecosystems.[24]

In Fig. 1, we illustrated an agroecosystem as potentially involving a lot of different heterogeneous elements like tractors, cows, knowledge, etc. And it is easy to comprehend that there is a potential surplus of elements that can be included in an agroecosystem, in terms of different plant and animal breeds, different technologies, etc. But each included element also offers a surplus of possibilities, e.g. a computer can be used to collect and process enormous amount of production data, or it can be used as a means of searching technical information on the internet or as a means of communication. The elements offer the possibilities, but they do define which possibilities are actualised; this must be defined by the agroecosystem (contingency). Like each word offers a surplus of meaning, but it is the sentence, the text or maybe the wider discursive context that defines what the meaning is or how it is meant to be actualised. In Fig. 1, the circle with two arrows illustrates this dynamic self-organising process of selection of elements and possibilities.

The basis of this approach to see a farm as a self-organising system is that the system can be observed as a continuous process of decision-making forming a more or less coherent strategy. From a systems theoretical point of view, there are two important dimensions of a decision. One is the systems closure dimension, in terms

[23] Luhmann (1984).

[24] Noe and Alrøe (2006).

of what options belong to the system and what potential options are excluded from the system. The other is the time dimension, that every decision needs to mark its present with a past and future.

To deepen the first dimension, systems closure, decisions should here be understood in relation to a semiotic understanding of contingency. In a Peircean terminology, every (dynamic) object has a surplus of meaning, not only in terms of actualised meaning but also in unknown potential meaning in relation to some interpretant.[25] To give an example, if we look at "cow" as a dynamic object and the farming system as the interpretant, there is a surplus of possibilities of how a cow can be interpreted in the farming system. It can be interpreted as a dairy cow yielding milk, a beef cow yielding beef, a grazing cow performing nature conservation, etc. The point is that not all these possibilities can be actualised at the same time and that the (dynamic) object in itself does not define what is realised and what is not; in a semiotic understanding, this belongs to the immediate object and the interpretation as parts of a triadic sign relation.[26] From this viewpoint, the way an organisation system creates itself, by closure, is through a continuous process of negotiation and decisions on what possibilities (interpretations) belong to the system and what possibilities are excluded.[27] These selections or interpretations are faced with contingency, in the sense that, on the one hand, the organisation system is forced to decide and, on the other hand, it could have made other choices of interpretation.

To deepen the time dimension of decision-making, every decision marks a before and after. For example, an agreement or contract is not valid without a date, which marks that now the decision has been made and will frame the future differently than the past. Introducing time supports two important insights. The first is that an organisation system needs to be seen as a flow of decisions continually marking both an inside and outside of the system and a before and after. The other insight is that an organisation system is forced to take a development pathway, not determined by the surroundings of the systems but determined by the decisions made by the system in reaction to its surrounding world.[28]

From this autopoietic understanding of a farm as an autopoietic self-organising system follows some other features. The continuous flow of decisions (interpretations) must be systems internal operations. It is only the system itself that can define what interpretations or immediate objects belong to the system. A way to understand this is that the organisation system has to produce and reproduce its own schema or logic, which connects the flow of decisions. Such different production logics can easily be observed empirically, and detailed studies have shown that there are several different viable development strategies.[29] Underlying these

[25] Alrøe and Noe (2014).

[26] Peirce (1994).

[27] Noe and Alrøe (2006), p. 45.

[28] Alrøe and Noe (2012), pp. 39–52.

[29] See, e.g., van der Ploeg and Long (1994).

systems logics is meaning, in the sense of Viktor Frankl's notion of meaning understood as meaningfulness.[30] Instead of understanding farming systems as primarily goal-orientated systems, we see autopoiesis as meaning driven. The meaning is inherent to the system and is not an option for negotiation. Farming systems may therefore also be understood as logo-poietic systems, self-organising and autopoietic systems that are driven by the will to meaning.

This autopoietic understanding also means that it is the organisation system itself that has to observe and react to changes in the surrounding world. For instance, if the price of milk is decreasing, it is up to the dairy farming system to recognise this as a difference it has to react to. And how to react to these changes is linked to the system's logic, e.g. either to increase milk yield or to expand herd size. No organisation systems are able to be sensitive to all changes in their surroundings, so often it can be observed that the systems are sensitive only to the changes that seem most important to their strategies, and thereby they reduce the complexity of their "Umwelt" or phenomenological world.[31]

This means that how an agroecosystem is organised and how it reacts on changes in the surrounding environment and perturbations of the system cannot be understood from a universal logic of the social, technical, economic or biologic systems but needs to be understood from the internal logic of the agroecosystem as an autopoietic self-organising system.

5 Implications for Understanding the Regulation of Agroecosystems

The systems theoretical understanding of an agroecosystem as a self-organising system and the understanding of agroecology as a platform for multidisciplinary second-order observations from a sustainability perspective have a range of implications for how we can understand and develop an integrative platform for sustainable agroecological regulation and thereby for how to bridge between agroecology and law. We will start with the implications of the autopoietic understanding of agroecosystems as object for regulation and how agroecology as second-order multidisciplinary platform can serve also as a platform for the integration of law and regulation from the perspective of agroecology.

[30] Frankl (1962).

[31] Alrøe and Noe (2012) and Noe and Alrøe (2015).

5.1 The Challenges of Legal Regulation of Agroecosystems

The autopoietic understanding of agroecosystems has strong implications for our understanding of the conditions for regulation. Firstly, the operational closure means that there isn't any direct access to the autopoiesis. Regulative measures can only perturb, disturb or irritate the autopoiesis of the agroecosystems, no matter whether it is prescriptive, economic or normative measures that are applied. All regulation of autopoietic systems depends on self-regulation. Secondly, there is no one-to-one causal relationship between the intended logic behind the measures and the reactions in the agroecosystems. It is the ability of the agroecosystem to observe the disturbance and its internal schema of logic that will define the reaction. To illustrate this, we again use pesticide tax as an example. How an agroecosystem reacts on a certain tax cannot be calculated solely as a cause–effect relation. If the tax is high, it of course has an effect in irritating or disturbing the agroecosystem, but from a social systems theoretical understanding the effect is not defined by the tax but by the agroecosystem. An agroecosystem continually needs to react on changes in the encompassing world, and any form of regulation can be seen as changes in the environmental setting of the agroecosystem. But the reactions depend on the system's logic and related values and thereby on how the agroecosystem has reduced the complexity of its environment. If it is a very market-orientated agroecosystem, it will likely react on even very small price fluctuations, e.g. by changing crop rotation or changing input factors. If it is a very production-orientated agroecosystem, it may adopt new breeds or technologies even with an expectation of small increases in yield.

In Table 1, we have used the understanding of agroecosystems as autopoietic systems as a framework to analyse the system reactions and pros and cons for three different forms of regulation. None of these forms of regulation target the agroecosystem as a whole. They work through technical, biological or behavioural regulation (perturbation) of the systems based on an underlying understanding of how the systems are functioning. Different types of regulative measures have different pros and cons, and as indicated in the table the reaction of the systems and the effects of the measures in the individual agroecosystems will depend on the system's logic and values.

In the case of prescriptive regulation, it is possible to ensure and control that the intended actions are taken, e.g. that 60 % of the fields are covered with catch crops. But the effects of this regulation are highly dependent on the agroecosystem due to, for instance, different soil types, crop rotations and management practices; for example, are the catch crops established properly and in time to be able to retain nutrients?

Looking at economical incitements, even small changes in taxes on, e.g., pesticides may have an effect on market-orientated agroecosystems, while even very high taxes will have no effect on the internal organisation of agroecosystems orientated towards high productivity, although they may affect the economy of the systems notably. Normative measures like voluntary schemes, support to extension,

Table 1 Advantages and disadvantages of different forms of regulation analysed from an agroecosystems perspective

Forms of regulation	Examples of measures	System reactions		Pros	Cons
		System logic	System values		
Legal: injunction/ prohibition	Green catch crops	The effect of the catch crops is very dependent on how the system is organised. If the rationale is not shared, the reaction can be contrary.	System values play only an indirect role.	It is possible to control.	The real effect is unknown, and the side effect to the system is unpredictable.
Incitements: taxes/ subsidies	Pesticides tax	The sensitivity to taxes is dependent on both the values and logic of the system.		It regulates directly on the target; there is less disturbance of the autopoiesis of the agroecosystem.	High taxes are imposed to make all agroecosystems react; it leads to permanent dependence on taxes.
Normative: campaigns/ information	Voluntary agreement on pesticide reduction	It will only be a part of the system's logic if it becomes incorporated in the system's values.	Sensitivity is very dependent on values; some react very strongly in the intended direction, while others react against.	There is cheap and little control. It is co-constructive with the autopoiesis of the systems.	The agroecosystems that do not share the intention behind the campaign may react contrary.

and recommendations for good farming practices may have an effect on agroecosystems that already hold a strong focus on sustainability but can have the opposite effect on agroecosystems based on other values, reflected in statements like "As long it is legal to spray, I will do so and need to do so to be competitive with other farmers". And the more and stronger normative measures used, the more frustrated the farmers already dedicated become, and the less the other farmers care.

As demonstrated, the autopoietic understanding of an agroecosystem exposes the challenges to sustainable regulation because it cannot be foreseen how the system will react to exposure of different regulation measures based on one perspective only. Here we argue that law and regulation are facing the same challenge to obtain a platform for agroecological regulation that agroecological research is facing to study the sustainability of agroecosystems.

5.2 The Polyocular Understanding of Agroecological Regulation

Agroecological regulation does not have its own first-order perspective from where it can observe an agroecosystem as such (see Fig. 2). Law systems in general must necessarily be based on second-order observations. In any kind of regulation, the law perspective has to build on other perspectives observing the agroecosystem, either as an economical system through the lens of economy, a biological system through the lens of biology, an agronomical system through the lens of agronomy, etc. If the agroecosystem is understood as a technical system, law and regulation will be targeting technical matters such as the handling of liquid manure or the requested space for the animals. If it is seen as an economical system, law and regulation will be targeting taxes and subsidies. The choice of scientific perspective has consequences not only for the measures that are applied by the law systems but also for how the effects are measured and, more importantly, for what effects are not measured.

To deal with the increased complexity created by the differentiation of scientific perspectives and the institutional differentiation in policy and administration, the law systems have also undergone a differentiation. In our understanding, the differentiated branches of law systems are facing the same challenge in producing a coherent and sustainable regulation of agroecosystems as the different research perspectives do in establishing a coherent agroecological research perspective. Our contention is that there is not only a need for cooperation between law and science in agroecology but that the both lawmakers and researchers need to meet on the second-order platform of agroecology to obtain a polyocular view on the potentials for sustainable regulation and support of the development of agroecosystems.

But how to organise such platform in practice is not an easy case. It needs to be institutionalised in some way avoiding to be just another differentiation of science and law perspectives. Here we have a few principles to follow:

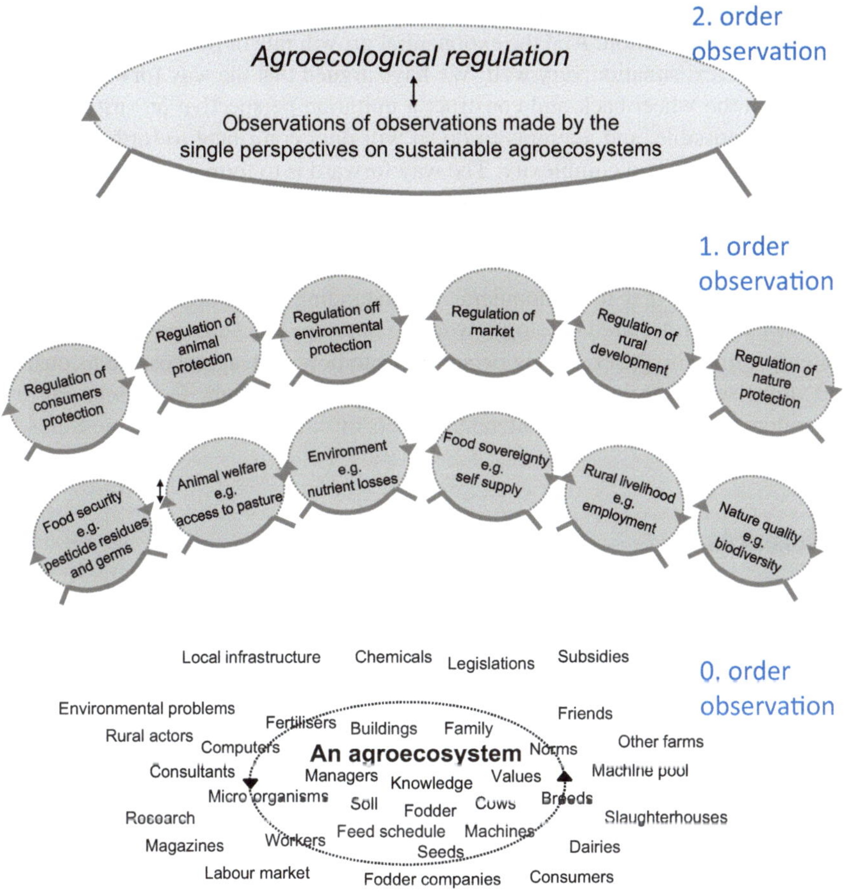

Fig. 2 Illustration of the second-order polyocular platform of agroecological regulation

- There must be time, room and resources for meeting and participating in this polyocular communication.
- Polyocular communication can only be organised around a specific problematic.
- The need for polyocular communication to be acknowledged by the different science and law perspectives involved.

6 Concluding Remarks

We have argued that the platform of agroecological regulation is fundamental for scientific perspectives and law perspectives to meet and communicate about sustainable development and regulation of agroecology. The increased complexity created by specialisation in science, law and farming practice[32] cannot be dealt with by further specialisation. And the exponential growth rate of positive law illustrates this paradoxical situation very well. We have argued that the way forward is not to try to turn the wheel back and construct a unitarian perspective on sustainability; this is not possible, and attempts to do so will only contribute to further differentiation and increase in complexity. The way forward is to form a new ground where different perspectives can meet and join in a second-order polyocular communication. But it requires that each perspective acknowledges its own blind spots and acknowledges that the agroecosystem can be seen from many other perspectives. We have argued that the autopoietic understanding of agroecosystems helps to establish an useful shared ontology (working ontology) that changes focus from the systems formed by the observing perspectives to how each agroecosystems observe and organise themselves. It also serves as a pivotal insight for discussing and observing the intended and unintended effects of different regulative measures.

References

Alrøe HF, Noe E (2008) What makes organic agriculture move: protest, meaning or market? A polyocular approach to the dynamics and governance of organic agriculture. Int J Agric Resour Gov Ecol 7(1/2):5–22

Alrøe HF, Noe E (2011) The paradox of scientific expertise: a perspectivist approach to knowledge asymmetries. Fachsprache (Int J Specialized Commun) XXXIV(3–4):152–167

Alrøe HF, Noe E (2012) Observing environments. Constructivist Foundations 8(1):39–52

Alrøe HF, Noe E (2014) Second-order science of interdisciplinary research: A polyocular framework for wicked problems. Constructivist Foundations 10(1):65–95

Altieri MA (1987) Agroecology: the scientific basis of alternative agriculture. Westview Press, Boulder

Altieri MA (1995) Agroecology: the science of sustainable agriculture. Westview Press, Boulder

Antle JM (1996) Efficient food safety regulation in the food manufacturing sector. Am J Agric Econ 78(5):1242–1247

Bateson G (1979) Mind and nature. A necessary unity. E.P. Dutton, New York

Dalgaard T, Hutchings NJ, Porter JR (2003) Agroecology, scaling and interdisciplinarity. Agr Ecosyst Environ 100(1):39–51

Eriksen J, Hermansen JE, Strudsholm K, Kristensen K (2006) Potential loss of nutrients from different rearing strategies for fattening pigs on pasture. Soil Use Manag 22(3):256–266

Frankl VF (1962) Man's search for meaning – an introduction to logotherapy. Simon & Schuster, New York

Luhmann N (1984) Soziale Systeme: Grundriß einer allgemeinen Theorie. Suhrkamp, Frankfurt [English edition: Luhmann N (1995) Social Systems. Stanford University Press, Stanford]

Monteduro M (2013) Environmental law and agroecology. Transdisciplinary approach to public ecosystem services as a new challenge for environmental legal doctrine. Eur Energy Environ Law Rev 22(1):2–11

Noe E, Alrøe HF (2006) Combining Luhmann and actor-network theory to see farm enterprises as self-organizing systems. Cybern Hum Knowing 13(1):34–48

[32] Noe and Alrøe (2015).

Noe E, Alrøe HF (2015) Sustainable agriculture issues explained by differentiation and structural coupling using social systems analysis. Agron Sustain Dev 35(1):133–144

Noe E, Alrøe HF, Langvad AMS (2008) A polyocular framework for research on multifunctional farming and rural development. Sociol Rural 48(1):1–15

Peirce CS (1994) The collected papers of Charles Sanders Peirce: electronic edition. Intelex Corporation, Charlottesville [reproducing Hartshorne C, Weiss P (eds) (1931-1935) Volumes I-Iv. Harvard University Press, Cambridge, and Burks A (ed) (1958) Volumes Vii-Viii. Harvard University Press, Cambridge]

Pretty J (2008) Agricultural sustainability: concepts, principles and evidence. Philos Trans R Soc B 363(1491):447–465

Schwarz G, Noe E, Saggau V (2012) Comparison of bioenergy policies in Denmark and Germany. In: Almås R, Campbell H (eds) Rethinking agricultural policy regimes: food security, climate change and the future resilience of global agriculture (Research in rural sociology and development, vol 18). Emerald Publishing, Bingley, pp 235–262

van der Ploeg JD, Long A (eds) (1994) Born from within: practice and perspectives of endogenous rural development. Van Gorcum, Assen

Wezel A, Bellon S, Doré T, Francis C, Vallod D, David C (2009) Agroecology as a science, a movement and a practice. A review. Agron Sustain Dev 29(4):503–515

Addressing Law and Agroecosystems, Sovereignty and Sustainability from a Legal Pluralistic Perspective

O. Hospes

Abstract This paper wants to contribute to the debate on the complex relationships between law and agroecosystems from a legal pluralistic perspective. For this purpose, it first explains what is legal pluralism, and then this notion is used to conceptualize law, the relationship between law and social fields, and sovereignty. Second, the paper critically reviews the concept of agroecosystems and explains how a legal pluralistic perspective can enrich agroecology as a transdisciplinary field of studies. Finally, to operationalize how the complex relationships between law and agroecosystems can be studied from a legal pluralistic perspective, it describes the rise of nonstate sovereigns over sustainability.

Keywords Agroecosystems • Law • Legal pluralism • Sovereignty • Sustainability

1 Introduction

This paper wants to contribute to the debate on the complex relationships between law and agroecosystems from a legal pluralistic perspective. For this purpose, I will first explain what legal pluralism is and then use this notion to conceptualize law, the relationship between law and social fields, and sovereignty. Second, I will critically review the concept of agroecosystems and explain how a legal pluralistic perspective can enrich agroecology as a transdisciplinary field of studies. Finally, to operationalize how the complex relationships between law and agroecosystems can be studied from a legal pluralistic perspective, I will describe the rise of nonstate sovereigns over sustainability. To this end, I will briefly review two legal orders for governing food systems and sustainability: food sovereignty and multistakeholder governance of sustainable cropping. Each of them can be seen as a counterframe to challenge neoliberal thinking, discourses and organization of politics, economics and agriculture in which food and nature are basically treated as commodities. My conclusion is that a legal pluralistic perspective on law and agroecosystems can be

O. Hospes (✉)
Department of Social Sciences, Public Administration and Policy Group, Wageningen University, Wageningen, The Netherlands
e-mail: Otto.Hospes@wur.nl, http://www.wageningenur.nl/en/Persons/Otto-Hospes.htm

M. Monteduro et al. (eds.), *Law and Agroecology*,
DOI 10.1007/978-3-662-46617-9_3

very helpful to identify different legal repertoires for governing food systems and sustainability. Two challenges can be distinguished in studies on the relationships between law and agroecosystems: first, to acknowledge the potential of legal pluralism and not to conduct legal "monocropping," focusing on the subjection of agroecosystems under one legal order, and, second, to identify and study interlegalities, or the intersection of different legal orders, as a space and opportunity for combining different nonstate and/or state laws, principles, norms.

2 A Legal Pluralistic Perspective on Law, Social Fields, and Sovereignty

In his seminal article on "What is legal pluralism?" Griffith explains that legal pluralism refers to "the presence in a social field of more than one legal order."[1] In a similar vein, Merry defined legal pluralism as "a situation in which two or more legal systems coexist in the same social field," asserting that "plural normative orders are found in virtually all societies."[2] In this way, legal pluralism "challenges a perceived monopoly of the state in making and administering law."[3] A legal pluralistic perspective on law holds that both state and nonstate actors are making law, conceived as a set of principles, norms, ideas, and rules, and that state and nonstate laws coexist in social fields and society. Nonstate actors include (representatives of) multinational companies, industry associations, farmer organizations, religious societies, family networks, charitable foundations, environmental NGOs, food sovereignty movements, etc.

From a legal pluralistic perspective, there is no one-way relationship between law as the independent variable and social fields as the dependent one. Social fields both are subject to external norms and have the capacity to generate own rules and to digest external norms. To qualify this property, the concept of the "semi-autonomous social field" was coined by Moore,[4] who has been "unanimously applauded among legal pluralists for having provided the appropriate locus of law in socio-legal research."[5] Social fields, for example, can be "an arena in which a number of corporate groups deal with each other," but, "Also, the corporate groups themselves may each constitute a semi-autonomous social field."[6] A social field can also be a farmer community, cultural entity, or neighborhood.

[1] Griffith (1986), pp. 1–55.

[2] Merry (1988), pp. 869–896, exp. 870 and 873.

[3] Michaels (2009), pp. 1–35, exp. 3.

[4] Moore (1973), pp. 719–746.

[5] Dupret (2007).

[6] Moore (1973), p. 722.

The concept of legal pluralism has traditionally been used by legal anthropologists to analyze the intersections of European, religious, and customary laws in former colonies. Merry called this "classical legal pluralism."[7] At the end of the 1970s, this classical notion gave way to "new legal pluralism": scholars of legal pluralism broadened the geographical focus of their studies to include industrialized societies in Europe and the USA and began to investigate relations between dominant groups and subordinate groups, such as religious, ethnic, or cultural minorities, immigrant groups. A third and dramatic step has been the shift from studying legal pluralism in a "small field observable to an anthropologist"[8] at the local level to exploring "global legal pluralism"[9]: global legal pluralism, first of all, refers to the coexistence of legal orders that span the globe or large parts of it, including transnational, supranational, and international laws; second, global legal pluralism is about the coexistence of differently scaled legal orders in one social field.[10] The social field can be a small field observable to an anthropologist and also an international network or transnational chain of economic transactions that requires the anthropologist to observe at different places and levels. Generally put, processes of globalization have given an enormous boost to legal pluralism as an object of study. According to Michaels,[11] legal pluralism "has recently moved into the mainstream of legal discourse. The most important reason is globalization."

A distinctive characteristic and driver of globalization have been that "more and more non-state institutions, from corporations to cultural communities and churches to criminal organizations, are asserting sovereignty of greater or lesser scale."[12] From a legal pluralistic perspective, the definition of state sovereignty as the supreme authority within a territory is too limited. As a result of globalization, different types of sovereignties have evolved: "Sovereignties over terrains and their inhabitants; sovereignties over transactional spheres, networks of relations, regimes of property; sovereignties over people conjoined in faith or culture."[13] Sovereigns can be both state and nonstate actors, producing different kinds of law in terms of scale, projection, and symbols.[14]

[7] Merry (1988).

[8] Moore (1973), p. 720.

[9] Berman (2007), pp. 1155–1238; Michaels (2009).

[10] Michaels (2009).

[11] Michaels (2009), p. 1.

[12] Comaroff and Comaroff (2009), pp. 31–59, exp. p 39.

[13] Comaroff and Comaroff (2009), pp. 39.

[14] de Souza-Santos (1987), pp. 279–302.

3 A Critical Review of the Concept of Agroecosystems from a Legal Pluralistic Perspective

An agroecosystem has been defined as "an ecosystem under agricultural management, connected to other ecosystems"[15]; "the organisms and environment of an agricultural area considered as an ecosystem" (Merriam-Webster Dictionary); or, simply, "an ecosystem on agricultural land" (Oxford Dictionary). The concept suggests that the organization of one particular kind of human activity (agriculture), living organisms (like plants and animals), and nonliving components (like air, water, and mineral soils) are interconnected and together form an integrated whole, which fulfils different functions or provides different services. Ecosystem services are defined as "the benefits people obtain from ecosystems,"[16] including provisioning (for instance, of food and water), regulating (for instance, of climate and diseases), supporting (such as nutrient cycling and crop pollination), and cultural services (like cultural diversity and spiritual benefits).

My first critique on the concept of agroecosystems is that it neglects the role of legal orders and social fields in defining benefits, services, values. Whereas the concept is holistic in paying attention to interconnectedness between agriculture and ecological processes in a particular area, the concept is reductionistic in leaving out the interconnectedness between agroecosystems and plural legal orders and ignoring the capacity of social fields to generate principles, norms, values.

My second critique is that the concept suggests something like harmony or a fit between agriculture and ecosystems, whereas the relationship may be or become full of tensions: agricultural practices may destroy ecosystems, and changing ecosystems may lead to the demise of agriculture as a practice and system. Law may help to address these tensions but also may fuel them. In a similar vein, legal pluralism can serve as a repertoire to address these tensions but can also be a source of tension itself.

My third critique is that the concept is biased towards a notion of territorial synchrony of agriculture and ecosystems and, more generally, a territory-based notion of both agriculture and ecosystems. However, an ecosystem may include different agricultural systems, and vice versa an agricultural system may trespass the boundaries of different ecosystems. Also, agriculture in a specific area may be connected to other, more or less distant, ecosystems through provision of inputs, sale of marketable surplus, or disposal of waste. "Distant laws" of state and nonstate actors that regulate these flows may affect the use of land, water, plants, and trees in an agroecosystem.

Whereas agroecology has been widely acclaimed as a transdisciplinary field of studies (see CAPORALI, BUONGIORNO and many others in this volume), my contention

[15] OECD (2003).

[16] Wallace (2007), pp. 235–246.

is that this field still can benefit enormously from the use of a legal pluralistic perspective on law, social fields, and sovereignty in several ways.

The first way is to explore the interconnectedness of different legal systems and normative orders with agroecosystems. Different state and nonstate actors generate rules, principles, and norms that serve as more or less incompatible normative frameworks for use and conservation of land, water, and/or soils. Expanding the idea of exploring a kind of symbiosis between agriculture and ecosystems to form complex agroecosystems, I propose to investigate how legal and agroecosystems form an integrated whole, in which the benefits that people obtain from ecosystems are defined not just by the material properties and dynamics of an agroecosystem but also by values, principles, rules, and norms as immaterial components of legal systems.

The second way is to study how agroecosystems are linked to one or more semiautonomous social fields, which may exist within the geographical boundaries of an agroecosystem and also may cross these boundaries to cover (parts of) different agroecosystems. This means that the question is not only how different external rules, principles, and norms affect an agroecosystem and its services but also how rules, principles, and norms generated by social fields within an agroecosystem affect the use and conservation of land, water, and soils. As semiautonomy can also be considered a property of agroecosystems, agroecologists and legal pluralists face a common methodological challenge in studying their object.

The third way is to consider agroecosystems as part of legal orders that span the globe or large parts of it. Legal pluralism is referring not only to coexistence of state and nonstate laws and to coexistence of different laws in terms of projection or subject matter (biodiversity, agriculture, environment, food security, property rights, etc.) but also to coexistence of differently scaled legal systems in or linked to (different components of) agroecosystems. From a pluralistic perspective on law, global laws can coexist in agroecosystems and social fields, next to local, regional, and national laws—together forming a complex of normative frameworks for use and conservation of land, water, and soils.

The fourth way is to acknowledge a plurality of sovereignties over an agroecosystem, seen as a complex system with material (land, crops, etc.) and immaterial dimensions (transactions, relations, property rights, culture, etc.). Different state and nonstate actors may claim to be the politically legitimate and supreme authority over different dimensions of an agroecosystem: some may consider themselves to be the sovereign over territory, others over a particular category of "inhabitants" (for instance, agricultural producers), and again others over culture. Each of them may use a particular kind of law to govern a particular component of an agroecosystem.

4 The Rise of Nonstate Sovereigns over Sustainability

There is no widely shared or universal definition of sustainability. At a general level, at least four kinds of definitions of sustainability can be distinguished: sustainability may refer to a property, an ability, a value or norm, or a desired situation. Some assume that sustainability is about a system that will remain the same forever; others use the concept to explore who or what makes a system able to survive, to resist or mitigate threats from outside and within, and to adapt. Some use sustainability in one particular sense (economic, environmental, social); again, others emphasize that sustainability is typically multidimensional. In any case, time and scale matter much in definitions and debates on sustainability.

From a legal pluralistic perspective, sustainability is not so much a characteristic or capacity of a system but rather refers to plurality of laws, values, and norms that define sustainability as a desired situation of a system, consisting of processes and practices. Such a value or norm can be generated by actors that live in or are part of a system and also by outsiders, be they state or nonstate actors. The term “outsiders” is a bit misleading here because these actors may consider the system to be situated within the confines of their territory or jurisdiction or to be (part of) a common good that needs to be protected from overexploitation. From a legal pluralistic perspective, an agroecosystem can be an appropriate object of study to analyze how different normative frameworks, being generated by different state and nonstate actors in social fields within or outside this system, coexist. Updating Moore’s concept of semiautonomous social field to the new era of globalization, it is important to emphasize that a social field can also refer to a policy arena and sustainability diplomacy at the national or international level.

One of the key drivers of globalization processes has been the spread of neoliberal values, thinking, and discourse in organizing politics, economics, and agriculture. This spread has led to the rise of nonstate sovereigns over sustainability: neoliberal policies like privatization, deregulation, and free trade gave room to nonstate actors to assume new roles and responsibilities in organizing politics, economics, and agriculture. Somewhat paradoxically or unexpectedly, different nonstate actors took this opportunity to challenge the neoliberal architecture of globalization processes. Expecting or witnessing detrimental effects of monoculture cropping on ecosystems and local food systems, they began to establish or mobilize alternative legal orders. Not only international NGOs, farmer movements, and human rights activists but also multinationals emerged as new sovereigns over sustainability, adopting noneconomic values as core values and promoting sustainability or agroecology as new principles or norms for organizing food systems and agriculture.

I will briefly review two legal orders for governing food systems and sustainability: food sovereignty and multistakeholder governance of sustainable cropping. Each of them can be seen as a counterframe to challenge neoliberal thinking, discourses, and organization of politics, economics, and agriculture in which food and nature are basically treated as commodities. Every review consists of a brief introduction, characterization, and identification of two or three critical issues.

4.1 *Food Sovereignty as an Alternative Legal Order*

The concept of food sovereignty was launched in 1996 by the international movement La Via Campesina as the "right of each nation to maintain and develop their own capacity to produce foods that are crucial to national and community food security, respecting cultural diversity and diversity of production methods"[17]; the Food Information and Action Network (FIAN) presented food sovereignty as "democracy in localized food systems."[18] The concept was used to express concerns about the lack of "voice of farmers, pastoralists, fisherfolk, food workers, and indigenous people"[19] in national and international policy debates on the future of food, farming, and development. Advocates of food sovereignty distinguish four principles of food sovereignty: practicing agroecological models of food production, securing the human right to food, promoting continued access of small food producers to productive resources, and strengthening of local markets and equitable trade policies.[20]

As an alternative legal order, food sovereignty has been established by a wide variety of nonstate actors but not including the private sector. Given the centrality of both agroecology and human rights as core principles, one could also say that food sovereignty is a legal order that addresses both man–nature and state–man relationships. Echoing concerns of local communities, food sovereignty can be seen as a legal order that is constructed bottom up. As food sovereignty addresses the links between agriculture and environment, it can be considered a form of agroecosystem law. Finally, as food sovereignty is declared by nonstate actors and at the same time embraces the human right to food that has been declared by states, food sovereignty can also be considered as a hybrid legal order.

Three issues are critical for the future development of food sovereignty as an alternative legal order. The first issue is how and to what extent food sovereignty fundamentally challenges the sovereignty of the state and can be connected or integrated in official laws. So far, very few public authorities at the national and international levels have adopted food sovereignty as a normative basis for alternative agriculture and food policies.[21] The second and related issue is about who is supposed or obliged (on what legal grounds) to respect or recognize food sovereignty. It is unclear what the consequences are of not respecting and what measures

[17] NGO (1996).

[18] Windfuhr and Jonsén (2005).

[19] Pimbert (2009).

[20] Windfuhr and Jonsén (2005), pp. 14–15.

[21] Hospes (2014a), pp. 119–130.

can be taken by whom to address lack of respect. The third issue is whether food sovereignty as a legal order can or has to substitute the set of values, principles, and norms that structure neoliberal policies and institutions or should be reconciled with such a set. Can the WTO and multinational food business adopt food sovereignty as a legal order?

4.2 Multistakeholder Sustainability Governance as an Alternative Legal Order

The 1990s and even more so the 2000s have been the era characterized by the emergence of a new legal order in the field of sustainable production of global commodities like palm oil, soy, timber, coffee, etc. Multinational businesses and environmental NGOs established global private partnerships and global principles and criteria for sustainable forestry, fishery, and cropping. These partnerships have created legitimacy through multistakeholder consultation and can be seen as new sovereigns over sustainability. Examples are the Forest Stewardship Council (FSC), Marine Stewardship Council (MSC), Roundtable on Sustainable Palm Oil (RSPO), Roundtable on Responsible Soy (RTRS), Better Cotton Initiative (BCI), Better Sugar Initiative (BSI), Sustainable Beef Roundtable (SBR), and Aquaculture Stewardship Council (ASC).[22]

As an alternative legal order, the sustainability principles and criteria of the partnerships have been established by a wide variety of nonstate actors, including business and civil society actors. The principles and criteria of the partnerships specify what economic, agricultural, social, and legal practices should be adopted by forestry companies, fishing businesses, and agricultural producers to protect biodiversity of ecosystems and respect local communities. All partnerships have been initiated by European and/or USA-based companies and environmental NGOs; the values, principles, and criteria of these sustainability partnerships can be seen as elements of a global legal order or, to be more precise, a western nonstate legal order that defines what is sustainable forestry or cropping on the territory of states in the South. Because the partnerships provide principles, norms, and criteria for best practices in agricultural and environmental management, their standards can be considered as forms of agroecosystem law. Finally, one of the key principles of every partnership is compliance with existing international and national laws. As such, the partnerships recognize and reinforce official laws.

Two issues are critical for the future development of global private partnerships as an alternative legal order. The first issue is how and to what extent public authorities at the national and local level in production countries in the South will respond to the proliferation of global standards that may easily be perceived to be

[22] Glasbergen (2008), pp. 15–25; Hospes (2011), pp. 38–46; Hospes et al. (2012), pp. 29 52; Hospes (2014b); Schouten and Glasbergen (2011), pp. 1891–1899.

western interventions in national affair. For example, the government of Indonesia has taken a legal initiative to halt the power of the RSPO as a new sovereign over palm oil plantations on its territory.[23] The second issue is how legitimate global private partnerships can be or remain when, for instance, the loss of forest and biodiversity due to agricultural expansion is not halted and even certified producers reportedly do not always comply with the voluntary sustainability standard.

5 Conclusion and Discussion

Agroecology as a transdisciplinary field of studies of agroecosystems can benefit enormously from the use of a legal pluralistic perspective on law, social fields, and sovereignty in four different ways: by exploring the interconnectedness of different legal systems and normative orders with agroecosystems, by studying how agroecosystems are linked to one or more semiautonomous social fields, by considering that agroecosystems are part of legal orders that span the globe or large parts of it, and by acknowledging a plurality of sovereignties over an agroecosystem, seen as a complex system with material and immaterial dimensions.

A legal pluralistic perspective can also be extremely useful as a contribution to debates on law and sustainability. Such a perspective will bring us to explore the coexistence of nonstate legal orders based on food sovereignty, agroecology, human rights, and sustainability as core values or principles, with state or corporatist legal orders based on neoliberal values, principles, norms to organize politics, economics, and agriculture. Of course, this "coexistence" is not without tensions. A very important task of legal pluralists would be not to take sides and not to conduct legal "monocropping," that is, to focus on the subjection of agroecosystems under one legal order. Just like agro-ecologists emphasize plurality and diversity of agroecosystems to be an asset, legal pluralists should emphasize plurality and diversity of legal systems to form a broad legal repertoire and resource base for governing food systems and sustainability.

From a legal pluralistic perspective on law and agroecosystems, the identification and study of interlegalities, that is, intersections of different legal orders,[24] form a major challenge. The study of interlegalities is not only extremely opportune from a legal-scientific perspective in a globalizing world characterized by legal pluralism all over the place but can also contribute to policy debate and insights on how to reconcile and combine different nonstate and/or state laws, principles, norms.

[23] Hospes (2014b).

[24] de Souza-Santos (1987).

References

Berman PS (2007) Global legal pluralism. South Calif Rev 80:1155–1238

Comaroff JL, Comaroff J (2009) Reflections on the anthropology of law, governance and sovereignty. In: von Benda-Beckmann F, von Benda-Beckmann K, Eckert J (eds) Rules of law and laws of ruling: on the governance of law. Ashgate, Farnham, pp 31–59

de Souza-Santos B (1987) Law: a map of misreading. Toward a postmodern conception of law. J Law Soc 14(3):279–302

Dupret B (2007) Legal pluralism, plurality of laws, and legal practices. Eur J Legal Stud 1(1). http://cadmus.eui.eu/handle/1814/6852. Accessed 24 Aug 2014

Glasbergen P (2008) Mondiale partnerships en collectieve actie. Bestuurskunde 4:15–25

Griffith J (1986) What is legal pluralism? J Legal Pluralism 24:1–55

Hospes O (2011) Besturen tussen globale en nationale schaal: De casus van de Ronde Tafel over Duurzame Palmolie. Bestuurskunde 4:38–46

Hospes O (2014a) Food sovereignty: the debate, the deadlock and some suggestions for a detour. Agric Hum Values 31:119–130

Hospes O (2014b) Marking the success or end of global multi-stakeholder governance: the rise of national sustainability standards in Indonesia and Brazil for palm oil and soy. Agric Hum Values. doi:10.1007/s10460-014-9511-9

Hospes O, van der Valk OMC, van der Mheen-Sluijer J (2012) Parallel development of five partnerships to promote sustainable soy in Brazil: solution or part of wicked problems? Int Food Agribus Manag Rev 15(B):29–52

Merry SE (1988) Legal pluralism. Law Soc Rev 22(5):869–896

Michaels R (2009) Global legal pluralism. Annu Rev Law Soc Sci 5:1–35

Moore SF (1973) Law and social change: the semi-autonomous social field as an appropriate subject of study. Law Soc Rev 7(4):719–746

NGO (1996) La Via Campesina, "Food Sovereignty: A Future Without Hunger," Statement by the NGO Forum to the World Food Summit. http://www.voiceoftheturtle.org/library/1996DeclarationofFoodSovereignty.pdf. Accessed 12 May 2014

OECD (2003) Definition of agro-ecosystem. Glossary of statistical terms. https://stats.oecd.org/glossary/detail.asp?ID=82. Accessed 9 May 2014

Pimbert M (2009) Towards food sovereignty. Gatekeeper 141:11–20. http://pubs.iied.org/pdfs/14585IIED.pdf. Accessed 22 Oct 2014

Schouten G, Glasbergen P (2011) Creating legitimacy in global private governance: the case of the roundtable on sustainable palm oil. Ecol Econ 70(11):1891–1899

Wallace KJ (2007) Classification of ecosystem services: problems and solutions. Biol Conserv 139 (3–4):235–246

Windfuhr M, Jonsén J (2005) Food sovereignty: towards democracy in localized food systems. FIAN/ITDG Publishing, Rugby. http://www.ukabc.org/foodovpaper.htm. Accessed 9 May 2014

From Agroecology and Law to Agroecological Law? Exploring Integration Between *Scientia Ruris* and *Scientia Iuris*

M. Monteduro

Abstract *Rus*, the rural phenomenon understood in its entirety, marks the plurality and the interdependence of different complex systems that are based jointly on the land as a central point of reference. "Rural" expresses a *quid pluris* as compared to "agricultural": if agriculture is understood traditionally as an activity aimed at exploiting the land for the production of material goods for use, consumption, and private exchange, rurality marks the reintegration of agriculture into a wider sphere, not only productive but also social and cultural; not only material but also ideal, relational, historic, and symbolic; not only private but also public. The natural and social sciences (*scientia ruris*), in approaching *rus*, at first became specialized, multiplied, and compartmentalized in a plurality of "first-order" disciplines; later, above all in recent decades, they have set up a process of integration into agroecology as a "second-order" polyocular, transdisciplinary, and common platform. The law (*scientia iuris*) seems instead to be frozen at the first stage. Following a reductionistic and hyperspecialized approach, the law has deconstructed and shattered the complex universe of *rus* into disjointed legal elementary particles, multiplying the planes of analysis and regulation (agricultural law, business law, environmental law, landscape law, town planning law, etc.), without caring to construct linkage platforms among the various legal fields. In this chapter, after examining some important experiences underway internationally, it is asserted that *scientia iuris* should experiment with the development of an agroecological law, like that which agroecology is today for *scientia ruris*. Agroecological law should counteract the antinomic interlegalities (among the various legal fields that deal with rural phenomena) through tools of negative coordination and favor instead compatible interlegalities through tools of positive coordination. In the conclusions are proposed by way of example four types of coordination tools: agroecological information collecting and sharing (AICS), agroecological zoning (AZ), agroecological planning (AP), and agroecological impact assessment (AIA).

M. Monteduro (✉)
Department of Legal Sciences, University of Salento, Lecce, Italy
e-mail: massimo.monteduro@unisalento.it

M. Monteduro et al. (eds.), *Law and Agroecology*,
DOI 10.1007/978-3-662-46617-9_4

Keywords Agriculture • Agroecological Impact Assessment • Agroecological Information Collecting and Sharing • Agroecological Law • Agroecological Planning • Agroecological Zoning • Agroecology • Coordination • Interlegalities • *Rus*

1 Introduction: *Rus* and Agriculture

Imagining having to converse with a third party observer, who does not take sides with one or the other of the specialists in each academic field, how would one try to describe the current physiognomy of the complex relations between *rurality, science*, and *law*?

At the upper vertex of this ideal triangle is positioned *rus*, that is, rural phenomena understood in the broadest sense.

Most contemporary scholars seem to agree, converging from the different academic fields to which they belong, on one point: the adjective "rural," placed next to a noun such as "space," "environment," "development," *et alia*, marks *the plurality and the interdependence of different complex systems that are based jointly on the land as a central point of reference.*[1]

"Rural" expresses, namely, a *quid pluris* as compared to "agricultural" in its restricted meaning (so to say, agro-centric[2]): if agriculture is traditionally understood as an activity focused on the exploitation of the land for the production of material goods for private use, consumption, and exchange, rurality marks the reinsertion of agriculture into a larger sphere, not only productive but also social and cultural; not only material but also ideal, relational, historical, symbolic; not only private but also public.

From here derives the inestimable richness of the rural "world," of the peasant "civilization," of the villages, of the heritage of wisdom tied to the rhythms and rites of the countryside, to the traditions[3] (see in this book the contribution of DE NITTO

[1] See, e.g., for a plurality of perspectives, Iacoponi (1998), pp. 51 et seqq.; Albisinni (1998), pp. 139 et seqq.; Albisinni (2000), pp. 421 et seqq.; Gray (2000), pp. 30–52; Esposti and Sotte (2002); Friedland (2002), pp. 350–371; Basile and Romano (2002); Marsden (2003); Buller (2004), pp. 101–119; Brouwer and van der Heide (2009); Martinez (2010), pp. 1–16; Bryden et al. (2011); Agnoletti (2013); Westlund and Kobayashi (2013); Camaioni et al. (2013); Lukić (2013), pp. 356–376; Bosworth and Somerville (2014).

[2] Buller (2004); Sturiale (2001), pp. 161–195, 161. As can be read in OECD (2009), "the new rural paradigm should promote the complementarity between agricultural and rural policy, that is there must be common aspects and a dynamic interaction, overcoming both the 'agrocentric' paradigm characterized by the complete coincidence between agricultural and rural policy, and that based on the 'divorce' between the two types of policy" (author's translation).

[3] Over 70 years ago, Serpieri (1940) declared in his Corso di economia e politica agraria, vol I. G. Barbera, Firenze, p. 42, that "we can succinctly call rurality" a "complex of feelings, customs, ways of life" that "neatly distinguish the agricultural world from the urban-industrial one". Agriculture, understood in the reductive sense of activity of production of food and fiber through exploitation of the land, can be "seen both as a threat to and a caretaker of cultural heritage,"

and his refined analyses of the lemmas of rurality, as well as the contribution of BUONGIORNO for the comparison with the perception of *rus* in Roman law): this richness is not *produced* according to linear and precise transformative schemes of cause and effect in the discontinuous rhythm of the production of individual assets, but it is *built* chaotically in the continual interaction between men and lands over centuries, through historical strata and sediments; it is not *marketed*, and it cannot be sold or consumed by single individuals, but it is *lived* in the communities; it can be destroyed, but only with the complete destruction of the latter, regardless of the processes of production of material agricultural individual assets.

The preferred term of contemporary specialists is the adjective "multifunctional," associated with the noun agriculture and in opposition to "monofunctional."[4] Multifunctional agriculture, as distinguished from monofunctional agriculture, is not limited to producing material goods for private use, consumption, and exchange on the market, but it also furnishes to the collectivity fundamental ecosystem services, whose value is not entirely monetizable: for example, it designs the countryside; protects the fertility of the soil; contributes to the integrity of the hydrological cycles, to the management of water resources and flood control through hydrological adjustment; maintains biodiversity; ensures natural recycling of nutrients; preserves the functioning of the natural carbon sink (terrestrial vegetation) and contributes, thus, to the fight against climatic changes induced by greenhouse gases; guarantees safety, healthiness, and food quality, even those of traditional local products; allows the socio-economic survival of rural communities and gives value to the human labor of nuclear family farmers with respect to artificial capital; educates for the rurality maintaining the historical roots of the relation between city and countryside; guards the cultural identity of the territory; favors the development of agro-ecotourism and the enjoyment of nature for educational and recreational purposes.[5]

Rurality and multifunctional agriculture are not, however, interchangeable synonyms: their relation is rather that between structure and flow, between organization and action. Multifunctional agriculture, released from its exclusive relation to the material production of *things* and tied also to the plurality of *interconnected services, values, expertises, and experiences* in which *rus* is articulated, is (metaphorically) the sap that flows constantly within the tree of rurality, sustaining its metabolism and historical evolution.

assuming thus a "double role" with respect to rurality understood in its cultural dimension, as noted by Daugstada et al. (2006), pp. 67–81.

[4] Mazzarino and Pagella (2003); Van Huylenbroeck and Durand (2003); Henke (2004); Brouwer (2004); Contò (2005); Wilson (2007); Russo (2007), pp. 231–245; OECD (2008); Carbone (2009), pp. 133–144; Milone (2009); Wilson (2010), pp. 364–381; Potter and Thomson (2011), pp. 213–223; Bonnal et al. (2012); Westhoek et al. (2013), pp. 5–13; Adam (2014).

[5] Monteduro (2013), pp. 2–11.

2 *Rus* and *Scientia Ruris*

First to look to *rus* are the natural and social sciences, which are involved in various ways with multifunctional agriculture. To refer to them collectively and succinctly, in this chapter a deliberately broad umbrella term will be used: *scientia ruris*.

At first glance, the plurality of the disciplines involved in the study of the "polytope" represented by *rus* is striking. The landscape that is shown to the observer has many different sciences with equally many different viewing angles, planes of analysis, principles, methods, techniques, findings: for example, agronomy, soil science, plant pathology, horticulture, genetics, food science, entomology, animal science, forest science, ecology, rural sociology, agricultural economics, rural geography, agricultural engineering, anthropology, environmental philosophy. They all revolve around a center of gravity represented by rural phenomena, and they each capture a fragment.

As has been observed by NOE & ALRØE in this book, each of these disciplines configures itself as a "first-order perspective" that represents only a partial and limited "point of observing an agroecosystem": though the agroecosystem be "the shared object" of all the sciences that intercept the rural phenomenon, "none of the individual perspectives can observe the agroecosystem as such." The agroecosystems must be considered in their wholeness and complexity as social-ecological systems, autopoietic and self-organizing: as such, they cannot be observed from only one viewing angle.

This leads on to the innovation that has characterized the scientific panorama of research on rural phenomena in the last decade: the emergence and consolidation of a transdisciplinary research platform called agroecology.

As underlined in the contribution of CAPORALI in this book, "the emergent characteristic of Agroecology is that of a transdiscipline as it integrates other fields of knowledge into the concept of agroecosystem viewed as a socio-ecological system."

Initially, agroecology was born from the key idea of linking two sciences that heretofore had been separate: agronomy and ecology. Inspired in particular by Odum's systemic ecology, the seminal studies of agroecology aimed to integrate the principles of ecology into the redefinition of agronomy. Taking as the object of scientific analysis the concept of "agro-ecosystem," these studies tried to identify theoretical principles and practical techniques for a sustainable agriculture, able to mimic natural processes and aimed at the creation of favorable biological synergies and interactions among the biotic and abiotic components of the agroecosystems. In a successive phase, the analytical field of agroecology broadened to include the study of processes of construction, organization, management, and development of food systems: this evolution scientifically integrated into agroecology new perspectives from sociology, economics, engineering, political sciences, history. Finally, agroecology incorporated points of view of the philosophical, bioethical, and demoethno-anthropological sciences. From all this comes the full transdisciplinarity of agroecology in the contemporary context.[6]

Nevertheless, as observed by NOE & ALRØE, "Agroecology is a polyocular platform of second-order observations." It is not a new scientific discipline that substitutes for others or that juxtaposes itself with them, subtracting spaces in part from one and in part from the other; neither is it an algebraic sum of sciences. It could be defined as a metadiscipline, a "science of sciences" that establishes a second level of observation, in which the different disciplines can meet and compare—preserving them—the different points of view.

The benefit of this second-level platform is given by the fact that the cultural diversity among the academic fields becomes a transformative factor of coevolution, rather than a regressive push towards isolation, among different sciences: borrowing from the lexicon of ecology, the aggregate level of the studies is raised from scientific *populations* (composed of scholars of the same disciplinary sub-field, e.g., the soil sciences) to scientific *communities* (composed of scholars of different disciplinary sub-fields, belonging, however, to the same macro-field, e.g., the life sciences) up to those that, with a metaphorical image, could be defined as scientific *ecosystems* (composed by scholars of different macro-fields that establish structured relations of coexistence in a *tòpos* of common and shared research, such as is agroecology): at each level of aggregation is found not a sum but a synthesis that generates emerging properties in the scientific research, namely, principles and new methodologies that the preceding level would not have been able to exhibit in isolation.

3 *Rus* and *Scientia Iuris*

Legal sciences also look to *rus*, and here they are gathered under the broad umbrella label of *scientia iuris*.

---

[6] Altieri (1983); Altieri (1987); Altieri (1989), pp. 37–46; Altieri (1991); Gliessman (1990); Carroll et al. (1990); Caporali (1991); Flora (2001); Uphoff (2002); Francis et al. (2003), pp. 99–118; Dalgaard et al. (2003), pp. 39–51; Gliessman (2006); Ruiz Rosado (2006), pp. 140–145; Warner (2007); Uphoff (2007), pp. 218–236; Wojtkowski (2008); Snapp and Pounds (2008); Bland and Bell (2008), pp. 280–294; Wezel et al. (2009), pp. 503–515; Bohlen and House (2009); Wezel and Jauneau (2011), pp. 1–25; Wezel and Soldat (2009), pp. 3–18; Caporali et al. (2010); Tomich et al. (2011), pp. 193–222; Caporali (2011), pp. 1–72; Van Dam et al. (2012); Lichtfouse (2013); Martin and Sauerborn (2013); Sevilla Guzmán and Woodgate (2013), pp. 32–44; Gonzalez de Molina (2013), pp. 45–59; Vandermeer and Perfecto (2013), pp. 76–89. See also Cleveland (2013): "Agroecology is defined as a comprehensive perspective of agrifood systems including the relationships between the biophysical and sociocultural components and between agrifood systems and the larger biophysical and sociocultural context in which they are embedded. As such, agroecology includes the internal ecology of agroecosystems, their social and cultural components including nutrition and food sovereignty, crop genotype-by-environment interactions including those of transgenic crop varieties, and the positive (ecosystem services) and negative (ecosystems degradation) effects of agroecosystems on the larger environment, especially climate. This is a broad view of agroecology that does not limit the term to the traditional discipline of ecology applied to agricultural production systems."

The relation between *ius* and *rus*, nevertheless, has followed a different evolutionary trajectory from that which characterized the natural and social sciences.

The latter, in approaching rural phenomena, first became specialized, multiplied, and compartmentalized in a plurality of "first-order" disciplines; later, above all in recent decades, they have set up a process of integration into agroecology as a "second-order" polyocular, transdisciplinary, and common research platform.

The law seems instead to be frozen at the first stage.

A first legal discipline that takes its own *nomen* from agriculture is agricultural law: this branch of the law, however, offers only a partial perspective on *rus* because its main object of study is represented today by the regulation of markets of agricultural and agri-food products.

As has been recently reiterated, "agricultural law regulates mainly *the production* obtained through raising plants and animals and the sale of the results of those activities"[7]; "agricultural law has its essence *in production* [...] agricultural law focuses on the regulation of agro-biological production activities while other legal disciplines deal with other productions or activities."[8] Food law is derived from agricultural law.[9]

Then there is environmental law, which intercepts the many ecological profiles linked to agricultural activities: for example, biodiversity in agriculture, protected animal and plant species, agricultural wastes, reclamation of contaminated lands, agricultural water use, relations between agricultural and animal husbandry activities and climate changes, organic farming, energy production by agricultural biomass as renewable resources, agro-forestry. Legal doctrine, notwithstanding the many interferences between the two disciplinary fields,[10] has preferred to keep separate agricultural law and environmental law (see, for example, the contributions of CRISTIANI, HERMON, SZILÁGYI, and DOOLEY in this book).[11]

---

[7] Costato (2008a), p. 6 (author's translation from Italian).

[8] Pastorino (2012), p. 55 (author's translation from Spanish).

[9] Russo (2012), pp. 141 et seqq. For a different perspective, see recently Perfetti (2014), pp. 3–20. About the relation between precautionary principle and food law, see Giliberti (2013), pp. 1 et seqq.

[10] D'Addezio (2008), pp. 9–34; Carmignani (2012).

[11] In this sense, the experience of Italian legal doctrine is emblematic. It boasts a great tradition in agricultural law. Italian scholars, while recognizing the interference between agricultural law and environmental law, have always proclaimed the scientific autonomy of agricultural law. The environment has been understood, for example, as a *limit* on the exercise of agricultural activities ("polluting" and "polluted" agriculture), as the *form* of agriculture (environmental constraints on the agro-forestry territory) or as the *product* shaped by the exercise of agriculture (with reference to the new role that the European CAP has assigned to agricultural undertakings and to the services that they can perform for the care of the environment): see Francario (1993), p. 519. For arguments that agricultural law can neither be fused nor confused with environmental law, see Carrozza (1994), pp. 151–172; Costato (2008b), pp. 15–24; Cristiani (2008), pp. 464–479. The autonomy of agricultural law with respect to environmental law is an issue that has been addressed also in the legal doctrine of other European countries. For example, for France, see Hernandez Zakine (1998), pp. 133–155; Hudault (1987); Doussan (2002); Hudault (2006), pp. 247–260. For Spain, see

In addition, the doctrine that has discussed "agri-environmental law" has done so, so far, in the perspective of a sub-field of agricultural law, contained in the latter.[12]

The phenomena linked to rurality are intercepted, in addition, by many other fields of law.

Taking the example of Italy, business law is interested in the legal definition and regulation of the agricultural entrepreneur[13]; intellectual property law (called "industrial law" in Italy, which today revolves around the new Industrial Property Code approved by Legislative Decree 30/2005) deals with topics such as collective trademarks, protected geographical indications, or protected designations of origin[14]; private law continues to study agrarian

---

Martinez De Marigorta Andreu (1987), pp. 19–30; de los Desamparados Llombart Bosch (1999), pp. 217–226; Navarro Fernández (2010). For Germany and Austria, see Winkler (1994), pp. 173–189; Winkler (2002), pp. 5–18; Welan (2002), pp. 48–53. For Hungary, see Szilágyi (2009), pp. 41–55. For the debate in the legal doctrine of the United States of America, see, e.g., Hamilton (1999), pp. 41–58; Schneider (2010), pp. 935–963. For the Latin American experience, see Zeledón Zeledón (2009a), pp. 9–26; Prado de Albuquerque (2007), pp. 69–82 (see particularly 79); Zeledón Zeledón (2009b); Pastorino (2009), pp. 3–14, 39–52, 151–164; González Linares (2011).

[12] See, e.g., Costato et al. (2011); Merusi (2007), pp. 495–501; Pastorino (2012), pp. 50–59; Massart and Sánchez Hernández (2001).

[13] See Cossu (2003), pp. 73–100, according to whom (p. 97) "it appears always less justifiable to subtract from the whole the agri-food sector of the *lex mercatoria*" (author's translation); Battista Ferri (2005), pp. 1–15; Jannarelli and Vecchione (2008).

[14] Of the "industrial law," it has been said that "it is a special law within the special law (commercial law)" (author's translation): Caruso (2011), p. 7. The Industrial Property Code approved in Italy by Legislative Decree 30/2005 regulates:

- at Art. 11, Para. 1, the "collective trademark" registered by persons who have "the function of guaranteeing the origin, nature, or quality of specified products or services"; Art. 11, Para. 4, specifies that "notwithstanding Art. 13, Para. 1, a collective trademark can consist of signs or indications that in commerce can be used to designate the geographical provenance of products or services. In that case, moreover, the Italian Office of Patents and Trademarks can refuse, with a reasoned decision, the registration when the trademarks requested could create situations of unjustified privilege or anyway prejudice the development of other analogous initiatives in the region. The Italian Office of Patents and Trademarks can request the opinion of the public administrations, categories and interested or qualified bodies. The successful registration of the collective trademark constituted by geographical name does not authorize the owner to prohibit third parties from using the same name in commerce, provided that this use complies with the principles of professional propriety";
- at Arts. 29 and 30, "the use of geographical labels and denominations of origin that identify a town, region, or locality, when they are adopted to designate a product which originates from and whose quality, reputation, or characteristics are owed exclusively or essentially to the geographic area of origin, including natural, human, and traditional factors [...] is prohibited, when it is likely to deceive the public or when it entails an undue exploitation of the protected denomination, the use of geographic labels and denominations of origin, as well as the use of any means in the designation or presentation of a product that indicate or suggest that the product itself comes from a locality that is different from its true place of origin, or else that the product presents the qualities that belong to the products that come from a locality designated by a geographic label."

On the protection of trademarks (collective, territorial, of quality), of protected geographic labels, and of protected denominations of origin, see Giacomini et al. (2007); Ubertazzi and Muñiz Espada (2009); Angelicchio (2014), pp. 345–386; Caforio (2014).

property[15]; criminal law analyzes crimes linked to agricultural production activities (as, e.g., adulteration and counterfeiting of agri-food products)[16] and the penal protection of trademarks and indications of provenance, of origin, and of quality, including those relative to agri-food products (e.g., olive oil)[17]; labour law regulates labor in agriculture (e.g., work health and safety, employment contracts)[18]; landscape law deals with rural landscapes[19]; cultural property law covers the protection of the material and immaterial rural cultural heritage[20]; town planning law deals with planning uses in the rural territory (besides the urban) and legal designation of some areas as

[15] Alpa et al. (2001), pp. 412 et seqq.; Moscarini (2009), pp. 177 et seqq.

[16] Gargani (2013), pp. 273 et seqq.

[17] Cingari (2008), particularly pp. 139 et seqq.; Mazzanti (2013), pp. 561–582. Art. 4, Paras. 49 and 49-bis, of Law 350/2003 establishes that "the importation and exportation for commercial ends, i.e. commercialization, or the commission of acts directed in an unequivocal way at commercialization of products bearing false or misleading indications of origin or source constitutes a crime and is punishable within the meaning of Art. 517 of the Penal Code. A false indication is constituted by stamping "made in Italy" on products and goods that do not originate in Italy within the meaning of the European regulation on origin; a false indication is also constituted, even when foreign origin and provenance of products or goods are indicated, by the use of signs, figures, or other things that might induce a consumer to believe that the product or good is of Italian origin including the false or misleading use of business trademarks within the meaning of the regulation of deceptive trade practices [. . .] The offenses are committed with the presentation of the products or goods in customs for release for consumption or in free circulation or in retail sales. The false labeling of goods can be remedied on the administrative level with the removal by and at the expense of the offender of the signs or figures of whatever else might create a belief that it is a product of Italian origin. The false labeling of origin or provenance of products or goods can be remedied on the administrative level through the correct indication of the origin or the removal of the "made in Italy" printing. A false labeling is constituted by the use of a brand, by the owner or the licensee, in such a way as to lead a consumer to believe that the product or good is of Italian origin within the meaning of European regulation on origin, without the same being accompanied by precise and evident indications of the foreign origin or provenance or in any event sufficient to avoid any misunderstanding by the consumer on the real origin of the product, or without being accompanied by the attestation, made by the owner or licensee of the brand, about the information, that by him, will be made during commercialization on the real foreign origin of the product. For food products, for real origin is meant the place of cultivation or breeding of the agricultural raw materials used in the production and preparation of the products and the place in which substantial transformation took place [. . .] Subject to the provisions of Para. 49-*ter* and subject to the sanctions referred to in Art. 16, Para. 4, of Legislative Decree 135/2009, amended, with modifications, by Law 166/2009, false labeling in the use of brand, referred to in Para. 49-*bis*, is punishable, as regarding virgin olive oil, within the meaning of Art. 517 of the Penal Code."

[18] Pelliccia (2011); D'Imperio (2011), pp. 1195–1198.

[19] See the contributions of BROCCA and BUIA & ANTONUCCI in this book.

[20] See the contribution of DENUZZO in this book and also De Giorgi Cezzi (2005), pp. 2955 et seqq.; Istituto Nazionale di Economia Agraria (2001).

agricultural zones[21]; constitutional law faces the problem of the distribution of legislative powers between State and Regions in matters of agriculture, rural development, agri-environmental measures, agri-food markets[22]; tourism law is involved in subjects such as rural tourism, family rural hospitality, farm stays (in Italian, "agriturismi"), country houses, and camping sites.[23]

The list could go on, but it behooves us to stop and ask a few questions.

Is the deconstruction/fragmentation of the complex universe of *rus* into elementary and disjointed legal particles, accomplished by the *scientia iuris*, an inevitable landfall?

What are the results of this reductionistic and hyperspecialistic approach that, multiplying the level of analysis and legal regulation, does not bother to construct platforms connecting the different disciplinary fields to surmount the barriers and elevate the point of view?

---

[21] Jannarelli (2004), pp. 592 et seqq.; Mengoli (2009), especially pp. 189 et seqq.; Portaluri (2011), pp. 241–255; Urbani (2011), pp. 597 et seqq.; Russo (2013), pp. 163–174.

[22] See the contribution of TROISI in this book.

[23] Santagata De Castro (2012), pp. 96 et seqq. and 186–187; Righi (2013), pp. 129 et seqq.; Busti (2013), 198 et seqq.; La Torre (2013), 271 et seqq. The Code of Tourism (The Italian Tourism Act, Legislative Decree 79/2011) regulates:
- at Art. 12, Para. 9, and Art. 9, Para. 1, "the lodgings within the area of agro-tourist activities," which "are local sites in rural buildings managed by agricultural entrepreneurs," and "agro-tourism" referring to Art. 3 of Legislative Decree 228/2001 and to Law 228/2006 (which establish in detail the regulation of agro-tourism);
- at Art. 12, Para. 9, "accommodations in rural residences or country houses," which "are facilities located in country villas or rural buildings to be used for sports or recreation entertainment composed of rooms with kitchenette possible, that have food service open to the public";
- at Art. 13, "camping" in general and "camping within the area of agro-tourist activities" in particular;
- at Art. 23, "local tourist systems," which are homogenous or integrated tourist contexts, including territorial areas belonging even to different regions, characterized by the integrated offering of cultural heritage, environmental resources, and tourist attractions, "including typical products of agriculture" and local crafts, or by the widespread presence of individual or associated tourist enterprises;
- at Art. 29, Para 2, "nature tourism," which includes hospitality, recreational, didactic, and cultural activities and services aimed at the proper use and enhancement of natural resources, of wildlife and aquatic heritage, and of routes of recovery of the "bridleways" (horse trails) and of the "ancient rural roads."
"Family rural hospitality" is instead regulated by Art. 23 of Law 122/2001, according to which "the Regions, in the area of initiatives aimed at rural development and enhancement of the multifunctionality of the businesses, can regulate the activity providing service of lodging and breakfast in one's own home. Should said activities have a professional and continuing character and be undertaken by agricultural entrepreneurs, they become part of agro-tourist activities. The Regions [...] determine, with their own laws, the characteristics of real estate that can be used [...] as well as the characteristics of professionality and continuity of the activity. No physical person can be the owner of more than one authorization for the exercise of this activity. The requirement of the prevalence of one's own products and of products of agricultural businesses of the area in the meals provided in the agro-tourist activities is applicable also for rural hospitality activities."

Is it possible to seek an alternative path that leads (in the future) the *scientia iuris* to experiment with the elaboration of an *agroecological law* in the likeness of that which *agroecology* represents (today) for the *scientia ruris*? Or is it only a utopia, an illusion, a red herring impracticable for a legal scholar who aspires to be rigorous in the method he applies when "he does his craft?"

And, if perchance it is not a utopia, what characteristic traits could this new *agroecological law* present?

4 Agroecological Law: Utopia or Overlooked Possibility?

The answer to the foregoing questions, in our opinion, is the following: for the present, the divisive approach used by the law heretofore is not ineluctable[24] and, for the future, the gradual construction of a new *agroecological law* is not a utopia.

Instead, it is a concrete and underestimated possibility that challenges legal scholars and commits them to renew deeply their theoretical models, to inspire legislation and jurisprudence able to put into dialog, on one hand, areas of law that have heretofore been separated and, on the other hand, law and agroecology.

The examination of meaningful experiences in the course of experimentation at the international level confirms that agroecological law is practicable, concrete, present, and urgent.

A few examples suffice.

Nicaragua recently approved the *Ley de Fomento a la Producción Agroecológica u Orgánica* (Law 765/2011).[25] This law is of considerable interest, inasmuch as:

- it strives to furnish a legal definition of "*agroecosistemas*" (agroecosystems),[26] as well as other concepts like "*bienes naturales*" (natural resources)[27] or "*sistema sucesional*" (successional systems)[28];

[24] Recently, some legal scholarship is exploring the possibility of building a "new law" based on systematic, integrated, and comprehensive understanding of social-ecological systems, by rethinking the idea of rule of law, which could evolve into "ecological rule of law" or "rule of law for nature," and introducing a legal concept of "ecological public order." For some references, see Monteduro (2014), pp. 1–44.

[25] Approved 14 April 2011 and published in the Gaceta n. 124 of 5 July 2011.

[26] Art. 3, Para. 1, Law 765/2011: "Agroecosistemas: Sistema ecológico que cuenta con una o más poblaciones de utilidad agrícola y el ambiente con el cual interactúa, cuyos componentes principales son los subsistemas de cultivos o de producción animal, identificados con las parcelas o áreas de la finca donde se tienen cultivos y sus asociaciones o las unidades de producción pecuarias."

[27] Art. 3, Para. 2, Law 765/2011: "Bienes naturales: Bienes comunes y servicios que proporciona la naturaleza sin alteración por parte del ser humano que contribuyen al bienestar y desarrollo de la vida en la tierra."

[28] Art. 3, Para. 7, Law 765/2011: "Sistema sucesional: Sistemas agroforestales que consiste en el asocio masivo de cultivos anuales y perennes con especies arbóreas de diferentes hábitos de crecimiento, usos y beneficios, que imitan la estructura y dinámica sucesional del bosque natural."

- it constructs around the agroecosystems a whole fabric of regulations intended to promote "*producción agroecológica*" (agroecological farming), defined as the process of production based on the synergic management of local resources of the agroecosystems through the use of practices that favor the biological and ecological complexity of the latter,[29] together with "*producción orgánica*" (organic farming), defined as the process of holistic production that applies organic methods rejecting the use of synthetic products[30];
- it establishes eleven legal principles[31] that represent the pillars of this regulatory complex: the *principle of sustainability* (duty to reach an overall result represented by the harmonic relationship between the factors of production and the ecosystems with their natural cycles, protecting biodiversity and respecting life in all its manifestations); the *principle of food sovereignty and safety* (protection of the individual and collective right to production, distribution, and consumption of food with quality and safety verifiable along the entire food chain); the *principle of healthiness* (requirement for production, conservation, processing, distribution, and consumption of products according to criteria of preventive health); the *principle of competition* (freedom to produce food and other products in a sustainable way for local and international markets, with quality, added value, and in a work setting that is safe, fair, and ecologically acceptable); the *principle of sustainable land management* (requirement to favor uses and productive practices in harmony with the spontaneous aptitudes and natural predispositions of ecosystems and agroecosystems, that they be able to reverse processes of degradation of soil and vegetation, erosion, loss of topsoil and fertile ground in arid, semiarid, and subhumid dry zones, caused mainly by inadequate human activities and climatic changes); the *principle of protection* (duty to apply activities, practices, and processes that are able to protect the integrity both of the ecosystems and of the human beings involved in production); the *principle of recognition* (duty to recognize, teach, and revitalize traditional and autochthonous knowledge in agricultural practices, reconciling the advancement of technological progress with the different conditions of each zone of production and its actors); the *principle of precaution* (duty to adopt, in the processes of agroecological and organic production, measures aimed at evaluating the social impacts together with the ecological ones in order to face the risks of irreversible damage to the ecosystems); the *principle of prevention* (duty to adopt, in the processes of agroecological and organic production, measures to minimize the negative impacts on the ecosystems and on human

[29] Art. 3, Para. 4, Law 765/2011: "Producción Agroecológica: Proceso productivo donde se aprovechan al máximo los recursos locales y la sinergia de los procesos a nivel del agroecosistema, utiliza prácticas que favorecen su complejidad, adoptando el control biológico y la nutrición orgánica de manera óptima en el manejo del sistema de producción o la finca."

[30] Art. 3, Para. 5, Law 765/2011: "Producción Orgánica: Sistema de producción holístico, que emplea al máximo los recursos de la finca mediante prácticas de gestión interna, aplicando métodos biológicos y descartando el empleo de productos sintéticos."

[31] Art. 4, Law 765/2011.

health); the *principle of fairness* (duty to assure a fair division of responsibilities and benefits deriving from access and use of natural resources in production processes); the *principle of participation* (duty to include—within legal and administrative procedures for decision, development, execution, and evaluation of the policies and strategies relative to agroecological production—public and private entities, institutions, companies, unions, civil society organizations, indigenous populations, ethnic groups, and communities of African origin);

- it entrusts to the *Ministerio Agropecuario y Forestal*, identified as "*Autoridad de aplicación*" of Law 765/2011,[32] a series of tasks, including conservation of genetic heritage and protection of the right of all producers to access, use, exchange, propagation, and protection of original germplasm; certification of agroecological and organic production systems on the basis of precise technical standards; promotion of training and instruction at all levels in agroecological and organic production, in coordination with competent institutions; validation of integrated and diversified production systems that involve both farmers and indigenes; protection of the immaterial cultural heritage represented by the traditional knowledge and wisdom of the indigenous populations[33];
- it gives to the *Ministerio Agropecuario y Forestal* and the *Ministerio de Fomento, Industria y Comercio* the task of informing the citizens and of sensitizing them to the consumption of agroecological and organic products; more generally, both Ministries are required by the law to promote, with apposite actions, the commercialization of agroecological and organic products both in domestic and foreign markets[34];
- it institutes a national Register of agroecological and organic producers[35];
- it institutes both a Specialized Unit (a governmental office) for the certification of agroecological and organic production systems and a Register of nongovernmental bodies empowered to certify, both nationally and internationally, agroecological or organic production systems[36];
- it institutes a national committee for reference and consultation on agroecological policies called the *Consejo de la Producción Agroecológica u Orgánica* (COPAGRO), which is participated in by representatives of the *Ministerio Agropecuario y Forestal*; the *Ministerio del Ambiente y de los Recursos Naturales*; the *Ministerio de Fomento, Industria y Comercio*; the *Instituto de Desarrollo Rural*; the *Instituto Nicaragüense de Tecnología Agropecuaria*; the *Consejos Regionales de las Regiones Autónomas de la Costa Atlántica*; the Municipalities; the public and private Universities with scientific research programs in agroecology; all the segments of the production and distribution lines;

[32] Art. 5, Law 765/2011.

[33] Art. 6, Law 765/2011.

[34] Art. 22, Paras. 2 and 3, Law 765/2011.

[35] Art. 8, Law 765/2011.

[36] Arts. 9–11, Law 765/2011.

agroecological and organic producers; and nongovernmental bodies involved in agroecological programs or projects;

- it envisions, in principle, a true and proper agroecological zoning and planning of the entire national territory, through recognition of agroecological and organic production zones (established in correspondence with the typologies and natural aptitudes of the soil and the agricultural productions correlated to them) and the successive planning of agroecological and organic production zones within the national territory.[37]

Of great importance also are Decree 2/2012 for the execution of the cited Law 765/2011[38] and, above all, the "*Norma Técnica Obligatoria Nicaragüense*" NTON 11 037-12/2013, having as its object the "*Caracterización, Regulación y Certificación de Unidades de Producción Agroecológica.*"[39]

In Venezuela, Law Decree 6129/2008 entitled "*Ley de Salud Agrícola Integral*"[40] defines "*salud agrícola integral*" (integral agricultural health) as the primary health of animals, plants, products, and byproducts of animal or vegetable origin, soil, water, air, human beings, and the close relations between them: Law Decree 6129/2008 expressly proclaims the necessity for the legislature and the administrative authorities to act "*incorporando principios de la ciencia agroecológica*" within the legal regulations,[41] according to an approach based not on mandatory requirements or coercive sanctions but rather on measures of promotion, monitoring, and information that are adequately justified scientifically.[42] According to this Venezuelan law, developing agroecology as a science is indis-

---

[37] Art. 22, Paras. 4 and 5, Law 765/2011: "promover [...] la declaratoria de zonas de producción agroecológica u orgánica, garantizando que se establezcan en correspondencia al tipo y vocación de suelo, según el uso en la producción de que se trate; y promover el ordenamiento territorial de las zonas de producción agroecológica u orgánica en el territorio nacional."

[38] "Reglamento General de la Ley nº 765, Ley de Fomento a la Producción Agroecológica u Orgánica, Decreto no. 02-2012", approved 23 January 2012, published in the Gaceta no. 15 of 25 January 2012.

[39] "Norma Técnica Obligatoria Nicaragüense NTON 11 037 – 12 Caracterización, Regulación y Certificación de Unidades de Producción Agroecológica", approved 30 April 2013 and published in the Gaceta n. 123 of 3 July 2013. http://legislacion.asamblea.gob.ni/normaweb.nsf/b92aaea87dac762406257265005d21f7/32d6ad99d191b0fe06257bc200799142?OpenDocument. See Salazar-Centeno (2013), pp. 58–65.

[40] "Decreto n° 6.129, con Rango, Valor y Fuerza de Ley de Salud Agrícola Integral" (n. 5.890 Extraordinary of the Gaceta Oficial de la República Bolivariana de Venezuela, 31 July 2008).

[41] Art. 1, Law Decree 6129/2008.

[42] "Exposicion de motivos", Law Decree 6129/2008: "los principios de la agricultura lo más sana posible por medio de las prácticas agroecológicas [...] no pueden transformarse en normas jurídicas puras, que como tales implican coerción, obligatoriedad y sanción, pero que como principios metas y objetivos deben quedar insertas en la nueva ley, a fin de impregnar esta nueva cultura agraria a las normativas, procedimientos y actos del propuesto Instituto Nacional de Salud Agrícola Integral [...] el Titulo III, referido a la Agroecología, establece políticas, definiciones y objetivos, pero no normas coercitivas."

pensable for the goal of guaranteeing food safety and sovereignty.[43] So also is assuring popular participation through involvement of city, village, and indigenous community councils and any other form of social organization whose principal activities are tied to the rural world.[44] The entire Title III of Law Decree 6129/2008 (Arts. 48–51) is dedicated to "*la Agroecología*," defined as a science whose principles are based on ancestral wisdom of respect, conservation, and preservation of all the natural components of the sustainable agroecosystems, of any scale and dimension.[45] The Central Government is expressly tasked with applying agroecology as the scientific basis for sustainable tropical agriculture in order to transform the economic and social model of the Nation, developing agroecological projects to stimulate food production processes of good biological quality and sufficient quantity for the population, promoting instruction and training for learning agroecological practices.[46] In order to apply agroecology, the Central Government, in cooperation with local councils, populations, indigenous communities, and other communities, must examine the various problems of agricultural health provoked by ecologically unsustainable models of agricultural production; it must propose, for each problem identified, agroecological projects to reconcile agricultural production with the environmental and cultural context; it must gather and process all correlated statistical information in order to survey and direct organizational assets to agroecological production.[47] Finally, the *Instituto Nacional de Salud Agrícola Integral* (INSAI, a public body directed by the *Ministerio del Poder Popular* and organized in regional and local administrative units corresponding to the various socio-bio-regional areas of the national territory[48]) is required to adopt strategies, plans, measures, and projects for agricultural health "*sobre la base fundamental de los principios agroecológicos*[49];" within INSAI is constituted, for these purposes, an apposite *Dirección de Agroecología y Participación Popular*.[50]

Brazil has focused instead on agroecological zoning ("*Zoneamento AgroEcológico – ZAE*").

In particular, with Federal Decree 6961/2009, agroecological zoning for sugarcane expansion (ZAE Cana)[51] was established. The general goal of agroecological zoning, entrusted to the *Ministério da Agricultura, Pecuária e Abastecimento*,

---

[43] Art. 2, Para. 1, Law Decree 6129/2008.

[44] Art. 2, Para. 7, Law Decree 6129/2008.

[45] Art. 48, Law Decree 6129/2008.

[46] Art. 49, Law Decree 6129/2008.

[47] Arts. 49 and 50, Law Decree 6129/2008.

[48] Art. 52, Law Decree 6129/2008.

[49] Art. 56, Para. 7, Law Decree 6129/2008.

[50] Art. 63, Law Decree 6129/2008.

[51] Decreto Presidencial n. 6961 of 17 September 2009. https://www.planalto.gov.br/ccivil_03/_ato2007-2010/2009/decreto/d6961.htm. See Almeida (2012), also for explanations on the position of the Federal Decrees (normative acts of the Executive) within the hierarchy of the sources of Brazilian law.

together with the *Ministério do Meio Ambiente*, is that of furnishing technical support for the formulation of public policies directed at the expansion and sustainable production of a specific crop. For the ZAE Cana, this involves the strategic necessity of evaluating, indicating, and spatializing the potential of the soil suitable for the expansion of the production of cane sugar crops (in rainforest conditions) for the production of bioethanol and sugar, as a basis for a comprehensive planning for the sustainable use of the territory in harmony with biodiversity. The ZAE allows, for example, the provision of sustainable economical alternatives to farmers, information for planning future development centers in rural zones, and useful data for coordinating rural development policies and energy policies.

The main indicators considered in the development of agroecological zoning are the vulnerability of the territory, climatic risk, potential for sustainable agricultural production, and existing environmental laws. Research is conducted to evaluate for each zone: *climatic suitability* (through a probability analysis of climatic risk), *pedological suitability* (through an estimate of the potential for agricultural production of a given crop in a particular model of crop management, on the basis of the classification of lands for physical and physiographic characteristics), *pedo-climatic suitability* (intersecting the results of the climatic and soil analyses),[52] and *use of the territory* (through mapping present uses and plant cover of the national territory, done with satellite imagery[53]).

The agroecological criteria introduced with the ZAE are important because they create a duty at the national level for all financial and credit institutions: namely, before financial institutions will issue loans (essential for large cultivation companies), they have to verify the compatibility of individual projects with agroecological zoning.[54]

[52] The pedo-climatic suitability gives rise to the classification of the soils in classes that are assigned certain letters: P, areas with high agricultural potential; R, areas with medium agricultural potential; MS, areas with low agricultural potential; ISC, areas not suitable because of the combination of soil and climate; IC, areas not suitable because of the climate, for thermal deficits or high risk of freezing; ID, areas not suitable because of the climate, by reason of unavoidable necessity of intensive irrigation; IE, areas not suitable because of the climate by reason of excess of water with prejudice to maturation and harvest; ICIS, areas not suitable both because of the climate and of the soil; IS, areas not suitable solely because of the soil.

[53] The legend of the uses in the territory is articulated in the abbreviations Ap (cultivated pastures), Ag (lands for agropastoral uses), and Ac (lands for agricultural use).

[54] See Almeida (2012), p. 33: "ZAE Cana is implemented through the Federal Decree 6.961 of 2009, which specifies the areas where sugarcane can be cropped and allows subsidised public and private financing only to existing or new sugarcane producers who expand within this zone. This financing is controlled by the National Monetary Council, which formulates policies for the Central Bank of Brazil. In November 2009, the National Monetary Council made the Rule 3.814, which prohibits public and private financing to sugarcane companies that produce sugar and/or ethanol and plan to expand outside ZAE Cana. ZAE Cana may also be implemented in the future by rules set up in the Resource Consent Bill 6.077 of 2009. This bill still needs to be approved by the House of Representatives and the Senate, and finally receive the presidential assent, to take legal effect. According to this bill, resource consents and the possibility to impose administrative, civil, and criminal penalties for illegal sugarcane expansions could become additional tools in the implementation of ZAE Cana." See also Oliveira Jr and Silva (2010),

Also, Federal Decree 7172/2010 follows a similar approach (though on the basis of different classifications of lands) by establishing the agroecological zoning for palm cultivation (*ZAE Palma de Óleo*),[55] in order to plan the expansion of Brazilian production of palm oil on a technical-scientific basis and to guarantee its sustainability economically, socially, and environmentally.

In Africa, similar attention has been given to agroecological zoning, for example, in Mali by Law 06-045/2006 ("*Loi d'orientation agricole*," promulgated by the *Président de la République du Mali* on 5 September 2006). After recognizing the importance for agricultural law of "knowledge" regarding "agroecological potential"[56] and "agroecological diversities,"[57] Law 06-045/2006 of Mali expressly establishes the principle by which local collectives must regulate their plans and management programs within the territory according to the different "agroecological zones of the Nation."[58] To this end, the local collectivities are required to identify, in their territorial planning projects, the "aptitudes of the lands" and the "types of production that best fit the potentials of each agroecological zone."[59] These territorial planning projects are then submitted for the opinion of the *Comité Exécutif Régional*, and then approved by the State, in order to assure integration with the strategies for interlocal and interregional land management.[60] In addition, Law 06-045/2006 requires that mandatory contributions or taxes to guarantee the

pp. 6343–6351. http://www.conpedi.org.br/manaus/arquivos/anais/fortaleza/3225.pdf; Strapasson et al. (2012), pp. 48–65.

[55] Decreto Presidencial 7.172/2010. http://www.planalto.gov.br/ccivil_03/_Ato2007-2010/2010/Decreto/D7172.htm.

[56] Art. 3, Law 06-045/2006: "La politique de développement agricole a pour but de promouvoir une agriculture durable, moderne et compétitive reposant, prioritairement sur les exploitations familiales agricoles reconnues, sécurisées, à travers la valorisation maximale du potentiel agro-écologique et des savoir-faire agricoles du pays."

[57] Art. 4, Law 06-045/2006: "La politique de développement agricole prend en compte les objectifs de la décentralisation et intègre les diversités agro-écologiques et la situation spécifique de chaque région du pays afin de déterminer les moyens à mettre en œuvre pour réaliser les objectifs visés. Elle intègre les stratégies et objectifs nationaux de lutte contre la pauvreté fixés dans le Cadre Stratégique de Lutte contre la Pauvreté."

[58] Art. 67, Law 06-045/2006: "La stratégie d'aménagement du territoire privilégie la gestion durable des ressources naturelles en conformité avec les engagements internationaux et la réduction des disparités inter et intra régionales. Elle tient compte des réalités des différentes zones agro-écologiques du pays dans le sens d'une responsabilisation effective des Collectivités territoriales, des exploitants agricoles et de leurs organisations. La stratégie d'aménagement du territoire intègre les contraintes majeures liées à l'aridité du pays périodiquement aggravée par les aléas climatiques."

[59] Art. 70, Law 06-045/2006: "Les Collectivités territoriales élaborent les schémas et programmes d'aménagement de leur ressort territorial qui sont soumis à l'approbation préalable de la tutelle après avis consultatif du Comité Exécutif Régional prévu à l'Article 190. Ces schémas précisent les vocations des terres et orientent les exploitants Agricoles vers les types de productions les plus conformes aux potentialités de chaque zone agro-écologique."

[60] Art. 70, Law 06-045/2006.

sustainability of agriculture must be differentiated based on different agroecological zones.[61]

As far as Europe is concerned, the most interesting legal experiment is taking place in France, on the initiative of Minister Le Foll. It is a bill for agriculture, food, and forests, already approved by the *Sénat* (on the first reading) on the evening between Tuesday 15 and Wednesday 16 April 2014, after 40 h of discussion, with 175 votes in favor and 134 against. It is now being examined by the *Assemblée nationale* (in its second reading).[62] This bill introduces important changes within the French *Code Rural* precisely in order to realize an agroecological law that is able to integrate agroecology into the law (see, on this subject, the contribution of HERMON in this book).

In particular, this bill inserts into the *Code Rural* a *Livre Préliminaire*, dedicated to the fundamental objectives of the public policies on agriculture, food, and maritime fishing, within which is the new Art. L.1 of the *Code Rural*,[63] according to which:

---

[61] Art. 74, Law 06-045/2006: "Les Collectivités territoriales peuvent prélever des redevances et taxes sur les aménagements et les infrastructures réalisés de leur ressort en vue d'assurer leur durabilité. L'assiette, le taux et les modalités de recouvrement des redevances et taxes sont déterminés par la législation, en tenant compte des spécificités régionales et agro-écologiques."

[62] *Assemblée nationale*, n. 1892, "Projet de Loi modifié par le Sénat, d'avenir pour l'agriculture, l'alimentation et la forêt", registered by the *Présidence de l'Assemblée nationale* 17 April 2014.

[63] "Projet de Loi modifié par le Sénat, d'avenir pour l'agriculture, l'alimentation et la forêt", Art. 1: " I. – Avant le livre Ier du Code Rural et de la Pêche Maritime, il est inséré un Livre Préliminaire ainsi rédigé: LIVRE PRÉLIMINAIRE. OBJECTIFS DE LA POLITIQUE EN FAVEUR DE L'AGRICULTURE, DE L'ALIMENTATION ET DE LA PÊCHE MARITIME. Art. L.1. – I. – La politique en faveur de l'agriculture et de l'alimentation, dans sa triple dimension européenne, nationale et territoriale, a pour finalités: 1° Dans le cadre de la politique de l'alimentation définie par le Gouvernement, d'assurer à la population, dans des conditions économiquement et socialement acceptables par tous et en quantité suffisante, l'accès à une alimentation sûre et saine, diversifiée et de bonne qualité, produite dans des conditions favorisant l'emploi, le respect des normes sociales, la protection de l'environnement et des paysages et contribuant à l'atténuation et à l'adaptation aux effets du changement climatique; 1° bis De répondre à l'accroissement démographique, en rééquilibrant les termes des échanges en matière de denrées alimentaires entre pays, dans un cadre européen et de coopérations internationales fondées sur le respect des principes de la souveraineté alimentaire permettant un développement durable et équitable; 2° De soutenir le revenu et de développer l'emploi des agriculteurs et des salariés, notamment par un meilleur partage de la valeur ajoutée et en renforçant la compétitivité et l'innovation des différentes filières de production, de transformation et de commercialisation. Elle préserve le caractère familial de l'agriculture et d'autonomie et de responsabilité individuelle de l'exploitant. Elle vise à améliorer la qualité de vie des agriculteurs; 3° De contribuer à la protection de la santé publique, de veiller au bien-être et à la santé des animaux, à la santé des végétaux et à la prévention des zoonoses; 3° bis De promouvoir l'information des consommateurs quant aux lieux et modes de production et de transformation des produits agricoles et agroalimentaires; 4° De participer au développement des territoires de façon équilibrée, diversifiée et durable; 4° bis De prendre en compte les situations spécifiques à chaque région. Elle valorise en particulier les services écosystémiques; 4° ter De rechercher des équilibres sociaux justes et équitables; 5° De développer la valeur ajoutée dans chacune des filières agricoles et alimentaires et de renforcer la capacité exportatrice de la France; 5° bis D'encourager la diversité des produits, le développement des productions sous signes de qualité et d'origine, la

- the new public law of agriculture must be founded "on the practices of agro-ecology" and on "agroecological production systems," so that they may protect the autonomy of farmers, reconcile agricultural competition and profitability with reduction of consumption of energy, water, fertilizers, phytopharmaceutical products, and veterinary medicines, using biological interactions and natural potential found in water, biodiversity, photosynthesis, soils, and air, maintaining their capacity of renewal quantitatively and qualitatively, and favoring adaptation to the effects of climate change;
- the State has the duty to facilitate the recourse of farmers to "innovative cultivation practices and systems according to an agroecological approach" and to sustain the professional actors in the development of "biocontrol" solutions (namely, control measures of insect infestations based not on chemically synthesized products but rather on microorganisms or natural pathogenic agents), accelerating the procedures for evaluation and authorization into commerce of products that use agroecological biocontrol;
- the State must intervene to "facilitate interactions between social sciences and agronomic sciences in order to make possible the production and transfer of knowledge necessary for the transition to agroecological models."[64]

transformation sur zone ainsi que les circuits courts; 5° bis De promouvoir la conversion et le développement de l'agriculture et des filières biologiques au sens de l'article L. 641-13; 6° De concourir à la transition énergétique, en contribuant aux économies d'énergie dans le secteur agricole, au développement des énergies renouvelables et à l'indépendance énergétique de la Nation, notamment par une valorisation optimale et durable des sous-produits d'origine agricole et agroalimentaire dans une perspective d'économie circulaire; 7° De développer l'aide alimentaire; 8° De lutter contre la faim dans le monde, dans le respect des agricultures et des économies des pays en développement et en cohérence avec les politiques de développement et de solidarité internationale française et communautaire. La politique d'aménagement rural définie à l'article L. 111-2 et les dispositions particulières aux professions agricoles en matière de protection sociale et de droit du travail prévues au livre VII contribuent à ces finalités."

[64] "Projet de Loi modifié par le Sénat, d'avenir pour l'agriculture, l'alimentation et la forêt", Art. 1: " [...] Art. L.1. [...] II. – Afin d'atteindre les objectifs mentionnés au I du présent article, la politique conduite par l'État favorise: 1° L'ancrage territorial de la production et de la transformation agricoles ainsi que de la commercialisation des produits agricoles y compris par la promotion de circuits courts; 2° Le développement de filières de production et de transformation alliant performance économique, haut niveau de protection sociale, performance sanitaire et performance environnementale, capables de relever le double défi de la compétition internationale et de la transition écologique, en mettant sur le marché une production innovante et de qualité, en soutenant le développement des filières des énergies renouvelables, des produits biosourcés et de la chimie végétale; 3° La recherche, l'innovation et le développement; 4° L'organisation collective des acteurs; 5° Le développement des dispositifs de prévention et de gestion des risques; 6° Les actions contributives réalisées par l'agriculture et la sylviculture en faveur de l'atténuation et de l'adaptation au changement climatique; 7° L'équilibre des relations commerciales; 8° La protection des terres agricoles. Les politiques publiques visent à promouvoir et à pérenniser les systèmes de production agricole et les pratiques agronomiques permettant d'associer la performance économique, la performance sociale et la performance environnementale. Elles privilégient les démarches collectives et s'appuient sur les pratiques de l'agro-écologie, dont le mode de production biologique fait partie. Les systèmes de production agro-écologiques privilégient l'autonomie des exploitations agricoles et l'amélioration de leur compétitivité en maintenant ou en augmentant

Also, very interesting is the close link between agroecology and the "*Groupement d'Intérêt Économique et Environnemental*" (GIEE), introduced by this bill. According to the bill, the State representative in the region shall legally qualify a group as GIEE at the outcome of a selection: nevertheless, an indispensable condition to obtain legal recognition as GIEE is that of presenting a multiannual project that proposes "relevant actions of agroecology able to improve the economic, social, and environmental performances of agricultural productions, in particular favouring the technical, organizational, or social innovation of the agricultural experiments." Hence, there is no GIEE without projects scientifically based on agroecology.[65]

---

la rentabilité économique, en améliorant la valeur ajoutée des productions, et en économisant la consommation d'énergie, d'eau, d'engrais, de produits phytopharmaceutiques et de médicaments vétérinaires, en particulier les antibiotiques. Ils sont fondés sur les interactions biologiques et l'utilisation des potentiels offerts par les ressources naturelles, en particulier les ressources en eau, la biodiversité, la photosynthèse, les sols et l'air, en maintenant leur capacité de renouvellement du point de vue qualitatif et quantitatif. Ils contribuent à l'atténuation et à l'adaptation aux effets du changement climatique. L'État veille aussi à faciliter le recours par les agriculteurs à des pratiques et à des systèmes de cultures innovants dans une démarche agro-écologique. À ce titre, il soutient les acteurs professionnels dans le développement des solutions de biocontrôle et veille à ce que les processus d'évaluation et d'autorisation de mise sur le marché de ces produits soient accélérés. L'État veille à faciliter les interactions entre sciences sociales et sciences agronomiques pour faciliter la production et le transfert de connaissances nécessaire à la transition vers des modèles agro-écologiques. Les politiques publiques visent à promouvoir et à pérenniser les systèmes de production agricole et les pratiques agronomiques permettant d'associer la performance économique, la performance sociale et la performance environnementale. Elles privilégient les démarches collectives et s'appuient sur les pratiques de l'agro-écologie, dont le mode de production biologique fait partie. Les systèmes de production agro écologiques privilégient l'autonomie des exploitations agricoles et l'amélioration de leur compétitivité en maintenant ou en augmentant la rentabilité économique, en améliorant la valeur ajoutée des productions, et en économisant la consommation d'énergie, d'eau, d'engrais, de produits phytopharmaceutiques et de médicaments vétérinaires, en particulier les antibiotiques. Ils sont fondés sur les interactions biologiques et l'utilisation des potentiels offerts par les ressources naturelles, en particulier les ressources en eau, la biodiversité, la photosynthèse, les sols et l'air, en maintenant leur capacité de renouvellement du point de vue qualitatif et quantitatif. Ils contribuent à l'atténuation et à l'adaptation aux effets du changement climatique. L'État veille aussi à faciliter le recours par les agriculteurs à des pratiques et à des systèmes de cultures innovants dans une démarche agro-écologique. À ce titre, il soutient les acteurs professionnels dans le développement des solutions de biocontrôle et veille à ce que les processus d'évaluation et d'autorisation de mise sur le marché de ces produits soient accélérés. L'État veille à faciliter les interactions entre sciences sociales et sciences agronomiques pour faciliter la production et le transfert de connaissances nécessaire à la transition vers des modèles agro-écologiques [...] IV. – La politique d'installation et de transmission en agriculture a pour objectifs: 1° De favoriser la création, l'adaptation et la transmission des exploitations agricoles dans un cadre familial et hors cadre familial; 2° De promouvoir la diversité des systèmes de production sur les territoires, en particulier ceux générateurs d'emplois et de valeur ajoutée et ceux permettant d'associer performance économique, haut niveau de protection sociale, performance sanitaire et performance environnementale, notamment ceux relevant de l'agro-écologie [...]."

[65] "Projet de Loi modifié par le Sénat, d'avenir pour l'agriculture, l'alimentation et la forêt", Art. 3: "Le Code Rural et de la Pêche Maritime est ainsi modifié. 1° Le chapitre Ier du titre Ier du livre III est complété par des articles L. 311-4 à L. 311-5-1, L. 311-6 et L. 311-7 ainsi rédigés. Art.

Finally, this French bill establishes that agroecology must be integrated into educational programs of the public system of education, professional training, development, and research in agriculture, agronomy, and veterinary sciences.[66]

In Switzerland, it suffices to mention the "*Règlement 910.21.1 sur l'Agroécologie (RAgrEco)*" of 15 December 2010 approved by the Canton of Vaud, which became applicable in January 2011. It contains "*les modalités d'exécution des dispositions relatives à l'agroécologie de la Loi sur l'agriculture vaudoise*" (LVLAgr) of 7 September 2010. These are legislative measures and regulations whose principal objective is to address the public economic and financial subsidies to farmers, with respect to agroecological objectives, such as the "promotion of voluntary ecological measures" by farmers;[67] the realization of "collective agri-environmental projects;"[68] the "maintenance of the fertility of the soil" through safeguarding and increasing "lawns"[69] and "pilot projects of cultivation by direct sowing;"[70] the

L. 311-4. – Peut être reconnue comme groupement d'intérêt économique et environnemental toute personne morale dont les membres portent collectivement un projet pluriannuel de modification ou de consolidation de leurs systèmes ou modes de production agricole et de leurs pratiques agronomiques en visant une performance à la fois économique, sociale et environnementale. Le projet pluriannuel contribue à renforcer la performance sociale en mettant en œuvre des mesures de nature à améliorer les conditions de travail des membres du groupement et de leurs salariés, à favoriser l'emploi ou à lutter contre l'isolement en milieu rural. Cette personne morale doit comprendre plusieurs exploitants agricoles et peut comporter d'autres personnes physiques ou morales, privées ou publiques. Les exploitants agricoles doivent détenir ensemble la majorité des voix au sein des instances du groupement. La reconnaissance de la qualité de groupement d'intérêt économique et environnemental est accordée par le représentant de l'État dans la région à l'issue d'une sélection. Le suivi, la diffusion des innovations ou l'accompagnement des groupements d'intérêt économique et environnemental relèvent de l'article L. 820-2. La qualité de groupement d'intérêt économique et environnemental est reconnue pour la durée du projet pluriannuel. Art. L. 311-5. – Pour permettre la reconnaissance d'un groupement comme groupement d'intérêt économique et environnemental, le projet pluriannuel mentionné à l'article L. 311-4 doit: 1° Associer plusieurs exploitations agricoles sur un territoire cohérent leur permettant de favoriser des synergies; 2° Proposer des actions relevant de l'agro-écologie permettant d'améliorer les performances économique, sociale et environnementale de ces exploitations, notamment en favorisant l'innovation technique, organisationnelle ou sociale et l'expérimentation agricoles; 3° Répondre aux enjeux économiques, sociaux et environnementaux du territoire où sont situées les exploitations agricoles concernées, notamment ceux identifiés dans le plan régional de l'agriculture durable mentionné à l'article L. 111-2-1 et en cohérence avec les projets territoriaux de développement local existants; 4° Prévoir les modalités de regroupement, de diffusion et de réutilisation des résultats obtenus sur les plans économique, environnemental et social. L'accompagnement, le suivi, la capitalisation et la diffusion des innovations des groupements d'intérêt économique et environnemental sont assurés par les organismes de développement agricole, dont les têtes de réseau ont conclu avec l'État un contrat d'objectifs ou un programme pluriannuel de développement agricole et rural dans des conditions définies par décret."

[66] "Projet de Loi modifié par le Sénat, d'avenir pour l'agriculture, l'alimentation et la forêt", Art. 26 and Art. 27.

[67] Art. 9, "Règlement 910.21.1 sur l'Agroécologie (RAgrEco)" of 15 December 2010.

[68] Chapter III, RAgrEco.

[69] Art. 18, RAgrEco.

[70] Art. 19, RAgrEco.

protection of the "biodiversity and the diversity of the countryside;"[71] the creation of a "network of compensatory ecological surfaces,"[72] etc.

The discussion on the necessity of overcoming barriers between environmental law and agricultural law and constructing a new "agroecological law" is in embryo also in Asia, in great countries such as China. Even though they have not yet adopted specific laws on the subject, the debate is under way.[73]

According to researchers involved in this discussion, what is needed is to conceive of an "agro-eco-environment legislation" that is not merely a conglomeration of environmental and agricultural laws and regulations heaped up in disorder, indifferently, confusedly, and in conflict with one another but rather "a systemized, interdependent organic whole classified according to definite standards"[74] that is centered on "ecological interest supremacy" and agroecology.[75] A

---

[71] Chapter VI, RAgrEco.

[72] Arts. 26–33, RAgrEco: "Le service, en collaboration avec le service en charge de la protection de la nature, détermine les exigences d'appréciation en matière de qualité biologique particulière et de mise en réseau des surfaces de compensation écologique (ci-après: réseau), conformément aux exigences minimales fixées par l'ordonnance fédérale sur la qualité écologique (ci-après: OQE) et par les instructions de la Confédération. Il requiert l'approbation de la Confédération. Ces exigences sont régulièrement mises à jour en fonction de l'évolution des connaissances [...] Le réseau doit permettre le développement de la flore et de la faune spécifiques de la région concernée. Il doit être constitué de surfaces de compensation écologique, au sens de l'ordonnance fédérale sur les paiements directs, en relation avec d'autres milieux naturels, tels que biotopes, forêts ou cours d'eau. Il doit tenir compte des inventaires nationaux, régionaux ou locaux, de documents scientifiques ou de plans directeurs publiés, et respecter d'autres projets de préservation des écosystèmes existants dans le périmètre [...] Le projet de réseau doit notamment indiquer: a. le promoteur et les partenaires du projet; b. le professionnel qualifié qui conseille les exploitants bénéficiaires du projet; c. le périmètre concerné; d. un descriptif de l'état initial des milieux naturels; e. la liste des inventaires et données de base prises en compte; f. les objectifs et les synergies avec d'autres projets; g. les types de mesures mises en place sur le terrain; h. les dispositions d'évaluation et de suivi du projet; i. le financement du projet [...] Le réseau doit couvrir au minimum 100 hectares de surface agricole utile ou impliquer, en tout ou partie, au moins 5 exploitations agricoles. Le service en charge de la protection de la nature peut demander une extension du périmètre d'un projet lorsque les objectifs en matière de biodiversité et de liaisons biologiques l'imposent ou lorsque la complémentarité est nécessaire avec un autre projet [...]."

[73] Lin (2010), pp. 1261–1265; Jin-hua et al. (2010), pp. 19465–19467; Legislation based on agro-ecological and environmental protection (June 10, 2014). http://www.nt20.com/index.php/archives/4859. Accessed 30 Sept 2014.

[74] Lin (2010), p. 1261: "The existing problems of agro-eco-environment legislation. First, poor match coordination. As a system, the internal structure of agro-eco-environment law composed of law and regulations are not piled up together with disorder, but a systemized, interdependent organic whole classified according to definite standard. Up to now, there exist in the frame system conflict and confusion at the level of legal validity [...] the intercross between different agricultural environmental laws and regulations, combined with the immature legislative techniques, leads to contradiction and conflict phenomenon among these laws and regulations and there is still some gap to fill."

[75] Lin (2010), pp. 1262–1263: "To improve further the path choice of ecological agricultural environmental legislation [...] Ecological interest supremacy. Ecological interest should be put

future horizon for Chinese legislation could be, therefore, "to formulate a unified agricultural ecological environmental protection law."[76]

5 Interlegalities and Coordination in Agroecological Law

The rapid examination of experiences under way throughout the world strengthens the conviction that it is possible to work on the idea of agroecological law, even if the way seems long and hard.

This does not require imagining a *super-law* that, top-down and hierarchically, purports to *incorporate* and *replace* the existing legal fields with their specializations (e.g., agricultural law and environmental law). On the contrary, it requires constructing a *trans-law* that, bottom-up and progressively, attempts to *link* and *coordinate* regulatory measures between different legal fields, respecting their autonomy and distinction but, at the same time, emphasizing their common roots in *rus*.

From this point of view, the concept of "interlegality," understood as "an intersection of different legal orders,"[77] is very useful. The contribution of HOSPES in this book refers to this notion.

Many legal fields, gravitating around the universe of rurality and gathering each a single fragment of agroecosystem regulation, create many "legal force fields" that interfere with each other. The "interference zones" between different legal force fields represent "interlegal niches" in which can be manifested both repulsive type interferences (which give way to disturbances, attrition, and noise, if left to themselves) and attractive type interferences (which synergically involve the forces of each sector multiplying its regulatory power, if channeled through appropriate coordination tools).

The task of constructing agroecological law, in this perspective, is twofold:

- counteract the *antinomic interlegalities* between the various legal fields appurtenant to *rus*, through tools of *negative coordination*;

---

the supreme place when enacting agro-eco-environmental legislation, because as a part of the ecosphere, human development can't surpass the ecological allowed limit. With ecological interest supremacy principle, the enactors are required to ensure economic growth on the basis of ecology while enacting laws. The sustainable agricultural development guided and achieved by ecology standard demands to abide by ecological law, like biodiversity rule, ecosystem cycle and regeneration law and ecological balance rule and so on. Soil and water loss, desertification, violent sandstorm in recent years in some districts of our country, they are all punishment nature return to human for violating ecological rule. In fact, the basic theory of ecology is the basic principle that we must obey today while dealing with the environmental problems and is the theoretical basis enacting environment and natural resources law."

[76] Legislation based on agro-ecological and environmental protection (June 10, 2014). http://www.nt20.com/index.php/archives/4859. Accessed 30 Sept 2014.

[77] Santos (1987), pp. 279–302, especially pp. 297–298; Santos (2002); Darian-Smith (2013), pp. 168 et seqq.; Tuory (2014), pp. 41 et seqq.

– catalyze instead the *compatible interlegalities*, through tools of *positive coordination*.[78]

6 Concluding Remarks

What are the main categories of operative tools that agroecological law could use to coordinate "rural interlegalities" through a horizontal platform shared by different legal fields?

Some can be suggested, with the warning that this is an open list, exemplifying, not exhaustive:

I) agroecological information collecting and sharing (AICS);
II) agroecological zoning (AZ);
III) agroecological planning (AP);
IV) agroecological impact assessment (AIA).

The first category of tools (AICS) requires that the public authorities (from the local level to the national level) be required to perform structured and systematic "readings" of rural territories, aimed at acquiring and continuously updating data and information on the characteristics of the various agroecosystems. This public survey requires the involvement and participation of the rural communities: not only producers, workers, and consumers but also, more generally, the inhabitants, starting with the nuclear farmer families, including also the scientific communities and the researchers of the rural traditions of those places. This means "photographing" and "mapping" the various agroecosystems present in the territory: the difference compared to the current public registers is that the AICS looks at the agroecosystems as social-ecological systems, whose characteristics and boundaries depend not only on material parameters of biophysics, agronomy, or economy but also on immaterial parameters dealing with historical, cultural, and social identity (for example, typical products of the agri-food traditions should be considered). The AICS consists therefore in the creation on a local, regional, national, and European scale of "agroecological cadastres," which should be hosted on open-access public media platforms, with the right for all to consult them and to propose to the competent authorities any corrections, integrations, improvements, and updates that reflect more accurately the reality of the surveyed and described agroecosystems.

The second category of tools (AZ) moves from the information gathered on the various agroecosystems through the AICS and aims to subdivide the territory into agroecological zones (or "rural districts," to use the terminology already present in

[78] On positive and negative coordination, see Scharpf (1994), pp. 27–53; Bobbio (1996), pp. 83–85.

the Italian legislation: see the contribution of BUIA & ANTONUCCI in this book).[79] The agroecological zones must be delimited intersecting the data relative to the agricultural aptitudes of the lands, the soil and climate parameters, the historical processes of settling and growth of the rural communities present in the territories, their social composition, the specific conditions of the agri-food market, the traditions of the autochthonous rural civilization, and so on. The purpose of the AZ is to order the rural territory, prohibiting types and modes of agricultural production that are incompatible with the characteristics of each agroecological zone and allowing, instead, those that respect the identity and uniqueness of each zone.

The third category of tools (AP) has the purpose of rendering consistent the numerous administrative planning acts that affect the agroecological zones delimited with the AZ. The coordination of the "first level" heterogeneous plans (about the environment, urban spaces, socio-economic activities, infrastructural development, tourism, coastal zones, etc.) can take place through "second level" agroecological plans (metaplanning) that analyze the "first level" plans and their impacts on rural territories, focus on all the points of convergence and divergence, and establish measures (including financial ones, through disbursements or denials of public subsidies) that help minimize antinomic interferences and maximize synergic interferences, in order to respect the characteristics of each agroecological zone.

Finally, the fourth category of tools (AIA) must provide an administrative procedure (co-managed through forms of consultation among the various competent public authorities for the different legal fields) that subordinates the realization of any *project of transformation of a rural territory* to the preventive evaluation of its *agroecological impact*. This is not an evaluation only of the *environmental* impact but of the interrelated complex of impacts that are ecological, economic, occupational, social, and cultural for that given rural area, to ensure compliance to the provisions of the AZ and AP so that an agroecological degradation or collapse is not caused.

The AIA should protect not only the land considered as a collection of ecological systems and as a resource for future generations: it also should preserve the equilibrium of the *human-rural environment*. More generally, to use the expression of DE NITTO in this book, the object of the inviolable rights protected by agroecological law is, ultimately, the "humanity of land": the heritage of past generations fills the forms and flavors of the land, marks the identity and the welfare of present generations, becomes the genetic heritage of future generations in a perspective of continuity of knowledge and at the same time the potential for evolutionary diversification, thanks to the formidable treasure chest of the biological and cultural

[79] The "rural districts" are defined Italy by Art. 13 of Legislative Decree 228/2001 as "the local production systems [. . .] characterised by a homogeneous historical and territorial identity deriving from the integration between agricultural activity and other local activities, as well as by the production of goods or services of particular specificity, coherent with the natural and territorial traditions and vocations."

diversities that reside not only in nature but also in the history of man's cohabitation with it.

References

Adam E (2014) Multifunctionality of agriculture and international trade. In: Eckinger J, Kaplan DM (eds) Encyclopedia of food and agricultural ethics. Springer, Dordrecht

Agnoletti M (ed) (2013) Italian historical rural landscapes: cultural values for the environment and rural development. Springer, Dordrecht

Albisinni F (1998) Lo spazio rurale come elemento d'impresa – Note per un diritto rurale. In: Various Authors, Agricoltura e Ruralità. Quaderni dell'Accademia dei Georgofili, vol 7. Studio editoriale fiorentino, Firenze, p 139

Albisinni F (2000) Ruralità come regola di diritto per uno sviluppo sostenibile. In: Various Authors, Agricoltura e diritto. Scritti in onore di Emilio Romagnoli, vol I. Giuffrè, Milano, p 421

Almeida M (2012) Analysing the Brazilian sugarcane agroecological zoning: is this government policy capable of avoiding adverse effects from land-use change? MSc dissertation, School of Geography, Environment and Earth Sciences, Victoria University of Wellington, Wellington. http://researcharchive.vuw.ac.nz/bitstream/handle/10063/2418/thesis.pdf?sequence=1. Accessed 30 Sept 2014

Alpa G, Bessone M, Fusaro A (2001) Poteri dei privati e statuto della proprietà, vol 1. Nozione e rilevanza costituzionale. SEAM, Roma, p 412

Altieri MA (1983) Agroecology. University of California Press, Berkeley

Altieri MA (1987) Agroecology: the scientific basis of alternative agriculture, 2nd edn. Westview Press, Boulder

Altieri MA (1989) Agroecology: a new research and development paradigm for world agriculture. Agric Ecosyst Environ 27(1–4):37–46

Altieri MA (1991) Agroecologia: prospettive scientifiche per una nuova agricoltura (trans: Angelini P). Muzzio, Padova

Angelicchio G (2014) Indicazioni geografiche e denominazioni di origine. In: Scuffi M, Franzosi M (eds) Diritto industriale italiano. Tomo I: Diritto sostanziale. Cedam, Padova, pp 345–386

Basile E, Romano D (eds) (2002) Sviluppo rurale: società, territorio, impresa. Franco Angeli, Milano

Battista Ferri G (2005) La "nuova" impresa agricola. Diritto e giurisprudenza 120(1):1–15

Bland WL, Bell MM (2008) A Holon approach to agroecology. Int J Agric Sustain 5(4):280–294

Bobbio L (1996) La democrazia non abita a Gordio: studio sui processi decisionali politico-amministrativi. Franco Angeli, Milano

Bohlen PJ, House G (eds) (2009) Sustainable agroecosystem management. Integrating ecology, economics and society. CRC, Boca Raton

Bonnal P, Bonin M, Aznar O (2012) Les évolutions inversées de la multifonctionnalité de l'agriculture et des services environnementaux. VertigO – La revue électronique en sciences de l'environnement 12(3). doi:10.4000/vertigo.12882

Bosworth G, Somerville P (eds) (2014) Interpreting rurality: multidisciplinary approaches. Routledge, London

Brouwer F (ed) (2004) Sustaining agriculture and the rural environment: governance, policy and multifunctionality. Edward Elgar, Cheltenham

Brouwer F, van der Heide CM (eds) (2009) Multifunctional rural land management: economics and policies. Earthscan, London

Bryden JM et al (eds) (2011) Towards sustainable rural regions in Europe: exploring inter-relationships between rural policies, farming, environment, demographics, regional economies and quality of life using system dynamics. Routledge, New York

Buller H (2004) The 'espace productif', the 'théâtre de la nature' and the 'territoires de développement local': the opposing rationales of contemporary French rural development policy. Int Plan Stud 9(2–3):101–119

Busti S (2013) L'agriturismo. In: Franceschelli V, Morandi F (eds) Manuale del diritto del turismo, 5th edn. Giappichelli, Torino, p 198

Caforio G (2014) La tutela delle tipicità appartenenti alla pubblica amministrazione. Aedon 17(1). http://www.aedon.mulino.it/archivio/2014/1/caforio.htm. Accessed 20 Oct 2014

Camaioni B et al (2013) Quanto è "rurale" la Politica di sviluppo rurale? Agriregionieuropa 9(34). http://agriregionieuropa.univpm.it/content/article/31/34/quanto-e-rurale-la-politica-di-sviluppo-rurale. Accessed 30 July 2014

Caporali F (1991) Ecologia per l'agricoltura. Teoria e Pratica. Utet, Torino

Caporali F (2011) Agroecology as a transdisciplinary science for a sustainable agriculture. In: Lichtfouse E (ed) Biodiversity, biofuels, agroforestry and conservation agriculture. Springer, Dordrecht, p 1

Caporali F, Campiglia E, Mancinelli R (2010) Agroecologia: teoria e pratica degli agroecosistemi. Città studi, Grugliasco, IT

Carbone A (2009) Agriculture multifunctionality: the contribution that the European primary sector can give to preserve and possibly to reinforce local cultural identities. Agricoltura Istituzioni Mercati 6(1–2):133–144

Carmignani S (2012) Agricoltura e ambiente. Le reciproche implicazioni. Giappichelli, Torino

Carroll CR, Vandermeer JH, Rosset PM (eds) (1990) Agroecology. McGraw-Hill, New York

Carrozza A (1994) Lineamenti di un diritto agrario ambientale. I materiali possibili. I leganti disponibili. Rivista di diritto agrario 73(2):151–172

Caruso MA (2011) Temi di diritto dei beni immateriali e della concorrenza. Giuffré, Milano

Cingari F (2008) La tutela penale dei marchi e dei segni distintivi. Wolters Kluwer Italia, Milano

Cleveland DA (2013) Agroecology. In: Gibson D (ed) Oxford bibliographies online: ecology. Oxford University Press, New York. doi:10.1093/obo/9780199830060-0117

Contò F (ed) (2005) Olivicoltura e ambiente: un nuovo equilibrio tra marginalità, condizionalità, multifunzionalità e sviluppo del territorio. Franco Angeli, Milano

Cossu M (2003) La "nuova" impresa agricola tra diritto agrario e diritto commerciale. Rivista di diritto civile 49(1):73–100

Costato L (2008a) Corso di diritto agrario italiano e comunitario. Giuffrè, Milano

Costato L (2008b) La lezione di Antonio Carrozza (e qualche possibile fraintendimento). Rivista di diritto agrario 87(1):15–24

Costato L, Rook Basile E, Germanò A (eds) (2011) Trattato di diritto agrario, vol II. Il diritto agroambientale. Utet, Torino

Cristiani E (2008) Le riviste agraristiche italiane e i nuovi contenuti del diritto agrario. Rivista di diritto agrario 87(4):464–479

D'Addezio M (2008) Le attività di servizi agroambientali nell'ordinamento giuridico sopranazionale italiano e comunitario: questioni di qualificazione. Agricoltura Istituzioni Mercati 5(2):9–34

D'Imperio G (2011) Sicurezza sul lavoro e utilizzo dei macchinari in agricoltura. Diritto e pratica del lavoro 28(20):1195–1198

Dalgaard T, Hutchings NJ, Porter JR (2003) Agroecology, scaling and interdisciplinarity. Agr Ecosyst Environ 100(1):39–51

Darian-Smith E (2013) Laws and societies in global contexts: contemporary approaches. Cambridge University Press, Cambridge

Daugstada K, Rønningena K, Skarb B (2006) Agriculture as an upholder of cultural heritage? Conceptualizations and value judgements – a Norwegian perspective in international context. J Rural Stud 22(1):67–81

De Giorgi Cezzi G (2005) Tutela dell'identità dei luoghi e rilevanza "monumentale" di un ulivo secolare. Foro Amministrativo – TAR 4(9):2955 et seqq
de los Desamparados Llombart Bosch M (1999) Las interrelaciones del Derecho Agrario y el Derecho Ambiental. In: Ballesteros Hernández LM, Oliván del Cacho J (eds) Actas del Congreso español de Derecho Agrario y ordenación rural. Instituto de Derecho Agrario de la Universidad de Zaragoza, Zaragoza, p 217
Doussan I (2002) Activité agricole et droit de l'environnement, l'impossible conciliation? L'Harmattan, Paris
Esposti R, Sotte F (eds) (2002) Le dinamiche del rurale: letture del caso italiano. Franco Angeli, Milano
Flora C (ed) (2001) Interactions between agroecosystems and rural communities. CRC, Boca Raton
Francario L (1993) Agricoltura e ambiente: nuovi stimoli per l'approccio giuridico. Diritto e giurisprudenza agraria e dell'ambiente 2(10):517 et seqq
Francis C et al (2003) Agroecology: the ecology of food systems. J Sustain Agric 22(3):99–118
Friedland WH (2002) Agriculture and rurality: beginning the 'Final Separation'? Rural Sociol 67 (3):350–371
Gargani A (2013) Reati contro l'incolumità pubblica. Tomo II: Reati di comune pericolo mediante frode. In: Grosso CF, Padovani T, Pagliaro A (eds) Trattato di diritto penale. Parte speciale, vol IX. Giuffré, Milano, p 273
Giacomini C et al (2007) Lo sviluppo dei marchi geografici collettivi e dei segni distintivi per tutelare e valorizzare i prodotti freschissimi. Franco Angeli, Milano
Giliberti B (2013) Il principio di precauzione nel diritto alimentare e farmaceutico. GiustAmm.it: Rivista di diritto pubblico 10(2):1 et seqq. http://www.giustamm.it/
Gliessman SR (ed) (1990) Agroecology: researching the ecological basis for sustainable agriculture. Springer, New York
Gliessman SR (2006) Agroecology: the ecology of sustainable food systems, 2nd edn. CRC, Boca Raton
Gonzalez de Molina M (2013) Agroecology and politics. How to get sustainability? About the necessity for a political agroecology. Agroecol Sustain Food Syst 37(1):45–59
González Linares N (2011) Derecho agrario contemporáneo y derecho civil. San Marcos, Lima
Gray J (2000) The common agricultural policy and the re-invention of the rural in the European community. Sociologia Ruralis 40(1):30–52
Hamilton ND (1999) A changing agricultural law for a changing agriculture. Drake J Agric Law 4 (1):41–58
Henke R (ed) (2004) Verso il riconoscimento di una agricoltura multifunzionale: teorie, politiche, strumenti. Edizioni Scientifiche Italiane, Napoli
Hernandez Zakine C (1998) L'influence du droit de l'environnement sur le droit rural. Conservation de la faune sauvage, droit communautaire, droit National. Ruralia, Revue de l'Association des ruralistes français 3:133–155
Hudault J (1987) Droit rural, droit de l'exploitation agricole. Dalloz, Paris
Hudault J (2006) L'evolution et les fondements actuels du droit rural. Rivista di diritto agrario 85 (3):247–260
Iacoponi L (1998) La sfida della moderna ruralità: sviluppo rurale, ambiente e territorio. In: Regazzi D (ed) L'agricoltura italiana tra prospettiva mediterranea e continentale. Atti del 33° Convegno di studi della Società Italiana di Economia Agraria – SIDEA, Napoli, 26–28 September 1996. SIDEA-Officine grafiche Grafitalia, Cercola, Napoli, p 51
Istituto Nazionale di Economia Agraria (2001) Beni culturali: una risorsa per lo sviluppo rurale. INEA Working Paper, Roma. http://inea.it/reteleader/pubblica/working/cultura.PDF, Accessed 20 Oct 2014
Jannarelli A (2004) Concessione edilizia tra destinazione "agricola" della zona e qualificazione "agricola" dell'attività: un rapporto da chiarire. Diritto e giurisprudenza agraria e dell'ambiente 13(9):592 et seqq

Jannarelli A, Vecchione A (2008) L'impresa agricola. In: Buonocore V (ed) Trattato di diritto commerciale, Sezione I, vol 2.2. Giappichelli, Torino
Jin-hua et al (2010) Legal safeguard of agro-ecological environment. China J Anhui Agric Sci 34:19465–19467
La Torre M (2013) I contratti di ospitalità. In: Franceschelli V, Morandi F (eds) Manuale del diritto del turismo, 5th edn. Giappichelli, Torino, p 271
Lichtfouse E (ed) (2013) Agroecology and strategies for climate change. Springer, Dordrecht
Lin H (2010) The path choice of legislation on agro-eco-environment in China. Adv Mater Res 113–116:1261–1265
Lukić A (2013) Tourism, farm diversification and plurality of rurality: case study of Croatia. Eur Countryside 5(4):356–376
Marsden T (2003) The condition of rural sustainability. Royal Van Gorcum, Assen
Martin K, Sauerborn J (eds) (2013) Agroecology. Springer, Dordrecht
Martinez MJ (2010) Nueva ruralidad, la remake del término pluriactividad. Nómadas. Revista Crítica de Ciencias Sociales y Jurídicas 26(2):1–16
Martinez De Marigorta Andreu J (1987) El problema del medio ambiente en el marco del Derecho agrario español. Revista de derecho agrario y alimentario 3(8):19–30
Massart A, Sánchez Hernández A (eds) (2001) Manual de instituciones de derecho agroambiental euro-latinoamericano. ETS, Pisa
Mazzanti E (2013) Nuove disposizioni penali in materia di olio d'oliva: uso fallace di marchi, responsabilità dell'ente per frodi alimentari, sanzioni accessorie (L. 14.1.2013 n. 9 "Norme sulla qualità e la trasparenza della filiera degli oli di oliva vergini"). La Legislazione Penale 33 (3):561–582
Mazzarino S, Pagella M (2003) Agricoltura e mondo rurale tra competitività e multifunzionalità. Franco Angeli, Milano
Mengoli P (2009) Manuale di diritto urbanistico, 6th edn. Giuffré, Milano
Merusi F (2007) Il diritto ambientale. Rivista di diritto agrario 86(4):495–501
Milone P (2009) Agricoltura in transizione. Donzelli, Roma
Monteduro M (2013) Environmental law and agroecology: a transdisciplinary approach to public ecosystem services as a new challenge for environmental legal doctrine. Eur Energy Environ Law Rev 22(1):2–11
Monteduro M (2014) Per una "nuova alleanza" tra diritto ed ecologia: attraverso e oltre le "aree naturali protette". GiustAmm.it: Rivista di diritto pubblico 11(6):1–44. http://www.giustamm.it/
Moscarini LV (2009) Proprietà terriera, autonomia privata e impresa nella questione dei "patti agrari". In: Moscarini LV (ed) Diritto privato e interessi pubblici. Saggi di diritto civile (2001–2008). Cedam, Padova, p 177
Navarro Fernández JA (ed) (2010) Introducción al Derecho agrario. La organización jurídica de las explotaciones agrarias. Supuestos prácticos y materiales. Tirant Lo Blanch, Valencia
OECD (2008) Multifunctionality in agriculture: evaluating the degree of jointness, policy implications. OECD Publishing, Paris
OECD (2009) Summary of the conference "Policy Forum on Agriculture and Rural Development", Paris. Rete Rurale Nazionale italiana. http://www.reterurale.it/flex/cm/pages/ServeBLOB.php/L/IT/IDPagina/2414
Oliveira CB Jr, Silva CSG (2010) Zoneamento Agroecológico da Cana-de-açúcar como Mecanismo de Efetivação do Princípio Constitucional do Desenvolvimento Sustentável. In: Anais do XIX Encontro Nacional do Conselho Nacional de Pesquisa e Pós-Graduação em Direito (Conpedi). Fundação Boiteux, Fortaleza, p 6343
Pastorino LF (2009) Derecho agrario argentino. Abeledo-Perrot, Buenos Aires
Pastorino LF (2012) Objeto y extensión del derecho agrario. Anales de la Facultad de Ciencias Jurídicas y Sociales de la Universidad Nacional de La Plata 9(42):50–59. http://sedici.unlp.edu.ar/handle/10915/26993. Accessed 28 July 2014
Pelliccia L (2011) Lavoro e legislazione sociale in agricoltura. Maggioli, Santarcangelo di Romagna (RN)

Perfetti LR (2014) Principi della disciplina pubblicistica dell'alimentazione: premesse ad un diritto amministrativo dell'alimentazione. Rivista di diritto agrario 93(1):3–20
Portaluri PL (2011) L'ambiente e i piani urbanistici. In: Rossi G (ed) Diritto dell'ambiente, 2nd edn. Giappichelli, Torino, pp 241–255
Potter C, Thomson K (2011) Agricultural multifunctionality and Europe's new land debate. In: Oskam A, Meester G, Silvis H (eds) EU policy for agriculture, food and rural areas, 2nd edn. Wageningen Academic, Wageningen, p 213
Prado de Albuquerque M (2007) O conteúdo do Direito Agrário Brasileiro na doutrina jusagrarista. Revista Amazônia Legal de estudos sócio-jurídico-ambientais 1(1):69–82
Righi L (2013) Le strutture ricettive. In: Franceschelli V, Morandi F (eds) Manuale del diritto del turismo, 5th edn. Giappichelli, Torino, p 129
Ruiz Rosado O (2006) Agroecología: una disciplina que tiende a la transdisciplina. Interciencia 31 (2):140–145. http://www.redalyc.org/pdf/339/33911311.pdf. Accessed 31 July 2014
Russo L (2007) La condizionalità da condizione a fine. Rivista di diritto agrario 86(2):231–245
Russo L (2012) Agricultural law and food law. In: Costato L, Albisinni F (eds) European food law. Cedam, Padova, p 141
Russo L (2013) Il consumo di suolo agricolo all'attenzione del legislatore. Aestimum 63:163–174
Salazar-Centeno DJ (2013) Nicaragua: potencial faro regional para el diseño y evaluación de agroecosistemas agroecológicos. La Calera 13(20):58–65
Santagata De Castro R (2012) Diritto del turismo, 2nd edn. Utet, Torino
Santos B de Sousa (1987) Law: a map of misreading. Toward a postmodern conception of law. J Law Soc 14(3):279–302
Santos B de Sousa (2002) Toward a new legal common sense: law, globalization, and emancipation, 2nd edn. Butterworths LexisNexis, London
Scharpf FP (1994) Games real actors could play: positive and negative coordination in embedded negotiations. J Theor Polit 6(1):27–53
Schneider SA (2010) A reconsideration of agricultural law: a call for the law of food, farming, and sustainability. William Mary Environ Law Policy Rev 34(3):935–963
Serpieri A (1940) Corso di economia e politica agraria, vol I. G. Barbera, Firenze
Sevilla Guzmán E, Woodgate G (2013) Agroecology: foundations in agrarian social thought and sociological theory. Agroecol Sustain Food Syst 37(1):32–44
Snapp SS, Pounds B (eds) (2008) Agricultural systems: agroecology and rural innovation for development. Elsevier Academic Press, Amsterdam
Strapasson AB et al (2012) Agro-ecological zoning and biofuels: the Brazilian experience and the potential application in Africa. In: Johnson FX, Seebaluck V (eds) Bioenergy for sustainable development and international competitiveness: the role of sugar cane in Africa. Routledge, London, p 48
Sturiale L (2001) Possibili modelli di valutazione della compatibilità tra sviluppo e tutela con particolare attenzione alle aree protette. In: Bove E, Gajo P (eds) Gestione delle risorse naturali nei territori rurali e nelle aree protette: aspetti economici, giuridici ed estimativi. XXX Incontro di studio del Centro Studi di Estimo e di Economia Territoriale – Ce.S.E.T, Potenza, 5–6 ottobre 2000. Ce.S.E.T., Firenze, p 161
Szilágyi JE (2009) The dogmatics of agricultural law in Hungary from an aspect of EC law. Eur Integration Stud (University of Miskolc) 7(1):41–55
Tomich TP et al (2011) Agroecology: a review from a global-change perspective. Annu Rev Environ Resour 36(2011):193–222
Tuory K (2014) Transnational law: on legal hybrids and legal perspectivism. In: Maduro M, Tuori K, Sankari S (eds) Transnational law. Rethinking European law and legal thinking. Cambridge University Press, Cambridge, p 41
Ubertazzi B, Muñiz Espada E (eds) (2009) Le indicazioni di qualità degli alimenti: diritto internazionale ed europeo. Giuffré, Milano
Uphoff NT (ed) (2002) Agroecological innovations: increasing food production with participatory development. Earthscan, London, Sterling

Uphoff N (2007) Agroecological alternatives: capitalising on existing genetic potentials. J Dev Stud 43(1):218–236

Urbani P (2011) La disciplina urbanistica delle aree agricole. In: Costato L, Rook Basile E, Germanò A (eds) Trattato di diritto agrario, vol II. Il diritto agroambientale. Utet, Torino, p 597

Van Dam D, Nizet J, Streith M, Stassart P (eds) (2012) Agroécologie: Entre pratiques et sciences sociales. Éducagri Éditions, Dijon

Van Huylenbroeck G, Durand G (2003) Multifunctional agriculture: a new paradigm for European agriculture and rural development. Ashgate, Aldershot

Vandermeer J, Perfecto I (2013) Complex traditions: intersecting theoretical frameworks in agroecological research. Agroecol Sustain Food Syst 37(1):76–89

Warner KD (2007) Agroecology in action: extending alternative agriculture through social networks. MIT Press, Boston

Welan M (2002) Agrarrecht versus Umweltrecht. Agrarische Rundschau 56(3):48–53

Westhoek HJ, Overmars KP, van Zeijts H (2013) The provision of public goods by agriculture: critical questions for effective and efficient policy making. Environ Sci Policy 32:5–13

Westlund H, Kobayashi K (eds) (2013) Social capital and rural development in the knowledge society. Edward Elgar, Cheltenham

Wezel A, Jauneau JC (2011) Agroecology – interpretations, approaches and their links to nature conservation, rural development and ecotourism. In: Campbell WB, López Ortíz S (eds) Integrating agriculture, conservation and ecotourism: examples from the field. Springer, Dordrecht, p 1

Wezel A, Soldat V (2009) A quantitative and qualitative historical analysis of the scientific discipline of agroecology. J Agric Sustain 7(1):3–18

Wezel A et al (2009) Agroecology as a science, a movement and a practice. A review. Agron Sustain Dev 29(4):503–515

Wilson GA (2007) Multifunctional agriculture: a transition theory perspective. CABI, Wallingford

Wilson G (2010) Multifunctional 'Quality' and rural community resilience. Trans Inst Br Geogr 35 (3):364–381

Winkler W (1994) L'impact du droit de l'environnement sur le droit agraire en Allemagne. Rivista di Diritto Agrario 73(32):173–189

Winkler W (2002) Agrarrecht – Umweltrecht: Gegensatz oder Ergänzung? Agrarische Rundschau 56(3):5–18

Wojtkowski P (2008) Agroecological economics. Sustainability and biodiversity. Elsevier, Amsterdam

Zeledón Zeledón R (2009a) Derecho Agrario contemporáneo y Derecho Agrario AAA (agricultura, ambiente y alimentación). Revista Estudios Agrarios (México, Procuraduría Agraria) 15(40):9–26

Zeledón Zeledón R (2009b) Derecho agrario contemporáneo. Juruá Editora, Curitiba, Paraná, Brasil

Agriculture, Environment and Law Between Ancient Experiences and Present Knowledge: Some Remarks

P. Buongiorno

Abstract This paper is an attempt to outline the points of contact between agroecology and the sciences of antiquity, with particular regard to the contribution that the reflection on the models of ancient (especially Roman) law can give agroecological research. In this regard, some examples are outlined, such as the regulation of deserted lands (*agri deserti*) between ancient and modern experiences, the complex problem of the commons, the epistemological debate to the contribution of history to the other research fields.

Keywords Agroecology • Ancient law • Commons • Deserted lands • Historicism

1 Agroecology, Classics and Law: A "Transdisciplinary" Dialogue?

The scholarly encounter between agroecology and the Classics (especially the study of Ancient Law) may seem at first to give way to an intricate and hopeless mish-mash of knowledge. The *Altertumswissenschaft* alone can already be considered a cutting-edge, forefront field of research since it incorporates and implies many different disciplines and forms of knowledge—from law to philology to the different subfields of history, not to mention "ancillary" sciences such as epigraphy, papirology, archaeology and antiquarianism.

This complex mix of knowledges and sciences may discourage from establishing or even trying to pursue a possible dialogue between the Classics and agroecology, but, in this instance, I prefer to follow K. Popper's maxim, just as well-known scholars such as S. Mazzarino have already done, for which: "There are no subject

This paper resumes, with a set of essential notes and some extension, my lecture at the International Workshop on "Agroecology and Law: a Transdisciplinary Dialogue" (Lecce, 15 November 2013). I thank Massimo Miglietta, Massimo Monteduro, Annarosa Gallo for valuable suggestions and Sara Saba for the help with the English version of this text.

P. Buongiorno (✉)
Dipartimento di Scienze Giuridiche, Università del Salento Lecce, Italy

Institute for the History of Law, University of Münster, Münster, Germany
e-mail: buongior@uni-muenster.de

M. Monteduro et al. (eds.), *Law and Agroecology*,
DOI 10.1007/978-3-662-46617-9_5

matters . . .: there are only problems . . ."[1] Complex problems cannot be explored and analyzed, adopting one single point of view. Instead, they need multiple perspectives that can therefore allow a thorough analysis of all their parts, as they belong to a unit.[2] This very method already helps to reduce the distance between agroecology and the Classics. Both sciences use, in fact, a multidisciplinary approach to their material, which this volume tries to reproduce and adopt also as a reaction to the many impulses that recent research has sent. In other words, this procedure helps to acquire *nouveaus yeux* with which, as Proust says, we can go through *le véritable voyage de découverte*.[3] The contribution to this form of research of the Classics in general, and in particular of the history of law and economics, must be considered then because of their methodology and content.

Ancient societies, and Rome in particular, had a special relation with land, as space (by which I mean the environment or the ecosystem) and as production center. Agroecological research looks also at the land from these two perspectives, and this common approach leads to ask whether (and eventually how) ancient solutions for problems, which today we would ascribe to the field of agroecology, can help modern research as terms of comparisons or heuristic tools.

The breath of the question I ask in this contribution is such that finding actual answers is unlikely here. But the aim of these pages is a different one, namely to stimulate and inform as much as possible a future debate by presenting a few reflections substantiated by examples taken from ancient history.

2 The Ancient Landscape, Between Agriculture and Environment

The many references to M. Porcius Cato that can be found in the work of a well-known agronomist like Fabio Caporali at first may come as a surprise (see CAPORALI in this volume). The work of that ancient author is normally considered very specific, and it is mostly Latin linguists who devote their attention to it rather than students of agronomy. Still, Cato should be a major reference for the field of

---

[1] "There are no subject matters; no branches of learning – or, rather, of inquiry: there are only problems, and the urge to solve them. A science (. . .) is, I contend, merely an administrative unit." Popper (1983), p. 3.

[2] Another famous maxim particularly liked by the Classicist Walter Otto describes this approach: *ek mérous manthánomen* (transl: "I learn from the detail"). On this expression, which now scholars read as invitation to become a multidisciplinary and "multitasking" scholar and is especially true for the approach to the Classics in Italy and France that have already attempted to start an interdisciplinary dialogue, see now Giardina (2009), pp. 72 et seqq. About the *ricerca pluriprospettica*, see now Di Nuoscio (2006), pp. 91 et seqq. (exp. p. 92).

[3] "Le seul véritable voyage, le seul bain de Jouvence, ce ne serait pas d'aller vers de nouveaux paysages, mais d'avoir d'autres yeux, de voir l'univers avec les yeux d'un autre, de cent autres, de voir les cent univers que chacun d'eux voit, que chacun d'eux est." Proust (1923), p. 69.

agronomy, if we were to give real historical depth to our research work on agroecology. For the Romans, an agrarian problem was also an environmental (and a juridical) problem since the environment was perceived and understood as the space in which men lived and produced. The best example for this approach is the *villa catoniana*.[4] According to one of the best known and most refined scholars of ancient agri-environmental studies, Luigi Capogrossi-Colognesi (who studied, among others, with Emilio Sereni), "the *villa catoniana* and its land do certainly shape anew the landscape, but are also a tremendously important means to preserve and renew the ancient covenant between men and his land."[5] The fundamental contribution of this covenant was to made possible the birth of human civilization from its very beginnings.

Agrarian problems then become environmental problems. But the opposite is also true: environmental problems include and imply agrarian issues.[6]

A couple of examples can help better explain this statement. In the first book of the *Annales*, Tacitus records a debate (15 AD), whose source is the *acta senates*, *i.e.* the official record of senatorial meetings, on the floods of the Tiber that threatened Rome.[7] Two senators of consular rank were in charge of reporting on the issue, L. Arruntius and the jurist C. Ateius Capito, the first maybe as member of the *curatores riparum et alvei Tiberis* the second as chief of the *curatores aquarum*. The presence of a jurist was certainly a pointed choice. Be that as it may, they proposed to deviate the rivers and the lakes' streams that fed the Tiber so that the Roman authorities could better control and regulate the floods of the river Tiber. The involved communities, however, were given a hearing in the Senate and brought forward several arguments (also of religious nature) against the plan of the two senators. The most forceful of the arguments stated that "the nature provided in the best way to the good of the men (*optume rebus mortalium consuluisse naturam*), giving to the rivers sources, flows, springs, mouths."

The idea of "violence against nature" plays a central role, but there is more to it.[8] The Italian communities whose survival and subsistence depended on the use of the land pointed out that deviating the tributary rivers of the Tiber meant changing the Italian landscape as well as the economy and the social (and religious) equilibrium of the region.[9] The real violence against nature then would have taken place through

[4] An introduction to (human and economic) geography in the Roman World in Traina (1990); for the so-called *villa catoniana*, see now Marzano (2007), *passim*, with further bibliography.

[5] Author's translation of Capogrossi Colognesi (2014), p. 58.

[6] In general on this, the still useful Sereni (1955).

[7] Tacitus, *Annales*, 1.76 and 1.79. On the floods of the Tiber, see now also Aldrete (2007).

[8] This topic is well known in the ancient sources starting from Herodotus up to the second century AD authors. See Traina (1987), pp. 40–49.

[9] *Actum deinde in senatu ab Arruntio et Ateio an ob moderandas Tiberis exundationes verterentur flumina et lacus, per quos augescit; auditaeque municipiorum et coloniarum legationes, orantibus Florentinis ne Clanis solito alveo demotus in amnem Arnum transferretur idque ipsis perniciem adferret. congruentia his Interamnates disseruere: pessum ituros fecundissimos Italiae campos, si amnis Nar (id enim parabatur) in rivos diductus supersta gnavisset. nec Reatini silebant, Velinum*

the actual menace to the very subsistence of Roman citizens or, in other words, with a nonsustainable intervention on the landscape. For this reason, the Senate decided not to act.[10]

A few decades later (42–53 AD), another debate took place on the reclamation of the riverbed of the Fucino, but this time the sources preserve memory of a different perception of the matter. The Senate and the emperor Claudius thought that this intervention would have reinforced and enhanced local economy rather than damaging it.[11] In order to reclaim the Fucino, Claudius was ready to employ 30,000 workers for 11 years to build the necessary drain. Local owners, however, took upon themselves the cost of the entire operation (leaving the glory of it to the emperor) in order to have the right to use the reclaimed land. This was only one of the numerous hydraulic projects that Claudius had planned and that encountered the favor of his contemporaries because they had the merit of being useful.[12]

3 *Res Communes Omnium*: Towards Common Goods

These few examples show already how rich the ancient past is in terms of information, experience and themes pertaining to general topic of land and its administration. This particular field of research is often on the forefront because its scholars are accustomed to and have used, for a long time now, the principles of transdisciplinary work. The new challenge, however, is to reach a broader audience that goes well beyond the small group of experts to whom this type of work is mostly addressed so that specialists of other fields can make a good and productive use of its results.

The utility of ancient, technical literature is by now well known also to nonclassicists.[13] It is enough to think of the amount of information preserved in

lacum, qua in Narem effunditur, obstrui recusantes, quippe in adiacentia erupturum; optume rebus mortalium consuluisse naturam, quae sua ora fluminibus, suos cursus utque originem, ita finis dederit; spectandas etiam religiones sociorum, qui sacra et lucos et aras patriis amnibus dicaverint: quin ipsum Tiberim nolle prorsus accolis fluviis orbatum minore gloria fluere. seu preces coloniarum seu difficultas operum sive superstitio valuit, ut in sententiam Pisonis concederetur, qui nil mutandum censuerat.

[10] Thus, I interpret the three elements *preces coloniarum*, *difficultas operum*, *superstitio* instead of arguing with the *difficultas operum* alone (as Kornemann 1963, p. 253, does, observing that "entscheidend wird wohl die Überzeugung gewesen sein, daß die Pläne technisch noch nicht ausgereift waren während die beiden anderen Gründe nur sekundäre Bedeutung besaßen").

[11] See, in particular, Cassius Dio 60.11.5 ("Claudius furthermore desired to make an outlet into the Liris for the Fucine Lake in the Marsian country, in order not only that the land around it might be tilled but also that the river might be made more navigable"), but see also Suetonius, *De Vita XII Caesarum*, *Claudius*, 20.2 (*Fucinum adgressus est non minus compendii spe quam gloriae, cum quidam privato sumptu emissuros se repromitterent, si sibi siccati agri concederentur*), and Plinius the Elder, *Naturalis Historia*, 36.124.

[12] See Thornton and Thornton (1985), pp. 105–120; Leveau (1993), pp. 3–16; Traina (2000), p. 37.

[13] See n. 4 *above*.

authors like Cato, Varro, Columella or Virgil in the Georgics (enriched by Servius Marius Onoratus' commentary or that of the so-called Servius Danielinus—another commentator of Virgil, who maybe added to the latter's work).

There are other ways too in which we can profit from ancient knowledge in modern research. A good example is represented by the recent and ongoing debate—to which, among others, contributed Ugo Mattei or Stefano Rodotà—on common goods.[14] The topic that scholars hotly discuss concerns the couple public–private, and even second-guesses the basis of article 810 of the Italian Civil Code.[15] The contribution that Roman law has given to this debate cannot be ignored. This originates from and pertains to the ill-known *res in publico uso* (goods that had to be at everybody's disposal; in a way, these are similar to modern common goods) up to the *res communes omnium*, namely those things that, according to Talamanca, "belong to all men no matter where they are from: air, running water, the sea, the coast, following the listing on the third book of the *Institutiones* by the third century A.D. jurist Elius Marcianus."[16]

The *communis opinio* reads the sources as if they testify that these goods could not belong to anybody, but everyone could use them. In other words, they are thought of as being "public," which means that scholars normally follow the generally accepted principle of the *ius gentium* (meaning, all free men). The distinction between *res publicae*,[17] *res in publico uso* and *res communes omnium,* which had already gone largely lost in the systematic recording of Justinian work, is still in the process of being recovered by scholarship in all its aspects. This is indeed the only way to avoid modern misinterpretations and enrich the current debate. In this context, it is necessary to quote Paolo Maddalena, emeritus vice president of the Italian Constitutional Court[18]:

> Molti scritti recenti dedicati ai beni comuni presentano questa categoria come una "scoperta nuova" della scienza giuridica e, proprio per questo, cioè proprio perché partono dal presente senza pensare che già nel passato si erano esercitati sull'argomento i giureconsulti romani, cadono in talune contraddizioni, come quella, a nostro avviso fondamentale, secondo la quale non è importante in questa materia il profilo dell'appartenenza, ma la "destinazione", e cioè la "disciplina d'uso" del bene di cui si parla. Decidere

---

[14] Mattei (2011) and Rodotà (2013).

[15] Art. 810 of Italian Civil Code: "Sono beni le cose che possono formare oggetto di diritti."

[16] Author's translation of Talamanca (2013), p. 203. For the texts of Marcianus, cf. *Digesta Iustiniani Augusti* 1.8.2 and 4. The equilibrium point of *res in usu publico* and *res communes omnium*, revolving around the node of the evolution of the processual protection, is examined, also historiographically, by Di Porto (2013), *passim.*

[17] It means *res populi Romani*, as in the sense of "*res* of the people of Rome" as far as—*abusive*—in the sense of "*res* of the single municipalities." For this distinction, see the opinion of the third-century AD jurist Ulpian in his tenth book *ad edictum*: *Bona civitatis abusive 'publica' dicta sunt: sola enim ea publica sunt, quae populi Romani sunt.* The distinction had become necessary because of the continuous procedural changes. See, for example, *CIL* X, 1018, which refers directly to an edict by Vespasian but then also mentions a *res publica Pompeianorum* (!). Cf. Müllejans (1961), pp. 47 et seqq., exp. p. 48.

[18] Maddalena (2012).

> dell'appartenenza, cioè stabilire se si tratta di beni in proprietà dei singoli ovvero in proprietà comune e collettiva di tutti, è invece importantissimo. Infatti, è evidente che non esiste nessun'altra possibilità di difendere un bene di uso comune, se non quella di considerarlo fuori commercio, siccome appartenente alla collettività e non nel patrimonio di alcuno, "*nullius in bonis, sed universitatis*", come affermava Gaio (Gai. *Inst.* 2.11), poiché, se lo si considerasse una *res nullius*, nel significato moderno della parola, a parte la considerazione che, come in seguito vedremo, questa categoria autonoma delle *res nullius* fu una creazione postclassica, sarebbe immediatamente occupabile da chiunque e non avrebbe nessuna garanzia di difesa.

The modern debate on the evolution of the juridical status of the *res* is therefore essential for an inter- and transdisciplinary work on agriculture and law: the basic elements for anybody's existence *terra, aer, acqua*, the *res par excellence*, are crucial reference points for both disciplines and part of the discussion I have just mentioned.

4 The "Deserted Lands"

The surviving regulations pertaining to land and water allow presenting observations and clarifications, which thus derive (and benefit) from the use of ancient models. The regulation that disciplines the recovery and best use of the so-called *agri deserti* stands out among all others. Ancient Rome faced repeatedly the issue of countryside's depopulation and the abandonment of the land, which the authorities tried to solve with the enactment of numerous regulations over the centuries. During the time of the Republic, the general approach to fighting land abandonment and its neglecting was the imposition of fines.[19] Later on, however, in the imperial time, the authorities tended to use a different attitude towards fining (probably this change is concomitant with the economic crisis under Tiberius).[20] With the provisions that date to these centuries, the authorities tried to force the richer strata of society to use the land or even assigned anew abandoned lots. Especially, this latter system was meant to curb the effects of the progressive depopulation of the countryside, *agri deserti* (deserted lands), that will become a very pressing problem in Late Antiquity.[21]

---

[19] See Aulus Gellius *Noctes Atticae*, 4.12pr., 1, 3: *Notae et animadversiones censoriae in veteribus monumentis repertae memoria dignae. 1.* Si quis agrum suum passus fuerat sordescere eumque indiligenter curabat *ac neque araverat neque purgaverat, sive quis arborem suam vineamque habuerat derelictui, non id sine poena fuit, sed erat opus censorium,* censoresque aerarium faciebant. ... *3. Cuius rei ... auctoritates sunt, et M. Cato id saepenumero adtestatus est.*

[20] Suetonius *De vita XII Caesarum, Tiberius*, 48.1 (see also Tacitus, *Annales*, 6.17.2): *Quorum alterum magna difficultate nummaria populo auxilium flagitante coactus est facere, cum* per senatus consultum sanxisset, ut faeneratores duas patrimonii partes in solo collocarent, *debitores totidem aeris alieni statim solverent, nec res expediretur.*

[21] For a traditional view on the issue, see Whittaker (1976), pp. 137–200 (with references); now also Tarozzi (2013).

The historian Herodianus preserves in his work one of the oldest provisions on this problem that dates to 193 AD under Pertinax. According to his testimony: "whoever wanted and could, (this) was allowed to take possession of deserted and untilled land, both in Italy and in the Provinces, even if they were part of the imperial possessions. These (men) were to own (*despoteía*) the land granted that they would take care of it and cultivate it. Furthermore he thus granted a 10-year long tax exemption and absolute ownership of those territories forever."[22] The text poses numerous juridical and historical problems,[23] but, if we put those aside for a moment, we can say that Herodianus' testimony shows unmistakably that the issue motivating the approval of this provision was the progressive depopulation of the countryside with its main consequence, namely the negative impact on agricultural production.

This ancient edict testifies to problems that have come back over and over in the past two millennia and for which solutions have been adopted that, at first glance, are not very dissimilar among themselves. It is therefore worth of consideration to carry out a diachronic study of all these solutions, even if with the caution due to such type of research. Such a study could indeed help to better grasp ideological continuity and discontinuity for these solutions, at least as far as their agrarian and environmental aspects are concerned.

For example, in regard to the question of the depopulation of the countryside and the abandonment of the land, we could cite a recent initiative—with its actual provisions—of the Italian region Tuscany that aims at implementing the principles of an old, national law[24] through a regional one (80/2012). The latter law authorizes the reassignment and use of untilled land through the so-called *Banca della terra* (Land Bank). This unusual bank is operated by the authority of the *Terre regionali Toscane* (which works according to a local regulation and with the help of the ARTEA[25]). The main goal that the *Banca della terra* pursues is "promoting the good use of the public and private land also through their exploitation." In order to attain this result, "the bank has a complete and up-to-date record of all lots of terrain, be they public of private, that can be rented or sold" and also of agricultural forest land (to be used for forestry) that can also be let or sold (art. 3). The authority grants the necessary permissions in agreement with the owners of the land, be they the Region or individuals registered in the lists of the Bank. These permissions

---

22 Herodianus, *Regnum post Marcum*, 2.4.6 (edition of Lucarini 2005).

23 The most pressing questions pertain to the definition of *despoteía*, to the regulation of property in *terra Italia* and the provinces and to the juridical statute of the recipients of this provision. I will return to these problems in some future papers (part of a bigger project on the juridical and social profiles of Herodianus' work; first issues of this project are my papers: Buongiorno 2014, pp. 81–89; Buongiorno 2015).

24 I am referring to the Italian law L. 440/1978, "Norme per l'utilizzazione delle terre incolte, abbandonate o insufficientemente coltivate" ("Standards for the use of uncultivated land, abandoned or inadequately cultivated").

25 ARTEA means *Agenzia Regionale Toscana per le Erogazioni in Agricoltura* (Tuscany Regional Agency for Disbursement in Agriculture).

establish the conditions, length and modes of use of the land. Also, it is in these permits that the authorities register the established rent or the price of the land (art. 4). The goals pursed by the project can be summarized as follows: "to help reclaim deserted land, curb local degradation, protect the land and the hydrogeological equilibrium, stop fires, help to promote the structure and arrangement of the territory through the most appropriate activities."[26] The region of Tuscany promotes and protects the use of deserted land by taking into account social, environmental and economic interests of local communities.

Even if we cannot state with any certainty or even probability that in 2012 the lawmaker knew either Herodianus or the development of the Roman legislation on the *agri deserti*, we must note that—if we look back to our past—we have at our disposal a great amount of knowledge, on which we should consciously (so without simplifications or generalizations) reflect, given the similarity of the problems that in antiquity and today afflict us.[27] Such a conscious approach to the techniques and ideologies used in the antiquity could have, even today, a stimulating function for the research.

5 "Blind Without History": An Ancient Debate

To recapitulate, before drawing our conclusions, we cannot ignore ancient law, ancient socio-economic phenomena and, what is more, the surviving evidence. The ancient sources indeed belong to that essential cultural heritage that helps interpret and administer the use and the very existence of the environment (see CAPORALI in this volume). This is not a new approach. Even if not continuously, we indeed find this same approach from the beginning of the modern age starting with the teaching of Jacques Cujas (sixteenth century) onwards.[28]

For example, the humanist and jurist Matteo Gribaldi Mofa was well aware of it, as he long insisted—in his *De methodo ac ratione studenti libri tres* (Lugduni 1541) that appeared in the middle of the sixteenth century—"on the close necessity of an historical-philological foundation of the interpretation." In it, he referred to the

[26] Author's translations of Arts. 3 and 4 of the Tuscany Regional Law 80/2012.

[27] We could bring forward other examples, such as the use, protection and reclaim of marshy areas; on this, see Traina (1988), pp. 93–108, now revisited in some forthcoming papers by Salvatore Vacante (DAI-AEK München).

[28] Roman law has never had a "continuous" success, so that a "famous" jurist, J.W. Goethe, could define Roman law with a metaphor as a duck "sometimes . . . prominent, swimming on the surface of the water; at other times . . . hidden from view, diving amid the depths. But . . . always there." So Stein (1999), p. 116. Eckermann (1837), p. 109 (6 April 1829) says: "Auch das römische Recht, als ein fortlebendes, das, gleich einer untertauchenden Ente, sich zwar von Zeit zu Zeit verbirgt, aber nie ganz verloren geht, und immer einmal wieder lebendig hervortritt." On this in the modern discussion of Roman law, see the contributions by Cassese (2011), pp. 26 et seqq.; and Capogrossi Colognesi (2011), pp. 43 et seqq.

maxim of Lorenzo Valla,[29] who said that *sine Latinitate, caeca omnis doctrina est, et illiberalis, praesertim in iure civili.*[30]

The same attitude appears again in the work by Francois Baudoin that was published a little later than Mofa's work. For Baudoin, *sine historia caeca est jurisprudentia*, a saying that has influenced in several ways different fields, among which is that of economics.

The German tradition of the history of economics of the middle of the nineteenth century also deserves a note and, with it, the new direction showed for these studies by Francesco Trinchera, an unjustly forgotten historian of economics of Neapolitan and then "Savoy" training. Trinchera authored a *Storia critica dell'economia pubblica,* which was never completed because of his untimely death. The section devoted to the *Età antica* appeared in 1873 as the tenth volume of the *Atti della Reale Accademia di Scienze Morali e Politiche di Napoli,*[31] and in it the historian went back to the principles expounded by Gribaldi Mofa and Baudouin, writing *caeca sine historia oeconomia* and thus referring in his introduction (pp. 11 et seqq.) to historicism. Trinchera thought this approach was necessary to give a historical depth to the discourse then taking place in his branch of study of the social science, especially when the most recent themes were object of debate. From this point of view, the part of Trinchera's work devoted to the ancient world shows that he focused on agriculture and its role before even approaching other themes such as commerce and trade. Agriculture was for him an economic and social phenomenon: "the most honorable and important occupation of the ancient Romans," so that "the eulogy ... of 'good farmer', *bonus colonus vel agricola*, was equivalent to 'good man', *vir bonus.*"[32] Furthermore, he dwelled on Roman agricultural procedures describing them in detail to then move to a careful analysis of the ancient sources that contributed to the birth of an agrarian discipline with principles and reference points that are still worthy of consideration.[33]

---

[29] On the interpretation of this maxim in Lorenzo Valla's thought (where it is inserted in a lexicographic research on the meaning of words), see Mantovani (2007), pp. 143–208, exp. pp. 180 et seqq. This approach can be found also in Poliziano, as Bellocci (1995), pp. 345–354 shows.

[30] *Latinitas* here does not only mean *eloquentia*, but it rather refers to the cultural approach to knowledge, esp. in the law. A discussion and further bibliography in Discussione e bibliografia in Quaglioni (2002), pp. 345–349. See also Quaglioni (2008), pp. 347–357.

[31] Trinchera (1873), pp. 1–424. On this work, which I intend to publish anew with commentary, see Di Taranto (2000), pp. 203–218, exp. 216 et seqq.

[32] As Fiori (2013), p. 25, notes "un'analisi degli usi dell'aggettivo *bonus* mostra che esso è legato innanzi tutto all'idea di abbondanza, di prosperità; è termine del linguaggio agricolo, indicando beni produttivi; applicato alle persone denota la qualità di compiere perfettamente la propria funzione."

[33] Trinchera (1873) writes in the *Summary* of his work (p. 421): "Le ricche esperienze formate nell'esercizio dell'agricoltura formarono un ramo di dottrina che ebbe un carattere tutto originale ed un'impronta tutta romana, come la si ravvisa nelle opere ... che ci lasciarono in rassegna M. Porcio Catone, Cn. Terenzio Scrofa, Terenzio Varrone, L. Giunio Moderato Columella, Palladio Rutilio Tauro, Crescenzio, Gargilio Marziale, Tiziano, Fabiano, ecc..."

6 Conclusions

These pages should have shown, among other things, that the theme of the impact of socio-economic models and ancient regulations on modern behavior is not new. But we must ask how much and what kind of influence the ancient evidence has and how our approach has changed because of them and still evolves.

As far as the study of the law is concerned, I believe we can at least try to give an answer to these questions. The necessary premise is, of course, that the law is a product of the activity of men, and its aim is to regulate conflicts that have been caused by opposite socio-economic interests. In a world where the human race is itself a threat to the environment and its equilibrium, the study of ancient regulations may not influence directly our behaviors but may (or better has to) show the right way towards more informed and sustainable actions.[34]

References

Aldrete GS (2007) The floods of the Tiber in ancient Rome. The Johns Hopkins University Press, Baltimore

Bellocci N (1995) Caeca sine historia iurisprudentia. Spunti di riflessione sul Poliziano e lo studio del diritto romano. Studi Senesi 107:345–354

Buongiorno P (2014) Erodiano 4.12.4 e i poteri di Flavio Materniano nell'anno 217 d.C. Fundamina 20:81–89

[34] On this, see also the exhortation of Pope Francis (who's actually working on an Enciclic on ecological themes) in 2013 during the World Environment Day proclaimed by the United Nations: "Today I would like to reflect on the issue of the environment, as I have already had an opportunity to do on various occasions. I was also prompted to think about this because of today's World Environment Day, sponsored by the United Nations, which is launching a pressing appeal for the need to eliminate waste and the destruction of food. When we talk about the environment, about creation, my thoughts go to the first pages of the Bible, to the Book of Genesis, where it says that God puts men and women on the earth to till it and keep it (cf. Gen. 2:15). And these questions occur to me: What does cultivating and preserving the earth mean? Are we truly cultivating and caring for creation? Or are we exploiting and neglecting it? The verb "cultivate" reminds me of the care a farmer takes to ensure that his land will be productive and that his produce will be shared. What great attention, enthusiasm and dedication! (...) However "cultivating and caring" do not only entail the relationship between us and the environment, between man and creation. They also concern human relations. The popes have spoken of a *human ecology*, closely connected with *environmental ecology*. We are living in a time of crisis; we see it in the environment, but above all we see it in men and women. The human person is in danger: this much is certain—the human person is in danger today, hence the urgent need for human ecology! And the peril is grave, because the cause of the problem is not superficial but deeply rooted. It is not merely a question of economics but of ethics and anthropology. (...) I would therefore like us all to make the serious commitment to respect and care for creation, to pay attention to every person, to combat the culture of waste and of throwing out so as to foster a culture of solidarity and encounter." See http://w2.vatican.va/content/francesco/en/audiences/2013/documents/papa-francesco_20130605_udienza-generale.html. Accessed 31 Aug 2014.

Buongiorno P (2015) “Il senso della crisi”. Ritual und Legitimität der kaiserlichen Macht nach Herodian. In: Babusiaux U, Kolb A (eds) Das Recht der Soldatenkaiser – rechtliche Stabilität in Zeiten politischen Umbruchs? De Gruyter, Berlin, pp 65–87
Capogrossi Colognesi L (2011) Un futuro senza storia? Index 39:43–57
Capogrossi Colognesi L (2014) La disciplina giuridica del rapporto tra società e ambiente naturale nel contesto romano. Quaderni Lupiensi di Storia e Diritto 4:53–59
Cassese S (2011) L’anatra di Goethe. Index 39:26–30
Di Nuoscio E (2006) Il mestiere dello scienziato sociale. Un’introduzione all’epistemologia delle scienze sociali. Liguori, Napoli
Di Porto A (2013) Res in usu publico e ‘beni comuni’. Giappichelli, Torino
Di Taranto G (2000) Deduttivismo e pragmatismo nell’Accademia delle Scienze morali e politiche di Napoli. In: Augello MM, Guidi MEL (eds) Associazionismo economico e diffusione dell’economia politica nell’Italia dell’Ottocento. II. Dalle società economico-agrarie alle associazioni di economisti. Franco Angeli, Milano, pp 203–218
Eckermann JP (1837) Gespräche mit Goethe in den letzten Jahren seines Lebens (1823–1832), II. Brodhaus, Leipzig
Fiori R (2013) Il vir bonus tra filosofia greca e tradizioni romane nel de officiis di Cicerone. In: Lovato A (ed) Vir bonus. Un modello ermeneutico della riflessione giuridica antica. La matrice, Bari, pp 19–38
Giardina A (2009) Il maestro, il discepolo e gli altri maestri: un percorso nella storia romana. Anabases 10:61–73
Kornemann E (1963) Tacitus. Annalen, I. Buch 1-3. Carl Winter Verlag, Heidelberg
Leveau P (1993) Mentalité économique et grands travaux: le drainage du Lac Fucin. Aux origines d’un modèle. Annales ESC 1:3–16
Lucarini CM (2005) Herodianus. Regnum post Marcum. Teubner, München
Maddalena P (2012) I beni comuni nel diritto romano: qualche valida idea per gli studiosi odierni. Federalismi.it(14). http://www.federalismi.it/ApplMostraDoc.cfm?Artid=20444&content_auth=#.U_uUO7x_u4M. Accessed 31 Aug 2014
Mantovani D (2007) “Per quotidianam lectionem Digestorum semper incolumis et in honore fuit lingua Romana”. L’elogio dei giuristi romani nel proemio al III libro delle Elegantiae di Lorenzo Valla. In: Studi per Giovanni Nicosia, V. Giuffrè, Milano, pp 143–208
Marzano A (2007) Roman villas in central Italy. A social and economic history. Brill, Leiden
Mattei U (2011) Beni comuni. Un manifesto. Laterza, Bari-Rome
Müllejans H (1961) Publicus und Privatus im römischen Recht und im alteren kanonischen Recht unter besonderer Berücksichtigung der Unterscheidung Ius publicum und Ius privatum (Doktorarbeit). Max Hüber, München
Popper K (1983) Realism and the aim of science (Galley proofs for the lectures at the London School of Economics, academic year 1956/57). Hutchinson, London
Proust M (1923) À la recherche du temps perdu. XII. La prisonnière. Gallimard, Paris
Quaglioni D (2002) Gribaldi Moffa. In: Dizionario Biografico degli Italiani, LIX. Treccani, Roma, pp 345–349
Quaglioni D (2008) L’education du juriste face au pouvoir: la Methodus de Matteo Gribaldi Mofa (1541). In: Krynen J, Stolleis M (eds) Science politique et droit public dans les facultés de droit européennes (XIIIe – XVIIIe siècle). Vittorio Klostermann, Frankfurt a.M., pp 347–357
Rodotà S (2013) Il terribile diritto. Studi sulla proprietà privata e i beni comuni. Il Mulino, Bologna
Sereni E (1955) Comunità rurali nell’Italia antica. L’Erma di Bretschneider, Roma
Stein P (1999) Roman law in European history. Cambridge University Press, Cambridge
Talamanca M (2013) Elementi di diritto privato romano, 2nd edn. In: Capogrossi Colognesi L (ed). Giuffrè, Milano
Tarozzi S (2013) Riforma dello Stato e gestione della terra. La questione degli agri deserti nella prospettiva dei Codici tardo antichi. Alcuni spunti di riflessione. Mel. École Franç. de Rome – Antiquité 125(2). http://mefra.revues.org/1861. Accessed 31 Aug 2014

Thornton MK, Thornton RL (1985) The draining of the Fucine Lake: a quantitative analysis. Anc World 12:105–120

Traina G (1987) L'impossibile taglio dell'istmo (Ps.-Lucian. Nero 1-5). Riv Fil Istr Class 115 (1):40–49

Traina G (1988) Paludi e bonifiche del mondo antico. Saggio di archeologia geografica. L'Erma di Bretschneider, Roma

Traina G (1990) Ambiente e paesaggi di Roma antica. NIS, Rome

Traina G (2000) La tecnica in Grecia e a Roma, 2nd edn. Laterza, Bari

Trinchera F (1873) Storia critica dell'economia pubblica, I. Età antica. Atti dell'Accademia di Scienze morali e politiche di Napoli 10:1–424

Whittaker C (1976) Agri deserti. In: Finley MI (ed) Studies in Roman property. Cambridge University Press, Cambridge, pp 137–200

Agriculture and the Environment in International Law: Towards a New Legal Paradigm?

S. Di Benedetto

Abstract Since some decades, the international regulation of environmental issues can be represented in unitary terms, as international environmental law. Conversely, a legal discipline that might be called "international agricultural law" does not exist. Although there are several international rules and even treaties involving agricultural matters, these are disparate and heterogeneous. This fact shows the asymmetrical positions of agriculture and the environment in international law. As a consequence, this chapter focuses less on the environment than it does on agriculture. Moving from international rules and instruments regarding agriculture, it tries to show the existing interaction with environmental law. Finally, it offers some suggestions on a possible agroecological evolution in international law.

Keywords Agriculture • Agroecology • Environment • Fragmentation • International law

1 Introduction: An Asymmetric Relationship

In the current theoretical mainstream, international law is depicted as being composed of many sectorial parts. International trade law, human rights law, international environmental law and international maritime law are some examples of such thematic fields whose existence reveals a sort of fragmentation of the international legal order. Scholars have strongly discussed the latitude and normative impact of such fragmentation by focusing particularly on the question of the autonomy of each field (or subsystem) of international law, by and in itself and with respect to general rules and principles.[1] Even if the degree to which such autonomy is

[1] The issue of fragmentation of international law is broadly dealt with by international scholars. One of the most important studies is that of the International Law Commission, 58th session, *Fragmentation of International Law: Difficulties Arising from the Diversification and Expansion of International Law*, Report of the Study Group of the International Law Commission finalized by

S. Di Benedetto (✉)
Department of Legal Sciences, University of Salento, Lecce, Italy
e-mail: saverio.dibenedetto@unisalento.it

M. Monteduro et al. (eds.), *Law and Agroecology*,
DOI 10.1007/978-3-662-46617-9_6

effective is still being debated, in terms of the concept of self-contained regimes, and varies according to the different "laws" under scrutiny, these fields of international law may arguably be ordered according to some unifying concepts and principles.[2]

Since some decades, the international regulation of environmental issues can be represented in such unitary terms. International environmental law (IEL) is composed of a great number of legal instruments that sharply differ in both subjective dimension and normative effectiveness. The ability to speak of "a law," however, is due to the possibility of singling out common features and objectives. This unitary construction of IEL is particularly important from a normative viewpoint as the existence of common objects and concepts is fundamental to the interpretation and application of different rules and standards.

Conversely, a legal discipline that might be called "international agricultural law" virtually does not exist.[3] Although there are several international rules and even treaties[4] involving agricultural matters, to all appearances these are disparate and heterogeneous. This point marks a crucial difference not only in the international regulation of environmental issues but also in the existence of the autonomous discipline of agricultural law in the legal tradition of western countries.

The foregoing involves the asymmetrical positions of agriculture and the environment in international law. For this reason, this chapter focuses less on the environment than it does on agriculture. It firstly tries to reconstruct international rules and instruments that pertain to agriculture by highlighting the different ideological premises that underlie them. By doing so, it is possible not only to compare how international law deals with the two subject matters but also to understand the extent to which environmental protection and the regulation of agriculture are strictly intertwined. The second section analyzes how IEL as a legal discipline addresses and influences agricultural matters, conceived as both source of environmental threat and object of environmental protection. The last two

Martti Koskenniemi, A/CN.4/L.682, 13 April 2006. According to the ILC Report, such regimes "are often identified as 'special' in the sense that rules of general international law are assumed to be modified or even excluded in their administration."

[2] Some international regimes, such as trade law or maritime law, present a unity as well as an internal coherence primarily based on a formal common structure. Indeed, international trade law corresponds for the most part to the law of the World Trade Organization, with its many covered agreements, while a similar reasoning is possible for international maritime law and the UN Convention on the Law of the Sea. In other cases, the supposed regime is composed of a patchwork of legally separate and autonomous legal instruments, which, however, share common principles and concepts. This is the case, for instance, in international investment law. In this case, a major unifying function is played by the jurisprudence of arbitral tribunals.

[3] "International agricultural law" does not exist as an autonomous legal discipline. We may find very sporadic references to it among legal scholars usually not specifically studying international law (by searching for "international agricultural law" in Google Scholars, we find a handful of quotations, usually of national agrarian law scholars).

[4] See, in particular, the WTO Agreement on Agriculture (*infra*, Sect. 2) and the FAO Treaty on Plant Genetic Resources (*infra*, Sect. 4).

sections attempt to give some insight into how a new idea of ecology and the recent discipline of agroecology might be relevant to international law by and in itself and particularly in relation to a reconceptualization of IEL itself.

2 Agriculture in International Law

The field of agriculture is of major interest to countries at the economic, political and security levels. To all appearances, however, this has not led to the development of a comprehensive and coherent set of international rules, principles and instruments on the matter, as has been the case in several other fields. This section follows a historical perspective in offering a general picture of the international rules and instruments that affect agriculture and, far more than constructing legal principles from too-heterogeneous practices, seeks to highlight both their main underlying ideological patterns and their interactions with environmental issues.

2.1 The Common Roots of Agriculture and the Environment in International Law (A Productivist Discourse)

It is quite surprising to note that the first examples of environmental cases and treaties referred to by authors involve agricultural matters in many cases. International cooperation to combat a vine pest was launched by the 1878 International Convention against Phylloxera,[5] which is referred to as a predecessor of environmental conventions that protect plants.[6] From among the early environmental treaties, scholars include the 1902 Convention for the Protection of the Birds Useful to Agriculture[7] and the 1929 Plant Protection Convention.[8]

As for early environmental cases, a prestigious report on environmental decisions singles out the Helmand River case,[9] where two arbitral tribunals decided on a

[5] *Convention relative aux mesures à prendre contre le phylloxéra vasastrix*, Berne, 17 September 1878, signed by 7 European states. Its text may significantly be found in Ruster and Simma (1975), pp. 1565–1570.

[6] Sand (2007), p. 33.

[7] Text in Ruster and Simma (1975), pp. 1615–1630; cf. Sands and Peel (2012), p. 24. The different approach from the IEL of today is attested to by the distinction made by this Convention between useful and detrimental birds.

[8] Sand (2007), p. 33. Sands and Peel (2012), p. 25, also includes conventions to avoid plant and animal diseases in his survey of early environmental instruments. Conversely, Ruster and Simma explicitly excluded the "international legal measures in the area of control and prevention of plant disease" and "promotion of agriculture" from their collection (*Introduction* to Ruster and Simma 1975, VI).

[9] *Helmand River Delta Case*, Arbitral Awards of 19 August 1872, in Robb (1999).

transboundary dispute between Afghanistan and Persia. The key point of the award was clearly the prohibition against state activity interference in the supply of irrigation water for nearby lands. The famous Trail Smelter dispute, a cornerstone of international environmental case law, concerned a case of transboundary pollution[10] primarily affecting cultivated lands and livestock, which led the Tribunal to order Canada to pay damages.[11] Moreover, the Tribunal explicitly summed up the main rationale behind its decisions as the exigency of giving equal recognition to the interests of both the agricultural community and industry.[12]

This important correspondence between early environmental protection and agricultural matters may be explained by looking at the existing approach to natural resources in the first part of the twentieth century. The international cases and instruments referred to above somehow qualify as being environmental since they were variously aimed at the conservation of natural resources. However, save for rare exceptions,[13] the logic underlying this kind of international cooperation during this period was chiefly based on economic reasons. That legal experience was permeated by a productivist (or industrialist) paradigm[14] whereby natural resources were internationally protected because of their utility for human economic activities. This functional premise mainly resulted in the defence of sovereign and property rights, related to the practice of entrepreneurial activities.

In sum, while the concrete objectives of international cooperation (and sometimes, significantly, the legal instruments employed[15]) were comparable to the current vision of environmental protection—thus justifying their inclusion in the

[10] The case is primarily well-known for having stated the prohibition, as a matter of principle in international law, against a state causing injury through fumes to the territory of another, including properties therein located (*Trail Smelter Case,* Award of Arbitral Tribunal, 11 March 1941, in Robb (1999), pp. 278 et seqq).

[11] In its first award (*Trail Smelter case*, Award of Arbitral Tribunal, 16 April 1938, in Robb 1999, pp. 248 et seqq.), the Tribunal eventually decided to compensate only damage to farming lands (in particular land used for crops) and, to a lesser degree, timber lands (259-70), while the supposed damage to livestock was encompassed within the damage resulting from reduced crop yield or grazing (271). This shows the strict agricultural nature of the affected goods.

[12] Indeed, neither “industrial effort should be prevented by exaggerating the interests of the agricultural community,” nor “agricultural community should be oppressed to advance the interest of industry” (Trail Smelter case, 279).

[13] Actually, even in the “traditional era” of international environmental protection (cf. Sand 2007, pp. 31–33), there are cases of resource protection without a strict productivist logic: see the 1933 London Convention Relative to the Preservation of Fauna and Flora in their Natural State.

[14] Cf. Dryzek (2007), p. 48. Other scholars speak of an “anthropocentric outlook” (Bodansky et al. 2007, p. 3).

[15] Two authors interestingly note that some techniques of protection set forth by the 1902 Convention for the Protection of the Birds Useful to Agriculture are still used today (Sands and Peel 2012, p. 24).

history of environmental protection[16]—the underlying logic, though nuanced, was productivist, and thus a proper conceptualization of the environment did lack. This fact explains the significant correspondence between *ante litteram* international environmental protection and legal instruments aimed at protecting and enhancing a fundamental economic activity such as agriculture.

2.2 International Cooperation on Agriculture from the Birth of the FAO to the End of the Cold War: Production, Development and Food Security

The international legal relevance of agricultural matters before the Second World War was thus substantially limited to the conservation of land natural resources for productive aims, by means of international treaties and arbitral decisions that relied on recognized general principles. On the other hand, some cases of international cooperation on agricultural matters are attested between the two world wars. The situation changed in the 1940s as the new legal order following the world conflict was based on a renewed and broader role for international organizations at both the political and economic levels. This changed climate, dominated by the United Nations (UN), provides the roots for the creation in October 1945 of the Food and Agriculture Organization (FAO).

In the early FAO experience, the objective of protecting and conserving natural resources used for agriculture, though remaining secondary to the overall policy architecture of the institution, did not in the least disappear. Indeed, among the functions of the FAO, there was and is that of making recommendations, *inter alia*, on "the conservation of natural resources and the adoption of improved methods of agricultural production."[17] In this period, however, international cooperation on food and agriculture fundamentally had two purposes: the enhancement of world food production and the (consequent) raising of the standard of living of populations (in particular, freeing people from hunger). These two objectives were affirmed in the Preamble of the 1945 FAO Constitution.[18] These objectives

[16] Indeed, scholars' reconstruction of the logic of environmental law origins has been construed as a form of management of natural resources, in accordance with an industrialist paradigm (Sand 2007, p. 31). Cf. also Sands and Peel: "a growing awareness that the exploitation of natural resources could not occur on an unlimited basis," p. 30.

[17] FAO Constitution, Art. 1.2 c). Moreover, in its first years of activity, the FAO approved the International Plant Protection Convention and the International Convention on the Protection of New Plant Varieties, both in 1951.

[18] Indeed, the first two points set out in the Preamble are "raising levels of nutrition and standards of living of the peoples" and "securing improvements in the efficiency of the production and distribution of all food and agricultural products." The third point sets out a further objective, strictly connected to that of raising living standards: "bettering the condition of rural populations." The latter point is a clear synthesis of the two objectives (in the aforesaid sense of consequentiality) from a broader perspective: "thus contributing towards an expanding world economy and ensuring humanity's freedom from hunger."

were clearly interrelated, in the sense that the former was a substantial (but not exclusive) requirement for the latter.[19]

While the purpose of conserving natural resources was progressively marginalized, the pursuit of the two main objectives of the UN food system was largely undertaken by the FAO in its early years through technical and political actions such as research programmes, information exchange and technical assistance.[20] In fact, the objectives of enhancing food production and fighting hunger directly touched legislative sectors that strictly belonged to state domestic jurisdiction, particularly with respect to the legal statute of land property, the organization of internal production and the management of agricultural output.[21] The only normative attempt the organization made to more directly address the question of reducing hunger and famines concerned the so-called "surplus disposal," yet the FAO's adoption of quasinormative instruments in its efforts to push developed states into directing their food surplus towards developing countries ultimately failed due to the opposition of western countries.[22]

The overall international scenario concerning food and agriculture changed in the 1960s under the impetus of a new season at the UN since, after the massive decolonization process, a large number of new states had become members of the Organization, with all claiming a different model of development.[23] The changed political equilibrium within the UN underlay the creation in 1961 of the UN World Food Programme, which today still plays a major role in managing and distributing food help throughout the world by looking at both urgent needs and development strategies.[24]

---

[19] It is worth noting that the connection between food production and the aim of raising human standards of living is reflected in the 1948 UN Declarations of Human Rights, in Art. 25.1: "Everyone has *the right to a standard of living adequate* for the health and well-being of himself and of his family, *including food*, clothing, housing and medical care and necessary social services" (emphasis added).

[20] Marchisio and De Blase (1991), p. 61. It is worth noting that technical assistance was implemented according to a north–south approach, in terms of the supply of technology and commodities from developed to developing countries.

[21] More analytically, the issues of land tenure and property regulation were outside the quasinormative competences of the FAO. As to the issue of country public policies in agriculture (in terms of strategies and management), Art. 1. 2 of FAO Constitution (on the power to make recommendations) just refers to the "adoption of improved methods of agricultural production" (Lett. c) and "the improvement of the processing, marketing and distribution of food and agricultural products" (Lett. d).

[22] Cf. Marchisio and De Blase (1991), pp. 25–32. The initial ambitious idea of creating a world reserve of excess food to be directed to poor countries was blocked by the opposition of the United States, the main producer of excess food that intended to autonomously manage its distribution as an instrument of political influence (cf. Mahiou and Snyder 2006, pp. 21–22).

[23] Their major political claim concerned the achievement of an effective process of economic independence. This position principally resulted in the demand for (full) sovereignty over their natural resources, such claim being in itself highly likely to strongly affect agriculture (from the perspective of the regime of foreign investments in land and farms).

[24] Mahiou and Snyder (2006), pp. 25–27. Cf. also Marchisio and De Blase (1991), p. 56.

To fully understand the underlying legal values of the international action on food and agriculture that were conceived during those years, it suffices to read the final declaration of the first World Food Congress of 1963.[25] Even if the declaration was a nonbinding instrument, it recognized important normative principles on food and malnutrition by ultimately sketching out a model of rural development. The declaration singles out two main legal bases underlying the fight against hunger. The first is perhaps intuitive but fundamental: human dignity, being the major principle in the human rights framework. The second is an expression of changed times, namely, hunger as a threat to social and international peace. This crucial point, which is clearly related to the "recent attainment of political independence" by millions of people, contains the idea of food security that would characterize the years following. Yet the suggested concrete approach to obtaining freedom from hunger is still more significant for the understanding of the implied model of agriculture.

The ideological discourse underlying the declaration—and the proposed means of freeing humanity from hunger—is productivism. This is evident when the declaration refers to the need for an "adequate increase in productivity" or relates the education of the rural population to the capability of "applying modern techniques and systems." The declaration even hints at the question of reversing negative effects on trade, which has become a fundamental issue for agriculture in our globalized era. But the major evidence of a productivist approach lies precisely in the model of rural development that the declaration provides. Freedom from hunger "can only be accomplished if all the available human and natural resources of the world are mobilized to this end through balanced economic and social development." Various crucial points are concentrated in this sentence. Natural resources are conceived of as a mere object of human development, as an instrument through which to pursue an economic goal, which should be entirely (all) employed in this sense. On the other hand, development is considered only from an economic and social perspective, with no reference to environmental protection and even to the objective of conserving natural resources.

This is a key point to understand the international approach to agricultural and rural matters during those years. The combination of human rights (of the first generation) and peace-related issues (food security) led to the expunction of the conservation perspective from the international political and legal agenda on food and agriculture, even in contrast to early FAO practice. This is probably the decisive factor in the fracture between agriculture as a productive activity and agriculture as land control. The latter idea of agriculture will be absorbed by the rise of international environmental law.

Subsequent developments in the 1960s and the 1970s legitimized the centrality of food security in the agricultural policies of both the UN and the FAO. At the

---

[25] Declaration of the World Food Congress of 18 June 1963. The Congress was organised by the FAO and the UN General Assembly during the UN Freedom from Hunger Campaign. http://legacy.library.ucsf.edu/documentStore/n/d/u/ndu74c00/Sndu74c00.pdf. Accessed 28 Sept 2014.

World Food Conference of 1974,[26] states thoroughly discussed methods and actions to cope with the food crisis, decided to create a fund to finance agricultural development,[27] approved several resolutions on different fields of cooperation and adopted a final declaration. Analysis of the text of the final declaration shows that the model of rural development substantially remained that of increasing production by fully utilizing productive factors, *in primis* natural resources. However, Point 9 affirmed the necessity for states "[t]o assure the proper conservation of natural resources [...] utilized for food production" and the recommendation that "all countries must collaborate in order to facilitate the preservation of the environment."

Another crucial issue for the FAO initiatives during the 1970s and early 1980s concerned the question of agrarian reform. This was connected to food security, and as a major objective underlying the initiatives for agrarian reforms was a better utilization of land in terms of productivity. At the same time, the value of equity, in relation to the individual right to food, was a factor that equally shaped proposals for agrarian reforms. In terms of international law, the output of such FAO initiatives was still less compulsory than that for the international redistribution of agricultural products, as this issue touched a delicate sector of state sovereignty.[28] The substantial failure of international attempts to promote agrarian reforms, and more generally the absence of international rules or effective practices on land tenure and property rights, is crucial in explaining the absence at the international level of a unitary discipline on agricultural (or agrarian) law,[29] unlike in national legal systems.

In sum, the FAO experience from its birth to the 1980s was characterized by the coexistence of many initiatives on different issues such as surplus disposal, food security, south–south cooperation, agrarian reform and trade. The most important activities carried out by the Organization were of a technical nature and were aimed

[26] The conference was jointly organized by the UN and the FAO.

[27] The International Fund for Agricultural Development (IFAD) is a specialised agency of the UN created in 1976. According to Art. 2 of its agreement, IFAD "shall provide financing primarily for projects and programmes specifically designed to *introduce, expand or improve food production* systems [...], taking into consideration: the *need to increase food production* in the poorest food deficit countries; the *potential for increasing food production* in other developing countries; and the importance of improving the nutritional level of the poorest" (emphasis added). The initial logic pursued by IFAD was thus mainly productivist.

[28] In particular, developing countries, to which FAO initiatives were mainly addressed, feared that such initiatives could jeopardize the difficult process of economic independence that was at the core of their international claims. See Marchisio and De Blase (1991), pp. 98–102.

[29] From a broader ideological viewpoint, the demand to reform land tenure in terms of the wider participation of populations in agricultural activities and management is not really important with respect to the dialectics between conservation and production. In fact, broader access by populations to lands, in particular to uncultivated land, may raise as much concern in the protection of the environment as the maintenance of large landed estates (see, for instance, the broad program of land access and distribution made by Brazil in the 1970s and 1980s in the Amazonian state of Rondonia, which has led to the clearing of most of Rondonia's rainforest).

at assisting developing countries and favoring collaboration between states. From a legal viewpoint, a number of various soft instruments were adopted, some of which were very important in embodying real legal principles and influencing state conduct. However, the key point seems to be the unwillingness of states to create binding rules that limit their sovereignty in a field as sensitive as the regulation of agriculture and land tenure, a fact that explains the difficulties in thinking about agriculture in a unitary sense from the perspective of international law.

While the different fields of initiatives were inspired by a number of legal values and principles—human dignity, equity, peace, state equality—with respect to the model of rural development the FAO's actions ultimately relied on an industrialist paradigm and resulted in the effort to enhance and increase food production. Against this background, the initiatives for the conservation of natural resources seem to have had a secondary importance and developed in a way that did not contradict the main productivist approach.[30] Some environmental initiatives were adopted in the 1980s[31] and opened the way for a change of perspective in subsequent decades.

2.3 *Agriculture in the Era of Globalization: Multifunctionality and Plurality of Institutional Poles*

In the above illustration of the FAO experience, we may identify the multiple issues that likewise characterize the current international legal experience on agriculture. Food security, technical assistance for agriculture, the promotion of rural investments, trade in food products, the conservation and protection of natural resources; these are all issues that were then, as they are today, the object of cooperation between states. A major difference between the present age of globalization and the period of the Cold War principally concerns the plurality of international institutions and regimes variously interested in agriculture.

Among the international institutions that are variously involved in food and agriculture, a major role is played by the WTO. One of the most important changes from the regime of the 1947 General Agreement on Tariff and Trade (GATT) to the current WTO legal system was the inclusion, among the so-called Marrakesh Agreements, of an Agreement on Agriculture.[32] The very title of the agreement

[30] The initiatives for conserving natural resources in the 1950s and 1960s, such as the 1951 Treaty on Plant Protection, could be connected to the early forms of environmental protection (see *supra* A), as national resources were internationally protected because of their utility for human economic activities.

[31] Cf. the Code on the use of pesticides adopted in 1985: see Marchisio and De Blase (1991), pp. 105–106. See also the Undertaking on Plant Genetic Resources, included in the Resolution 8/83 of 23 November 1983 (*infra*, Sect. 4).

[32] The WTO became operational on the first day of 1995. It administers a body of international agreements that were adopted at Marrakesh on 15 April 1994 (Marrakesh Agreements). Its birth

might well generate the idea that a specific regulation for agriculture in international law does exist today. In actual fact, the WTO Agreement on Agriculture is—though aware of the multifunctional dimension of agriculture[33]—a trade agreement whose adoption was the result of years of negotiations during the Uruguay Round, as well as decades of discussions under the old GATT 1947 regime.[34]

Under the impetus of developing countries, one of the major issues discussed during the Uruguay Round was the effective liberalization of trade in agricultural goods. The solutions reached with the adoption of the Agreement on Agriculture were just the beginning of a liberalization process that is still ongoing. In fact, the Agreement provides for crucial liberalizing rules—on market access, domestic support and export subsidies—but the obligations are to be progressively implemented by member states. By the same token, Art. 20 required members to continue negotiations after a 6-year period of initial implementation. This further process started during the Doha Round in 2001 and is still ongoing, thus testifying to the difficulty in regulating matters that have a highly political relevance for states. Among the more debated questions, we find the north–south relationship (in terms of both the liberalization of developed countries' markets and more favorable concessions for developing states) and the construction of agriculture as a social phenomenon that goes beyond mere trade relevance, in line with its multifunctional character.[35]

The impact of the WTO legal system on agricultural matters extends to two other covered agreements, namely those on Sanitary and Phytosanitary Measures (the SPS Agreement) and on Trade-Related Intellectual Property Rights (known as the TRIPS Agreement). The SPS Agreement significantly links trade interests and nature conservation concerns where both involve agricultural matters.[36] The objective of the Agreement is to regulate such measures as a kind of exception to

originates in the 8-year negotiation round that was held during the regime of the GATT 1947 (Uruguay Round).

[33] According to its Preamble, the WTO Agreement should be applied "in an equitable way among all Members, having regard to non-trade concerns, including food security and the need to protect the environment."

[34] Even if the GATT 1947 had set out a comprehensive regime to liberalize trade, member states, particularly developed countries, strongly derogated from rules of this regime as to agricultural products by maintaining, or even increasing, barriers to trade and export subsidies for food and agricultural products. Reasons for this mainly lay in internal political reasons, namely the will of supporting agricultural classes and the interest in maintaining a sufficient level of internal food production as a guarantee of sovereignty and independence. Cf. Picone and Ligustro (2002), pp. 149–150.

[35] Since the beginning of negotiations, the EU has been claiming for the multifunctional character of agriculture (in particular food security and environmental protection, as developed in the EU ACP) to justify some of its trade-restrictive measures (cf. Smith 2000, pp. 707–713).

[36] For an analysis on the typology of SPS measures (in accordance with both a classical conservative logic and an ecological one), see *infra*, Sect. 3), note 68.

GATT-WTO liberalizing principles[37]: trade-restrictive SPS measures are allowed if they respect some requirements set forth by the Agreement.[38]

The TRIPS Agreement contains a detailed regulation on geographical indications (GI), namely trade indications "which identify a good as originating in the territory [. . .] where a given quality, reputation or other characteristic of the good is essentially attributable to its geographical origin" (Art. 22). The Agreement sets out a cluster of obligations aimed at creating a legal system to prevent the abusive use of such indications in each member state. The structural relationship between property right and territory of origin explains the particular importance of GI for agriculture,[39] in line with a legal model of protection that ultimately rewards the distinctive features of different territories so as to favor the diversity of cultures and uphold traditional rural practices.

Along with the WTO, another international economic institution has played a growing role in agriculture, namely the World Bank (WB).[40] Among the original objectives of the WB is the financing of investments for the development of less developed countries, in accordance with an obvious industrialist logic. However, since the 1990s, the Bank has integrated other spheres of values, such as environmental protection, into its actions. A great number of WB projects in developing countries have concerned agricultural activities.[41] Reducing global poverty is one of the main objectives of the organization as a whole.[42] Another institution of the World Bank Group (WBG), the International Finance Corporation (IFC), today plays a crucial role in the sector. Indeed, the private sector arm of the World Bank, whose mandate is to encourage development by both investing and providing technical assistance, has made agriculture lending a priority because of its potential

[37] Indeed, the entire SPS Agreement is a sort of specific development of Art. XX GATT Lett. b), which qualifies as general exceptions those state measures "necessary to protect human, animal or plant life or health." This claim is confirmed at the end of the Preamble of the SPS Agreement ("*Desiring* therefore to elaborate rules for the application of the provisions of GATT 1994 which relate to the use of sanitary or phytosanitary measures, in particular the provisions of Article XX (b)").

[38] The importance of this point of conjunction between trade and nontrade concerns is exemplified by Art. 5.7, which is shaped by the precautionary principle, even if the debate concerning the concrete scope of the provision remains open.

[39] It is worth noting that particular protection is afforded to GI for wines and spirits (Art. 23).

[40] The WB encompasses the International Bank for Reconstruction and Development (IBRD) and the International Development Association (IDA).

[41] The first project that the IBRD financed in a developing country, in 1947, concerned precisely the realization of irrigation facilities in Chile (cf. Viterbo 2013, p. 75). As for the IDA, its statutory role is to finance economic development of the least developed countries, where agriculture often represents the first economic activity.

[42] As 75 % of the world's poor live in rural areas, agricultural development is a key source of economic growth. For the IBRD/IDA, agricultural assistance has risen from an average of 9 % of total lending to 12 % from 2010 to 2013. In Sub-Saharan Africa, where agriculture accounts for three-quarters of employment, the WB mobilized $1.4 million, recording a 35 % increase compared to 2010–2012. See WBG Agriculture Profile Results. http://www.worldbank.org/en/results/2013/04/15/agriculture-results-profile. Accessed 4 Oct 2014.

for broad impact.[43] Altogether, the IBRD, IDA and IFC need to make their commitments in such a way that the projection of the global WBG Agriculture Action Plan is met.[44] The employment of approaches based on biodiversity protection, green growth, food security, as well as landscape, climate-smart and community-driven development, shows a deep commitment to the responsible use of agro-investments. Overall, the nature of these operations[45] bears testament to the growing awareness of the multifunctional character of agriculture among international institutions.

On the other hand, current international law is characterized by the flourishing development of internal subsystems with autonomous characteristics and aims. Some of them regulate matters that partially cross the agricultural sector. The rapid development of IEL has led to the adoption of multilateral agreements that directly concern agriculture and rural landscapes. The influence of this body of law on the legal status of agriculture in international law is fundamental and is specifically investigated in the next section.

Since the beginning of the century, another field of international law has been consolidating in unitary terms, namely international investment law. Like IEL (and unlike WTO law), this legal experience is characterized by a great number of treaties not formally correlated with each other but sharing common features as to the concrete models of regulation. Moreover, arbitral tribunals applying such treaties have had a fundamental unifying action on this field. The essential function of this body of international law is to limit sovereign prerogatives in favor of foreign investment protection. Such limitations may affect state exercise of fundamental public policy prerogatives such as those on land and territory. This issue is debated mainly in terms of human rights[46] and environmental issues[47] but may also involve agricultural matters.[48]

---

[43] The IFC operates on a commercial basis and invests only in for-profit projects, mainly in developing countries.

[44] See the latest World Bank Group Agriculture Action Plan. The strategy refers to years 2013–2015. http://www-wds.worldbank.org. Accessed 4 Oct 2014.

[45] Viterbo (2013), p. 75. The IFC's approach to social and environmental sustainability has been codified by creating a comprehensive set of rules that IFC borrowers—primarily corporations and states—have to comply with in order to qualify for project funding: cf. IFC, *Performance Standards on Environmental and Social Sustainability*, January 1, 2012. On this point, see Solidoro (2013).

[46] Cf. Dupuy et al. (2009).

[47] Cf. Di Benedetto (2013).

[48] This limitation basically operates from an indirect point of view: tribunal decisions do not prohibit regulatory measures *per se* but award (full) compensation when investment treaty rules are violated, thus discouraging the adoption or implementation of such policies. For instance, two different arbitral awards ordered Mexico to pay compensation for having closed two landfills for hazardous wastes controlled by two foreign companies. The main legal basis for the decision was the violation of the obligation to grant fair and equitable treatment to foreign investors. While these decisions mainly raised environmental concerns, they also negatively impacted on agricultural development (as demonstrated by the massive protests of the local population against landfills): cf. Di Benedetto (2013), pp. 121–123.

2.4 First Conclusion Concerning Agriculture in International Law: Fragmentation and Coexistence of Different Ideological Patterns

International rules and policies regarding agriculture are scattered around several fields.[49] Each international legal field affecting agriculture has a conceptual unity (trade, finance/development, environment, foreign investments) that tends to prevent a possible autonomous, crosscutting discourse on agriculture from occurring. This point is particularly evident in IEL, as will be examined in the following section, but is also valid for WTO law, international investment law and WB practice. Even the legal experience of the FAO, despite its obvious focus on agriculture, risks being partly absorbed in the categories of food security (primarily related to the political dimension of state cooperation rather than to models of rural development) and right to food (which belongs to the field of human rights).

3 International Environmental Law and Agriculture

3.1 A Definitional Premise: (The) Environment and Agriculture

A single definition of "the environment" is very difficult to find, as international legal scholars addressing such a task commonly acknowledge.[50] Perhaps the same "definitional" approach conceals an intrinsic limit particularly from a legal viewpoint. However, what we are trying to do here, in speaking of agriculture and the environment in international law, is to relate the two concepts by starting with a very simple, literal definition (thus intrinsically limited and not definitive) of both.

The basic definition of "environment" is relational: it concerns the relationship between human beings and their surrounding reality. "Environment" is what is around the subject,[51] principally around human beings. When referring to "the

[49] A representation of such a fragmentation is provided by UN Millennium Development Goals: among its eight objectives, at least four involve agricultural matters—Objective 1 on eradicating extreme poverty and hunger and also Objectives 6 (against diseases), 7 (environmental sustainability), 8 (partnership for development)—but none explicitly deals with agriculture.

[50] Cf. Birnie et al. (2009), pp. 4–5. Maljean-Dubois (2010), pp. 12–13, underlines the plurality of meanings embedded in the term "the environment."

[51] Accordingly, the New Oxford Dictionary of English (1998) defines "environment" as "the surroundings or conditions in which a person, animal or plant lives or operates." Interestingly, the French word "environnement" in the current meaning is ultimately derived from the English word (though the latter was based in turn on the French "environ"): see Le Grand Robert de la Langue Francaise, 10th edn., vol IV.

environment," the definite article isolates a particular meaning that thus restricts the semantic scope of "environment." "The environment" is therefore a specific, essential portion of what is termed "environment." This essential portion of reality that surrounds human beings, defined in English as "the environment," has its core meaning in "nature," with an important nuance on protective terms.[52] This basic definition is confirmed in its fundamental features by IEL authors addressing such a task.[53] Any further analysis of the meaning of "the environment" should move from the concrete disciplines to which it pertains, thus also IEL.[54]

The meaning of "agriculture" is perhaps more intuitive as it concerns a determinate human activity, that of farming (growing crops and rearing animals).[55] Around this core meaning, "agriculture" may encompass other related activities such as those of harvesting wood products or fishing in rivers. More generally, "agriculture" may refer to the whole of human knowledge and practice concerning these activities.

These two elementary definitions show the inherent relationship between agriculture and the environment. Each agricultural activity implies as "object" a portion of the natural world that is composed of living natural organisms interacting with physical surroundings.[56] In this way agriculture overlaps with the basic notion of "the environment" as encompassing living and nonliving natural resources. Of course, this is only a partial overlapping as the environment encompasses many natural resources that are not directly affected by agriculture.[57]

So far, this definitional approach, in delimiting a relational field, affords us only a very generic picture. A deeper understanding requires us to say something more on the environment, but this would mean going further within clusters of theories developed by natural and social scientists, in particular ecological theories. For the moment, we will try to speak of agriculture and "the environment" by moving from the history of how international environmental legal practice has variously affected agriculture. The following sections will attempt to add further to the ecological background underlying environmental matters.

---

[52] The New Oxford Dictionary of English defines "the environment" as "the natural world, as a whole or in a particular geographical area, especially as affected by human activity."

[53] Cf. Fitzmaurice (2001), pp. 22 et seqq; Birnie et al. (2009), p. 5.

[54] Cf. in this sense Birnie et al. (2009), p. 185.

[55] See the *New Oxford Dictionary of English*. In other languages, such as Italian, the core meaning of the corresponding word (It.: agricoltura) is simply that of growing crops (in accordance with the Latin etymology).

[56] The biological character of the main object of agricultural activities is unsurprising if one thinks that agriculture is basically aimed at producing food. However, the natural interaction of crops and domestic animals with other living organisms as well as with chemical-physical processes highlights the ecological dimension of agriculture. On this point, which basically explains the discipline of agroecology, see Sects. 4 and 5.

[57] On the other hand, a broader meaning of "agriculture" also concerns collateral knowledge and activities that are autonomous in themselves in relation to the environment.

3.2 A Brief Picture of the Intersections Between International Environmental Regulation and Agriculture

As noted in the previous section, early IEL frequently concerned cases involving the protection of natural resources with a prevailing or exclusively agricultural use, according to an industrialist paradigm. As for the international protection of natural resources, such industrialist paradigm was called into question during the 1960s,[58] and authors usually pinpoint 1972—the year of the UN Conference on the Human Environment—as the symbolic moment of transition into the modern era of environmental law.[59] Environmental protection today is mainly characterized by a concern with the overall worth of nature for present and future human generations, beyond its economic and market value. This revolution in perspective was basically driven by the natural sciences and particularly by ecology.

This paradigm shift in the consideration of natural resources, which led to the foundation of IEL as a discipline, marked a decisive, axiological difference with respect to agricultural matters. The cluster of customs, practices and knowledge that characterized farming implied the traditional idea of nature as an entity that played the role of antagonist to human beings and that should be progressively eroded and modified by human activities. While the protection of natural resources as economic utilities progressively lose its importance in IEL, modern agricultural practices have even been targeted by critics for their harmfulness to nature. Such a different axiological basis may explain the current absence of an autonomous consideration of agriculture within the discipline of IEL.

This point appears quite evident when looking at how IEL scholars speak of agriculture in general terms, particularly in handbooks and treatises. Structurally, none of the chapter headings in any of the IEL books analyzed for this research[60] refers to agriculture.[61] Of course, many chapters treat matters *related to* agriculture such as toxic substances, biological diversity and waste management, in line with a coherent and unitary practice in international law that typically revolves around

[58] Even if the current IEL is far from that model, the industrialist paradigm remains fundamental to understand the evolution of the law and its new foundations from a theoretical viewpoint (cf. Dryzek 2007, p. 48). This is still more important as the influence of IEL on agricultural matters is under scrutiny, for agriculture is essentially an economic activity where productivist aspects obviously remain central.

[59] Sand (2007), p. 33; Sands and Peel (2012), pp. 30–32.

[60] Birnie et al. (2009), Bodansky et al. (2007), Sands and Peel (2012), Bodansky (2010), Desai (2014), Kiss and Beurier (2010), Kiss and Shelton (2004) and Fitzmaurice et al. (2010).

[61] An apparent exception is provided by a collected book on sustainable development (Schrijver and Weiss 2004), where two chapters do concern agricultural issues (namely, agriculture and WTO, and plant genetic resources; for the latter, see *infra*, Sect. 4). Nevertheless, this should be set on a different level from the other books since the law of sustainable development is distinct from IEL (*infra*, Sect. 5).

environmental agreements. But none of the chapters tries to do a transversal analysis on the relevance of agriculture for IEL.[62]

Conversely, the same books make sporadic references to agriculture and farming activities simply because international environmental rules and policies do so. This interaction between environmental rules and agriculture occurs in two different, and contradictory, sets of hypotheses. On the one hand, agriculture is seen as a potential source of risk and damage to the environment.[63] Such environmental threats are posed both by agricultural activities in general (to ecosystems, and in particular to tropical forests, due to the expansion of farming[64]) and by specific polluting practices (e.g., pesticides and fertilizers). Accordingly, IEL may regulate agriculture with the aim of preventing the risks it poses to the environment.

On the other hand, agriculture also qualifies as an (indirect) object of protection under environmental rules. Indeed, the rules aimed at conserving natural resources with an economic value often serve a protective function for agriculture. By the same token, rules generally concerning prevention and compensation for environmental harm may also protect agricultural goods.[65] Furthermore, some environmental conventions have positive spillovers for agriculture. This is particularly important for soil protection: the UN Desertification Convention[66] and the African Convention on the Conservation of Nature[67] are clear examples of this. As for the regulation of genetic resources, the CBD affords an important discipline that also involves the critical issue of plant genetic resources and seed regimes.

In sum, agricultural activities have a double (and opposite) relationship to the current ecological paradigm underlying IEL, as they may be a source of environmental threat and also an object of protection under environmental regimes.[68] This

---

[62] Even no section of the chapters explicitly mentions agriculture in their heading. Instead, an environmental handbook has a subsection named "agriculture": Sands and Peel (2012), p. 550. It significantly occurs under the book's section dealing with "other hazardous activities," gathering items that have no structural relationship to each other and thus confirming the marginalisation of the relationship agriculture–environment from the perspective of IEL. This is due, as the authors note, to the fact that "agriculture is not subject to a coordinate regime of legal obligations which apply specific rules at the regional or global level, and which might prepare and implement strategies to use agricultural lands optimally" (Sands and Peel 2012, pp. 550–551).

[63] Very significantly, at the beginning of an IEL handbook, the authors indicate "agricultural practices" as one of the six products and behaviors that are "particularly harmful for the environment" (Sands and Peel 2012, p. 5).

[64] See, e.g., Doos (1991), p. 44.

[65] We have seen that the leading case on state responsibility for transboundary environmental harm—the *Trail Smelter* case—essentially dealt with the negative effects on farming activities.

[66] UN Convention to Combat Desertification in those Countries Experiencing Serious Draught and Desertification, particularly in Africa, adopted on 17 June 1994. Cf. Kiss and Shelton (2004), pp. 445–449.

[67] African Convention on Conservation of Nature and Natural Resources. Its Art. 6 requires states to adopt "effective measures to prevent land degradation." In particular, this article calls for state measures improving soil conservation and promoting sustainable farming and forestry.

[68] This double character of agriculture is also reflected in the SPS Agreement. Art. 1 of Annex A, in defining a sanitary or phytosanitary measure, refers to both "risks for agriculture from pests and diseases" and "risks for humans and animals from additives and contaminant." The latter implies

quite contradictory relationship might provide an initial explanation for the lack of a theoretical consideration of agriculture by IEL scholars. However, the main point is that in international law there is currently no environmental concept or regime that directly addresses agriculture as such (unlike biodiversity, for example), simply because agriculture is not conceived in ecological terms. This point, however, could evolve according to the currently prevailing ideas of ecology and ecosystems, whereby human activities are not in themselves excluded from the scope of ecosystem theory, as it is argued in next two sections.

4 Ecosystem Ecology, Agriculture and IEL

This section addresses the relationship between agriculture and the environment in international law from the different perspective that a renewed idea of ecology and the new discipline of agroecology have opened up.

4.1 Ecology and Ecosystems

Since its founding in the early twentieth century, ecology as a discipline has been in continuous evolution. Born as a branch of biology and concerned with the study of the relationships among individuals in a population or a biological community, ecology has progressively accentuated its character as a *sui generis* discipline. A first, a basic definition of ecology that should not disappoint the majority of ecologists could be that of a natural science that studies living organisms gathered in populations and communities as being intrinsically interactive with each other and with surrounding physical-chemical processes.[69] This first definition implies a fundamental idea, whereby natural world is composed of basic entities, which are made up of a living community and their physical environment and may unitarily be represented as a system, namely an ecological system, or an ecosystem.[70]

Nevertheless, there is a highly controversial debate on the importance and latitude of ecosystems in terms of ecology as a science.[71] This could be roughly synthesized as a debate opposing traditional (or mainstream) ecology and new (or ecosystem) ecology. This debate is of primary importance for the issue being

an ecological perspective, whereby agricultural activities may provoke pollution; the former follows a logic similar to that of the early plant protection conventions, whereby legal values are related to agriculture.

[69] Cf. Odum (1988), pp. 1–4.

[70] An early definition of "ecosystem" may be found in Tansley (1935). Some authors have indicated Forbes (1887), pp. 537–550, as the first *ante litteram* construction of an ecosystem (cf. Odum 1988, pp. 12, 19). See also the definition of Willis (1997), pp. 268–271. According to Art. 1 of CBD, "'*Ecosystem*' means a dynamic complex of plant, animal and micro-organism communities and their non-living environment interacting as a functional unit."

[71] This subsection is essentially based on the collected book Raffaelli and Frid (2010). Yet important references may also be found in some IEL books.

dealt with here since the concrete policy and legal choices concerning nature protection are highly influenced by scientific explanations of natural phenomena, on both national and international levels.[72]

On the one hand, "traditional ecology," although it accepts and develops the concept of ecosystem, substantially maintains a reductionist approach whereby ecology—as a science—should rely on strong, effective linkages between the elements that compose an ecosystem.[73] This has led to a main focus on species population dynamics and to privilege small-scale ecosystems[74] that are constructed as having an internally stable balance.[75] This kind of approach to ecosystems tends to exclude human and social factors from their components.

On the other hand, some ecologists have promoted further evolution of their discipline in line with a holistic—and interdisciplinary—perspective. With this approach, ecology is mainly conceived as an ecosystem ecology where ecosystems are holistically constructed, from the smallest to the broadest levels, as complex and dynamic systems and as being reciprocally interactive along different scales.[76] From this perspective, rather than simply qualifying human activities as being negative in terms of ecological values,[77] the importance of considering the presence and impact of human societies in ecological models and representations becomes fundamental.[78] Therefore, this background has opened the door to an effective interdisciplinary approach that also involves, along with biological and chemical-physical disciplines, economic and social matters.[79]

As may be easily understood, the holistic perspective, and its deep interdisciplinary dimension, may have fundamental repercussions in the construction of environmental regimes,[80] in particular on the international level where global environmental concerns involve different ecosystems and are required to cope with highly complex interactions.[81] As for agricultural matters, this could lead to

[72] Cf. Tarlock (2007). See also Birnie et al. (2009), pp. 585–586.

[73] Cf. Preface of Raffaelli and Frid (2010).

[74] Raffaelli and Frid (2010).

[75] Tarlock (2007), pp. 578–579.

[76] Raffaelli and Frid (2010). They underline that ecosystem ecology concerns "the ecosystem as rich in ecological linkages, some of which may be strong but many of which will be individually weak," while "a reductionist approach [. . .] may fail to correctly understand the system's topology and dynamics" (Preface of Raffaelli and Frid 2010, vii). For an attempt to bridge the two different perspectives, by enhancing mathematical models also in the holistic approaches, see Fenton and Spencer (2010).

[77] Raffaelli and Frid (2010).

[78] Cf. also Sands and Peel (2012), p. 13.

[79] "There is a clear imperative, therefore, to understand what motivate people, whether they be individuals, communities, organisations or nation states, to manage ecosystems in the way they do, *and to incorporate this knowledge into ecosystem models*" Raffaelli and Frid (2010), p. 14.

[80] Cf. also Tarlock (2007).

[81] Significantly, ecologists' argumentation in favor of this further evolution of ecology relies on international cooperation and even legal instruments, such as the Millennium Ecosystem Assessment promoted by the UN, the Intergovernmental Panel on Climate Change reports, the practice developed within the CBD regime (Raffaelli and Frid (2010), pp. 4–6).

the acknowledgment of their ecological dimension, thus allowing them an autonomous room in IEL.

4.2 *Old and New Ecological Approaches in IEL*

The effective extent of the impact of ecological ideas and models in the making of environmental agreements is controversial. Yet IEL references to ecological concepts, such as endangered species, populations and ecosystems, are very common and in some regimes have implemented ecological values initially marginalized, as for the UNESCO Convention.[82] What remains controversial is the degree of the incidence of ecological ideas, that is, how effective the influence of ecological patterns and models has been on the law and policy making within international environmental regimes.[83] The percolation of new ecology ideas into IEL is still more controversial. An author has expressed scepticism regarding the actual influence of the renewed meaning of ecosystems on current international environmental regimes.[84] While this view seems to be substantially correct, several regimes provide important normative elements that rely on concepts and ideas of ecosystem ecology.

At its outset, the CBD Preamble affirms that contracting parties are conscious "of the importance of biological diversity for evolution and for maintaining life-sustaining systems of the biosphere." Such a holistic approach[85] is reaffirmed in Art. 1, which defines biodiversity in broad terms as encompassing "terrestrial, marine and other aquatic ecosystems and the ecological complexes of which they are part."[86]

The evolution of the Antarctic regime provides an important indication of the ability of ecological concepts to increasingly influence IEL. The 1959 Antarctic Treaty has built up a singular regime over a vast area of the earth,[87] where state prerogatives have been effectively limited in favor of a conservation strategy. Environmental protection has been one of the objectives of the regime since the beginning,[88] but the 1991 Protocol on the Environment[89] expanded environmental

[82] The original values of the UNESCO World Heritage Convention: science, conservation, natural beauty and, subsequently, natural sites that are also "examples representing significant on-going ecological and biological processes in the evolution of terrestrial, fresh water, coastal and marine ecosystems and communities of plants and animals" (Operational Guidelines 2002, referred to by Gillespie (2007), p. 77).

[83] Cf. Haas (2007), pp. 798–802.

[84] Tarlock (2007).

[85] Birnie et al. (2009), p. 8.

[86] According to Birnie et al. (2009), p. 616, this concept of biodiversity "could become the 'organizing' or at least the 'integrating' concept for relating relevant existing agreements."

[87] The treaty applies to the area south of the 60° South line of latitude, according to Art. VI.

[88] Sands and Peel (2012), pp. 579–580.

[89] Protocol on Environmental Protection to the Antarctic Treaty, adopted in Madrid on 4 October 1991.

objectives according to a fully ecological perspective by referring to the "protection of the Antarctic environment and dependent and associated ecosystems and the intrinsic value of Antarctica, including its wilderness and aesthetic values" (Art. 3.1).[90]

Another environmental regime provides a still more important sign of the influence of a holistic ecological approach, as including human activities among the components of some ecosystems. The 1971 Ramsar Convention is aimed at protecting wetlands[91] and since the Preamble member states have recognized the "fundamental ecological functions of wetlands as regulators of water regimes and as habitats supporting a characteristic flora and fauna." While at the beginning protected areas were essentially habitats of waterfowl populations, the Ramsar regime has progressively enlarged its scope by considering a broader set of ecological values.[92] Moreover, among the three major types of wetlands that are now protected, there are "human made wetlands," which thus extends the ecological approach to land also used by people according to the perspective of ecosystem ecology. Significantly, they also include aquaculture ponds, irrigated land and seasonally flooded agricultural lands, which thus demonstrates that the ecological perspective may also involve agricultural territories.[93]

4.3 Two Different Regimes on Plant Genetic Resources and Agrobiodiversity

The Ramsar Convention is an important though quite isolated case of a treaty dealing with human and natural elements blended together in the same ecological model, particularly with respect to agricultural areas. Conversely, the underlying logic of current international environmental regimes—even when aimed at protecting large-scale ecosystems and relying on modern ecological patterns—is that of protecting environmental values from human activities, mainly the economic ones, in line with the mainstream ecological vision that diffidently sees the direct relevance of social factors within ecosystem models.[94] Arguably, the latter function of IEL remains essential. The point is whether, along with the perception of human economic activities as dangerous for the environment, there is room for another assessment of them as intrinsic components of a number of ecosystems on

[90] Cf. Birnie et al. (2009), p. 186: the Protocol is the "most comprehensive, and significant example in which an entire continent and the surrounding marine environment have been protected on such an ecosystem basis."

[91] Ramsar Convention on Wetlands of International Importance especially as Waterfowl Habitat, February 1971.

[92] Cf. Gillespie (2007), pp. 64–65.

[93] Gillespie (2007), p. 73.

[94] Raffaelli and Frid (2010).

earth. This is a sensitive point where agriculture is concerned. A complex matter that has different approaches on just this point is that of the international regulation of plant genetic resources.

In an early approach, genetic resources were considered part of the "common heritage of humankind."[95] As a consequence, no state would be able to affirm sovereign rights on such resources and access to them would remain free.[96] However, this approach seems to be merely theoretical, as international practice has not confirmed the existence of a customary rule in this sense[97] due mainly to the strong opposition by industrialized states.[98] On the other hand, two different treaty models are at stake, one developed within the CBD regime, the other set forth by a FAO Convention.

One of the three main objectives of the CBD is "the fair and equitable sharing of the benefits arising out of the utilization of genetic resources," including access to them (Art. 1), but within the framework of the principle of state sovereignty over natural resources, which is understood to include genetic resources. This principle is strongly affirmed by Art. 15.1 on Access to Genetic Resources, whereby "[r] ecognizing the sovereign rights of States over their natural resources, the authority to determine access to genetic resources rests with the national governments and is subject to national legislation." States may exercise their sovereign right by providing their genetic resources to another state by allowing the said state access to resources (Art. 15.5), ultimately according to a contractual mechanism.[99]

The key concept of such a regime is that of "country of origin" of the genetic resources .[100] The state being country of origin has sovereign rights to genetic resources and, accordingly, decides on the access to genetic resources of another state. As an author has admirably remarked, the concept of "country of origin" becomes highly problematic when applied to the genetic resources of domesticated species, particularly vegetable ones.[101] The country of origin "is not always the

[95] This model in particular was affirmed by the FAO in its 1983 *Undertaking on Plant Genetic Resources*.

[96] Rayfuse (2007), p. 378.

[97] Cf. Mgbeoji (2003); Francioni (2006), pp. 9–11.

[98] Industrial states—and breeder associations—"were all concerned that the common heritage would pose an unacceptable threat to proprietary rights over plant varieties and their potential improvement by way of biotechnological manipulation," Francioni (2006), p. 9.

[99] The concrete dispositions set out by the 2010 Nagoya Protocol has confirmed the centrality of state sovereignty and this access mechanism (Nagoya Protocol on Access to Genetic Resources and the Fair and Equitable Sharing of Benefits Arising from their Utilization to the Convention on Biological Diversity, adopted at the tenth meeting of the CBD Conference of the Parties on 29 October 2010, in Nagoya).

[100] According to Art. 2, "'*Country of origin of genetic resources*' means the country which possesses those genetic resources in *in-situ* conditions," that is, "conditions where genetic resources exist within ecosystems and natural habitats, and, in the case of domesticated or cultivated species, in the surroundings where they have developed their distinctive properties."

[101] Santilli (2012), pp. 115–117.

same as the country in which the species developed its distinctive properties,"[102] and attribution of the right thus risks being arbitrary. Furthermore, the long and continuous process of species evolution that has been guided by farmers' communities risks being terminated by a specific attribution of sovereign rights on a genetic resource, which therefore jeopardizes the further diversification of cultivated species.[103]

The FAO International Treaty on Plant Genetic Resources,[104] which concerns only the genetic resources of plants used for food and agriculture, offers a rather different regulation of the matter.[105] A further restriction of its scope is realized by a mechanism, whereby only the plants listed in Annex I are subject to the specific treaty system.[106] The treaty reaffirms the principle of state sovereignty over its genetic resources, including its general right to determine access to such resources.[107] The distinctive feature of the regime, however, concerns access to and benefit sharing of the plant genetic resources listed in Annex I, provided that they "are under the management and control of the Contracting Parties and in the public domain" (Art. 11.2). Instead of the bilateral logic that characterizes the CBD regulation, the FAO Treaty sets out a multilateral system involving all the contracting parties (Art. 10-13). States shall take the necessary measures to provide access to genetic resources to the other parties—as well as to their legal and natural persons—in an expeditious manner and substantially free of charge (Art. 12.2-3). As for benefit sharing, the Treaty firstly recognizes that a major common benefit of the system is the regime of access, but it sets out further duties for states on information exchange and monetary benefits (Art. 13).

Moreover, the Treaty sets forth important rules on the conservation and sustainable use of all agricultural plant genetic resources, and therefore not only those listed in Annex I. Duties of inventory, collection, monitoring and conservation of genetic resources are set out by Art. 5, while Art. 6 provides a set of rules on sustainable use that is basically aimed at the protection, promotion and enhancement of agricultural biodiversity. In both dispositions, in particular in Art. 6, the Treaty underlines the major role of local communities of farmers as promoters and stewards of agrobiodiversity.

Finally, the CBD and FAO regimes follow quite different perspectives but are not in opposition. Their relationship is based on complementariness: the FAO

[102] Santilli (2012), p. 115. The author refers to important scientific studies from the nineteenth century to today that testify to such difficulties in determining centers of origins and centers of diversity.

[103] Santilli (2012), p. 117.

[104] International Treaty on Plant Genetic Resources for Food and Agriculture, Rome, 3 November 2001.

[105] The FAO Treaty would tend to qualify Plant Genetic Resources as a "common concern of humankind": Footer (2004), pp. 433–466.

[106] They are 35 food crops and 29 forage crops. However, very important crops are excluded by the list, such as soybean, tomatoes, coffee, cacao (see Santilli 2012, p. 123).

[107] Art. 10.1, which uses the same terms as Art. 15.1 of CBD.

regime prevails for the plants listed in Annex I of the Treaty, provided they belong to the public domain, while the bilateral system set forth by the CBD applies to the other genetic resources. As for the models of land use and rural development underlying the two regimes, both reaffirm the centrality of state sovereignty over its own natural resources.

However, in its general rules on conservation and sustainable use, the FAO Treaty sets out a few decisive elements capable of integrating a new and different model of the relationship between agriculture and natural resources. Along with the reference to agrobiodiversity,[108] the FAO Treaty establishes local communities of farmers as the decisive actors in the sustainable use of plant genetic resources. From this perspective, the dimension of agriculture and that of conservation and protection of the agricultural environment merge. This change of model is well represented by Art. 6.2 b), which underlines the particular role of those farmers "who generate and use their own varieties and apply ecological principles in maintaining soil fertility and in combating diseases, weeds and pests." This disposition could well be deemed to comply with an agroecological approach.

5 Conclusion: Is There Room for Agroecology in International Law?

The FAO Treaty on plant genetic resources provides an important path towards the concrete integration of agricultural activities and environmental protection in accordance with a unitary ecological model. This opens the door to a broader consideration of how such a model, namely an agroecological model, might influence international rules or even inspire a new international legal regime.

As noted in Sect. 3, semantically "agriculture" and "the environment" have an important overlap as both directly concern living natural resources and their physical surroundings. Besides, agricultural lands are vast, are multiform, integrate noncultivated zones and also host wild animals. A great number of ecological processes and functions are developed in such lands, some of which may have an important impact at a global scale. Modern ecology has recognized these crucial points, which are precisely addressed by the new discipline of agroecology.[109]

Considering agroecology in international law would require a radical and difficult change of paradigm, principally because, unlike national systems, this law does not deal unitarily with agriculture matters, thus preventing a possible further evolution towards an "agroecology law." Besides, as noted in Sect. 3,

[108] It is worth noting that the Nagoya Protocol in its Preamble recognizes "the special nature of agricultural biodiversity, its distinctive features and problems needing distinctive solutions."

[109] Agroecology is specifically dealt with in several chapters of this book (see in particular CAPORALI), and the relationship between agroecology and law is more broadly investigated in some of them.

environmental law likewise lacks awareness of the possible autonomous role of agriculture within its fields. Against this background, only some suggestions about the possibility of considering agroecology in international law may be offered.[110] Such a first attempt could perhaps prelude to a more systematic construction of a part of international ecosystem law as being intrinsically characterized by human activities.

The first point to be addressed is the possible relationship between agroecology and sustainable development. While the legal status of sustainable development in international law remains highly controversial,[111] ranging from a mere ordering category or policy to a normative concept and, finally, to a real legal principle,[112] its tripartite meaning is quite commonly accepted today, wherein the economic, social and environmental dimensions[113] should be integrated together according to an intergenerational perspective.[114] In other words, rules and policies concerning one of the three dimensions should take into account (i.e., assess or balance with) the other two, according to the legal force of each value and the interest that is concretely involved.[115]

---

[110] For a pioneer study on agroecology and law, cf. Monteduro (2013).

[111] See, for instance, Boyle and Freestone (1999), pp. 16–18. This uncertainty concerns both the way treaties qualify sustainable development and the recognition of its status in general international law. Cf. Barstow Magraw and Hawke (2007), pp. 622–626.

[112] In its Separate Opinion of the ICJ Judgment in *Gabcikovo-Nagymaros case*, Judge Weeramantry claims that sustainable development would be "a principle with normative value." The thesis of the legal principle is also espoused by Voigt, who speaks of sustainable development as a general principle of law. The view of Lowe (1999) is more complex: he substantially qualifies sustainable development as a policy goal, thus incapable of assuming a normative character as a custom of international law; however, it would play the role of a "modifying" or "interstitial" norm, influencing the application (*in primis* by judges) of primary rules.

[113] These three "pillars" of sustainable development are commonly recognised by the UN Declarations since the mid-1990s (cf. Copenhagen Declaration on Social Development, 19 April 1995; UN Millennium Declaration, GA Res. 55/2, 8 September 2000; World Summit Outcome, GA Res. 60/1, 24 October 2005; Declaration on the follow-up of Millennium Development Goals, GA Res. 65/1, 19 October 2010). At its origin, the two conceptual poles of sustainable development were simply economic development and environmental protection, and this perception has continued until the end of the past century (cf. Lowe 1999 and the famous 1997 *Gabcikovo-Nagymaros Judgment* of the ICJ, para. 141).

[114] Among international law scholars, for instance, the book of Voigt (2010), precisely interprets sustainable development as a "principle of integration" (pp. 145–186) and thus employs the principle to cope with the question of reconciling climate and trade law. Cf. also Cordonnier Segger and Khalfan (2004), pp. 50–53 (see *infra*, note 123). The same approach is also implied in the famous ICJ Judgment in the *Gabcikovo-Nagymaros case*: "This need to reconcile economic development with protection of the environment is aptly expressed in the concept of sustainable development" (para. 141). Even Lowe (1999), p. 26, though exposing a more articulated theory (cf. *supra* note 111), says that "[t]he idea that development and environmental protection must be reconciled is clearly central to the concept."

[115] This legal force (or weight, if referred to principles according to the Dworkinian theory) would depend on each regime, or legal order concretely involved, so explaining why similar questions might lead to different results even though all are framed under sustainable development.

Such an integrative perspective would seem to be the ideal framework to foster a mutually supportive relationship between agriculture (as an economic and social activity) and the environment and also to allow room in international law for agroecology. Arguably, the pair agroecology–sustainable development is potentially rich in concrete applications at the international level. The suggestion proposed here, however, tries to go a little further in relation to the notion of integration as being currently constructed within the framework of sustainable development. The point is that all the above-discussed construction of sustainable development in international law implies a structural separation of its three dimensions.[116] The same idea of balance entails the interaction of separate elements, each of which follows a different logic or serves different values.[117]

On the other hand, since the concept of agroecosystem is based on a combination of natural and human factors (the latter to be intended as both economic and social), or, more precisely, concerns ecosystems that are intrinsically modified and shaped by human action, it is quite impossible to disentangle the three dimensions from each other. This agroecological perspective does not simply concern the impact of human activities on the environment (in terms of both damage and protection) but also pertain to the same logic (or ontology) of agriculture. In this way the ecological dimension is transversal to human economy, human society and environmental resources.[118]

What might this mean for international law? At this moment, the concept of agroecology might be represented as a policy goal for a different method of cooperation in agricultural matters. This method would revolve around a real interdisciplinarity[119] in accordance with a holistic approach. In the long term, it could become the unifying concept that deals autonomously with agriculture in international law. It would be the conceptual framework to unitarily address rules and principles involving agriculture, which, as explained in Sect. 2, are today fragmented in different fields. Some characteristics might typify this agroecological perspective in international law, firstly, the notion of agricultural ecosystem service and, secondly, rural culture and traditional farming knowledge (all merging a human and an environmental dimension). In sum, the new idea of ecology, which

[116] This point is confirmed by Koskenniemi (2009), pp. 7–19: "notions such as 'sustainable development' [...] single out fragile compromises in areas where the struggle between opposing groups of experts and their preferences has not (yet) been taken to the end."

[117] For instance, according to Cordonier Segger and Khalfan (2004), p. 50, sustainable development entails "a balance between three intersecting systems of international law," i.e. international social, economic and environmental law. Therefore, "International sustainable development law is found at the intersection" of these three fields of international law (p. 51). Sustainable development should play a role "to reconcile economic, environmental and social treaties and regimes in case of conflicts between norms" (p. 53).

[118] In this way, the concept of integration evolves, being viewed "as an ontological link operating in the construction of agriculture as a human activity system" (see CAPORALI in this volume). Accordingly, the concept of sustainability, if holistically conceived, would likewise rather assume a shift in its meaning.

[119] Raffaelli and Frid (2010), p. 14.

accepts the role of human activities as a component of the ecosystem and is intrinsically interdisciplinary, would represent the backbone of this possible new field of international law.

References

Barstow Magraw D, Hawke LD (2007) Sustainable development. In: Bodansky D et al (eds) The Oxford handbook of international environmental law. Oxford University Press, Oxford, pp 622–626

Birnie P, Boyle A, Redgwell C (2009) International law and the environment, 3rd edn. Oxford University Press, Oxford

Bodansky D (2010) The art and craft of international environmental law. Harvard University Press, Cambridge

Bodansky D, Brunnée J, Hey E (eds) (2007) The Oxford handbook of international environmental law. Oxford University Press, Oxford

Boyle A, Freestone D (1999) Introduction. In: Boyle A, Freestone D (eds) International law and sustainable development. Past achievements and future challenges. Oxford University Press, Oxford, pp 1–21

Cordonier Segger MC, Khalfan A (2004) Sustainable development law: principles, practices, and prospects. Oxford University Press, Oxford

Desai BH (2014) International environmental governance. Brill Nijhoff, Leiden

Di Benedetto S (2013) International investment law and the environment. Edward Elgar, Cheltenham

Doos BR (1991) Environmental issues requiring international action. In: Lang W, Neuhold H, Zemanek K (eds) Environmental protection and international law. Martinus Nijhoff, London, pp 1–54

Dryzek JS (2007) Paradigms and discourses. In: Bodansky D, Brunnée J, Hey E (eds) The Oxford handbook of international environmental law. Oxford University Press, Oxford, pp 44–62

Dupuy P-M, Francioni F, Petersmann E-U (eds) (2009) Human rights in international investment law and arbitration. Oxford University Press, Oxford

Fenton A, Spencer M (2010) Linking population, community and ecosystem ecology within mainstream ecology. In: Raffaelli DG, Frid CLJ (eds) Ecosystem ecology – a new synthesis. Cambridge University Press, Cambridge

Fitzmaurice MA (2001) International protection of the environment. The Hague Academy Collected Courses, vol 293, pp 22 et seqq

Fitzmaurice M, Ong DM, Merkouris P (eds) (2010) Research handbook on international environmental law. Edward Elgar, Cheltenham

Footer ME (2004) Our agricultural heritage: sustainability, common heritage, and intergenerational equity. In: Schrijver N, Weiss F (eds) International law and sustainable development – principles and practice. Martinus Nijhoff, Leiden, pp 433–466

Forbes SA (1887) The lake as a microcosm. Bull Sc A Peoria [Reprinted in Ill Nat Hist Surv Bull] 15:537–550

Francioni F (2006) International law for biotechnology: basic principles. In: Francioni F, Scovazzi T (eds) Biotechnology and international law. Hart, Oxford, pp 3–27

Gillespie A (2007) Protected areas and international environmental law. Martinus Nijhoff, Leiden

Haas P (2007) Epistemic communities. In: Bodansky D et al (eds) The Oxford handbook of international environmental law. Oxford University Press, Oxford, pp 791–806

Kiss A, Beurier J-P (2010) International environmental law: fairness, effectiveness and world order. Cambridge University Press, Cambridge

Kiss A, Shelton D (2004) International environmental law, 3rd edn. Arsdley, New York

Koskenniemi M (2009) The politics of international law – 20 years later. Eur J Int Law 20(1):7–19
Lowe V (1999) Sustainable development and unsustainable arguments. In: Boyle A, Freestone D (eds) International law and sustainable development – past achievements and future challenges. Oxford University Press, Oxford
Mahiou A, Snyder FG (eds) (2006) La sécurité alimentaire – Food security and food safety. Martinus Nijhoff, Leiden
Maljean-Dubois S (2010) La "fabrication" du droit international au defi de la protection de l'environnement. In: Société Francaise pour le Droit International (ed) Le droit international face aux enjeux environnementaux – Colloque d'Aix-en-Provence. Pedone, Paris, pp 9–38
Marchisio S, De Blase A (1991) The Food and Agriculture Organization. Martinus Nijhoff, London
Mgbeoji I (2003) Beyond rhetoric: state sovereignty, common concern, and the inapplicability of the common heritage concept to plant genetic resources. Leiden J Int Law 16:821–837
Monteduro M (2013) Environmental law and agroecology. Transdisciplinary approach to public ecosystem services as a new challenge for environmental legal doctrine. Eur Energy Environ Law Rev 22:2–11
Odum EP (1988) Basi di ecologia. Piccin, Padova [Italian translation of Odum EP (1983) Basic ecology. CBS College Publishing]
Picone P, Ligustro A (2002) Diritto dell'Organizzazione Mondiale del Commercio. Cedam, Padova
Raffaelli DG, Frid CLJ (2010) The evolution of ecosystem ecology. In: Raffaelli DG, Frid CLJ (eds) Ecosystem ecology – a new synthesis. Cambridge University Press, Cambridge
Rayfuse R (2007) Biological resources. In: Bodansky D et al (eds) The Oxford handbook of international environmental law. Oxford University Press, Oxford, pp 362–393
Robb CAR (1999) International environmental law reports, vol 1 (Early decisions). Cambridge University Press, Cambridge
Ruster B, Simma B (1975) International protection of the environment – treaties and related documents. Oceana Publications Inc., Dobbs Ferry
Sand PH (2007) The evolution of international environmental law. In: Bodansky D et al (eds) The Oxford handbook of international environmental law. Oxford University Press, Oxford, pp 29–43
Sands P, Peel J (2012) Principles of international environmental law, 3rd edn. Cambridge University Press, Cambridge
Santilli J (2012) Agrobiodiversity and the law – regulating genetic resources, food security and cultural diversity. Earthscan, Abingdon
Schrijver N, Weiss F (eds) (2004) International law and sustainable development – principles and practice. Martinus Nijhoff, Leiden
Smith F (2000) Multifunctionality and non-trade concerns in the agriculture negotiations. J Int Econ Law 3(4):707–713
Solidoro S (2013) International human rights accountability mechanisms benchmarking. In: Costa A (ed) Banks and human rights: pathways to compliance. Lulu Press, pp 193–214
Tansley AG (1935) The use and abuse of vegetational concepts and terms. Ecology 16:284–307
Tarlock D (2007) Ecosystems. In: Bodansky D, et al (eds) The Oxford handbook of international environmental law. Oxford University Press, Oxford, pp 574–596
Viterbo A (2013) Fondo Monetario Internazionale e Banca Mondiale. In: Comba A (ed) Neoliberismo internazionale e global economic governance, 2nd edn. Giappichelli, Torino, pp 69–130
Voigt C (2010) Sustainable development as a principle of international law – resolving conflicts between climate measures and WTO law. Martinus Nijhoff, Leiden
Willis AJ (1997) The ecosystem: an evolving concept viewed historically. Funct Ecol 11:268–271

Ecosystem Services: European Agricultural Law and Rural Development

B. Jack

Abstract This chapter examines the concept of ecosystem services and analyses the role that the European Union's Common Agricultural Policy ("the CAP") is playing in supporting and protecting these services throughout the European Union. The chapter begins by exploring both the nature and condition of ecosystem services before going on to examine the commitments that have been made in international and European Union environmental policies to protect these services. The chapter then provides a brief examination of the severe impact that early CAP production policies had upon ecosystems producing important ecosystem services. Over recent decades, no less than six major reform packages have been introduced and have made fundamental changes to the CAP. The most recent of these occurred with the regulations introduced on December 2013, which are intended to govern European agriculture until the end of 2020. In the light of these reforms, the chapter analyses the modern CAP and questions whether the policy measures currently available provide the degree of protection and support for ecosystem services that is envisaged by the European Union's environmental policy. Finally, the chapter points out that, in addition to the CAP, private markets are also likely to develop, providing payments for landowners who help to protect ecosystem services.

Keywords Common agricultural policy • Ecosystem services • Private markets • Rural development

1 Introduction

The existence of the Common Agricultural Policy ("the CAP"), which has controlled European agriculture since the early 1960s, has often been linked to a compromise between France and Germany when the European Economic Community was being established.[1] If German industrial exports were to be granted free access to other markets, then France sought similar access for its agricultural

[1] See, for example, Grant (1997), p. 63.

B. Jack (✉)
School of Law, Queen's University of Belfast, Belfast, UK
e-mail: b.jack@qub.ac.uk, http://www.qub.ac.uk/schools/SchoolofLaw/Staff/DrBrianJack/

M. Monteduro et al. (eds.), *Law and Agroecology*,
DOI 10.1007/978-3-662-46617-9_7

exports. When viewed in terms of increased agricultural productivity, the CAP has been a major success. However, society has become increasingly aware of the environmental costs associated with the drive towards increased productivity that so characterised early CAP measures. In particular, biodiversity has suffered as agricultural intensification promoted land use changes that removed a broad range of wildlife habitats. These changes are part of the broader story of human development, which has had major environmental impact not just within the European Union but also across the world.[2] The impact has been so severe that scientists are warning that we may be triggering a global mass extinction of biodiversity.[3] At the same time, we have also become increasingly aware of the wide range of services, ecosystem services, that are provided by biodiversity and of the extent to which humanity is reliant upon them.

This chapter begins by examining the concept of ecosystem services, exploring both their nature and condition before going on to explore the commitments made to protect ecosystem services in international and European Union environmental policies. The chapter then provides a brief examination of the impact that the initial CAP production policies had upon ecosystems producing important ecosystem services. The modern CAP, however, is a very different policy instrument. Since the early 1990s, no less than six major reform packages have been introduced to make fundamental changes to the CAP. The most recent of these occurred with the introduction of a package of regulations in December 2013, intended to govern European agriculture until the end of 2020. The chapter therefore examines the modern CAP, in the light of these reforms, and questions whether the policy measures currently available provide the degree of protection and support for ecosystem services that is envisaged by the European Union's environmental policy. Finally, the chapter points out that, in addition to the CAP, private markets may increasingly develop to provide payments for landowners who help to protect ecosystem services.

2 Ecosystem Services

Several publications in the late 1990s drew attention to the extent to which humanity was reliant upon a range of ecosystem services provided by the natural world.[4] At that time, the global value of these services was estimated to be US $33 trillion per annum, some 1.8 times greater than the value then of the entire world economy.[5] However, it was really the publication of the *Millennium*

[2] See, variously, Rockström et al. (2009), pp. 472–475; Lynas (2011) and Ellis (2011), pp. 1010–1035.

[3] See, for example, Barnosky et al. (2011), pp. 51–57.

[4] See Daily (1997) and Baskin (1997).

[5] Costanza et al. (1997), pp. 253–260 and 259.

Ecosystem Assessment, in 2005, that drew particular attention to the importance of ecosystem services.[6] The *Millennium Ecosystem Assessment* had been established in 2001 at the request of the United Nations. Its task was to "assess the consequences of ecosystem change for human well-being and to establish the scientific basis for actions needed to enhance the conservation and sustainable use of ecosystems and their contribution to human well-being".[7] The final report identified twenty-four different ecosystem services, which it divided into four categories: provisioning services, regulating services, cultural services and supporting services.[8] Provisioning services are the products that humanity receives from functioning ecosystems, such as food, fresh water and raw materials (for example, timber, cotton and wool); regulating services, on the other hand, acknowledge the role that ecosystems play in maintaining the general environment within which humanity lives. This includes its role in regulating our climate, protecting air and water quality, pollinating plants and preventing soil erosion or flooding. Cultural services, in contrast, recognise the enjoyment that humans gain from natural spaces, whilst supporting services are concerned with the role that ecosystem services play in a range of fundamental areas that underpin all the other ecosystem services, such as soil formation and water cycling.

The *Millennium Ecosystem Assessment* also highlighted the degradation that human development had caused many of the ecosystems associated with essential ecosystem services over the last 50 years.[9] Although this had brought substantial gains in human well-being and economic development, the report identified a less certain future.[10] It predicted that ecosystem service degradation could grow significantly worse and that ecosystems would provide substantially fewer benefits for future generations.[11] Ultimately, the report stressed that it would be possible to at least partially reverse ecosystem degradation whilst meeting increased demand for ecosystem services.[12] However, it warned that this would require significant changes in policies, institutions and practices and cautioned that these changes were not yet being implemented.[13] Ultimately, the Millennium Ecosystem Assessment provided a global evaluation of the condition of ecosystem services. However, studies by the *European Academies Science Advisory Council* and by the *United Kingdom National Ecosystem Assessment* have similarly highlighted both the importance of ecosystem services within the European Union and the need to ensure that this is better reflected in decision-making.[14]

[6] Millennium Ecosystem Assessment (2005).

[7] Millennium Ecosystem Assessment (2005), preface.

[8] Millennium Ecosystem Assessment (2005), p. 7.

[9] Millennium Ecosystem Assessment (2005), p. 1.

[10] Millennium Ecosystem Assessment (2005), p. 1.

[11] Millennium Ecosystem Assessment (2005), p. 1.

[12] Millennium Ecosystem Assessment (2005), p. 1.

[13] Millennium Ecosystem Assessment (2005), p. 1.

[14] See the European Academies Science Advisory Council (2009); Fitter et al. (2010), and UK National Ecosystem Assessment (2011).

There has now been broad recognition of the need for action to protect ecosystems producing important ecosystem services. At an international level, the fifth meeting of the parties to the Convention on Biological Diversity, in 2000, first highlighted the importance of conserving ecosystem structures and functions in order to maintain ecosystem services as part of the development of an ecosystem approach towards the management of natural resources.[15] Subsequently, in adopting the Strategic Plan for Biodiversity 2011–2020, the parties to the Convention agreed "to take effective action to halt the loss of biodiversity in order to ensure that by 2020 ecosystems are resilient and continue to provide essential services, thereby securing the planet's variety of life, and contributing to human well-being, and poverty eradication".[16] In addition, they also agreed to the Aichi Biodiversity Targets, which establish the following targets of particular relevance to agriculture:[17]

Target 5: By 2020, the rate of loss of all natural habitats, including forests, is at least halved and where feasible brought close to zero and degradation and fragmentation is significantly reduced.

Target 7: By 2020, areas under agriculture, aquaculture and forestry are managed sustainably, ensuring conservation of biodiversity.

Target 8: By 2020, pollution, including from excess nutrients, has been brought to levels that are not detrimental to ecosystem function and biodiversity.

Target 14: By 2020, ecosystems that provide essential services, including services related to water, and contribute to health, livelihoods and well-being, are restored and safeguarded, taking into account the needs of women, indigenous and local communities and the poor and vulnerable.

Target 15: By 2020, ecosystem resilience and the contribution to carbon stocks has been enhanced, through conservation and restoration, including restoration of at least 15 % of degraded ecosystems, thereby contributing to climate change mitigation and adaption and to combating desertification.

Equally, the European Union, as a party to the Convention on Biological Diversity, has also made the protection of ecosystem services a central element of its own biodiversity strategy for the period up to 2020.[18] The strategy's headline target during this period is "halting the loss of biodiversity and the degradation of ecosystem services in the EU by 2020 and restoring them in so far as possible, while stepping up the EU contribution to averting global

[15] Conference of the Parties to the Convention on Biological Diversity, Decision V/6 on the ecosystem approach, principle 5. http://www.cbd.int/decision/cop/default.shtml?id=7148. Accessed 21 May 2014. See also Decision VII/11, adopted in 2004, on the ecosystem approach. http://www.cbd.int/decision/cop/default.shtml?id=7748. Accessed 21 May 2014.

[16] Conference of the Parties to the Convention on Biological Diversity, Decision X/2: "The Strategic Plan for Biodiversity 2011–2020 and the Aichi Biodiversity Targets", adopted in Oct 2010, Para. 12. http://www.cbd.int/decision/cop/default.shtml?id=12268. Accessed 21 May 2014.

[17] Conference of the Parties to the Convention on Biological Diversity, Decision X/2, Para 13.

[18] European Commission (2011a), Our Life Insurance, Our Natural Capital: An EU Biodiversity Strategy to 2020, COM (2011) 244 final. The Environment Council subsequently endorsed the Commission proposals on 21 June 2011.

biodiversity loss”.[19] In relation to agriculture, the strategy sets the following specific target for the European Union:[20]

> By 2020, maximise areas under agriculture across grasslands, arable land and permanent crops that are covered by biodiversity-related measures under the CAP so as to ensure the conservation of biodiversity and to bring about a measurable improvement in the conservation status of species and habitats that depend on or are affected by agriculture and in the provision of ecosystem services as compared to the EU2010 Baseline, thus contributing to enhance sustainable management.

3 Ecosystem Services and Agriculture

Both in the European Union and elsewhere, agriculture traditionally concentrated solely on the production of food and raw materials. Indeed, the beef mountains and wine lakes that so characterised the “the CAP” in the 1980s and early 1990s,[21] as the European Union wrestled with the problems caused by surplus production, were a testament to its success in so doing. However, as the *Millennium Ecosystem Assessment* noted, actions to increase one ecosystem service often cause the degradation of others.[22] In the case of the CAP, the environmental impact of policies that encouraged the sector to intensify in order to maximise its production capacity are now well known.[23] Their impact on other ecosystem services has been summarised, by the *European Academies Science Advisory Council*, as follows[24]:

> Large areas have been devoted to monocultures, with increasing use of fertilisers, fungicides and pesticides to maintain productivity. This process has prioritised production services, to the extent that other key services, in particular those associated with complex ecosystems or high biodiversity, have suffered. Soil carbon stores have declined, with implications for climate regulation, and the loss of species-rich lowland grasslands and wetlands has reduced biodiversity in many parts of Europe. The long-term consequences of this are likely to be severe. Sustaining production levels without recourse to natural processes for nutrient cycling and disease and pest regulation will be increasingly difficult and costly.

In addition, the abandonment of poorer quality agricultural land, where natural conditions discouraged intensification, has also had a damaging impact. This has often been in areas where low intensity farming had created high-nature-value farmland.[25] For example, it has been estimated that 32 habitats protected under the Habitats Directive are threatened by the abandonment of

[19] European Commission (2011a) Our Life Insurance, Our Natural Capital, 2.

[20] European Commission (2011a) Our Life Insurance, Our Natural Capital, 6.

[21] See, for example, Jack (2001), pp. 1–18.

[22] Millennium Ecosystem Assessment (2005), p. 6.

[23] See, variously, Jack (2009), Chap. 2; Harvey (1997); Shoard (1982).

[24] European Academies Science Advisory Council (2009), p. 1.

[25] See, for example, European Environment Agency (2004).

extensive grazing.[26] Overall, farmland bird and grassland butterfly populations provide a measure of agriculture's biodiversity impact. The *European Environment Agency* reports that European farmland bird populations declined sharply until the mid-1990s and remain low in comparison to levels recorded in the early1980s.[27] It also notes that grassland butterfly populations continue to decline sharply and have fallen by almost 70 % since 1990.[28]

Ultimately, the problem is one of market failure.[29] The *Millennium Ecosystem Assessment* reported that only four of the 24 ecosystem services it identified had become more plentiful over the last 50 years.[30] It is no coincidence that these four, crops, livestock, aquaculture and global carbon sequestration, are all ecosystem services for which well-identified markets have been established in which they can be bought and sold. In other cases, ecosystem services are generally freely available as public goods. Landowners receive no financial incentive to provide them and consequently have no incentive to continue protecting or supporting the ecosystems that produce them. The challenge for the European Union has therefore been twofold: firstly, to ensure that agricultural production policies discourage activities that damage ecosystems producing essential ecosystem services and, secondly, to encourage and support agricultural practices that benefit those ecosystems.

4 Ecosystem Services and the CAP

Reforming the CAP. The European Union's initial agricultural production policies encouraged widespread environmental damage as a result of the emphasis placed upon maximising production potential. However, the 2013 reforms to the CAP are one of several significant reforms that have been made to these policies over recent decades. These began with incremental changes within individual production sectors in the 1970s and 1980s and were followed by a number of reform packages, which introduced changes across a range of produce groups,[31] culminating in the 2013 reform package. Equally, the 1986 Single European Act introduced a treaty obligation to integrate environmental protection measures into all European policies. Today, this continues to be a core European Union objective. Article 11 of the Treaty on the Functioning of the European Union provides that "environmental protection requirements must be integrated into the definition and implementation

[26] European Environment Agency (2006b), p. 42, referring to habitats protected by Council Directive 92/43/EEC of 21 May 1992 on the conservation of natural habitats and of wild fauna and flora, [1992] OJ L206/7.

[27] European Environment Agency (2010), p. 6.

[28] European Environment Agency (2010), p. 6.

[29] See Ruhl et al. (2007), p. 63.

[30] Millennium Ecosystem Assessment (2005), p. 7.

[31] See, Jack (2009), pp. 51–62 and Cardwell (2004).

of the Union policies and activities, in particular with a view to promoting sustainable development". However, the effective integration of environmental measures remains elusive. The *European Environment Agency* has pointed out that key challenges continue to exist in dealing with issues such as soil erosion, water quality and quantity and the protection of biodiversity.[32] The reality is that environmental issues have had to compete with other issues at each reform round, such as the protection of the European Union budget, the liberalisation of world trade and preparations for European Union enlargement.[33]

Cross-compliance. One of the most obvious changes to agricultural production policy over recent decades has been the introduction of direct payments to farmers involved in major sectors such as dairy, beef, sheep and cereal farming. These payments, now known as "the basic payment scheme",[34] were originally introduced to compensate farmers for reforms that reduced commodity prices and market guarantees for farmers.[35] Viewed from an environmental perspective, these direct payments gave the European Union the opportunity to introduce cross-compliance obligations, which link farmers' eligibility to receive the payments to an obligation to comply with a number of environmental conditions.

Cross-compliance has been a compulsory element of farmer eligibility for direct payments since 2005.[36] The cross-compliance requirements are currently set out in Regulation 1306/2013.[37] They provide that farmers receiving basic payments must comply with the statutory management requirements imposed under 13 European Union regulations and directives concerned with environmental protection, public, animal and plant health and with animal welfare. In addition, they must also comply with standards of good agricultural and environmental condition (the "GAEC standards") that Member States have been required to develop from a common framework set out in the Regulation.

In terms of statutory management requirements, Regulation 1306/2013 requires farmers to comply with the obligations imposed upon landowners and occupiers by

[32] See European Environment Agency (2006a) and European Environment Agency (2005).

[33] See Jack (2001).

[34] Under Regulation 1307/2013 of the European Parliament and the Council of 17 December 2013 establishing rules for direct payments to farmers under support schemes within the framework of the common agricultural policy, OJ [2013] L347/608.

[35] See Jack (2009), p. 54.

[36] It became compulsory under Council Regulation 1782/2003 of 29 Sept 2003 establishing common rules for support schemes under the common agricultural policy, OJ [2003] L270/1.

[37] Regulation 1306/2013 of the European Parliament and the Council of 17 December 2013 on the financing, management and monitoring of the common agricultural policy, OJ [2013] L347/549.

national laws implementing each of the following European Union environmental law directives:

- the 1991 Nitrates Directive;[38]
- the 2009 Wild Birds Directive;[39]
- the 1992 Habitats Directive.[40]

Regulation 1306/2013 also requires compliance with GAEC standards introduced by Member States on the following environmental issues:

GAEC 1: Establishing buffer strips along water courses;
GAEC 2: Where use of water for irrigation is subject to authorisation, compliance with authorisation procedures;
GAEC 3: Prohibiting direct discharges into groundwater and preventing indirect pollution of groundwater through discharge on the ground and percolation through the soil of dangerous substances, as listed in the Annex to Directive (EEC) 80/68 in its version in force on the last day of its validity, as far as it relates to agricultural activities;
GAEC 4: Maintaining minimum soil cover;
GAEC 5: Conducting minimum land management, reflecting site specific conditions, to limit erosion;
GAEC 6: Maintaining soil organic matter level through appropriate practices including a ban on burning arable stubble, except for plant health reasons;
GEAC 7: Retaining landscape features, including where appropriate, hedges, ponds, ditches, trees in line, in group or isolated, field margins and terraces, and including a ban on cutting hedges and trees during the bird breeding and rearing season and, as an option, measures for avoiding invasive plant species.

Taken at face value, these measures have potential to contribute towards environmental protection, in general, and the protection of ecosystem services, in particular. Closer inspection reveals important gaps in this protection.[41] In particular, the Water Framework Directive is an important absentee from the statutory management requirements.[42] The European Parliament and Council merely annexed a statement to Regulation 1306/2013 inviting the Commission to monitor the transposition and implementation of both the Water Framework Directive and the Pesticides Directive,[43] with a view to including them within future cross-

[38] Council Directive 91/676 of 12 December 1991 concerning the protection of waters against pollution caused by nitrates from agricultural sources, OJ [1991] L375/1.

[39] Directive 2009/147 of the European Parliament and of the Council of 30 Nov 2009 on the conservation of wild birds, OJ [2010] L20/7.

[40] Council Directive 1992/43 of 21 May 1992 on the conservation of natural habitats and of wild flora and fauna, OJ [1992] L206/7.

[41] See further Jack (2012), pp. 258–273 and 263.

[42] Directive 2000/60 of the European Parliament and of the Council of 23 Oct 2000 establishing a framework for Community action in the field of water policy, OJ [2000] L327/1.

[43] Directive 2009/128 of the European Parliament and of the Council of 21 Oct 2009 establishing a framework for Community action to achieve the sustainable use of pesticides, OJ [2009] L309/71.

compliance conditions when they have been implemented in all Member States and the obligations directly imposed on farmers have been identified. In the case of the Pesticides Directive, not all obligations have fallen due. For example, it provides for pesticide application equipment to be inspected by December 2016.[44] However, Member States ought to have had all measures in place to implement the Water Framework Directive by December 2012. Its exclusion from cross-compliance is therefore particularly disappointing. Regulation 1306/2013 has also removed compliance with the Sewage Sludge Directive from the list of statutory management requirements.[45] Additionally, Regulation 1306/2013 also omits a Commission proposal that the GAEC standards should include requirements that farmers retain wetlands and carbon-rich soils,[46] which play an important role in supporting ecosystem services such a flood prevention and climate regulation and in protecting biodiversity.

Ultimately, cross-compliance is by no means a universal measure in European farming. It does not apply in sectors, such a pig and poultry farming, that have traditionally not benefited from direct payments under the CAP. Additionally, it does not apply to farmers receiving up to €1,250 per annum in direct payments and participating in the European Union's small farmer scheme, who are exempt from cross-compliance under Regulation 1306/2013.[47] These farmers must still comply with national laws implementing European Union environmental directives but have no obligation to meet the GAEC standards. In particular, therefore cross-compliance will have limited impact in areas where small farm structures predominate.

A number of other regulatory weaknesses also limit the effectiveness of cross-compliance as a tool to protect ecosystem services. In the first place, Regulation 1306/2013 merely requires Member States to ensure they conduct "a minimum level of on-the-spot checks for an effective management of the risks".[48] Previously, Member States were required to inspect only 1 % of those who claimed basic payments, to ensure compliance with the cross-compliance requirements.[49] It seems likely that this again will become the accepted minimum standard under Regulation 1306/2013. Where an inspection reveals noncompliance with the cross-

[44] Pesticides Directive, Art. 8(2).

[45] Council Directive 86/278 of 12 June 1986 on the protection of the environment, in particular the soil, when sewage sludge is used in agriculture, OJ [1986] L181/6.

[46] See European Commission (2011c), Proposal for a Regulation of the European Parliament and of the Council on the financing, management and monitoring of the common agricultural policy, COM (2011) 628 final/2 at Annex II to the proposed regulation.

[47] See Regulation 1306/2013, Art. 92. For details of the small farmer scheme see Council Regulation 1307/2013, Arts. 61-63.

[48] Regulation 1306/2013, Art. 59(5).

[49] Under Commission Regulation 1122/2009 of 30 Nov 2009 laying down detailed rules for the implementation of Council Regulation 73/2009 as regards cross compliance, modulation and the integrated administration and control system under the direct support schemes for farmers provided by that Regulation. OJ [2009] L316/65, Art. 50.

compliance conditions, then Member States were previously required, as a general rule, to withhold 3 % of the farmer's payments where this is considered to result from negligence and 20 % where it is considered to have been an intentional breach.[50] However, they had discretion to reduce this to 1 and 15 %, respectively, or increase it to 5 and 100 %, depending on the extent of the noncompliance, its severity and the degree of permanence of its effects. The Commission has proposed that similar provisions should again apply in relation to basic payments made under Regulation 1306/2013.[51] However, as the Court of Auditors has pointed out, these deductions reflect neither the cost of compliance nor the consequences of noncompliance.[52] Equally, as regards negligent noncompliance, the Court of Auditors noted that Member States it had audited tended to apply flat rate reductions of 1 %.[53] Member States have also been given discretion to decide not to apply cross-compliance penalties that do not exceed one hundred euros.[54] The Court of Auditors found that the average penalty in the Netherlands was one hundred euros and that 65 % of penalties in Finland and 94 % of those levied in Poland were less than this amount.[55] When combined with the low inspection rates, such limited financial penalties raise serious questions as to whether cross-compliance plays any practical role in protecting ecosystem services. Indeed, this is further emphasised by a serious shortcoming that the Court also uncovered in the implementation of cross-compliance requirements within Member States and in national inspection systems.[56]

The Greening Component. Arguably, the most innovative aspect of the 2013 CAP reform measures was the introduction of payments for agricultural practices beneficial for the climate and the environment, which is generally referred to as "the greening component". It was also the most contested issue in the reform process. In implementing the greening component, Regulation 1307/2013 introduces a mandatory requirement that all farmers receiving basic payments whose farms meet particular thresholds should engage in environmental practices beneficial for the

---

[50] Council Regulation 73/2009 of 19 Jan 2009 establishing common rules for direct support schemes for farmers under the common agricultural policy, OJ [2009] L30/16 Art. 6.

[51] European Commission (2014) Proposals for a Commission delegated regulation supplementing Regulation 1306/2013 of the European Parliament and the Council with regard to the integrated administration and control system and conditions for refusal or withdrawal of payments and administrative penalties applicable to direct payments, rural development support and cross-compliance, COM(2014), 1476 final, Art. 39.

[52] European Court of Auditors (2009) Is Cross Compliance an Effective Policy? Special Report 8/2008. Office for Official Publications of the European Communities, Luxembourg, Para 67.

[53] European Court of Auditors (2009), Para. 71. Based upon an audit conducted in Finland, France, Greece, the Netherlands, Poland, Portugal and Slovenia between Sept and Nov 2007. See similarly European Commission (2007), Report on the Application of the System of Cross-Compliance, COM (2007) 147 final, para 4.

[54] See Regulation 1306/2013, Art. 97(3).

[55] European Court of Auditors (2009), Para. 74.

[56] European Court of Auditors (2009), Paras 22–44 and 59–66. See also Birdlife International (2010).

climate and the environment.[57] In turn, one-third of the monies allocated to Member States by the European Union for basic payments are to be used to pay them for doing so. The Regulation provides that the greening component makes payments in return for "simple, generalised, non contractual and annual actions that go beyond cross-compliance".[58] It sets out three particular obligations, with farmers being required to comply with each that is relevant to their farm:[59]

(a) to comply with crop diversification requirements;
(b) to maintain permanent grassland;
(c) to have ecological focus areas on farmland.

Alternatively, Member States may exempt farmers from these obligations on the basis that their participation in an agri-environment-climate change scheme (see below) provides an equivalent or higher level of benefit for the climate and the environment.[60] Exemptions can also be granted on the basis that farmers are participating in national or regional environmental certification schemes that Member States have introduced, in accordance with Regulation 1307/2013, as an alternative to the greening component requirements and that, again, provide an equivalent or higher level of benefit for the climate and the environment.[61]

The greening component has a prominent role in ensuring that the European Union achieves its biodiversity objective of maximising "areas under agriculture across grasslands, arable land and permanent crops that are covered by biodiversity-related measures under the CAP".[62] Equally, it also has the potential to protect and support ecosystems producing important ecosystem services. In proposing the concept, the European Commission suggested that it would "ensure that all farms deliver environmental and climate benefits through the retention of soil carbon and grassland habitats associated with permanent pasture, the delivery of water and habitat protection by the establishment of ecological focus areas and the improvement of the resilience of soil and ecosystems through soil diversification".[63] In practice, however, a number of weaknesses seem likely to limit its effectiveness in achieving these goals.

To begin with, the greening component measures introduced by Regulation 1307/2013 are more limited in scope than those initially proposed by the Commission. The Commission had proposed that all farmers with more than 3 ha of arable land would be required to practice crop diversification by growing at least three

[57] Regulation 1307/2013, Arts. 43-47.

[58] Regulation 1307/2013, Recital 37.

[59] Regulation 1307/2013, Art. 43(2).

[60] Regulation 1307/2013, Art. 43(3).

[61] Regulation 1307/2013, Art. 43(3).

[62] European Commission (2011a), Our Life Insurance, Our Natural Capital, 6.

[63] European Commission (2011b) Proposal for a Regulation of the European Parliament and of the Council establishing rules for direct payments to farmers under support schemes within the framework of the common agricultural policy, COM(2011), 625 final/2, 3.

crops.[64] In contrast, Regulation 1307/2013 sets the threshold at 10 ha.[65] Farmers with between 10 and 30 ha of arable land must grow at least two different crops on this land, with the main crop not covering more than 75 %.[66] Those with more than 30 ha of arable land are required to grow at least three crops, with the main crop not covering more than 75 % of that land and the two main crops not covering more than 95 %.[67] The Commission had also proposed that farmers receiving basic payments should be required to retain all permanent grassland on their farms.[68] However, Regulation 1307/2013 has again introduced more limited measures.[69] Farmers are prohibited from converting or ploughing permanent grasslands within areas designated by Member States as being environmentally sensitive. Member States, in turn, are only obliged to designate those located within areas covered by the Habitats and Wild Birds Directives. They then have discretion to also designate other grasslands. Where they opt not to do so, the only protection these grasslands gain under Regulation 1307/2013 is a requirement that Member States ensure that the ratio of permanent grassland to agricultural area does not decrease by more than 5 %.[70] This, however, provides no specific protection for high-nature-value or carbon-rich areas that have not been designated under the Habitats or Wild Birds Directive. Finally, in relation to ecological focus areas, the Commission had proposed that all farmers should ensure that at least 7 % of their farmland was an ecological focus area.[71] Regulation 1307/2013, however, restricts the requirement to establish ecological focus areas to farmers who have more than 15 ha of arable land.[72] It also limits the extent of these areas to just 5 % of that arable area whilst providing for the possibility that this may in future be increased to 7 % following a Commission evaluation of the measure.[73] Far from applying to all farmers, as the Commission had imagined, the reality is that many farmers will actually be excluded from the scope of the measures introduced by Regulation 1307/2013.

The additional environmental benefits provided in return for the greening component payments are also unclear. Under Regulation 73/2009, for example, Member States had previously been required to implement a GAEC condition obliging

[64] European Commission (2011b), Proposal for a Regulation of the European Parliament and of the Council establishing rules for direct payments, proposed Art. 30.

[65] Regulation 1307/2013, Art. 44(1).

[66] Regulation 1307/2013, Art. 44(1).

[67] Regulation 1307/2013, Art. 44(1).

[68] European Commission (2011b), Proposal for a Regulation of the European Parliament and of the Council establishing rules for direct payments, proposed Art. 31(1).

[69] Regulation 1307/2013, Art. 45(1).

[70] Regulation 1307/2013, Art. 45(2) and (3).

[71] European Commission (2011b), Proposal for a Regulation of the European Parliament and of the Council establishing rules for direct payments, proposed Art. 32(1).

[72] Regulation 1307/2013, Art. 46(1).

[73] Regulation 1307/2013, Art. 46(1).

farmers to retain permanent pasture on their farms.[74] The greening component now pays farmers to provide potentially less protection to permanent grasslands. Equally, in relation to ecological focus areas, Regulation 1307/2013 identifies a range of farm areas that are eligible to be included within these areas.[75] These include, *inter alia*, fallow land, terraces, buffer strips and landscape features. Given that cross-compliance currently also includes a GAEC condition requiring farmers to retain landscape features, it is not clear what additional benefit will be achieved by also designating them as being ecological focus areas.

Questions can also be posed concerning the consequences of failing to comply with the greening component. Given that all farmers receiving basic payments who meet the thresholds set out above must participate in the scheme, it could be argued that noncompliance should place their entitlement to this basic payment at risk. In contrast, the Commission's legislative proposals indicate that farmers in this situation will only lose a proportion of the payment otherwise due to them under the greening component scheme.[76] As the *Institute for European Environmental Policy* has previously argued, this effectively separates the greening component from the basic payment and turns it, in practice, into a voluntary measure.[77]

Finally, the greening component is built around generalised environmental practices that can be applied across the European Union. In practice, greater environmental outcomes can be expected from measures that are tailored to the specific environmental situation existing in a particular region. In theory, the concept of national or regional certification schemes enables Member States to design such measures. Member States that choose to introduce such schemes are required to obtain Commission approval for their schemes. In principle, this should enable the Commission to ensure that Member States implement effective environmental requirements. However, Regulation 1307/2013 is silent as to how this evaluation should be conducted. It is by no means clear that the Commission has the necessary administrative capacity to evaluate a range of national certification schemes. Indeed, one study has suggested that there would be considerable practical difficulty involved in assessing the equivalence both of differing farming practices required under national schemes and of their environmental impact.[78] It concluded that[79] "while the concept of equivalence may sound like a reasonable and convenient approach in theory, the practical issues with its application are likely to lead to far greater administrative complexity and cost, both for Member

[74] Council Regulation 73/2009, Art. 6 and Annex III.

[75] Regulation 1307/2013, Art. 46(2).

[76] European Commission, Proposals for a Commission Delegated Regulation supplementing Regulation 1306/2013 of the European Parliament and of the Council, proposed Arts 24–27.

[77] Institute for European Environmental Policy (2013), Plenary Vote on CAP reform package: a bitter sweet victory for the environment, press release 19 March 2013. http://www.cap2020.ieep.eu. Accessed 28 May 2014.

[78] Hart and Menadue (2013), p. 26.

[79] Hart and Menadue (2013), p. 28.

States and within the Commission, with arguably little additional environmental benefit".

Rural Development Policy. As well as its agricultural production policies, the early CAP also introduced measures to help support farm incomes by promoting structural reforms across the farming sector and providing income support for farmers located on poor quality agricultural land. These measures stemmed initially from a number of directives adopted in the 1970s.[80] In more recent years, the realisation that agriculture was only a small part of a larger rural economy, and was no longer the predominant employer in most rural areas, stimulated the European Union to develop a more broadly based rural development policy.[81] Today, this rural development policy forms the second pillar of the modern CAP and is principally governed by Regulation 1305/2013.[82]

Regulation 1305/2013 stipules that "restoring, preserving and enhancing ecosystems related to agriculture" and "promoting resource efficiency and supporting the shift towards a low carbon and climate resilient economy" are two of the European Union's six policy priorities for its rural development policy in the period 2014–2020.[83] This shows the European Union's awareness of the role the policy can have in protecting ecosystem services, with the shift towards a climate-resilient economy potentially including measures supporting soil carbon storage.

Regulation 1305/2013 follows an identical path to its predecessor, Regulation 1698/2005,[84] in establishing a menu of twenty-six potential measures and requiring Member States to develop national or regional rural development plans that adopt and implement those that are most suited to their own rural areas. A number of these measures could potentially play a role in protecting and supporting ecosystems linked to essential ecosystem services. These include

- payments to support afforestation (the establishment of new forestry);
- payments for agro-forestry (the establishment and maintenance of trees on agricultural land);
- financial support for organic farming;
- Natura 2000/Water Framework Directive Payments (compensating farmers for costs and income loss associated with restrictions introduced by the Wild Birds, Habitats and/or Water Framework Directives);
- financial support for farmers in mountain areas and areas facing natural constraints (discouraging land abandonment through payments to farmers located on poor-quality agricultural land);

[80] See, Jack (2009), p. 8–12.

[81] See, Jack (2009), p. 12–14.

[82] Regulation 1305/2013 of the European Parliament and of the Council of 17 December 2013 on support for rural development by the European Agricultural Fund for Rural Development, OJ [2013] L347/487.

[83] Regulation 1305/2013, Art. 5.

[84] Council Regulation 1698/2005 of 20 Sept 2005 on support for rural development by the European Agriculture Fund for Rural Development, OJ [2005] L277/1.

- financial support through the agri-environment-climate-change scheme.

Of these measures, only the agri-environment-climate-change scheme is compulsory and must be included in every Member State rural development plan. It also accounts for one-third of the European Union's entire expenditure on rural development.[85] As such therefore, the agri-environment-climate-change scheme is the principal rural development measure available to protect and support ecosystem services.

The Agri-Environment-Climate-Change Scheme. The agri-environment-climate-change scheme ("the agri-environment scheme") has been a compulsory element of the European Union's rural development scheme since 1992.[86] Member States must therefore incorporate agri-environmental measures into their rural development programmes. At farm level, however, farmer participation is voluntary. The agri-environment scheme provides a mechanism through which national agriculture authorities offer farmers voluntary management agreements. Those who decide to participate enter a contractual arrangement under which they receive annual payments in return for providing environmental services over a period of 5–7 years.[87] The agri-environment scheme therefore provides a mechanism to address the problem of market failure highlighted earlier, by paying farmers to protect and restore ecosystems providing ecosystem services. This is acknowledged in the strategic guidelines for the European Union's 2007–2013 rural development programme:[88]

> European citizens expect farmers to respect mandatory standards. But many also agree that farmers should be remunerated for signing up to commitments which go further, delivering services that the market will not provide alone, particularly when focused on specific resources of particular importance in the context of agriculture and forestry, such as water and soil.

Some 21 % of farmland across the EU-27 is now enrolled in the scheme.[89] However, enrolment is higher in the EU-15, where 25 % of farmland is enrolled, as against only 9.7 % in the newer EU-12. Member States enjoy considerable flexibility in designing their agri-environmental schemes. Regulation 1305/2013 provides for them to do so "in accordance with their national, regional or local specific needs and priorities" and to make payments to farmers who "undertake operations consisting of one or more agri-environmental commitments".[90] In practice, the

[85] European Commission (2012) Rural Development in the European Union: Statistical and Economic Information Report 2012. Office for Official Publications of the European Union, Luxembourg, p. 369.

[86] Council Regulation 2078/92, OJ [1992] L215/1, first made it a compulsory element of every Member State rural development plan.

[87] See Regulation 1305/2013, Art. 28(5).

[88] Council Decision 2006/144 of 20 Feb 2006 on Community strategic guidelines for rural development (programming period 2007-2013), OJ [2006] L55/20, Para. 3.2. (i).

[89] European Commission (2011e), Rural Development in the European Union: Statistical and Economic Information Report 2011 (Commission, 2011), 27.

[90] Regulation 1305/2013, Arts. 28(1) and 28(2).

schemes themselves tend to fall into one of two general categories, being either broad (entry level) or narrow (higher level) schemes.[91] Broad schemes generally cover a wide area and are designed to attract a large number of entrants by making modest demands and providing relatively small payments in return. These account for a vast majority of farmers participating in the agri-environment scheme. In contrast, narrow schemes target site-specific environmental issues. They have fewer farmers enrolled within them but make more substantial demands and provide higher payments in return. Overall, the scheme has the potential to make a considerable contribution towards protecting ecosystem services by introducing management measures that protect landscapes, soil and water quality and biodiversity.[92] Unfortunately, however, its potential has been restricted by a number of important practical considerations.

In particular, although large areas of farmland have been enrolled, it is by no means clear that the agri-environment scheme has attracted the land that would be most valuable in terms of protecting ecosystem services. This can be illustrated by comparing enrolment rates within individual Member States against the amount of high-nature-value land that exists there. These are areas with predominantly low-intensity farming systems that support a diverse range of wildlife and can play an important role in protecting ecosystem services.[93] The European Commission has identified ten Member States in which more than one-third of farmland is considered to be of high nature conservation value: Austria, Bulgaria, Cyprus, Finland, Greece, Italy, Portugal, Romania, Spain and Slovenia.[94] Ninety per cent of farmland in Finland is enrolled in the agri-environment scheme, as is 70 and 40 % of farm land, respectively, in Austria and Slovenia.[95] In each of the other Member States, however, less than 10 % of farmland has been enrolled.[96] This illustrates that Member States have often failed to capitalise upon the scheme's potential as a means to protect and support ecosystem services. This, in turn, is due to a number of factors: firstly, poor design by Member States in failing to use national agri-environment schemes to target and address environmental priorities. The Court of Auditors, for example, has identified situations in which national agri-environmental programmes had not sought to address environmental pressures that had specifically been identified in the Member States' own rural development

[91] See Buller (2000), p. 233.

[92] See European Commission (2005) and Kantor Management Consultants (2011), 202. http://www.ec.europa.eu/agriculture/evaluation/rural-development-reports/ex-poste-evaluation-rdp-2000-2006_en.htm. Accessed 30 May 2014.

[93] See Baldock and Beaufoy (1993) or European Environment Agency (2004).

[94] European Commission (2011e) Rural Development in the European Union: Statistical and Economic Information Report 2011, 159.

[95] European Commission (2011e) Rural Development in the European Union: Statistical and Economic Information 2011, 27.

[96] European Commission (2011e) Rural Development in the European Union: Statistical and Economic Information 2011, 27.

programmes.[97] The European Union's biodiversity strategy provides for Member States to have mapped and assessed the state of the ecosystems and ecosystem services in their territory by 2014.[98] As the author has argued elsewhere, this could have provided a framework for Member State agri-environmental schemes, with Member States being required to construct those schemes around the core ecosystem services identified by their national assessment.[99] Secondly, unlike the basic payments and greening component payments, agri-environmental scheme payments are not fully funded by the European Union. Although the European Union will fund up to 85 % of rural development expenditure in less-developed regions, in most regions it provides a maximum of 53 % of eligible expenditure.[100] Consequently, participation rates in agri-environment schemes also reflect the extent to which Member States are prepared to provide matched funding to support the schemes. This situation is further compounded by the flexibility that Regulation 1307/2013 provided for Member States.[101] On the one hand, they were authorised to decide, by 31 December 2013, to use up to 15 % of the monies allocated to them for the basic payment and greening component schemes to increase their spending on rural development policy measures. On the other hand, they were also authorised to do the reverse, using 15 % of the monies allocated to them for rural development spending to increase direct payments. Indeed, twelve Member States (Bulgaria, Estonia, Spain, Latvia, Lithuania, Poland, Portugal, Romania, Slovakia, Finland, Sweden and the United Kingdom) are authorised to syphon up to 25 % of their rural development budget for this purpose. Although the former situation would help Member States to fund additional agri-environmental spending, the latter threatens to decimate the agri-environment scheme in individual Member States. The situation is particularly critical in Bulgaria, Spain, Romania and Slovakia, four Member States with large areas of high-nature-value land but low enrolment in the agri-environment scheme. In the past, various studies have recommended that direct payments, such as the current basic payment, should simply be abolished and replaced with payments for environmental services.[102] Given our knowledge now of the ecosystem service benefits that our environment provides, it may be time to revisit these suggestions.

Allied to the fact that Member State expenditure under the agri-environment scheme is only partially reimbursed by the European Union, an additional problem arises from the fact that Regulation 1305/2013 also requires that the payments made to farmers are based upon the additional costs they have incurred and income they

[97] Court of Auditors (2011) Is Agri-Environment Support Well Designed and Managed? Special Report 7/2011. Publications Office of the European Union, Luxembourg, Para 30.

[98] European Commission (2011a), Our Life Insurance: Our Natural Capital, Target 2, action 5.

[99] Jack (2012), p. 271.

[100] See Regulation 1305/2013, Art. 59.

[101] Regulation 1307/2013, Art. 14.

[102] See Buckwell et al. (1997), and Jenkins (1990).

have foregone as a result of enrolling in an agri-environmental scheme.[103] This is required under the GATT Agriculture Agreement, which stipulates that payments under government agri-environment schemes should be based upon the costs incurred and income foregone by their participants.[104] However, in practice, it values environmental resources in agricultural production terms. As is the case with greening component payments, agri-environment payments take no account of the value of the environmental resources that are being managed. In reality, areas with high environmental value often have limited agricultural productivity. Farm abandonment in these areas has become a major social and environmental threat, precisely because of the low incomes obtained from agriculture. Basing payments under the agri-environment scheme on costs and profit foregone limits its ability to address this issue. It would seem more effective for payments to be based upon the value of the ecosystem services that they were protecting.[105]

Finally, the question arises of the policy cohesion that exists between the requirements of cross-compliance and the greening component within agricultural production policy and the requirements introduced by the agri-environment scheme within rural development policy. On paper at least there is a clear separation between all three. The European Union's policy has consistently been that the cross-compliance provisions establish basic environmental standards that all farmers should be expected to provide without payment.[106] The greening component is then intended to build upon this, by introducing payments for "simple, generalised, non-contractual and annual actions that go beyond cross compliance".[107] In turn, Regulation 1305/2013 provides that the agri-environment scheme only provides payments for commitments that go beyond those introduced by both cross-compliance and the greening component.[108] It is not clear, however, that this separation always exists in practice. Inconsistencies have created the situation in which farmers in one Member State receive agri-environment payments for management actions that are cross-compliance requirements in others.[109] The Court of Auditors, for example, has highlighted the fact that agri-environment provisions concerning the introduction of buffer strips along water courses in Poland could have been implemented through cross-compliance.[110] Equally, as this author has previously highlighted, there also appears to be opportunity for duplication under the greening component of the basic payment scheme and the agri-environment

---

[103] See Regulation 1305/2013, Art. 28(6.)

[104] 1994 GATT Agriculture Agreement, Annex 2, Para. 12(b). http://www.wto.org/english/docs_e/legal_e/14-ag.pdf. Accessed 30 May 2014.

[105] Jack (2012), p. 272.

[106] See, for example, European Commission, Directions Towards Sustainable Agriculture, COM (99) 22 final, 28.

[107] Regulation 1307/2013, recital 37.

[108] Regulation 1305/2013, Art. 28(3).

[109] Jack (2015).

[110] Court of Auditors (2011), Para. 39.

scheme.[111] The Commission had proposed that each scheme should be kept distinct so that farmers would only receive agri-environment payments for measures that went beyond the requirements of the greening component.[112] The Council, however, suggested that farmers enrolled in agri-environment schemes should be deemed to be 'green by definition'.[113] This would have exempted them from further obligations under the greening component but would have enabled them to receive both the greening component and agri-environment payments. Ultimately, Regulation 1307/2013 enables Member States to treat commitments under the agri-environmental scheme as being equivalent to the practices required under the greening component and pay the greening component payment without asking farmers to meet additional environmental management requirements.[114] However, the Member States must then deduct these greening component payments from the payments the farmers would otherwise receive under the agri-environment scheme.[115] This avoids double payments and, in principle at least, maintains a separation between both schemes. But it appears highly bureaucratic and may prove difficult to police, with the result that overlaps actually occur in practice.

5 Ecosystem Services Beyond the CAP

Schemes making payments to landowners who provide ecosystem services are becoming increasingly common across the world.[116] Like the European Union's agri-environment-climate-change scheme, the vast majority of these schemes are government funded. However, a number of schemes funded by the beneficiaries of particular ecosystem services have also been introduced. In Panama, for example, insurance and shipping companies have financed payments to encourage farmers in the area surrounding the Panama Canal to reforest their land.[117] By both absorbing rainwater and preventing soil erosion, this reduces the risk that the canal will be closed by floodwaters or siltation.[118] The insurance companies gain by receiving fewer claims, whilst shipping companies benefit from paying lower insurance premiums.[119] Schemes to protect drinking water quality, by paying for upstream land management, also operate in a number of countries, including Brazil, Ecuador,

[111] Jack (2015).
[112] European Commission (2011d), Art. 27(3).
[113] See Jack (2013).
[114] Regulation 1307/2013, Art. 43(a).
[115] Regulation 1305/2013, Art. 28(6).
[116] See, for example, Salzman (2005), pp. 870–961 and Pagiola et al. (2004).
[117] Irwin and Ranganathan (2008), p. 21–69, 62.
[118] Irwin and Ranganathan (2008), p. 62.
[119] Irwin and Ranganathan (2008), p. 62.

Columbia and Japan, funded either by water companies themselves or through a charge on water supplies.[120]

Agriculture within the European Union similarly has the potential to produce a range of ecosystem services for private markets and is in fact already doing so. Recent severe flooding in the United Kingdom, for example, has served to highlight the role that upland heathlands play in absorbing rainwater and reducing downstream river levels. They also have the reverse function during droughts, releasing waters absorbed during wetter periods. In England, the South West Water company's Exmoor Mires project is paying farmers to block or remove upland drains in order to restore up to 2,000 ha of peat to its former condition.[121] Its principal objective in doing so is to maintain water levels during periods of drought, but the scheme is also likely to have associated benefits in flood prevention. Elsewhere, in France, Nestlé has also used a payments scheme to safeguard its Vittel mineral waters business, when nitrate and pesticide pollution threatened its water supply. The company signed all farmers in the water catchment to long-term contracts under which they received payments and technical support in return for avoiding practices that endangered water quality.[122] Water companies in England spent £189 million removing nitrates and £92 million removing pesticides from water supplies, between 2004–2005 and 2008–2009.[123] This raises the question of whether payments for ecosystem services might offer a cheaper long-term means of water purification. In turn, it also highlights the potential that exists outside the scope of the CAP for markets to develop to pay for the delivery of ecosystem services.

6 Concluding Remarks

The *Millennium Ecosystem Assessment* highlighted both the extent of humanity's reliance upon ecosystem services and also the detrimental impact that human development has had upon the ecosystems that produce them. Within the European Union, the CAP traditionally focused on the production of food and raw materials. However, this focus on maximising agricultural production potential left other ecosystem services to decline, to the extent that future agricultural productivity was also placed at risk.

The *Millennium Ecosystem Assessment* called for significant changes to be made in policies, institutions and practices to provide protection for ecosystem services. Internationally, this call is reflected in the goals set by the parties to the Convention on Biological Diversity. Within the European Union, it is similarly recognised by

[120] McNeely (2009), p 135–150, 142.

[121] See Couldrick (undated).

[122] Perrot-Maître (2006).

[123] National Audit Office (2010), p. 13.

the European Union's own biodiversity strategy. However, putting broad principles into practice creates a much greater challenge. In terms of the CAP, the task for European Union policymakers has been twofold: on the one hand, to ensure that agricultural production policies discourage activities that damage ecosystems producing important ecosystem services and, on the other hand, to encourage and support practices that benefit these ecosystems.

The modern CAP is very different from the initial policy that caused so much damage to ecosystem services. The recent 2013 reform package is just one of a number of reforms that have introduced change. At the same time, however, environmental issues have only been one of a number of policy considerations that have sought to influence these reforms. This raises the question of whether, amongst these competing issues, sufficient attention has been given to the imperative of protecting ecosystem services.

The CAP today is divided into two pillars. One concentrates on agricultural production policies, whilst the second is concerned with the broader rural development policy. Within agricultural production policy, the introduction of direct payments, now known as the "basic payment", was one of the most obvious changes introduced by past reforms. It gave the European Union the opportunity to link farmers' eligibility to receive these payments with the obligation to comply with various environmental conditions through cross-compliance. The introduction of the greening component, under Regulation 1307/2013, also adds an additional environmental component to the direct payment regime. In contrast, within rural development policy, the agri-environment-climate-change scheme provides a means to pay farmers to protect and restore ecosystems producing important ecosystem services.

On paper therefore CAP now contains a range of measures capable of protecting ecosystem services. In practice, however, the effectiveness of these measures in achieving this task appears much less certain. At European Union level, the failure to incorporate the Water Framework Directive within the cross-compliance statutory management requirements or require Member States to adopt GAEC standards to protect wetlands and carbon-rich soils undermines the value of the cross-compliance mechanism in protecting ecosystem services. This is further compounded by the fact that the small farmer exemption effectively means that cross-compliance measures will have little relevance in areas in which small farm structures predominate. At a Member State level, the governance failings exposed by the *Court of Auditors* also cast a deeper shadow over the fitness for purpose of cross-compliance as a whole. In contrast, it is perhaps too early to pass judgment on the greening component. However, the measures adopted by the European Parliament and the Council are much more limited than those initially proposed by the Commission, which can only mean that they will have less influence in practice in protecting ecosystem services. Equally, the fact that farmers who fail to comply with greening component obligations seem destined to only lose a portion of their greening component payments effectively turns the greening component into a voluntary measure. In turn, these factors will severely limit the ability of the greening component to serve its intended role as a central strand of the European

Union's strategy for protecting ecosystem services. Elsewhere, the agri-environment scheme will need to be better targeted towards the protection of key ecosystem services in each Member State if it too is to play a more effective role. There is also a need to address the funding disparities that witness the European Union budget fully underwriting basic payments and greening component payments providing limited environmental returns through cross-compliance and the greening component and only partially supporting agri-environment schemes that have greater potential to support ecosystems and ecosystem services. In some Member States, those who are unwilling or unable to provide adequate matched funding, this has resulted in the introduction of small-scale agri-environmental measures, providing limited support for ecosystem services. Additionally, the European Commission will need to be vigilant to ensure that the intended policy cohesion between the cross-compliance, greening component and agri-environment schemes is in fact achieved. Ultimately, whilst it provides greater support and protection than it has done in the past, more remains to be done to align the modern CAP with the European Union's vision of the role that it should play in protecting ecosystem services. Equally, as time progresses, private markets may also increasingly evolve to assist with this task.

References

Baldock D, Beaufoy G (1993) Nature conservation and new directions in the EC Common Agricultural Policy. Institute for European Environmental Policy, London

Barnosky A et al (2011) Has the earth's sixth mass extinction already arrived? Nature 471:51–57

Baskin Y (1997) The work of nature: how the diversity of life sustains us. Island Press, Washington DC

Birdlife International (2010) Through the green smokescreen: how is CAP cross compliance delivering for biodiversity? Birdlife International, Brussels

Buckwell A et al (1997) Towards a common agricultural and rural policy for Europe. Eur Econ 5

Buller H (2000) Regulation 2078: Patterns of Implementation. In: Buller H, Wilson G, Höll A (eds) Agri-environmental policy in the European Union. Ashgate Publishing Ltd, Aldershot, pp 219–253

Cardwell M (2004) The European model of agriculture. Oxford University Press, Oxford

Conference of the Parties to the Convention on Biological Diversity, Decision V/6 on the Ecosystem Approach. http://www.cbd.int/decision/cop/default.shtml?id=7148. Accessed 21 May 2014

Conference of the Parties to the Convention on Biological Diversity, Decision VII/11 on the Ecosystem Approach. http://www.cbd.int/decision/cop/default.shtml?id=7748. Accessed 21 May 2014

Conference of the Parties to the Convention on Biological Diversity, Decision X/2: "The strategic plan for biodiversity 2011–2020 and the Aichi Biodiversity Targets", adopted in Oct 2010, para. 12. http://www.cbd.int/decision/cop/default.shtml?id=12268. Accessed 21 May 2014

Costanza R et al (1997) The value of the world's ecosystem services and natural capital. Nature 387:253–260

Couldrick L (undated) Restoring river catchment function using payments for ecosystem services. West Country Rivers Trust, Callington

Daily G (ed) (1997) Nature's services: societal dependence on natural ecosystems. Island Press, Washington DC

Ellis E (2011) Anthropogenic transformation of the terrestrial biosphere. Philos Trans R Soc 369:1010–1035

European Academies Science Advisory Council (2009) Ecosystem services and biodiversity in Europe. Royal Society, London

European Commission (2005) Agri-environmental measures: overview on general principles, types of measures and application. European Commission, Brussels

European Commission (2007) Report on the application of the system of cross-compliance. COM (2007), 147 final

European Commission (2011a), Our life insurance, our natural capital: an EU Biodiversity Strategy to 2020. COM (2011), 244 final

European Commission (2011b), Proposal for a regulation of the European Parliament and of the Council establishing rules for direct payments to farmers under support schemes within the framework of the common agricultural policy. COM (2011), 280 final/2

European Commission (2011c), Proposal for a Regulation of the European Parliament and of the Council on the financing, management and monitoring of the common agricultural policy. COM (2011), 628 final/2

European Commission (2011d), Proposal for a regulation of the European Parliament and of the Council on support for rural development by the European Fund for Rural Development (EAFRD)

European Commission (2011e), Rural development in the European Union: statistical and economic information report 2011 (Commission, 2011), 27

European Commission (2012) Rural development in the European Union: statistical and economic information report 2012. Publications Office of the European Union, Luxembourg

European Commission (2014) Proposals for a commission delegated regulation supplementing regulation 1306/2013 of the European Parliament and the council with regard to the integrated administration and control system and conditions for refusal or withdrawal of payments and administrative penalties applicable to direct payments, rural development support and cross-compliance, COM(2014), 1476 final, Art. 39

European Commission, Directions towards sustainable agriculture. COM (99), 22 Final

European Commission, Proposals for a commission delegated regulation supplementing regulation 1306/2013 of the European Parliament and of the Council, proposed Arts 24–27.

European Court of Auditors (2009) Is cross compliance an effective policy? Special report 8/2008. Publications Office of the European Union, Luxembourg

European Court of Auditors (2011) Is agri-environment support well designed and managed? Special report 7/2011. Publications Office of the European Union, Luxembourg

European Environment Agency (2004) High nature value farmland: characteristics, trends and policy challenges. European Environment Agency, Copenhagen

European Environment Agency (2005) Agriculture and the environment in the EU-15: the IRENA indicator report. European Environment Agency, Copenhagen

European Environment Agency (2006a) Integration of Environment into EU Agriculture Policy- the IRENA indicator-based assessment report. European Environment Agency, Copenhagen

European Environment Agency (2006b) Progress towards halting the loss of biodiversity by 2010. European Environment Agency, Copenhagen

European Environment Agency (2010) 10 messages for 2010: agricultural ecosystems. European Environment Agency, Copenhagen

Fitter A et al (2010) An assessment of ecosystem services and biodiversity in Europe. In: Hester RE, Harrison RM (eds) Issues in environmental science and technology: ecosystem services. Royal Society of Chemistry, London, pp 1–28

Grant W (1997) The common agricultural policy. MacMillan, London

Hart K, Menadue H (2013) Equivalence mechanisms used for complying with greening requirements under the new common agricultural policy. Institute for European Environmental Policy, London

Harvey G (1997) The killing of the countryside. Jonathan Cape, London

Institute for European Environmental Policy (2013), Plenary vote on CAP reform package: a bitter sweet victory for the environment, press release 19 March 2013. http://www.cap2020.ieep.eu. Accessed 28 May 2014

Irwin F, Ranganathan J (2008) Action agenda for sustaining ecosystem services. In: Ranganathan J, Munasinghe M, Irwin F (eds) Policies for sustainable governance of global ecosystem services, Edward Elgar Publishing, pp 21–69

Jack B (2001) Economy and the environment: shaping the development of European agricultural law. Web J Curr Leg Issues 2:1–18

Jack B (2009) Agriculture and EU environmental law. Ashgate Publishing Ltd, Farnham

Jack B (2012) Ecosystem services and the rural environment: reforming European agricultural law. Eur Energ Environ Law Rev 21(6):258–273

Jack B (2013) The reform of the common agricultural policy: meeting the environmental challenge. Queen's University of Belfast Law Research Paper 17:1–18. http://ssrn.com/abstract=2263163. Accessed 30 May 2014

Jack B (2015) Research handbook on EU Agricultural Law. In: McMahon J, Cardwell M (eds) Research handbook on the common agricultural policy. Edward Elgar, Cheltenham

Jenkins TN (1990) Future harvests: the economics of farming and the environment- proposals for action. Council for the Protection of Rural England, London

Kantor Management Consultants (2011) Ex post evaluation of rural development programmes 2000–2006. Kantor Management Consultants, Athens

Lynas M (2011) The god species: how the planet can survive the age of humans. Fourth Estate, London

McNeely JA (2009) Payments for ecosystem services: an international perspective. In: Ninan KN (ed) Conserving and valuing ecosystem services and biodiversity: economic institutional and social challenges. Earthscan Ltd, London, pp 135–150

Millennium Ecosystem Assessment (2005) Ecosystems and human well-being: synthesis. Island Press, Washington DC

National Audit Office (2010) Tackling diffuse water pollution in England. The Stationary Office, London

Pagiola S et al (2004) Paying for biodiversity conservation services in agricultural landscapes. World Bank, Washington DC

Perrot-Maître D (2006) The Vittel payments for ecosystem services: a perfect PES scheme? International Institute for Environment and Development, London

Rockström J et al (2009) A safe operating space for humanity. Nature 461:472–475

Ruhl JB, Kraft SE, Lant CL (2007) The law and policy of ecosystem services. Island Press, Washington DC

Salzman J (2005) Creating markets for ecosystem services: notes from the field. N Y Univ Law Rev 80(3):870–961

Shoard M (1982) The theft of the countryside. Temple Smith, London

UK National Ecosystem Assessment (2011) The UK National Ecosystem Assessment: synthesis of the key findings. UNEP-WCMC, Cambridge

The Rural Development Programme (RDP) as a Strategic Tool for Linking Legal and Agroecological Perspectives

G. Buia and M. Antonucci

Abstract Starting from a reconstruction of the multidimensional concept of "rural development", this chapter intends to highlight the agroecological importance of an interesting (and still little investigated by the legal doctrine) administrative instrument of scheduling and planning, represented by the Rural Development Programmes (RDPs), also in the light of the new Regulation (EU) 1305/2013. The analysis is articulated in two parts. In the first part, the investigation focusses on the priorities established at EU level as regards rural development policies and the consequent transposition of these priorities into the RDPs, with particular reference to the Programme adopted, in Italy, by Regione Puglia. The second part examines the measures for safeguarding and developing the rural territories contained in the RDP 2007–2013 of Regione Puglia; these measures need to be compared to the protective actions envisaged by the Regional Landscape Plan ("Piano Paesaggistico Territoriale Regionale" or PPTR) recently adopted by Regione Puglia, in order to show the close interrelation between landscape and rural development in view of a new agroecologically oriented administrative planning process.

Keywords Agroecology • Landscape Plan • Rural Development Programme • Rural landscape • Sustainable development

1 The New EU Approach to Rural Development Policies

An analysis of the methodological approach recently adopted in Europe as regards rural development requires a preliminary consideration: the problem consists, above all, of defining what we mean by *rurality*.

The chapter represents the result of a joint research common to both authors; with reference to the distinction of the individual contributions, paragraphs 1–4 and 8 have been drawn up by Giuseppina Buia and paragraphs 5–7 by Mariacristina Antonucci.

G. Buia (✉) • M. Antonucci
Faculty of Law, Department of Legal Sciences, University of Salento, Lecce, Italy
e-mail: giusybuia@live.it; cristina.uni@libero.it

M. Monteduro et al. (eds.), *Law and Agroecology*,
DOI 10.1007/978-3-662-46617-9_8

The term, for many years identified with the supremacy of agriculture in a specific territory, now takes on a much more complex meaning; it concerns inherent elements of agriculture that were previously ignored, combined systematically. In the experience of mono-functional agriculture predominant until a few decades ago, agriculture was synonymous to production of goods for private use and consumption; following a relatively more recent reconstruction, rurality has been referred to territorial structures characterised by poor demographic densities and the absence of significant urban agglomerations.

Today, it is acknowledged, changing the interpretative standard, that agriculture is fully *multifunctional*: in other words, it is aimed at protecting even fundamental values of collective interest like the landscape, biological diversity and cultural diversity.

Multifunctional agriculture is characterised by *polymorphism*, by *complexity* and by *differentiation*, features that pertain to the integration between territory, nature, society and economy, on the one hand, and on the fair relation between services, industry and agriculture, on the other.

The new agriculture model, no longer exclusively connected to the economic aspect but also focussing on providing ecosystemic public services (see the contribution of JACK in this volume), consequently leads to consider the allotropic properties of *rus*.[1]

In this sense, therefore, *rurality* can be intended as the result of the balance and of the interaction between various types of capitals: the natural one, the social one, the human one and the artificial one.

Rurality, as just described, included in a context of *sustainable development*, outlines the notion of *rural development*.

Sustainable development, traditionally connected to the needs of environmental protection, tends to now assume a wider sphere of influence than in the past: no longer only *equity*, *economy* and *environment* but also *education*. The latter means pre-eminence of the duties to the future generations and implies an "educational" tendency of sustainability.[2]

The legal concept of rural development combines not only with the idea of a multifunctional agriculture, but also with the inherent transdisciplinarity of

---

[1] In this sense, agriculture "designs the landscape; protects the fertility of the soil; contributes to the integrity of the hydrogeological layout, to the management of the water resources and to the control of the flooding through the hydraulic layout. It also maintains biodiversity; ensures the natural recycling of the nutrients; preserves the operation of the natural carbon sink (the earth's vegetation) and thus contributes to the fight against climate change caused by greenhouse gases; guarantees the safety, heath and quality of food, also regarding typical products; permits the socio-economic survival of the rural communities and enhances the human labour of the farming families with respect to the artificial capital; it educates towards rurality by maintaining the historical roots of the relationship between town and country; guards the cultural identity of the territory; favours the development of agricultural ecotourism and the use of nature for teaching and recreational purposes": Monteduro (2013), pp. 2–3.

[2] Fracchia (2013).

agroecology, the platform of scientific research that has arisen simultaneously with the new model of agriculture (see CAPORALI'S contribution in this volume).

The expression *rural development* has also replaced what for a long time was called *development of the rural territory*: this name change incorporates a different way of understanding the territory, now considered in its proactive role and no longer only conservative one.[3] Rural development thus consists of "a long term strategy, aimed at maintaining the complexity and the balance between the two components, and at integrating the rural areas in an overall process of sustainable development".[4]

In this perspective, the territory operates as a true "binding agent" between agricultural holdings and the rural environment intended as the system (long lasting but at the same time flexible and mobile) of relations and is positioned at the centre of a new law that is the territorial one.[5]

The EU policy on rural development[6] represents the "second pillar" of the Common Agricultural Policy (CAP), established to develop the economy and promote the specific resources of the rural areas[7] (see the contribution of JACK again in this volume).

The CAP has undergone a radical change with the "medium-term reform" of 2003,[8] which, obviously, had significant repercussions in the current form of the EU policy on rural development. This reform, in particular, is responsible for the introduction of the following principles: *decoupling*, *cross-compliance* and *modulation*.

Decoupling, as is well known, consists of providing loans (allocated by the EU in favour of the farmers) independently from the quantity of products marketed by each farmer. The Single Farm Payment is based on the average of payments received under the main existing subsidy schemes during a historic reference period (as a general rule, 2000, 2001 and 2002). In this perspective, we are witnessing the change from *support to production* to *income support*. The coupled payments, however, can be evoked (as an exception to the rule) when there is a risk of abandoning production due to the excessive cost and the anti-economic nature for the operators of the sector.

The payment and the size of the payments, however, depend on other parameters, not connected to the production: on the one hand, respect of environmental legislation, care of food safety and care of plants and of animals (*obligatory management criteria*) and, on the other, care for the land so as to guarantee its

---

[3] Saija (2006), pp. 284–289.

[4] Sotte (2003), pp. 18–21 (author's translation).

[5] Saija (2006), p. 287.

[6] Hoffmann (2006).

[7] The "first pillar", on the other hand, aims at making agriculture more competitive on the internal and international markets, focussing on the stabilisation of the price level and product quality.

[8] The reform was achieved with Regulation (EC) 1782/2003, later replaced by Regulation (EC) 73/2009.

good state (*agricultural and environmental conditions*).[9] This is the so-called cross-compliance: financial aid depends on respect of the conditions of correct ecological, agricultural and agri-food management. If a farmer fails to comply with those rules through negligence, direct payments may be reduced; in the event of deliberate non-compliance, the producer may be completely excluded from receiving aid (but it may also be decided not to apply the reduction if the non-compliance is considered minor in terms of its seriousness, duration and persistence).

Modulation, finally, has the purpose of providing incentives to the small and medium-sized enterprises (SMEs), rather than to large operators or groups, reducing the payments to the large companies to increase the funds available for loans and improving the rural development policy.[10]

2 The Regulation (EC) 1698/2005 and the Regulation (EU) 1305/2013

The new idea of rurality can be immediately deduced by the general objectives of the Regulation (EC) 1698/2005,[11] on the basis of which the Rural Development Programmes (RDPs) 2007–2013 have been developed.

Indeed, Art. 4 of the Regulation (EC) 1698/2005 outlines a community policy focussing on increasing the *competitiveness* of the agricultural and forestry sector (Lett. a), on *promoting the environment and the natural space* (Lett. b), on improving the *quality of life* in the rural areas and promoting *diversification of the economic activities* (Lett. c).

A closer look highlights that all the objectives show the changing approach of the rural policies: not only those directly referring to the ecological profiles (promoting the environment and the natural space) but also those of an apparently economic nature (competitiveness, for example).

[9] The final purpose of decoupling should be that of remunerating the farmers for the positive external effects produced, rather than providing incentives for production. However, some disagree with this interpretation: see, e.g., Pennacchi (2006), pp. 11–16, who asks the provocative question (page 12): "why do the environmental, cultural and social services created through the agricultural activities have to be considered as goods combined with the traditional productive assets and, as such, treated as public goods for which an assessment is not defined by the market? And once again, why should the public resources destined to agricultural holdings and rural enterprises, which are capable of approaching extra-agricultural purposes shared by the whole community, exclusively affect the financial budget of the agricultural policy?" (author's translation).

[10] Angilieri (2009), pp. 44–46; Costato (2012), pp. 393–404. Regulation (EU) 1305/2013, coming into force on 1st January 2014 and referring to the programming period 2014–2020, spurred the Member States to give increasing impulse to the SMEs, providing them with greater resources.

[11] Regulation (EC) 1698/2005 succeeds the Delors I packet (which concerned the period 1988–1992), the Delors II packet (for the period 1993–1999) and Agenda 2000 (for the period 2000–2006).

Just consider that, as laid down by the already mentioned Art. 4 of the Regulation (EC) 1698/2005, competiveness must be achieved by "restructuring, development, innovation". Restructuring includes and implies the need to recover the historical roots and the cultural identity of the territory. The reference to the past (restructuring) thus combines harmoniously with the projection towards the future (development and innovation).

The same economic diversification comes within an "agroecosystemic" view: to stimulate the rural economy, it is necessary to focus not only on the increase in quantity of production but also on the connections between the variety of agri-food products and the many types of environment, landscape and history that distinguish each different rural region.

The reference to the quality of life in the rural areas finally opens the way to the social horizon of agroecology: an example is the so-called *Social Agriculture*, which consists of the vocation of the agricultural holding and rural enterprise to satisfy the needs of underprivileged categories or those risking social exclusion.[12]

The objectives indicated in Art. 4 provided the three "axes" of the EU rural development policy, to which the so-called *Leader* axis is added.

The inclusion of the latter initiative in the strategic community programme has completely changed the idea of *support* to rural development because with it the sectorial bottom-up type of approach, focussing on the individual figure of the farmer and the structural profile of agriculture, is abandoned. In place of this approach, the *Leader* axis introduced an *integrated* support to help the collective local abilities of a certain rural area and contributes to disseminating it elsewhere.[13] Some of the elements that the *Leader* approach is made up of suggest endorsement of this new support logic: the territorial strategies of local development, for instance, and also the local public–private partnership (the so-called local action groups), the multisectorial conception and actuation of the strategy based on the interaction between operators belonging to various sectors of the local economy.

Each one of the above-mentioned axes corresponds to specific *measures* to be distinguished according to the type: (a) measures aimed at promoting *human potential* (for example, the actions aimed at providing incentives to professional training and disseminating new information and new agricultural practices); (b) measures aimed at reconstructing *physical capital* (for example, the strategies for modernising agricultural holdings and rural enterprises, for re-qualifying forest areas, for cooperating in developing new food products or new production techniques and technologies); (c) measures regarding the *quality of production and of the agricultural products* (in particular, incentives to farmers who respect the standards coming from EU legislation and who participate in food quality systems); (d) measures for *supporting rural development* (like agri-environmental payments,[14] non-productive forestry investments also paid to promote forests and

[12] On this subject, Finuola and Pascale (2008).

[13] Crialesi (2009), pp. 6–10.

[14] Adornato (2011), pp. 567–595.

woods "in terms of public utility", actions for the maintenance, restoration and re-qualification of rural heritage).[15]

In the light of this framework, there is a clear emergence of co-penetration amongst environment, agriculture and rural communities: the above-described measures fully express the multifunctionality of agriculture and outline the change "from the right to land to the right of the land, intended as an intrinsic connection between the land as a productive factor and land as an asset and value to preserve".[16]

The innovations introduced by the Regulation (EC) 1698/2005 also move in another direction: that of the autonomy of the different State Programmes, whose maximum expression is in the RDPs.

Each Member State, on the basis of the EU strategic directions,[17] approves its own national strategic programme (NSP), which guarantees coherence and cohesion between the (potentially different) regional RDPs. In this case too, therefore, a multilevel regulation has been opted that responds to a logic of subsidiarity and proportionality.[18]

Thanks to the *vicinitas* with the individual territories, the regional RDPs easily manage to understand the particular characteristics of each rural macro-area and to respond to the relevant needs, thus reducing the "structural and natural disparity between the different agricultural regions".

The "differentiated" management of the territories, the adoption of equally specific production models and more flexible conditions to access EU loans represent significant innovations introduced by Regulation (EC) 1698/2005.[19]

The European Agricultural Fund for Rural Development (EAFRD)—in addition to the European Agricultural Guarantee Fund (EAGF)—is an important tool in order to achieve the goal of rural development.[20] The management of the EAFRD is carried out on the basis of partnership logics, in close consultation between the

[15] Regulation (EC) 1698/2005 on the support to rural development from the European Agricultural Fund for Rural Development (EAFRD).

[16] Adornato (2011), p. 554 (author's translation). The expression "from the right to land to the right of land" is by Graziani (2012), pp. 65–79. Graziani adopts this expression to indicate the change from land intended as "economic asset" to land as "ideal asset": see Graziani (2007), pp. 65–94.

[17] For the programming period 2007–2013, the EU strategic guidelines were aimed at determining the sectors that, thanks to the Community aid, could have provided greater added value on an EU level; at guaranteeing adequate attention for the EU priorities published at the European Council of Lisbon (2000) and of Gothenburg (2001); at harmonising the policy for the rural territories with the other EU policies and in particular with the cohesion and environment policy; at simplifying the actuation of the new CAP. See the Decision of the Council 144/2006 regarding the strategic EU guidelines for rural development (programming period 2007–2013).

[18] Adornato (2011), pp. 546–547.

[19] Magno (2007), pp. 226–230.

[20] Before 2005, both the Funds came within the European Agricultural Guidance and Guarantee Fund (EAGGF); now, however, the "Guarantee" section is absorbed in the EAGF, that of "Guidance" in the EAFRD: Adornato (2011), p. 545.

Commission, Member States, authorities and bodies determined by the Member States, including economic and social organisms.

Both the requests for loans made by private authorities to the EU and the corresponding administrative measures issued on a domestic level must demonstrate the respect of the principles aimed to protect the ecosystems and the principle of sustainable development. In particular, for the administrative measures of granting contributions, this demonstration must be contained in the motivation of the measure.[21]

The recent Regulation (EU) 1305/2013, abrogating Regulation (EC) 1698/2005, started the new 2014–2020 programming period, laying the basis for future RDPs to be prepared by each of the Member States.

The ways of applying Regulation (EU) 1305/2013 have been defined by the Commission Implementing Regulation (EU) 808/2014, which regulates the presentation of the RDPs, the procedure and the expiry dates for approval and modification of the RPDs and of the domestic legislation, the content of the domestic legislation, the information and the publicity regarding the RDPs, the actuation of certain measures of rural development, the monitoring, the assessment and the presentation of reports. In particular, the Commission Implementing Regulation (EU) 808/2014, in its own Annex I, establishes details of the presentation of the content of the RDPs indicated in Art. 27 of Regulation (EU) 1303/2013 and in Art. 8 of Regulation (EU) 1305/2013.

On the multifunctional and agroecological side, the overall framework of the new CAP (which includes several EU Regulations, apart from 1305/2013, like nos. 1303/2013, 1306/2013, 1307/2013, 1308/2013 and the Regulations for the transition between the two programming periods 2007–2013 and 2014–2020) guarantees continuity with the previous community policy but, at the same time, pushes it towards a further evolution.

Just consider, for instance, the new Regulation (EU) 1307/2013 that has replaced Regulation (EC) 73/2009, defining a new system of direct payments from 1 January 2015. In Regulation (EU) 1307/2013, there is a shift from "full decoupling" to "targeting". Targeting is a system of payments to the farmers in which the reference to prior historical periods (that is, the 3-year period 2000–2002) is usually excluded. A system of multipurpose payments, with seven components (the first three components are compulsory for Member States, while the last four are optional) is envisaged: (1) a basic payment per hectare; (2) a "greening" component, as additional support to compensate for the costs of providing environmental public goods not remunerated by the market; (3) an additional payment for 5 years for young farmers; (4) a redistributive payment whereby farmers may be granted additional support for their first hectares; (5) additional income support in areas

[21] Rampulla (2010), pp. 729–733.

with specific natural constraints; (6) coupled support for production, granted to certain areas or types of farming for economic and social reasons; (7) a simplified system available to small farmers.[22]

The priorities of the community rural development policy are also reprocessed in a significant manner: objectives like *employment* and *growth*, already declared in Lisbon and in Gothenburg and underlying the previous programming period, are revitalised by assimilating the interpretation proposed in the Communication of the Commission dated 3 March 2010, better known as "Europe 2020". Indeed, the first "Whereas" clause of Regulation (EU) 1305/2013 stands on a rural policy aimed at accompanying the market measures of the CAP and functional to an "intelligent growth" (thanks to more effective investments in education, research and innovation), which is "sustainable" (thanks to the choice for a low CO_2 emissions economy) and "inclusive" (aimed at creating more job places and reducing poverty).

As regards the intervention axes, replaced by *priorities* in the new Regulation (EU) 1305/2013, the following should be pointed out: as regards competitiveness, the quality criteria that the products must respect (Art. 16) are specified; as regards the exploitation of the environment and of the natural space, direct investments are envisaged to increase resilience, the natural merits and the forest ecosystems; the obligatory range of the agro-climatic environmental payments and their fundamental role are specified; support is given to organic farming (Arts. 28 and 29); in terms of improving the quality of life, the essential role of cultural and rural heritage for the growth of the rural areas is pointed out in view of sustainable development ("Whereas" clause 19, Art. 20); in terms of income support, the *Leader* approach is confirmed as a valid instrument for the "multi-sectorial needs of rural development" and a "LEADER start-up kit" is introduced ("Whereas" clause 31, Arts. 42 and 43), also envisaging *ad hoc* measures for managing the risks deriving from adverse weather conditions, from epidemics and plant diseases, from parasitic infestations, from environmental emergencies and no longer only for fire hazards (Art. 18). Finally, the fundamental role of the European network of rural development, which is directed towards the participation of *all* the interest bearers in terms of rural development, is highlighted (Art. 52).

As regards the programming, there is a specification of the intervention areas: having confirmed the choice for a multilevel governance, "Thematic sub-programmes" (Art. 7) have been introduced that respond to *specific needs*, in order to assure a prompt realisation of the EU priorities.

In order to guarantee the efficiency of the plans, apart from their *ex ante* and *ex post* assessment (envisaged by the old Regulation (EC) 1698/2005), a *during the programme* assessment is also envisaged: this replaces the old interim assessment every year, but it differs from it because the success of the strategy is *constantly* monitored. The efficiency of the whole programme must also be controlled, thanks

[22] Tropea (2014).

to indicators and targets that express the effective achievement of the objectives established by the Union in terms of quality and quantity.

A significant improvement aimed at ensuring the authentically "integrated" character of the strategy is represented by the choice of returning to unitary programming, but with an important difference with respect to the past: the independence of the EAFRD is retained, and it remains an independent Fund with respect to the other annexes to the single programming;[23] however, at the same time, the aim to join the rural development policy to the other EU policies (particularly, to the "first pillar" of the CAP and to the economic and social cohesion policy) continues. The joint articulation of the different EU Funds is guaranteed by a Common Strategic Framework (CSF), drawn up on a European level; it envisages the adoption, by each State, of a National Reform Programme (NRP), which operates as a programmatic agenda of the past and future interventions, and of a Stability Programme (SP), aimed at representing the economic-financial system, at indicating the future objectives, at managing the budget operations. Each Member State will consequently have its own Partnership Contract (PC): the PCs will basically correspond to the old National Strategy Plans (NSPs) and will guarantee, with more vigour, coordination between the various policies and their integration, in the light of the priorities of "Europe 2020".[24]

However, the new instruments to increase integration between the CAP, rural development policy and cohesion emerging in the EU law have raised some eyebrows. In particular, in the draft annexes to the Framework Regulation 2014 2020, the CSF was presented as a very general document, as such unsuitable for giving concrete replies to the need for integration and cohesion. Basically, the fear is that the only real points of contact between the policies will occur by means of the programmes adopted on domestic and regional levels.[25]

3 From *An* to *Quomodo*: The Assimilation of the European Objectives in the Italian Rural Development Programmes 2007–2013. In Particular: The Case of Regione Puglia

In the light of the preceding analysis, regarding the EU scheme for rural development, in this section we will seek to focus on if and how the Italian Regions have complied with the obligations imposed by the EU as regards rural development.

[23] Reference is made in particular to the European Regional Development Fund (ERDF), to the European Social Fund (ESF) and to the European Maritime and Fisheries Fund (EMFF), whose common and general provisions are now contained in Regulation (EU) 1303/2013: Mantino (2013), pp. 47–52.

[24] Mantino (2013), p. 48; Romito (2012), pp. 39–41.

[25] Mantino (2012), pp. 28–31, in particular 30.

The budget of the last programming period presents important positive aspects: however, there is no lack of criticalities, above all as regards the desired territorialisation of the RDPs.

The effort to come to a "differentiation of rurality in Italy" has been an undoubted success: many Italian Regions have known how to thoroughly investigate the rural reality of their own territories, and despite following general and abstract conditions dictated at EU and national levels, they have managed to determine "made to measure" solutions of each territory in terms of specifying the objectives, drawing up the policy and dividing the resources. On the other hand, the principle of subsidiarity, several times mentioned by virtue of the multilevel governance chosen for rural development, could only have blended with another fundamental principle of Italian legislation, which is specifically the principle of differentiation (Art. 118 of the Italian Constitution).

Furthermore, the regional RDPs have paid suitable attention to the measures destined to implement the human capital, and that is to training and to specialisation of the subjects involved, in the new multifunctional logic of agriculture. In this sense, above all the promotion of the establishment of Young Farmers produced positive results.[26]

As regards the types of investment, looking, for instance, at the case of Regione Puglia, the funds allocated for the first EU objective (competition), which represent 40 % of the resources of the Apulian RDP, 36 % was allocated to modernising the agricultural holdings and rural enterprises and 30.9 % to increasing the added value of the agricultural and forestry products.[27] It can, therefore, be stated that in the planning phase 2007–2013, much has been focussed on the "restructuring" component of competitiveness. This does not seem to be a negative piece of data, as long as the restructuring is also considered as recovery of historical-cultural identity and the emphasis on this component does not have a detrimental effect on the others (with particular regard to "innovation").

---

[26] Sotte (2009), pp. 237–238. The generation change in the agricultural and forestry sector is one of the main objectives of Law 80/2012 of Regione Toscana, which establishes the "Banca della Terra" ("Bank of Land"): it is a public inventory, managed through a web platform, which surveys all the uncultivated or abandoned land and the publicly or privately owned agricultural holdings and rural enterprises available for rent or concession, in order to make them preferentially available to young farmers who request it by promoting a specific development plan. The initiative favours those wanting to work but who do not have the availability of land and, at the same time, imposes as the condition to access the uncultivated land the positive contribution of the farmers to the market and to the territory by safeguarding biodiversity, protecting the landscape, protecting and maintaining the forestry resources, also to prevent hydrogeological imbalance and defend the mountain areas and populations from natural calamities. After the law of Regione Toscana, initiatives aimed at establishing regional "Banks of Land" are multiplying in other Italian Regions: for example, Sicilia, Art. 21, Regional Law 5/2014; Liguria, Regional Law 4/2014; Umbria, Art. 3, Regional Law 3/2014; Veneto, Regional Law 26/2014.

[27] Puglia, Rural Development Programme 2007–2013, http://svilupporurale.regione.puglia.it/portal/pls/portal/PSR_PORTALE.DYN_ASSE.show?p_arg_names=asse&p_arg_values=1. Accessed 30 July 2014.

The RDPs 2007–2013 adopted by the various Italian Regions, examined in their entirety, however, denote a series of problematic aspects, which is worthwhile highlighting.

To begin with, many regional RDPs have chosen an unbalanced policy to the detriment of territorialisation.[28] It is necessary to focus on this point to better explain the nature of the problem.

We have already seen how, on the basis of Regulation (EC) 1698/2005, the EU policy of rural development goes through three theme axes, to which a fourth is added: the first two axes (competition and exploiting the natural environment and space) constitute the *sectorial policy* of the EU, while the third and the fourth (quality of life and diversification of the economic activities, *Leader* axis) represent the *territorial policy*.[29]

As a consequence, from a domestic viewpoint, the Italian NSP 2007–2013 envisages three types of integrated actions, later assimilated by the 21 regional RDPs:

(1) the packages of measures for enterprise;
(2) integrated system projects;
(3) integrated territorial or area projects.

The plans of the type (3), unlike the first two, promote rural development through *intersectorial* investments, destined to areas whose historic, cultural and territorial identity was similar; these areas are distinguished not only for the products but also for the overall *services* offered, including the intangible ones.[30]

On a regional scale, most EU aid has been employed by the Italian Regions to finance types (1) and (2) measures, that is, enterprises and/or lines of enterprise: so a more "by sector" than "by territory" approach was used. This is not positive in terms of strategic approach and therefore in terms of multifunctionality.

Territorialisation and sectoriality, on the other hand, should live together in the rural development policy: their reciprocal integration[31] should not occur *ex post*,

---

[28] *Ex multis*: Sotte (2009), p. 243; Fugaro and Giuliodori (2006), pp. 22–25; Lucatelli (2006), pp. 8–10.

[29] Sotte (2009), pp. 243–244.

[30] Varotto (2007), pp. 571–576; Zumpano (2007), pp. 56–59.

[31] Monteleone (2010), pp. 45–48. However, even today there are those who are in favour of a clear separation between productivity support policies to the agriculture sector and policies of rural development in terms of territorial cohesion: see Moyano Estrada (2010), pp. 51–52, who asserts "the convenience of separating, on the one hand, the agriculture policies, whose objective must be that of favouring a new phase of modernising European agriculture, from the policies directed at developing and the cohesion of rural territories, on the other [. . .] If, in the backdrop of the current food situation, European agriculture must be once again reactivated in its productive dimension to satisfy the demand for food of the European population and must maintain its position in the world markets, I believe that it is necessary to again implement a guided agriculture policy in the EU territories with greatest productive potential with a logic directed at production and focussed above all on modernisation and competiveness of food systems [. . .] I would be of the opinion that it is necessary to improve an agricultural policy of clear productivistic vocation that revolves around a

but *ex ante*, at the moment the plans and the measures are processed, envisaging from the beginning a dialogue between the different *stakeholders*, which would prevent stalling in their application phase.

Another profile of criticality has been represented by the incredible delay with which the various RDPs have been adopted, party due to the adoption procedure of the RDPs imposed by EU law,[32] which is extremely complex and not flexible against the diversity of the individual countries. The node of the partnership, which is central in the monitoring and assessment phase of the RDPs, also remains problematic, considering that the absence of collaboration by the recipients could jeopardise the whole success of the plans.[33] For instance, in a study carried out on the RDP 2007–2013 of Regione Veneto, it was found that there are no defined rules on how the various consultations occur and on the effective incision of them on the rural development policy.[34]

Finally, there is a noticeable absence of a National Framework for Rural Development: this instrument, while being envisaged by Regulation (EC) 1698/2005, has not been adopted in Italy (the contrary of what has happened in Spain, for instance).[35]

The National Framework for Rural Development, unlike the NSP, which comes from a directive document for each of the Member States, would have been functional to the coordination of the various RDPs if, like in Italy, the presentation of a series of regional programmes had been opted for.

In particular, on the basis of Art. 15 of Regulation (EC) 1698/2005, in the event of multiple regional programmes, the States *could have* adopted the National Framework for Rural Development, which *should* have contained "the common elements of these programmes".

Summarising: insufficient territorialisation, evident delays and absence of coordination have had negative implications on the plan of assessing the RDPs in the

single axis and at a single base (concentrating the current two pillars of the CAP in a single one) [. . .] If it is not possible, due to the difficulties it involves, nor convenient, due to the characteristics of the new scenario, to integrate the agricultural and territorial visions in the backdrop of a common European policy, wouldn't it be better to focus on the separation of the agricultural policy and of the development policy of the rural territories, equipping each one with its own funds and its own actuation instruments, and establishing the necessary mechanisms of coordination? On the other side of that integration there is also an absence of necessary political and social support (considering the divergent positions of the agricultural, rural and environmentalist organisations)" (author's translation).

[32] Adornato (2011), p. 549.

[33] For the importance of the design of participatory monitoring and evaluation (PM&E) within rural development programmes, see, e.g., Parkinson (2009), pp. 229–237. The "absolute value" of the participation has been underlined in the introduction to the Deliberation of the Regional Government of Puglia 39/2013, with which, for the purposes of preparation of the Puglia RDP 2014–2020, the relevant procedure of Strategic Environmental Assessment (SEA) was established, a phase in which the preliminary consultation of the organisms of socio-economic representation, of the stakeholders and of the community assumes great importance.

[34] Secco et al. (2011), pp. 104–112.

[35] Sotte (2009), p. 242.

Italian experience, determining a dispersion of the responsibilities and of the initiatives for the purposes of the rural development.

In spite of the numerous steps forward carried out to increase the assessing capacity of the RDPs, the *feedforward* that would have come if the information collected with the monitoring and assessment procedures had been used regularly in the regional programmes, updating them constantly, has also been absent; furthermore, in this way, reallocation of the resources would have been constant and thus more efficient.[36]

4 The Priorities of the Rural Development Programmes 2014–2020: The Challenge of the Agroecological Approach

The issue of Regulation (EU) 1305/2013[37] definitively marked the completion of the programming period 2007–2013 and laid the important base for the new period of planning 2014–2020.

In the new perspective opened by the European legislator, it would seem possible to state that the Rural Development Programme is ready to be one of the most suitable legal instruments to establish fruitful connections in the future between law and agroecology.[38]

---

36 Camaioni (2009), pp. 51–54, in particular 54.

37 For the transitory provisions, see Regulation (EU) 1310/2013.

38 It is no coincidence that Regulation (EU) 1305/2013, in Art. 55, makes explicit reference to agroecology with reference to one of the essential instruments for rural development: the European Innovation Partnership (EIP) for agricultural productivity and sustainability. Art. 55 of Regulation (EU) 1305/2013 establishes the following:

"1. The EIP for agricultural productivity and sustainability shall: (a) promote a resource efficient, economically viable, productive, competitive, low emission, climate friendly and resilient agricultural and forestry sector, *working towards agro-ecological production systems and working in harmony with the essential natural resources on which farming* and forestry *depend*; (b) help deliver a steady and sustainable supply of food, feed and biomaterials, including existing and new types; (c) improve processes to preserve the environment, adapt to climate change and mitigate it; (d) build bridges between cutting-edge research knowledge and technology and farmers, forest managers, rural communities, businesses, NGOs and advisory services.

2. The EIP for agricultural productivity and sustainability shall seek to achieve its aims by: (a) creating added value by better linking research and farming practice and encouraging the wider use of available innovation measures; (b) promoting the faster and wider transposition of innovative solutions into practice; and (c) informing the scientific community about the research needs of farming practice.

3. The EAFRD shall contribute to the aims of the EIP for agricultural productivity and sustainability through support, in accordance with Article 35, of the EIP operational groups referred to in Article 56 and the EIP network referred to in Article 53."

According to the "Whereas" clause 44 of Regulation (EU) 1305/2013, RDPs and EIP are closely connected: "Rural development programmes should provide for innovative actions promoting a resource-efficient, productive and low-emission agricultural sector, with the

The new objectives selected by Regulation (EU) 1305/2013 for the programming period 2014–2020 demonstrate that the RDPs, while being created as legal instruments for defining rules, can be transformed, in the light of the new purposes imposed by the EU, into true "agro-ecological witnesses".

Each RDP actually becomes a bearer of multiple and interconnected interests like the interest in progress (intelligent growth), interest in environmental protection (sustainability) and interest in the reduction of poverty and of unemployment (inclusivity): the RDP thus incarnates an approach that (even if this term has not been explicitly used in Regulation (EU) 1305/2013) can *ante litteram* be defined as "agroecological", in the sense that it forces itself to integrate systematically from a legal viewpoint the economic, employment, social, ecological, landscape and cultural dimensions of *rus*, following the same pluri-dimensional and transdisciplinary approach that has been adopted, on the scientific level, by the agroecology (see, on the point, the contribution of MONTEDURO in this volume).

Regarding this, it is sufficient to highlight some of the innovations envisaged on an EU level for the current programming period.

We refer, for instance, to the "Agri-environment-climate payments" (AEC),[39] which, in the light of the criteria of decoupling, are amongst the main symptomatic indicators of an agroecological approach.

Then there is the increased importance of the *Leader* axis: it is an index of the multisectoriality of agriculture, and as such the original valence of horizontal instrument must be recovered. This means not only using it to support the third axis (*quality of life and diversification of the economic activities*) but also linking it with the first and second axis (*competitiveness and promoting the environment and the natural space*).

The agroecological valence of the RDPs could also emerge by further interventions in terms of governance.[40]

support of the EIP for agricultural productivity and sustainability. The EIP should aim to promote a faster and wider transposition of innovative solutions into practice. The EIP should create added value by enhancing the uptake and effectiveness of innovation-related instruments and enhancing synergies between them. The EIP should fill gaps by better linking research and practical farming."

[39] Art. 28, Regulation (EU) 1305/2013. The previous Regulation (EC) 1698/2005 simply referred to the "Agri-environmental payments". The increased attention paid to climate changes has led, in the new Regulation (EU) 1305/2013, to extending the payments to this new specific topic too.

[40] Regarding the assessment of the quality of the governance in a rural environment, it has been underlined that, in order to achieve a satisfactory assessment result, respect of the criteria of efficiency and efficacy of the various instruments must be backed up by respect for transparency, for participation, for sharing responsibility, for equity: see Secco et al. (2010), pp. 61–65.

In particular, it is essential to continue with promoting the differentiation of rural areas.[41] It is no coincidence that the European legislator has deemed it necessary to introduce "Thematic sub-programmes" for specific needs within the RDPs: these could allow the identity and inimitable characteristics of the different rural areas to be understood better than programming on a broader spectrum.

Even the "rural districts", established in Italy by Art. 13 of Legislative Decree 228/2001, could cover a crucial role: they are defined by the same Art. 13 as "the local production systems [...] characterised by a homogeneous historical and territorial identity deriving from the integration between agricultural activity and other local activities, as well as by the production of goods or services of particular specificity, coherent with the natural and territorial traditions and vocations".[42]

Regulation (EU) 1305/2013 and the future RDPs seem to be capable of including the "rural districts" indicated in Art. 13 of Legislative Decree 228/2001 in a more comprehensive manner within their range of application: this is because Regulation (EU) 1305/2013 significantly promotes the figures of the rural "clusters"[43] and "networks", the "EIP operational groups" and the "forms of co-operation" in a more general sense[44] (Arts. 35, 53 and 55–57, Regulation (EU) 1305/2013).

---

[41] Even regional centralism may, indeed, become a problem; it is important to favour an effective administrative decentralisation for the purposes of authentic rural development, as noted by McAreavey and McDonagh (2011), pp. 175–194.

[42] Varotto (2005), pp. 2–20; Albisinni (2010), pp. 21–25. Regarding the rural districts, see also Art. 1, para. 369, Law 266/2005 and the Decree of the Ministry of Agricultural, Food and Forestry Policies dated 21 April 2008.

[43] Art. 1, para. 1, Lett. q), Regulation (EU) 1305/2013, according to which the "cluster" is a "grouping of independent undertakings, including start-ups, small, medium and large undertakings as well as advisory bodies and/or research organisations - designed to stimulate economic/innovative activity by promoting intensive interactions, the sharing of facilities and the exchange of knowledge and expertise, as well as contributing effectively to knowledge transfer, networking and information dissemination among the undertakings in the cluster".

[44] "Whereas" clause 29 of Regulation (EU) 1305/2013: "During the 2007–2013 programming period the only type of co-operation which was explicitly supported under rural development policy was co-operation for the development of new products, processes and technologies in the agriculture and food sector and the forestry sector [...] it has become clear that supporting a much broader range of types of co-operation, with a wider range of beneficiaries, from smaller operators to larger ones, can contribute to achieving the objectives of rural development policy by helping operators in rural areas overcome the economic, environmental and other disadvantages of fragmentation. Therefore, that measure should be widened. Support to small operators for organising joint work processes and sharing facilities and resources should help them to be economically viable despite their small scale. Support for horizontal and vertical co-operation among actors in the supply chain, as well as for promotion activities in a local context, should catalyse the economically rational development of short supply chains, local markets and local food chains. Support for joint approaches to environmental projects and practices should help to produce greater and more consistent environmental and climate benefits than those which can be delivered by individual operators acting without reference to others (for example, through practices applied on larger, unbroken areas of land). Support should be provided in various forms. Clusters and networks are particularly relevant to the sharing of expertise as well as the development of new and specialised expertise, services and products. Pilot projects are important tools for

In order for the rural districts to not be considered counterproductive, for instance aggravating the isolation of the various rural realities, it would be useful to also consider the so-called *meta-districts*: they could operate as centres of connection between the various rural areas and intensify the processes of multifunctionality of which the RDP are promoters.[45]

5 The Measures of Protecting Rural Landscapes in the Rural Development Programme 2007–2013 of Regione Puglia

Within the rural development policies, a fundamental role is played by the actions of promoting and developing the rural territories.

Starting from the 1990s, the rural territories were overcome by a progressive process of "heritage-making". As scholars[46] have noted, the effort to stem the agriculture crisis has induced the promotion of quality local products, their direct sale, the organisation of rural and wine and food tourism, the rediscovery of traditions capable of transforming themselves into events, the recreating of "natural spaces" and the formation of parks.

From this viewpoint, an important contribution to protecting and developing rural territories has been provided by the measures that the regional RDPs have introduced to protect the *rural landscapes*: the theme is very important where it is considered that the "rural landscape" itself represents a critical conceptual articulation (see, in this volume, the contribution of BROCCA), in which agricultural law blends with cultural heritage law and the environment law, disciplines that are traditionally separated from a legal viewpoint.

Reflecting on the relationship between RDPs and legal protection of the rural landscape thus means interrogating oneself on the limits of the divisionistic and reductionistic model until now adhered to by the law and opening to the law the innovative horizon of agroecology as a "beyond the barriers" research platform,

testing the commercial applicability of technologies, techniques and practices in different contexts, and adapting them where necessary. Operational groups are a pivotal element of the European Innovation Partnership ("EIP") for agricultural productivity and sustainability. Another important tool lies in local development strategies operating outside the framework of LEADER local development – between public and private actors from rural and urban areas. Unlike under the LEADER approach, it is possible for such partnerships and strategies to be limited to one sector or to relatively specific development aims, including those mentioned above. Member States have the possibility to give priority to co-operation among entities involving primary producers. Inter-branch organisations should also be eligible for support under this measure. Such support should be limited to a period of seven years except for collective environmental and climate action in duly justified cases."

[45] Contò et al. (2012), pp. 82–94.

[46] Zerbi and Fiore (2009), p. 7.

capable of connecting the different faces of the "polytope" *rus* (see the contribution of MONTEDURO in this volume) into a system.

Indeed, the landscape is an essential element not only for protecting the cultural heritage (tangible and intangible) incorporated in the rural areas but also for the economic development of the territory of agricultural vocation, to promote the products with denomination of origin, for tourist development, for maintaining biodiversity, for improving the quality of life in the rural areas.[47]

Precisely for this reason, within the three axes that make up the strategy for rural development as it is designed by Regulation (EC) 1698/2005,[48] great attention has been focussed on landscape,[49] through measures aimed at protecting the rural landscape and its distinctive elements, at disseminating ecocompatible agroforestry practices, at promoting the historical-cultural heritage and the natural one, at disseminating territorial marketing actions aimed at connecting the tradition and the quality of the products to the production places and to the different natural and historical-cultural attractions of the rural territory and at developing activity connected to tourism in the rural areas.

In the new Regulation (EU) 1305/2013, first, and in the Italian National Strategic Plan for rural development 2007/2013, second, the protection of the rural landscape is, therefore, considered a *leitmotiv*, which in itself permeates all the actions to support the rural development.

The preservation and the exploitation of the agricultural landscape are, in Axis I, a stimulating factor in the growth of territorial competition of the agricultural products; in Axis II, an instrument for recovering a correct relationship between man and nature to protect biodiversity; in Axis III, an incentive for re-valuating and safeguarding the series of values, culture and traditions connected to the farming civilisation.[50]

---

47 Torquati (2007), pp. 41–43.

48 Indeed, Regulation (EC) 1698/2005, in Art. 4, defines the objectives of the EU rural development policies: Axis I) increase competitiveness of the agriculture and forestry sector, supporting restructuring, development and innovation; Axis II) promoting the environment and the rural space, supporting the management of the territory; Axis III) promoting the diversification of the rural economy and improving the quality of life in the rural areas; and, finally, the *Leader* approach aimed at improving the local governance and promoting the endogenic resources of the rural areas.

49 In this sense, we must point out how Regulation (EEC) 797/85 already established the adoption of economic instruments aimed at the need to preserve the rural landscape. Following the introduction of the agri-environmental measures contained in Regulation (EEC) 2078/92 and of the forestry measures of Regulation (EC) 2080/92, the importance of the landscape problem within environmental conservation and the acknowledgement of the role of the multifunctionality of agriculture have been strengthened. Subsequently, the EC interventions in favour of the rural landscape were concretised after the presentation of Agenda 2000, in Regulation (EC) 1257/99, by establishing agri-environmental measures, to be determined in the Rural Development Programmes that the Member States were called upon to draw up.

50 Torquati (2007), pp. 42–43.

The rural territories are identified as the bearers of a series of "identity resources",[51] which are not only physical-naturalistic (vegetation, habitat, landscape, ecosystems) but also historical-cultural ones (historical matrixes of the territory, literary and figurative representations), social and symbolic ones (places of collective identification, universes of meanings, perceptive characters), and, in this sense, become bearers, through uses, traditions, history and knowledge, of an "intangible cultural heritage", as defined by Art. 2 of the UNESCO convention for safeguarding the intangible cultural heritage.[52]

Devised in this way, the notion of rural landscape exceeds a purely productive dimension and becomes a place of co-production, where the relation between nature and society[53] is expressed, representing a *species* of the largest *genus* of the cultural landscapes.

This new way of conceiving the rural landscapes reflects a different way, which is innovative with respect to the past,[54] of considering the agricultural phenomenon

---

[51] Sonsini (2006), pp. 37–38, in particular 38.

[52] According to Art. 2 of the Convention, the intangible cultural heritage includes "the practices, representations, expressions, knowledge, skills – as well as the instruments, objects, artefacts and cultural spaces associated therewith – that communities, groups and, in some cases, individuals recognize as part of their cultural heritage. This intangible cultural heritage, transmitted from generation to generation, is constantly recreated by communities and groups in response to their environment, their interaction with nature and their history, and provides them with a sense of identity and continuity, thus promoting respect for cultural diversity and human creativity. For the purposes of this Convention, consideration will be given solely to such intangible cultural heritage as is compatible with existing international human rights instruments, as well as with the requirements of mutual respect among communities, groups and individuals, and of sustainable development. The "intangible cultural heritage", as defined in paragraph 1 above, is manifested inter alia in the following domains: (a) oral traditions and expressions, including language as a vehicle of the intangible cultural heritage; (b) performing arts; (c) social practices, rituals and festive events; (d) knowledge and practices concerning nature and the universe; (e) traditional craftsmanship."

[53] Cavazzani (2006), pp. 60–61, who quotes van der Ploeg (2006).

[54] The Rome Treaty of 1957 had already configured a diversified definition of agriculture. It is significant that in the texts of EU law, the term agriculture in many cases does not appear in its individuality since it is often replaced by other expressions like "agri-food systems", "rural development" or, as in the financial framework of the EU programming period 2007/2013, "compatible development". Starting from the second half of the 1960s, with the crisis of the Fordist model, a renewed interest for agriculture was asserted, so that from the 1980s the rural development policy became the second pillar of the CAP and one of the axes around which a new policy of qualification and differentiation rotates, concerning both the agricultural product and the rural spaces. See Adornato (2007), pp. 54–55, according to whom (author's translation) "we are in the presence of "the surrender" of the agricultural product, which has been the reference point of all the agricultural standards of the Rome Treaty and, at the same time, of the assertion on the one hand of the "centrality" of the territory compared to that of the enterprise and, on the other, of the plurality of subjective figures (not necessarily business ones and in a not necessarily productive relation with the land) [...] it is agriculture that promotes agriculture overall and its new distinctive profiles. Directives, moreover, supported on the constitutional level, by a more updated interpretation of Art. 44 of the Constitution, of which the more attentive readings have made it possible to interpret the formula "*rational exploitation of the land*" not only in a productive

overall, which now seems to be directed towards "a new geography of the (multi) functions, in which the rural areas are assigned several roles: not only a productive role, but also a residential, cultural and environmental one".[55]

According to some scholars, indeed, the objective of "keeping rural communities alive" can only pass through a "diversification of the economic activities", which is pursued, on the one hand, with the stimulation of endogenous development mechanisms and the exploitation of the local resources and, on the other, with the removal of obstacles, like the decline of agricultural activities, rural depopulation, population ageing, isolation, the weakness of infrastructures and services.[56]

Rural development is, therefore, no longer conceived as a "product" but is considered as a "locally rooted and socially controlled process which is determined on the basis of a specific combination of endogenous and exogenous factors, involves a series of actors and concerns several dimensions".[57]

From this viewpoint, the strategic national plan for regional development offers interesting opportunities for the development of the rural territories, whose concrete actuation modes, as has been previously evidenced,[58] are delegated to the choices of regional policy.

Amongst the actions protecting the rural landscape contained in the Rural Development Programme of Regione Puglia, those contemplated with Axis 2 must be considered ("Improvement of the environment and of the rural space"), as well as those of Axis 3 ("Quality of life in rural areas and diversification of the economy").

In particular, Axis 2 includes the priority objectives of "protecting the territory", articulating it in turn in a series of specific objectives like "Protecting the characteristic elements of the rural landscape", "Promoting the durability of sustainable agricultural activities in deprived areas", "Promoting agricultural and forestry systems aimed at protecting the land, tackling in particular the phenomenon of desertification and in the hilly areas also the phenomenon of erosion".

In the specific, the precise objective of "Protecting the territory" is satisfied by a combination of Measure 214 ("Agri-environmental payments"), with actions 1, "Organic farming", and 3, "Protecting the biodiversity", and Measure 216 ("Supporting non productive investments"), the Measures to protect the forest heritage (221-223-226-227) and, finally, the Measures aimed at favouring the durability of agricultural activities in deprived areas by paying subsidies,

manner, as it would seem from the presence of the (little opportune, but historically comprehensible) noun "*exploitation*", but also in the perspective of environmental protection, with an opening in this sense permitted by the presence of the adjective "*rational*", until a substantial formulation is reached which upturns the literal meaning and can be interpreted as "*optimal use of the territory*" in its most complex meaning"; Adornato quotes in this sense Desideri (1985) and Graziani (1985).

55 Fonte (2010), pp. 4–6 (author's translation).

56 Fonte (2010), p. 5.

57 Cavazzani (2006), p. 61 (author's translation).

58 See above, paragraph 2 of this chapter.

contributing to the maintenance of traditions, of knowledge and of the cultural identity of the landscapes (211-212).

Also, in order to ensure a minimum level of maintenance of the land and to avoid deterioration of the *habitats*, the regional RDP also establishes specific measures aimed at maintaining olive trees (Rule 4.3), by means of forbidding uprooting, imposing obligatory pruning, actuating farming techniques aimed at ensuring a balanced vegetative development according to the local habits and uses and at maintaining the characteristic elements of the landscape (Rule 4.4), as well as by forbidding the uprooting of century-old olive trees, prohibiting the elimination of the terracing marked by drystone walls, imposing the respect of regional measures of protecting the characteristic elements of the landscape and of Law 14/2007 of Regione Puglia, on the "Protection and exploitation of the landscape of the monumental olive trees of Puglia".[59]

More specifically aimed at promoting economic diversification of agricultural activity, at improving the quality of life in the rural areas and at increasing the level of attractiveness of agricultural territories, by exploiting the multifunctional role of the agricultural holding and rural enterprise, are the measures and actions established within Axis 3, entitled "Quality of life in the rural areas and diversification of the economy".

Axis 3, in particular, includes amongst its priority objectives the "Improvement of the attractiveness of the rural territories for agricultural holdings and rural enterprises and for the population", which develops into the following specific objectives: "Improving the offer and the use of services essential to the population, above all to the weaker layers and to the productive system"; "Re-qualifying the villages and the anthropogenic and landscape elements of the rural heritage"; "Promoting interventions for the care and maintenance of the territory, safeguarding the landscape, exploiting the cultural heritage". In relation to this latter aspect, Measure 323 ("Protection and re-qualification of the rural heritage") takes on significant importance, aimed at exploiting the archaeological, architectural, historical-artistic heritage and the landscape of the rural areas, in order to increase the tourist attractiveness of the same areas and to improve the quality of the life of the population.

These purposes are concretely sought by means of restoration interventions and by exploiting the regional cultural heritage, that of the rural villages and, particularly, that of property (privately and publically managed) of particular and proven artistic, historic and archaeological interest, which are an expression of the history, of the art and of the culture of the regional territory; extraordinary care, restoration and conservation of the typical and characteristic elements of the agricultural landscape and of the common spaces, even those of historical and

[59] Law 14/2007 of Regione Puglia reconnects the instruments of protecting and exploiting the monumental olive trees, as laid down by Art. 1, with the aim of safeguarding "their productive, ecological and hydrogeological defence functions, as well as particular and characterising elements of the regional history, culture and landscape".

religious importance, typical of the rural environment; setting up museums of the farming civilisation; maintenance work of the monumental olive trees as regulated by Regional Law 14/2007, surveyed and classified.[60]

Further measures envisaged within Axis 3 are also those aimed at providing incentives for the diversification into non-agricultural activities, connected to tourism and services (teaching, recreational activity and social assistance) like Measures 311, 313 and 321, as well as aimed at supporting the development of micro-enterprise like Measure 312, and at improving the level of knowledge and professional skills and the business skills of the local operators, concretely pursued through Measure 331.

In conformity with the general objectives determined by Regulation (EC) 1698/2005, the RDP of Regione Puglia 2007–2013 thus establishes a clear connection between agricultural dimension and territorial space, contemplating actions of re-qualification and protection of the characteristic elements of the landscape, of the culture and of the traditions, as well as creating infrastructures and services to favour the improvement of the quality of life of the rural areas.

Safeguarding and exploiting the characteristic elements of the Apulian regional territory become instruments, within Axis 2, for favouring the use of sustainable production practices by means of agri-environmental payments and incentives to organic farming and, in Axis 3, to increase the attractiveness of the rural territories, by creating and/or improving services and infrastructures, promoting actions aimed at supporting building restructuring work and re-qualifying villages and hamlets, disseminating the knowledge of rural values and rural trades, the training necessary for these purposes.

Amongst the actions of recovery and safeguarding of the rural heritage, the Apulian RDP assigns an important role to the re-valuation of the local knowledge, which has been subject to a process of erosion for a long time, derived from the marginalisation of the local rural communities in economic development. Furthermore, it is obvious how this knowledge has contributed in time to maintaining the biodiversity in the rural territories and to the survival of local cultures, food or craft products, local languages and traditions.

The process of recovery and reconstruction of the rural territories and identities expressed in the RDP of Regione Puglia re-valuates the local knowledge and the great heritage expressed in the farming, knowledge and traditions, restoring the evolutional and adapting capacity, as a fundamental link to ensure an authentically sustainable rural development.[61]

The rural territory, in the complexity of the matrices—naturalistic, historical-cultural and social-symbolic—which characterise it, thus becomes the fulcrum of the agricultural phenomenon and an essential element for rural development.

[60] As laid down by Arts. 4 and 5 of the already mentioned Law 14/2007 of Regione Puglia.

[61] Fonte (2009), pp. 13–15.

6 The Protection of Rural Landscapes in the Regional Landscape Plan ("Piano Paesaggistico Territoriale Regionale" or PPTR) of Regione Puglia and the Relationship with the Rural Development Programme 2007–2013

The question of protecting agricultural landscapes is also tackled within the recently adopted Regional Landscape Plan ("Piano Paesaggistico Territoriale Regionale" or PPTR) of Regione Puglia,[62] which poses "the heritage values of the Apulian landscape", including "historical rural landscapes", amongst the general objectives of the actions of exploiting and safeguarding the PPTR.

In Italian legislation, protecting agricultural landscapes—defined by Emilio Sereni, in a work that represents a milestone of studying the agricultural landscape, as "the shape that man consciously and systematically impresses on the natural landscape during and for the purposes of his agricultural productive activities"[63]—mainly goes along three lines.[64]

The first line is expressed by Art. 135, para. 4) of the Code on Cultural Heritage and Landscape (Legislative Decree 42/2004), which, in defining the prescriptions that landscape planning must necessarily establish, provides at Lett. d) to pay particular attention "to safeguarding rural landscapes".

The second is focussed on ensuring the protection of rural architecture, following Law 378/2003, "Provisions for protecting and exploiting rural architecture", whose purpose is to safeguard and exploit the types of rural architecture, like agricultural settlements, rural buildings or constructions present in the domestic territory, which bear witness to the traditional rural economy.

Finally, the third is discovered in the conservation policy of some characteristic elements of the rural landscape, like the "monumental trees", admired and protected as true "monuments" of the artistic heritage of the country and specifically within Art. 135, Lett. a) of the Code on Cultural Heritage and Landscape, as "individual beauties" of the national territory.

Following the first of the three lines, the PPTR of Regione Puglia, amongst the "further contexts" to be subjected to constraints and protection as laid down by Art. 38 of the Norme Tecniche di Attuazione (NTA, administrative and technical procedures for the implementation of the PPTR), expressly identified "rural landscapes", defining them then, at Art. 76, as "those parts of the rural territory whose landscape valence is linked to the particular integration between landscape identity of the territory and a material culture that in the long times of history has permitted the sedimentation of the characters".

[62] Adopted with DGR 1435/2013.

[63] Sereni (2006).

[64] Ferrucci (2011), pp. 202–215.

For this purpose, the PPTR envisages actions aimed at ensuring the safeguarding of the characteristics and the identity values of the anthropogenic and historical-cultural components, like restoring and recovering drystone products, maintaining and reusing buildings and rural heritage, exploiting and protecting monumental olive trees, by virtue of their importance for the identity of the landscape, the culture and the regional history, as well as their productive and ecological and hydrogeological defensive functions.[65] Furthermore, added to this, the protection of the assets spread out in the rural landscape, like drystone walls, bushes, terracing, minor drystone architecture, wells, rainwater channels.

They are elements of the territory that give the landscape a unique nature and that become recipients, within the PPTR, of a series of safeguarding measures and specific uses.

In particular, in Art. 83 of the NTA, it is established that it is forbidden to carry out all the interventions that jeopardise "the anthropogenic, semi-natural and natural elements characterising the agricultural landscape", which do not guarantee "the correct landscape insertion, respect of building types and of traditional agri-cultural landscapes, as well as the ecosystemic-environmental balances", or which involve "urban transformations...which change the long-term character of the settlement pattern".

On the other hand, after ascertaining their landscape compatibility, interventions aimed at the following are permitted: "realising systems for collecting rainwater, water-sewage networks and systems for treating wastewater"; "maintenance and restoring drystone walls in poor state of repair"; "creating gentle itineraries on existing roadways ...correctly inserted in the landscape"; "re-naturalisation, maintenance, restoring ...of the minor buildings and architecture".

In relation to the effectiveness of the "system of values" and of actions outlined in the PPTR, in the environmental Report annexed to the PPTR the opportunity was represented to proceed with an analysis of the interaction in relation to other plans and programmes and in particular with the RDP.

From the analysis performed by the Supervision Committee, it was found that the RDP fully contributes to the PPTR in achieving the general objectives aiming at re-qualifying and promoting historical rural landscapes, exploiting the cultural-settlement identity heritage, re-qualifying the decaying landscapes, exploiting the aesthetic-perceptive structure of the Apulian landscapes and the environmental quality of the territory.

Therefore, between the Assessment Group, the Technical Secretary of the PPTR and the Regional Departments of Agriculture and Territorial Structure, an interesting activity of collaboration and coordination was set up, which is capable of reorienting the planning strategies created by the RDP of Regione Puglia.

---

[65] For this purpose, Law 14/2007 of Regione Puglia expressly requires, in Art. 6, the respect of provisions stated by the Regional Landscape Plan, also acknowledging, in Art. 8, the importance of the promotion aspects "of the Apulian olive grove landscape, particularly the monumental olive groves and their productions, also for tourism purposes".

More specifically, the outcomes of this true process of "co-planning" led to the introduction of modifications in the text of the RDP, later approved by the Supervision Committee on June 2009, above all in relation to some measures envisaged within Axis 2, like Measure 214 on "Agri-environmental payments" regarding the protection and strengthening of the ecosystems connected to agricultural-forestry activity and Measure 216 on "Support to non-productive investments" as regards the buffer strip and the marshland.

In the Environmental Report annexed to the PPTR, furthermore, a detailed analysis has been carried out of the coherence between the objectives of the PPTR and the objectives and the measures/actions of the regional RDP: this has made it possible to find a positive interaction between the two planning instruments, not only as regards the maintenance of the hydromorphological balance and the development of the environmental quality of the territories but also with reference to the objectives of exploiting historical rural landscapes, the cultural-settlement heritage and re-qualifying decaying landscapes.

The results coming from the analysis performed have thus shown a complete integration between the RDP and the PPTR in relation to the protection of rural landscapes.

Both the planning and programming instruments,[66] despite following different directions—the promotion of rural development, for the RDP; protecting the landscape in general and particular types of landscape, for the PPTR—overlap and converge in creating a series of measures and actions focussed on exploiting the identity characters of the rural landscapes and their hydraulic, ecological and productive potential connected to the landscape, to promote rural re-population, to exploit rural buildings and constructions also for farmhouse tourism and to raise the quality of life of the populations by offering connecting services and infrastructures.

Some projects drawn up within the PPTR, like "Gentle itineraries" and the "City-Countryside pact", on the other hand, specifically point in this direction.

The former has the purpose of making the regional landscapes usable through an integrated network of cycle-pedestrian paths, train itineraries, panoramic roads and ancient paths to be recovered.

The latter responds to the need, also expressed in some measures of the RDP, to raise the environmental and landscape quality of both territories: the urban one, by clearly defining its margins, the public functions and spaces that characterise it on the historical level, improving the quality of buildings and of town planning; the

[66] According to prevailing Italian interpretation, in public law legal jargon the terms *plan* and *programme* must be considered interchangeable synonyms: see Giannini (1983), pp. 629 et seqq.; Stipo (1991), pp. 1 et seqq., which offers the following definition of *planning*: "a technique of each activity whose duration is projected in time and consists of having this activity carried out according to a pre-ordered pattern, within a time span, establishing certain contents and providing for certain financial means in view of the objectives to be achieved" (author's translation).

rural one, by restoring its specificity and functions. The "City-Countryside pact" comes from the firm belief that the degeneration process triggered by the excessive urbanisation of the countryside has ended up by determining a progressive decay of *both* these life environments of man, the urban and rural ones.[67]

---

[67] The "City-Countryside pact" in the PPTR of Regione Puglia is articulated in many directions (see the description on the webpage http://paesaggio.regione.puglia.it/index.php/lo-scenario-strategico/cinqueprogetti/il-patto-citta-campagna.html. Accessed 31 July 2014):

- the "confined countryside": is an area of agricultural territory around the city which surrounds its peripheral fringes with a greenbelt; in this area the construction of ancient "confines" is envisaged (on the limits of the current outskirts) to promote the re-proposal of an agricultural landscape well related to the city, like the "*ristretti*" (confines) were like in the past;
- the "multifunctional agricultural parks": are periurban, agri-urban or agri-environmental territories which are more extensive than the "confine"; they propose proximity agricultural forms which associate the externalities of multifunctional agriculture to agricultural activities producing not only quality agricultural products, hydrogeological safeguarding, landscape quality, ecological complexity and local closure of the cycles, but also good use of the rural space, exploitation of diffused and monumental rural building, activation of local economic systems; the agricultural park thus becomes a bearer of new ecological, social, cultural and symbolic values;
- the "CO_2 park": it is urban forestation of the productive or industrial areas as areas for environmental compensation; the proposal is that of covering the large parking areas and the open spaces on the outskirts of the industrial areas with large wooded areas; to create barriers to the noise and dust to protect the built-up edges; to contribute to energy production from biomass;
- the "urbanised countryside": to contrast the problems connected to the proliferation of low-density and de-contextualised urban settlements dispersed in the last few decades in the rural space (villas warehouses, shopping centres, etc.), making up the urban sprawl of the scattered city, the Apulian PPTR aims at blocking the further occupation of the agricultural land and at promoting the regeneration of the already existing settlement fabrics in order to integrate them harmoniously in the rural context or alternatively to connect them to the city as long as they become eco-compatible;
- the "lived-in countryside"; unlike the urbanised countryside, it consists of the diffusion of productive and residential fabrics which have maintained a relationship with the agricultural activities and ensure a positive functional link between the city and countryside; for these territories, the strategy of the PPTR aims at maintaining the physical and social manner of constructing the link between the residential community and agricultural-forestry activities, in order to the protect the quality of the scattered agricultural settlement, to improve services and infrastructures for the villages, to support the interventions on rural and drystone building work and provide incentives for farmhouse tourism;
- the "natural coastal park": the PPTR aims at stopping the mounting process of urbanisation of the Apulian coastal areas, that is a territory of high environmental value; for this purpose, in the delicate landscape of transition between the coast and the sea, the PPTR envisages a project for the coast with *nature* (pine woods, marshland, and so on) in place of *houses*; the areas of high natural importance are the landscapes which can be integrated with the coastal environment, from whose materials all the strategies for re-qualifying the coast have to start;
- the "agri-environmental coastal park": in order to ensure the important function of maintaining open breaches, marking the landscape rhythm of the coastal cities and guaranteeing the continuity of the open spaces between the countryside and the sea, the PPTR promotes agri-environmental activities as an alternative to coastal concrete, also by enhancing the historical landscapes.

This communion of intents between the RDP and PPTR, therefore, further strengthens the pursuit of the objectives of protecting agricultural landscapes, allowing financing to be directed and to promote projects that, in respect of the provisions and directives of the Landscape Plan, are aimed at guaranteeing re-qualification or reconstruction of quality landscapes.

7 Open Questions and Critical Surveys

In spite of the positive potential expressed by the RDP of the Regione Puglia 2007–2013 as regards safeguarding rural territories, it should still be pointed out that a series of critical nodes remains.

On the level of available resources, analysis of the economic data reported in the Section of the RDP of Regione Puglia 2007–2013 dedicated to "*ex ante* assessment" has highlighted a reduction of loans compared to the previous period 2000–2006 in precise relation to the strategic Measures for the rural territories, like Measure 323 on "Protection and re-qualification of the rural heritage" and also Measure 216 on "Support to the non-productive investments in the agricultural field".

From the economic data reported, it can be seen that a large part of the available resources have been directed at improving the competitiveness of the agri-food sectors (Axis I), while the incentives on interventions aimed at increasing the quality of the territories, improving the quality of life in the rural areas and the access to services (health assistance services, transport), creating new job opportunities and diversifying the economy in the rural areas has been less.

Predictive analyses on the economic impacts conducted after the approval of the RDP 2007–2013 have also highlighted that, on the total contribution to the estimated economic growth of the Apulian territory for each of the measures of the RDP 2007–2013, only 5.29 % derive from the actions included in Axis 3. In terms of the employment impact, only 2 % of new jobs derive from the measures indicated in Axis 3 and, even, a percentage close to 0 % from those envisaged in Axis 2.

This is a negative *trend* that was already recorded with the programming for the previous period (2000–2006), above all in relation to the investments aimed at creating new job opportunities and at improving life in the rural areas,[68] but that then, in the subsequent phase of programming, also extended to the investments more specifically aimed at protecting and re-qualifying the rural heritage.

[68] On the point, see Lucatelli (2006), pp. 8–10.

From this viewpoint, a recent contribution published by the Ministry of Agricultural, Food and Forestry Policies[69] assumes significant importance. In it a proposal for modifying Art. 143 of the Code on Cultural Heritage and Landscape was tendered, which defines the minimum essential content of the landscape plans, by predicting a specially dedicated part to the rural and agricultural dimension of a territory and to the preparation of rules of coordination between the landscape plans and the regional RDP.

This would allow overcoming the problem of coordination between planning measures and instruments, which often blocks a real and effective development of the territories.

Another interesting proposal in the mentioned contribution of the Minister is the one to modify the Cultural Heritage and Landscape of the Code in relation to the introduction of a protection of the "intangible cultural heritage", that is, practices, traditions, rituals that are passed down from generation to generation and that form the cultural heritage and the identity of a community. Art. 7-*bis* of the Code currently protects the "expressions of collective cultural identity" only when "they are represented by material evidence", thus excluding many expressions inherent of the farming community.

It is certain that the development prospects of the rural territories must measure themselves against the expectations of the new Regulation (EU) 1305/2013, which inaugurated the programming period 2014–2020.[70]

The new Regulation (EU) 1305/2013, from the "Whereas" clauses, highlights the need to ensure "the sustainable development of rural areas" through specific objectives, like transferring knowledge and innovation in the agricultural sector, safeguarding ecosystems, reducing poverty and economic development of the rural areas (point 4).

Following this line, Regulation (EU) 1305/2013 assigns an important weight to the development of services, of infrastructures and of connections, in order to promote the diversification of the agricultural economy also through extra-agricultural activity (point 17) and social inclusion, inverting the trends of decline and depopulation of the rural areas (point 19). Furthermore, the new directives of the policy of rural development, no longer articulated in Axes but in "Priorities", provide the investments for re-qualifying and exploiting the cultural and natural heritage of the villages and of the rural landscape as "essential elements of any effort to realise the growth potential and promote the sustainability of the rural areas" ("Whereas" clause 19).

The growth of the rural areas and a sustainable rural development cannot, therefore, disregard investments aimed at protecting and re-qualifying the rural

[69] Ministry of Agricultural, Food and Forestry Policies (2014) Tutela giuridica del paesaggio culturale rurale tradizionale (Roma: Ministry of Agricultural, Food and Forestry Policies). http://www.reterurale.it/flex/cm/pages/ServeAttachment.php/L/IT/D/u%252Fn%252Fi%252FD.2f93de7cfb74f2f38af5/P/BLOB%3AID%3D13591. Accessed 31 July 2014.

[70] See above paragraph 2.

landscape, in its many dimensions and potential: from actions aimed at supporting organic farming and biodiversity through services of assistance and consultancy to agricultural holdings and rural enterprises (Art. 15) and economic incentives like payments for environment and climate-friendly services (Arts. 28 and 29) to the measures of protecting the rural landscape by investing in restoring the cultural and natural heritage of the villages and, finally, to the measures aimed at training the agricultural producers ("Whereas" clause 12) also through the exploitation of that heritage of traditions and knowledge that is part of the cultural wealth of agricultural landscapes and that has permitted, in time, the survival of specific local resources and cultivations, typical food and craft products and also, more specifically, cultural resources like musical expressions or linguistic idioms.[71]

This imposes on the individual Regions, which are basically called upon to interpret and ensure the application of these measures, to consider that a balanced and sustainable rural development cannot avoid protecting the territory and the rural heritage, considered as a series of values, culture, history and traditions connected to the places of the farming civilisation, restoring their capacity for evolution and adaptation and exploiting the intrinsic human capital.[72]

From this viewpoint, the rural landscapes become part of the conception of heritage, which includes everything to which the community recognises a value, in that it bears witness to its identity and to which it feels a responsibility: *primarily* of conservation, and also of enhancement, a concept that is firmly linked to that of heritage and confers on the latter a dynamic character, making it an important factor for stimulating sustainable rural development.[73]

8 Concluding Remarks

The analysis carried out here has highlighted how the Rural Development Programme represents an important programming instrument capable of "balancing" the legal discipline of the various dimensions of the agroecosystems, intended as complex social-ecological systems.

It is a financial and incentive type of coordination in which the individual "measures" stimulate the actors of the rural systems to maintain certain behaviour; the groups of measures are ordered and grouped to the service of broader "priorities"; the overall pattern of the "measures" and of the "priorities" constitutes the programme.

Only at a programme level can it be understood that the individual measures look at the "part" with respect to their immediate scope, but at the "whole" compared to their indirect purpose; it is, therefore, a question of creating a mosaic

[71] Fonte (2009), p. 14.

[72] Fonte (2009), pp. 13–15.

[73] Zerbi and Fiore (2009), p. 6.

that, in its whole, forms a complete image, that of agroecosystems, no longer regulated in a mono-dimensional manner (into non-communicating compartments) but in a pluri-dimensional one.

The RDP is a legal instrument of integrated balancing, in which the freedom of the individual operators of the rurality is not compressed authoritatively with the imposition of unilateral obligations and bans, but directed by providing incentives in order to make it compatible with the necessity to cohabit the exercising of the different individual rights within the "common house" represented by the super-individual sphere of the agroecosystems, which are at the same time ecological systems, social systems, economic systems and cultural systems, in which the millenary interaction between human populations and nature has generated emerging properties.

The possible collapse of an agroecosystem, which could derive from the absence of instruments of managing the internal conflicts between its different dimensions, would, indeed, lead to the collapse of the same individual rights, which could not be exercised outside the living system that feeds them, sustains them and allows them to "breathe".

The RDP reflects the multifunctional nature of the agricultural phenomenon and thus becomes a legal instrument capable of bridging the gap for a precious and undoubtedly innovative dialogue between law and agroecology: despite the RDPs not yet making explicit reference to agroecology as a transdisciplinary scientific platform, it is the approach under the RDP that is already "agroecologically oriented" because it aims at coordinating differentiated measures that address the many components of the agroecosystems in a holistic perspective.

Paradigms of this new agroecological approach are also the provisions contained in the RDP of Regione Puglia as regards protection and enhancement of the rural landscape, whose discipline has shown a positive convergence with the standards of the recently approved Regional Territorial Landscape Plan (PPTR).

Analysis of the measures contained in the two planning instruments (RDP and PPTR) and the relation between them still highlights the need for a process of co-planning, in order to avoid antagonisms between programming instruments that would jeopardise a real and effective protection of the rural landscape, which—it should be remembered—represents the visible "shape" of the agroecosystems and manifests to the outside their internal state of health (or, otherwise, of deterioration).

References

Adornato F (2007) Oltre l'agricoltura? Osservazioni di un giurista. Agriregionieuropa 3(11): 54–55. http://agriregionieuropa.univpm.it/en/content/article/31/11/oltre-lagricoltura-osservazioni-di-un-giurista. Accessed 5 Apr 2014

Adornato F (2011) La disciplina comunitaria dei piani di sviluppo rurale e la contrattazione in agricoltura con la pubblica amministrazione. In: Costato L, Germanò A, Rook Basile E (eds) Trattato di diritto agrario, vol 1. UTET Giuridica, Torino, pp 545–595

Albisinni F (2010) Distretti e sviluppo rurale: elementi per una lettura delle regole di diritto. Agriregionieuropa 6(20):21–25. http://agriregionieuropa.univpm.it/content/article/31/20/distretti-e-sviluppo-rurale-elementi-una-lettura-delle-regole-di-diritto. Accessed 5 Apr 2014

Angilieri V (2009) Politica Agricola Comune e protezione del paesaggio. In: Zerbi MC, Fiore F (eds) Sviluppo sostenibile e risorse del territorio. Il ruolo del patrimonio rurale, 1st edn. Giappichelli Editore, Torino, pp 44–46

Camaioni B (2009) Lo stato di attuazione dei PSR. Agriregionieuropa 5(18):51–54. http://www.agriregionieuropa.univpm.it/content/article/31/18/lo-stato-di-attuazione-dei-psr. Accessed 5 Apr 2014

Cavazzani A (2006) Lo sviluppo rurale come superamento della modernizzazione Agricola. Agriregionieuropa 2(7):60–61. http://agriregionieuropa.univpm.it/content/article/31/7/lo-sviluppo-rurale-come-superamento-della-modernizzazione-agricola. Accessed 5 Apr 2014

Contò F, Fiore M, La Sala P (2012) The metadistrict as the territorial strategy: from set theory and matrix organization model hypothesis. Int J Food Syst Dyn 3(2):82–94

Costato L (2012) La PAC riformata, ovvero la rinuncia a una politica attiva. Rivista di diritto agrario 2:393–404

Crialesi G (2009) L'iniziativa Leader per lo sviluppo delle comunità locali. Finanziamenti su misura news 13(11):6–10

Desideri C (1985) Costituzione economica ed agricoltura. Nuovo Diritto Agrario 12(3):301 et seqq

Ferrucci N (2011) La tutela del paesaggio e il paesaggio agrario. In: Costato L, Germanò A, Rook Basile E (eds) Trattato di diritto agrario, vol 2. Utet Giuridica, Torino, pp 202–215

Finuola R, Pascale A (2008) L'agricoltura sociale nelle politiche pubbliche. Biemmegraf, Macerata

Fonte M (2009) La conoscenza locale, una componente negletta del capitale umano nelle aree rurali. Agriregionieuropa 5(16):13–15. http://agriregionieuropa.univpm.it/content/article/31/16/la-conoscenza-locale-una-componente-negletta-del-capitale-umano-nelle-aree. Accessed 3 Apr 2014

Fonte M (2010) C'era una volta il mondo rurale. Agriregionieuropa 6(20):4–6. http://agriregionieuropa.univpm.it/content/article/31/20/cera-una-volta-il-mondo-rurale. Accessed 5 Apr 2014

Fracchia F (2013) Introduzione allo studio del diritto dell'ambiente. Editoriale Scientifica, Napoli

Fugaro A, Giuliodori A (2006) Una visione integrata e strategica dei piani di sviluppo rurale 2007–2013. Agriregionieuropa 2(6):22–25. http://www.agriregionieuropa.univpm.it/content/article/31/6/una-visione-integrata-e-strategica-dei-piani-di-sviluppo-rurale-2007-2013. Accessed 3 Apr 2014

Giannini MS (1983) Planning. In: Enciclopedia del Diritto, vol 23. Giuffré, Milano, pp 629 et seqq

Graziani CA (1985) Sull'attualità dell'art. 44 della Costituzione. Nuovo Diritto Agrario 12(3):309 et seqq

Graziani CA (2007) Proprietà della terra e sviluppo rurale. Agricoltura Istituzioni Mercati 4(1):65–94

Graziani CA (2012) Dal diritto alla terra al diritto della terra. Agricoltura Istituzioni Mercati 9(2–3):65–79

Hoffmann A (2006) La nuova politica di sviluppo rurale. FrancoAngeli, Milano

Lucatelli S (2006) Lo sviluppo dei territori rurali nel Documento Strategico Preliminare Nazionale. Continuità, discontinuità, priorità per la politica regionale 2007–2013. Agriregionieuropa 2(4):8–10. http://www.agriregionieuropa.univpm.it/content/article/31/4/lo-sviluppo-dei-territori-rurali-nel-documento-strategico-preliminare-nazionale. Accessed 3 Apr 2014

Magno P (2007) Le linee comunitarie per lo sviluppo dell'agricoltura nel periodo 2007–2013. Diritto e giurisprudenza agraria alimentare e dell'ambiente 2(4):226–230

Mantino F (2012) Alla ricerca dell'integrazione perduta tra aiuti PAC, sviluppo rurale e politiche di coesione nella riforma 2014–2020. Agriregionieuropa 8(29):28–31. http://www.agriregionieuropa.univpm.it/content/article/31/29/alla-ricerca-dellintegrazione-perduta-tra-aiuti-pac-sviluppo-rurale-e. Accessed 3 Apr 2014

Mantino F (2013) La programmazione dello sviluppo rurale 2014–2020: il position paper e l'accordo di partenariato. Agriregionieuropa 9(32):39–41. http://www.agriregionieuropa.univpm.it/content/article/31/32/la-programmazione-dello-sviluppo-rurale-2014-2020-il-position-paper-e-laccordo. Accessed 1 Apr 2014

McAreavey R, McDonagh J (2011) Sustainable rural tourism: lessons for rural development. Sociol Rural 51(2):175–194

Monteduro M (2013) Environmental law and agroecology: a transdisciplinary approach to public ecosystem services as a new challenge for environmental legal doctrine. Eur Energy Environ Law Rev 22(1):2–3

Monteleone A (2010) La politica di sviluppo rurale in Italia tra presente e future. Agriregionieuropa 6(20):45–48. http://www.agriregionieuropa.univpm.it/content/article/31/20/la-politica-di-sviluppo-rurale-italia-tra-presente-e-futuro. Accessed 4 Apr 2014

Moyano Estrada E (2010) La Pac e lo sviluppo rurale dopo il 2013. Integrazione o separazione? Agriregionieuropa 6(20):51–52. http://agriregionieuropa.univpm.it/content/article/31/20/forum-sulla-pac-dopo-il-2013-parte-5. Accessed 3 Apr 2014

Parkinson S (2009) Power and perceptions in participatory monitoring and evaluation. Eval Program Plann 32(3):229–237

Pennacchi F (2006) Il ruolo della politica di sviluppo rurale per la crescita del territorio. Agricoltura e sviluppo rurale: scenario competitivo e modelli di sviluppo. Agriregionieuropa (4):11–16. http://agriregionieuropa.univpm.it/content/article/31/4/il-ruolo-della-politica-di-sviluppo-rurale-la-crescita-del-territorio. Accessed 4 Apr 2014

Rampulla FC (2010) Lo sviluppo sostenibile: una politica da perseguirsi in via procedimentale ed il caso dell'agricoltura. Diritto e giurisprudenza agraria, alimentare e dell'ambiente 5(12): 729–733

Romito G (2012) Gli strumenti di programmazione per la politica di sviluppo rurale 2014–2020. Agriregionieuropa 8(29):39–41. http://www.agriregionieuropa.univpm.it/content/article/31/29/gli-strumenti diprogrammazione-la-politica-di-sviluppo-rurale-2014-2020. Accessed 1 Apr 2014

Saija R (2006) Sviluppo rurale, *governance* locali e nuovo regionalismo in agricoltura dopo la riforma del Titolo V della Costituzione. Diritto e giurisprudenza agraria, alimentare e dell'ambiente 15(5):284–289

Secco L, Da Re R, Pettenella D, Cesaro L (2010) La valutazione della qualità della governance partecipativa in ambito rurale. Agriregionieuropa 6(23):61–65. http://www.reterurale.it/flex/cm/pages/ServeBLOB.php/L/IT/IDPagina/4523. Accessed 13 May 2014

Secco L, Pettenella D, Gatto P (2011) Forestry governance and collective learning process in Italy: likelihood or utopia? Forest Policy Econ 13(2):104–112

Sereni E (2006) Storia del paesaggio agrario italiano. Edizioni Laterza, Roma Bari

Sonsini A (2006) Per una nuova definizione urbanistico-architettonica dei territori rurali. Agriregionieuropa 2(5):37–38. http://agriregionieuropa.univpm.it/content/article/31/5/una-nuova-definizione-urbanistico-architettonica-dei-territori-rurali. Accessed 4 Apr 2014

Sotte F (2003) Introduzione. In: Arzeni A, Esposti R, Sotte F (eds) Politiche di sviluppo rurale tra programmazione e valutazione, 1st edn. FrancoAngeli, Milano, pp 18–21

Sotte F (2009) Considerazioni conclusive. In: Sotte F (ed) La politica di sviluppo rurale 2007–2013. Un primo bilancio per l'Italia, 1st edn. Edizioni Tellus, Roma, pp 237–238; 243–244

Stipo M (1991) State planning and regional planning. In: Enciclopedia Giuridica, vol 24. Treccani, Rome, pp 1 et seqq

Torquati B (2007) Il paesaggio nel Piano Strategico Nazionale per lo Sviluppo Rurale. Agriregionieuropa (8):41–43. http://agriregionieuropa.univpm.it/content/article/31/8/il-paesaggio-nel-piano-strategico-nazionale-lo-sviluppo-rurale. Accessed 4 Apr 2014

Tropea F (2014) First pillar of the cap: II—direct payments to farmers. http://www.europarl.europa.eu/aboutparliament/en/displayFtu.html?ftuId=FTU_5.2.5.html. Accessed 11 July 2014 van der Ploeg JD (2006) Oltre la modernizzazione. Processi di sviluppo rurale in Europa. Rubbettino, Soveria Mannelli, CZ

Varotto M (2005) Nuovi modelli di "*governance*" dello sviluppo locale: i distretti rurali. Azienditalia 12(5):2–20

Varotto M (2007) La progettazione integrata nelle aree rurali. Azienditalia 14(9):56–59

Zerbi MC, Fiore F (2009) Sviluppo sostenibile risorse del territorio. Il ruolo del patrimonio rurale. Giappichelli Editore, Torino

Zumpano C (2007) L'approccio integrato nelle politiche di sviluppo rurale: strumenti e modalità di attuazione. Agriregionieuropa 3(9):56–59. http://www.agriregionieuropa.univpm.it/content/article/31/9/lapproccio-integrato-nelle-politiche-di-sviluppo-rurale-strumenti-e-modalita-di. Accessed 3 Apr 2014

Part II
Multidimensional Relations Between Land, Agriculture and the Environment

The Common Agriculture Policy (CAP): Achievements and Future Prospects

A. Isoni

Abstract This paper aims to offer an analysis of the Common Agriculture Policy over the past 60 years. Taking the poor conditions of European agriculture between the two world wars as a starting point, this analysis considers the reasons that compelled the EU founding fathers to insert agriculture among the key sectors seen to further integration. Furthermore, this study aims to demonstrate how the protectionist attitude adopted by the European Commission actually arose from a long tradition of intervention by national governments in almost every European State.

As a matter of fact, the CAP was initially characterised, on the one hand, by a protectionist approach and, on the other hand, by a strong productivist attitude, in order to both guarantee European food independence and support farm incomes. These goals were attained, thanks to a price support system, which became very expensive with respect to the available EEC budget. In the 1980s, the European Commission came under the pressure of both national governments and economic globalisation, and consequently reviewed the CAP, thereby contributing towards a change of paradigm in the European agricultural sector. As a result of three main reforms—the MacSharry Reform (1992), the Agenda 2000 (1999) and the Fischler Reform (2003)—the CAP has become more centred on a multifunctional approach based on two principal pillars: firstly, aid towards food production, i.e. direct support to farmers, and, secondly, initiatives promoting the development of sustainable agriculture, according to an "agroecological" perspective, which allows for the protection of nature, as well as of regional cultures and traditions

Keywords Agroecology • Common agriculture policy • Multifunctional agriculture • Productivism • Protectionism

1 The Agricultural Exception

A consideration of the history of European agriculture over the last fifteen centuries reveals a clear moment of change in the second half of the nineteenth century. This was marked by the introduction of the first agricultural machinery and the

A. Isoni (✉)
Department of History, Society and Human Studies, University of Salento, Lecce, Italy
e-mail: alessandro.isoni@unisalento.it

M. Monteduro et al. (eds.), *Law and Agroecology*,
DOI 10.1007/978-3-662-46617-9_9

increasingly widespread use of chemical fertilisers.[1] Nonetheless, agriculture has continued to be a high-risk economic activity, subject not only to normal market fluctuations but also, above all, to weather uncertainties. Bearing in mind these precarious premodern conditions allows without doubt for a greater appreciation of the purposes of the Common Agriculture Policy and its less famous antecedents. According to the "prophecies" of Alexis de Tocqueville and Donoso Cortés, it was the introduction of Russian and American wheat into the world market at the end of the nineteenth century that initiated a process of economic globalisation, consequently obliging European States to find solutions to the cut-throat competition of these future superpowers.[2] While Great Britain continued with its free trade tradition, all the other governments adopted a protectionist approach, in order both to protect weak agricultural sectors and to avoid strong deficits in their trade balances.[3] Moreover, during the First World War, food requirements became one of the main battlefields over which the Central Powers lost the war, but only because France, Great Britain and Italy received extensive aid from the United States. A key feature of this aid was characterised by "Executives", an innovative administrative tool through which national governments could negotiate without parliamentary control.[4]

As a matter of fact, the war amply demonstrated that the agricultural sector was unable to feed all Europeans, thereby obliging every single government to find new ways of protecting their farmers and food production. In large parts of Europe, agriculture was primarily engaged in at a subsistence level, as farmers used traditional methods, while production was low and not market oriented. Aside from the war, other reasons contributed to this degenerating situation, namely, the generally low level of education among rural populations, widespread illiteracy and low standards of living. Policies centred on the well-established protectionist approach were adopted almost everywhere, and this tendency was strengthened during the 1930s by the outset of the Great Depression, resulting in the slowing down of agricultural reforms. Subsequently, European agriculture was characterised by marked underdevelopment on the eve of the Second World War, which confirmed the idea that agriculture had to be managed in exceptional ways. World War II heavily contributed to the ruin of the entire European economy, and its agricultural sectors were strongly affected by the general depression, leading to much concern among political leaders, particularly in the US. When new democratic States were re-established by the Allied Powers in Western Europe, every single government concerned decided to intervene openly in the agricultural sector, in order to produce more food and overcome the famine experienced during the war.[5] As a matter of fact, widespread opinions suggested that price instability

[1] Slicher Van Bath (1963).

[2] O'Rourke (1997), pp. 775–801.

[3] Foreman-Peck (1983).

[4] Laqua (2011).

[5] Milward (1984).

during the 1930s had provoked the establishment of fascist dictatorships in Europe. Meanwhile, in the postwar period, fear of a Communist revolution predominated, together with concerns about the financial imbalance related to trade between Europe and the USA, without much consideration being given to the electoral weight of farmers and their political organisations.[6]

In order to help the agricultural sector, almost every national government decided to buy agricultural products at fixed, high prices so that farmers' income could be protected from market fluctuations.[7] This political decision exemplified some of the main features of the first CAP paradigm: the idea of agricultural exception, the consequent need for protectionist policies and, last but not least, political concerns regarding a significant part of the electorate.

Discussions on the integration of agricultural policies in Europe began immediately after the Second World War had ended. Talks took place in the Council of Europe and the OEEC (Organisation for European Economic Co-operation) between seventeen nations, based on proposals from France, Britain and the Netherlands. Meetings were centred on two important issues: firstly, the security of food supplies, which was hard to guarantee as a result of the war and had key consequences, such as a decline in food consumption and a high level of dependency on food imports, and, secondly, the security of income for farmers, which was of considerable significance, seeing as important empirical studies had shown that farm incomes were lagging behind those of other sectors.[8]

Unfortunately, these negotiations on creating a common policy for agriculture, which took place between 1952 and 1954—during the same years in which the European Coal and Steel Community (ECSC) was taking its first steps—failed to bring about any kind of agreement. These discussions were paralysed by differing opinions between France and the Netherlands on the one side, both arguing for a supranational policy and strong community preferences, and Britain on the other side, opposing any form of supranationalism and strongly committed to maintaining relations with the Commonwealth.

Nonetheless, the discussions and, likewise the breakdown of these talks, served to identify differences in opinion between European countries, at least in relation to agricultural issues. It became evident that views on these issues varied, not only according to political preferences but above all depending on a wide range of reflections dating back to historic events and also concerning the traditions and specific culture of every single State.[9]

A historic tendency towards protectionism is only one of the reasons that may explain why, during the 1955 Conference of Messina, the six ECSC member States decided to insert agriculture in the future Treaty establishing the European Economic Community (EEC). Another reason for this decision can undoubtedly be

[6] Milward (1992).

[7] Johnson (1947).

[8] OECD (1961).

[9] Tracy (1989).

found in the awareness demonstrated by the European founding fathers, in considering the complexity of agricultural issues, without isolating these from more general economical concerns. Although agriculture represented only a minor part of the GNP of many European States, it was impossible to imagine a single market without involving agricultural goods.[10] According to Sicco Mansholt, there were at least four good reasons to include agriculture in the European integration process in 1957: first, the practical difficulty of excluding agriculture from being part of an integrated market, due to the impossibility of drawing a clear line between agricultural and industrial products; second, agriculture played a major role in the economies of the six countries concerned; third, food price levels and fluctuations were significantly influenced within each national economy by agricultural markets and policies; and, fourth, changes and adjustments in the agricultural sector were essential to general economic growth.[11]

Nonetheless, in spite of the decision to include agriculture in the process of European economic integration, the Treaty of Rome failed to provide clear regulations for the CAP, thereby referring agreements on a common agricultural policy to further negotiations among European States. The decision to wait for the Treaty to enter into force had some practical consequences, allowing every member State to introduce its own political traditions into the new communitarian policy. As a matter of fact, the implementation of an intergovernmental approach in establishing the new Common Agricultural Policy reveals the difficulties in understanding the CAP without adopting a long-term perspective, able to match the EEC legal framework with the cultural and political heritage linked to its agricultural milieu.[12]

2 The Birth of the Common Agriculture Policy

On 1 January 1958, the EEC Treaty, signed in Rome on 27 March 1957, came into being, driving European integration towards new and more ambitious goals following the innovative experience of the ECSC. Besides providing for the establishment of a common market, articles 38–47 of the Treaty covered the agricultural sector, offering, however, only some basic guidelines that required further negotiations in order to be implemented. Article 39, more specifically, presented a set of objectives for the CAP, such as the resolution to increase agricultural productivity by promoting technological progress, by ensuring the rational development of agricultural production and by optimising the use of factors influencing production, with particular regard to labour. As a consequence, European policymakers aimed to ensure a fair standard of living within the agricultural community, most specifically by increasing the individual earnings of those engaged in agriculture.

[10] Brandow (1977), pp. 209–294.

[11] Mansholt (1963).

[12] Tracy (1984), pp. 307–318.

Through the process of stabilising markets, the CAP aimed to ensure both the availability of supplies and reasonable prices for consumers. Article 40, in particular, set forth an agenda for the actualisation of the CAP. This was articulated in stages to be implemented over a 5-year transition period starting in 1962. Article 43, subsequently, made the Commission responsible for designing the actual policy, and the deadline for starting the CAP was fixed within 2 years.[13]

According to the new communitarian method, the European Commission held the power of initiative to set up the CAP framework: with this in mind, the European Commission invited national delegations to a conference in Stresa (Italy) from 3 to 12 July 1958, each composed of politicians, civil servants and representatives of organisations in the farm and food industries, in order to discuss key problems facing the agricultural sector. Discussions involved three working groups and led to conclusions, which influenced all future EEC decisions. The affirmation that agriculture had to be considered both as an effective part of the economy and as a basic factor of social life was taken as a starting point. According to this general principle, the new CAP aimed to promote both intra-community adjustments and fixed custom duties on agricultural goods coming from extra-EEC countries. In this context, the need to protect the single market was accompanied by a policy oriented towards the reformation of the entire European agricultural sector, with the goals of promoting productivity and increasing price levels while avoiding the risks of overproduction and also supporting less competitive regions.[14] However, the medium-term goal was to fill the gap between the agricultural and industrial sectors, in order to avoid the depopulation of rural areas and the breakup of secular traditions, well represented by family-owned farms.

The strategy adopted to attain these goals was centred on three pillars: first, the absolute centrality of agriculture in the general economic strategy of the Community; second, the protection of intra-Community trade in agricultural products against distortions from the world market; third, the provision of a market organisation based on price support, working in close relation to structural policy measures.

This latter point, together with the idea that family-run farms should be the cornerstone of European agriculture, represented the most important issue that emerged at Stresa: a common agricultural policy established on price support without any form of structural policy would never result in the achievement of the most important objective concerning farmers' income. On the other hand, some delegations argued that this kind of approach would create a wide range of problems for the Community, especially concerning the EEC budget and the long-term sustainability of European agriculture.[15]

In 1960, after 2 years of in-depth reflections and hard work, the European Commission presented an initial draft for the CAP. This was founded on the French

[13] Fearne (1997), pp. 11–55.

[14] Communauté européenne (1958), p. 250.

[15] Tracy (1994), pp. 357–374.

and Dutch points of view and contained three pivotal features: firstly, free intra-community trade, with no barriers and restrictions to trade in agricultural products between member States; secondly, a preference for Communitarian agricultural products; and, finally, common funding for the CAP.[16]

In order to understand this broader framework, it is vital to consider a question that concerns the CAP in all its different aspects: why did the EEC and its member States decide to adopt the continental approach, founded on artificial price fixing, instead of direct income support, drawing on the example of the United Kingdom? The reason may be found in the long tradition of direct interventions experienced by many European States, on the basis of fears of predictably heavy political and social consequences incidental to a productivist approach. However this may be, this proposal was implemented in the course of the 1960s and succeeded in dealing with numerous problems, most importantly concerning the price levels of agricultural products and, moreover, regarding who was financially responsible for the high costs required to ensure an adequate standard of living for agricultural populations, particularly through higher individual earnings.

In this sense, price fixing was at the heart of the struggle between France and Germany during the time between the Conference of Stresa and the launch of the CAP, in December 1964. As a matter of fact, the CAP was born as a French–German agreement, centred on the reciprocal exchange between the superiority of German manufacturing industries and French predominance in agricultural productions. In some way, CAP negotiations provide some of the most valuable insights to how European integration developed, characterised by recurring breaks between France and Germany, with the latter initially reluctant to accept the French proposal and eventually obliged to do so.[17]

France, with over 20 % of its population working in the agricultural sector and almost 10 % of the GNP deriving from farm production, had a real interest in the CAP and considered it to be an indispensable requisite for participation in European integration. According to De Gaulle, the entry into force of the CAP allowed France to modernise its agricultural sector, thereby solving extensive social problems, as well as helping France to keep economic parity with Germany, in order to safeguard its political predominance inside the Community. On the other side, Germany was worried about losing all traditional trade links with third countries, involving the buying of agricultural goods and the sale of industrial products. As a matter of fact, the perspective of having a single market for its enterprises convinced Germany to adopt the CAP, thereby allowing Dutch and French agricultural goods to enter a bigger market.[18]

---

[16] European Commission (1960) Proposals for the Working Out and Putting into Effect of the Common Agricultural Policy in Application of Article 43 of the Treaty Establishing the European Economic Community. DG VI/COM (60) 105, Brussels.

[17] Thiemeyer (2009), pp. 47–59.

[18] Germond (2010), pp. 25–44.

The new CAP was built on the principles of free intra-community trade, on an EEC scheme of preferences and on common financing, which meant that it would be funded by means of a European budget. This latter feature resulted in the establishment of the European Agricultural Guidance and Guarantee Fund (EAGGF), with the "Guarantee section" responsible for paying almost 100 % of expenditures for community organisations dealing with agricultural markets.

During the first years of implementation, the European Commission, together with its member States, was engaged in the creation of a complex framework, oriented towards ensuring reliable income sources for farmers. On the domestic side, a tariff union was set up as a prerequisite for a common market based on free trade, while the creation of market organisations for all agricultural products allowed for high institutional prices. On the external side, the protectionist origins of the various national agricultural policies resulted in the establishment of a system of import levies and export restitutions, in order to safeguard communitarian agricultural products against the competition of third country goods.[19]

Clearly, this approach protected the strongest producers inside the CAP, such as French and Dutch farmers. However, according to the principle of common financing, the CAP shared all costs and benefits between member States, thanks to the provision that these issues were to be handled through the Community budget.

Following a period of transition, the CAP was fully implemented in the summer of 1967, with the fixing of high institutional prices for some of its major products—like sugar and butter—as a result of long and hard negotiations, later to be known as "agricultural marathons", between the six countries involved. These high prices were the consequence of a strategy adopted by Germany and Luxembourg, the weakest countries from an agricultural point of view, which insisted that their price levels were converted into CAP price levels in order to protect their inefficient farms.[20]

From a constitutional point of view, the agreement envisaged that structural policies would fall under the responsibility of national governments, contradicting the European Commission, which argued that the best solution was for common structural policies to work together with price support policies at a communitarian level.[21] In the end, as was predicted, the increasing costs of market support and the high price policy immediately resulted in a budgetary crisis. This was combined with the inability of many farmers to receive good incomes, in spite of all the mechanisms provided for by the CAP. On December 1968, the need to address the negative effects of the CAP led the European Commission to present a memorandum, which was significantly entitled "Agriculture 1980". In this document, the communitarian policymakers outlined a wide set of solutions to the problems the CAP was facing after only a few years.[22] Inspired by the strong figure of Sicco

[19] Hill (1984).

[20] Neville-Rolfe (1984).

[21] Ludlow (2006).

[22] European Commission (1968) Memorandum of the reform of agriculture in the European Community. DG VI/COM (68) 1000, Brussels.

Mansholt, the Dutch Commissioner for Agriculture, the memorandum revealed how the gap between farmers' incomes and other productive sectors had not been bridged. This was both a consequence of the inability of the high-price policy to solve problems concerning productivity and a result of some structural problems, such as the small dimensions of many farms and the progressively ageing population in the agricultural sector.[23]

The memorandum was inspired by the desire to improve on the welfare programs for farmers and, for this purpose, advanced the following proposals: the first aimed at transforming European agriculture by means of structural modernisation and the transformation of farmers into businessmen; the second concerned prices, which had to play their classical market economy role, matching supply with demand.[24]

In order to proceed with price cuts and reform agricultural structures, the Mansholt Plan foresaw that, in the long term, farms had to become bigger, thereby reducing costs and enabling farmers to compete on the world market. With this purpose in mind, the European Commission provided for two kinds of farms: on the one hand, "Production Units" (PU), big individual farms or farms associated with others, working on areas of 80–120 ha, and, on the other hand, "Modern Agricultural Enterprises" (MAE), i.e. farms that decided to merge with other farms, in order to reach a critical mass of 80–120 ha. In both cases, the goal was to ensure adequate incomes for farmers while envisaging the possibility that these new productive units would, within 5 years, become the only beneficiaries of EEC aid.

In order to limit the increasingly big agricultural surplus, the Mansholt Plan aimed to reduce the existing number of European farmers by 5 million within 10 years, by way of a program of incentives, early retirements and other preferential treatments. At the same time, European technocrats sanctioned the reduction of 5 million ha of arable land.

According to this dirigiste approach, the entire Mansholt Plan revealed itself to be not only a wide set of measures concerning agricultural issues but above all an extensive study aimed at finding solutions to many social, political and economical problems concerning European agriculture.[25] On the other hand, throughout the 1960s the project of establishing a more advanced agricultural sector was the main topic on the communitarian agenda, involving technicians, agronomists, engineers, law scholars and sociologists in a long-lasting debate. In spite of this complex cultural background, agriculture was basically considered to be an economical problem. Some issues—such as the common use of financial funds and services and the merging of farms—clashed with the general opinion of EEC member States, with their preference for not considering their respective agricultural sectors within a long-term perspective. In other words, the Mansholt Plan was ahead of its time, while member States were extremely keen on maintaining socio-economical

[23] Seidel (2010), pp. 83–102.

[24] Knudsen (2009).

[25] Sheingate (2001).

issues under their control, leaving only the power to fix common prices to the European Commission. As a consequence, the more advanced proposals of the Mansholt Plan were not accepted and, on April 1972, the European Commission presented further directives, which outlined a program centred on three pillars: modernisation, early retirements and socio-economical assistance.

In spite of the violent opposition provoked by the dirigiste approach of the Mansholt Plan in farming circles throughout Europe, it provided anything but a straightforward analysis of some undeniable tendencies affecting Western Europe at the time, namely the passage from agrarian to postindustrial societies, marking the end of a century-old world based on farmers and landowners.[26]

Before moving on to analyse the numerous reforms adopted between the end of the twentieth century and the first decade of the twenty-first century, it is equally important to mention the Schuman Plan, which was also launched in 1951 to create a single market for coal and steel. However, after a few years, the High Authority of the European Coal and Steel Community was obliged to manage the closure of coal mines across Western Europe. Similarly, the EEC found itself faced with a process of extensive reorganisation in an economical sector, which was about to lose its centrality after at least 2000 years.

3 The Obsession for Reforms

As has been seen, the CAP price support policy turned out to be too expensive to both improve agricultural production and attain market stabilisation and was consequently identified as a kind of original CAP sin.[27] From the very beginning, many observers highlighted that if producers knew that all their products would be sold at a fixed price, they would expand their production up to a point where marginal costs matched guaranteed prices. As easily imagined, farmers immediately took advantage of this simple truth and production rates soared, leading to overproduction and a budgetary deficit.[28]

Moreover, during the first 30 years following its implementation, the CAP neglected structural policies while focusing on an agrarian and productivist strategy in order to increase farmers' income. As a result, it failed to reach one of its main goals, namely the structural reform of European agriculture, according to the "Guidance section" of the EAGGF, which co-financed measures to improve agricultural production and marketing structures, as well as compensatory allowances for less-favoured areas.

From 1969 onwards, the financial situation increasingly worsened, seeing as the EEC was obliged to abandon the common price policy, one of the pillars on which

[26] Ludlow (2005), pp, 347–371.

[27] Hofreither (2007), pp. 333–348.

[28] Josling (2009), pp. 115–176.

the entire CAP was built. The main reason for this decision was a strong re-evaluation of the Deutsche Mark (DEM) and a simultaneous devaluation of the French Franc (FF), which led to the introduction of the "Monetary Compensatory Amount" (MCA), which was able to balance differences in income between French and German producers. Many observers noted that the introduction of MCAs resulted in a new nationalisation of price policies, making a CAP reform even more difficult.[29]

In this sense, the role played by national organisations in establishing the CAP structure in each country turned out to be a major stumbling block to reforming the entire agricultural sector. The combined effect both of agricultural policies managed by national governments and of a corporatist approach resulted in decades of inertia, characterised by yearly meetings during which common prices were fixed.[30]

Along with the impossibility of proceeding with a deep structural reform of the European agricultural sector, the protectionist nature of the CAP provoked severe international disputes with the United States and third-world countries, all of which requested that their agricultural products be treated equally to communitarian ones. To tell the truth, the General Agreement on Tariffs and Trade (GATT), established in 1947 on the principles of non-discrimination, open markets and fair trade, did not provide for agricultural products, thus leaving the way clear for the EEC to rule the agricultural sector according to its original protectionist approach. However, over the years, the postwar international trade framework needed to be reformed, and during both the Kennedy Round (1964–1967) and the Tokyo Round (1973–1979), the CAP was accused of being an obstacle to free trade and of creating distortions within the market.[31] Aside from critiques coming from third countries, the CAP gained increasing disapproval within the EEC, especially because a large part of its budget was destined to the CAP, thus limiting the development of new policies in other fields. During the 1980s, the CAP approach became untenable, also due to the British crusade against the EEC budgetary policy, which was principally devoted to financing the CAP and, indirectly, the French, Dutch and German agricultural sectors. Up against Margaret Thatcher's request to "get her money back", the European Commission realised that it was high time to proceed with a wide set of reforms.

Apart from the entry into force, in 1985, of the new European Commission chaired by Jacques Delors, there were many other reasons in those years that led to the inauguration of a new CAP season. A primary reason undoubtedly concerned economic issues, such as the concentration of properties in Northern Europe and the enduring smaller dimensions of farms in Mediterranean countries. If on the microeconomic level the problem concerned the low profitability rate of many farms, especially in Southern Europe, on the macroeconomic level the main problem was

[29] Webber (1999), pp. 45–67.

[30] Spoerer (2010), pp. 143–162.

[31] Daugbjerg (2004).

the increasing budgetary deficit and the challenge of managing massive European overproduction, especially in the milk and dairy sector.

In this context, with the CAP literally drowning in a lake of milk, the start of the Uruguay Round in 1986 influenced the decision to proceed with a process of reforming the PAC, also due to the increasing pressure of the United States and the ACP countries, which were lobbying for the elimination of both EEC trade barriers and custom duties for third-country agricultural goods.[32] In this sense, the successful conclusion of the Uruguay Round, with the establishment of the World Trade Organization (WTO) in 1994 and the inclusion of agricultural goods in global trade regulations, marked a new step in the economic history of the CAP.[33]

As a matter of fact, the general dissatisfaction with the CAP not only was based on economic reasons but was also due to a cultural change that occurred between the late 1970s and mid-1980s. At this time, European societies experienced a wide range of cultural innovations, ranging from a new awareness of natural resources and sustainable development to a novel degree of consumerism requiring new regulations ensuring food safety. Meanwhile, debates on globalisation and the development of capitalism reinforced arguments that both domestic and EEC policies should not distort trading, thus handicapping less-developed countries.

The fight against protectionism was one of the main issues supported by economists, who underlined how the CAP gave rise to overproduction, thereby favouring hidden income transfers from consumers to producers, as well as having negative effects on income distribution in importing countries and on fair trade.[34] Other issues concerned the need to redefine price policies and inevitable doubts concerning the upkeep of a quota system in order to reduce and control overproduction, without, meanwhile, neglecting the welcome increase in overall environmental awareness.

To sum up, in the mid-1980s, widespread opinion stated that the first paradigm of CAP—centred both on the idea that agriculture had to be considered a protected sector and that the modernisation of agriculture had to be managed through State intervention—was in need of change. Above all, it was necessary to imagine new perspectives for economic activities, conceived until then only from a productivist point of view. In this sense, the European Commission took on the responsibility of changing the CAP, as it had been to date, namely, established on price policies, protectionism and the forced modernisation of agriculture.

In 1985, the new President of the Commission, Jacques Delors, inaugurated a well-structured path to reform this outdated agricultural sector, which was unable to ensure all the goals of the Rome Treaty and also a heavy burden on the EEC budget. Foremost, the new European Commission promoted two studies: a green paper with consultative purposes and a white paper focused on operational proposals and the need to base budgetary constraints and agricultural support on social and

[32] Meunier (1998), pp. 193–211.

[33] Josling and Tangermann (1999), pp. 371–388.

[34] Bullock (1992), pp. 59–67.

environmental grounds.[35] Although the proposals outlined in these EC studies were never implemented as reforms, they did contribute to a debate on the future of European agriculture, as demonstrated by the 1987 report presented by Frans Andriessen, the Dutch European Commissioner for Agriculture, centred on the need to safeguard the EEC budget. A year later, the EEC Commission presented a first reform, known as "Delors First Package", which—in the period 1989–1993—provided for an annual limit to agricultural expenditure growth and a mechanism of maximum guaranteed quantities, aimed at reducing price support, when quotas established in 1984 were surpassed. As to structural policies, these were strengthened by the EEC Commission, thanks to a reform of the Guidance section of the EAGGF, inspired by increased environmental awareness.

4 Reforming the CAP

Without wanting to adopt a deterministic approach, it is fair to say that the first successful attempt of reforming the CAP in 1992 was not by chance. Instead, it represented an out-and-out break in the European integration process. In the same year in which the European Union Treaty was signed in Maastricht, the European Commission launched an in-depth reform, which took the name of the Irish Agricultural Commissioner, Ray MacSharry.[36] In 1991, the European Commission had presented a report, which outlined the CAP weaknesses, clearly bringing to light the budgetary deficit deriving from the farm income support system. According to this EEC paper, the imbalance in CAP finances was caused by the guarantee placed on market prices rather than farm incomes, leading to detrimental consequences: first of all, high prices for European consumers and high costs for the EEC budget; second, increase in trade tensions with international partners, especially the United States and less-developed countries; and, last but not least, increase in environmental costs caused by soil over-exploitation.

In this sense, the 1991 paper proposed to reduce production incentives through a radical change of focus from market prices to direct income aid. Thanks to a second paper published some months later, the European Commission outlined a wide set of measures, in order to both reduce overproduction and provide an answer to the increase in environmental concerns, following the June 1992 Rio Conference on sustainable development.[37] Among the envisaged measures, the most important one was to reduce price support levels by 35 % over the next 3 years, with farmers being compensated for any resulting losses. However, compensation was dependent on

---

[35] European Commission (1991) The development and future of the CAP: reflections paper of the Commission, COM (91)100 final.

[36] Coleman and Tangerman (1999), pp. 385–405.

[37] Swinbank (1993), pp. 359–372.

participation in a set-aside program, which foresaw the reduction of arable land by 15 %.[38]

As a matter of fact, ever since 1981, the EEC had been experiencing an agonising reappraisal concerning the over-exploitation of the European environment: in the self-same year, the European Commission launched its first programs of integrated development, centred on a multisectorial approach, with agriculture being conceived as a key element influencing regional economies. Five years later, the Single European Act established the principles and goals of an EEC Environmental Policy, which concerned issues closely linked to the agricultural sector, such as the preservation and improvement of the environment and health protection, as well as a more functional exploitation of natural resources.[39] In other words, with the First Delors Commission, the EEC aimed to establish a new green paradigm, founded on the necessity to maintain social cohesion in rural areas while also safeguarding unique agricultural landscapes created over a period of two millennia. Beyond price support reductions and the introduction of direct payments managed by member States, the MacSharry Reform provided for the establishment of compulsory agri-environmental measures in aid of farmers, who were remunerated for their role in preserving rural landscapes. Besides the reforms, which affected some commodities, including the milk and livestock sectors, the most important proposals concerned the reinforcement of structural measures in line with the Guidance section of the EAGGF. On the one hand, the MacSharry reform further extended four measures originally envisaged at the time of the Mansholt Plan: (1) supporting the early retirement of farmers, (2) providing aid for mountainous and less-favoured areas, (3) promoting the agri-environmental approach, and (4) transforming agricultural land into forests. On the other hand, the new structural funds consisted of various measures, including the adjustment of agricultural structures, the development of rural areas, and the setting up of the LEADER initiative, in order to involve rural communities in the EEC development program, and thereby transforming farmers into key actors implementing the European cohesion policy.[40]

As a matter of fact, the year 1992 marked the redefinition of European agriculture on the basis of a number of key principles: foremost, agricultural exception was no longer reason enough to justify protectionism for a sector that was by now well obsolete. On the contrary, the need for a modern agricultural sector became highly evident in order both to prevent individual member States from establishing their own separate agricultural policies and to maintain political stability in the countryside.[41] This latter goal was aimed at creating a system of agricultural welfare, aligned with the new CAP environmental goals to resist social instability and mass migration from rural areas.

---

[38] Kay (1998).

[39] Jack (2013).

[40] Mahé and Roe (1996), pp. 1314–1323.

[41] Moyer and Josling (2002).

Nevertheless, the new European Union was about to face new challenges: externally, these included the end of the Uruguay Round and the establishment of the WTO in 1994, whereas domestically both the greater degree of integration and the envisaged enlargement of the EU to Central and Eastern European countries presented their own trials.[42]

As a consequence, only 3 years later, on December 1995, the new European Commission presented an "Agriculture Strategy Paper", which considered all the long-term questions concerning the CAP.[43] This paper outlined three different strategies for the future: firstly, a new reform of the CAP following a few years on from the MacSharry Reform; secondly, a development of the CAP following the 1992 reform; and, lastly, time to reflect on the repercussions of the 1992 reform.[44] Predictably, the new European Commission, chaired by Jacques Santer, decided to adopt an approach able to develop the MacSharry reform without dismissing the pillars on which it had been established, i.e., the direct income system and the link between income payments and the provision of agri-environmental services. Without doubt, all these measures revealed how the new challenges—economic globalisation, environmental issues and the subsidiarity principle introduced by article 5 of the EU Treaty—induced the European Commission to conceive a CAP reform able to make European agriculture more competitive, more devoted to structural policies and, last but not least, generally more simple in order to allow member States to implement EU decisions.[45]

In this line, the Austrian EU Commissioner, Franz Fischler, announced a conference on rural development in Cork (Ireland) on November 1996, at the end of which a declaration was adopted, which reaffirmed the need for establishing the CAP on sustainable principles of development, by means of a rural policy centred on a multidisciplinary and multisectorial approach, so as to take the complexity of different rural areas within Europe into account.[46] After 30 years, the EU Commission finally acknowledged the diverse qualities of rural areas and landscapes, by involving local communities in the implementation of CAP programmes. However, this new intervention strategy was not accepted by a number of member States—i.e. France, Germany and some Mediterranean countries—that feared that rural development would divert funds from direct income. In 1997, the European Commission presented a new paper "Agenda 2000: For a stronger and wider Europe", which outlined a broad overview of the future European Union with respect to various fields, such as enlargement, financial stability and the reform of key EU policies.[47]

[42] Daugbjerg and Swinbank (2004), pp. 99–119.

[43] European Commission (1995) Study on alternative strategies for the development of relations in the field of agriculture between the EU and the associated countries with a view to future accession of these countries (Agricultural Strategy Paper), CSE (95) 607.

[44] Tangermann (1998).

[45] Coclanis (2003), pp. 71–84.

[46] European Conference on Rural Development (1996).

[47] European Commission (1997).

It is impossible to dwell here on all the measures regarding the CAP; however, it is fair to say that the proposals made by the Commission can be summarised on the basis of two guidelines: on the one hand, financial sustainability was considered a primary goal of the CAP, thanks to strict limitations on the EU budget, and, on the other hand, the entire CAP was submitted to severe reassessments.[48] According to these EC proposals, the new CAP would be founded on two main pillars: market measures and price support embodied the first pillar, aimed at improving the domestic and external competitiveness of EU agriculture while also ensuring food safety and quality, as well as a fair standard of living for rural communities.[49] The second pillar recalled the results of the Cork negotiations on rural development by integrating environmental concerns into the CAP and turning European farmers into *stewards of the countryside*, as famously coined by Jacques Delors. The creation of new job opportunities for farmers and the social cohesion of rural areas had to be guaranteed by the Guidance section of the EAGGF, while structural funds were dedicated to reinforce rural heritage and the modernisation of the entire agricultural sector. Despite endless discussions and heated debates provoked by these proposals, the European Council approved the Commission plan on March 1999, and in June of the same year, the new reform came into force with some minor changes, particularly concerning price policies, due to strong opposition among member States.[50]

Differences of opinion between the European Commission and various member States led to some hesitance in implementing this reform, which failed to solve the budgetary question of the CAP. As a result, the new century started with a new menace on the horizon, represented by the foreseen EU enlargement towards Central and Eastern Europe, which according to many observers, jeopardised the endurance of the entire CAP budget.[51]

In order to avoid structural problems related to a potential financial crisis, "Agenda 2000" required ongoing reform inspections so as to monitor the implementation of measures planned in 1999. As a consequence, a "mid-term review" took place in 2003. This eventually resulted in the most radical reform of the CAP since its creation, thanks particularly to the pivotal role played by the Austrian EU Agricultural Commissioner, Franz Fischler, who was able to manage a complex process of bargaining among various political, economical and social actors, such as the European Commission, national governments and various stakeholders, including farmers, consumers and other parties involved.[52] Beyond domestic dimensions, the international context played a key role in this process, conditioned both by the start-up of the WTO Doha Round and the Iraqi War, which catapulted

[48] Galloway (1999), pp. 9–35.

[49] Ackrill (2000), pp. 343–353.

[50] Katranidis and Vakrou (2002), pp. 5–29.

[51] Van Meijl and Van Tongeren (2000), pp. 445–470.

[52] European Commission (2002) Mid-term review of the common agricultural policy, COM (2002) 394 final.

the EU towards a free-trade approach and opened its internal markets to agricultural goods coming from the USA and less-developed countries.[53]

Regarding the main innovations introduced by this reform, it is noteworthy here to underline that budgetary problems were not of key relevance in explaining the main provisions adopted. As a matter of fact, the reform introduced the Single Farm Payment (SFP), which decoupled a large share of CAP support from production, thanks to the two new instruments of "cross-compliance" and "modulation".[54] Modulation ensured the shift of funds from the first pillar—based on policy-fixed prices—to the rural development policies of the second pillar, thereby reducing financial transfers to larger farms and helping smaller farms.[55] The measure of cross-compliance made the SFP contingent upon a series of regulations relating to the environment and the multifunctional role of farms. Moreover, regarding food safety and the new environmental awareness, the Fischler Reform focused above all else on quality, thanks to the decision of decoupling farm support from production, thus indirectly favouring rural development. In reality, this reform presented some darker sides, as it was not only unable to influence the actual distribution of CAP benefits across countries and farms, but it was also ineffective in limiting its protectionist approach towards non-EU agricultural goods.[56] Likewise, the second CAP pillar was unsuccessful in reaching all the set goals, and at the end of Fischler's tenure, funds for rural development were, paradoxically, lower than in 2003. In Fischler's defence, it is fair to note that the 2003 reform marked a radical shift in the general CAP planning, which became more centred on the future role of agriculture, leaving behind the traditional pattern of day-by-day management, thanks to the powerful engagement demonstrated by the European Commission in reforming a sector of the economy that was lagging behind other productive sectors. As a matter of fact, a reform that began as a mid-term check revealed itself to be an important step in the redefinition of a new paradigm for European agriculture, directed towards holding a multifunctional role in twenty-first century Europe.

5 A New Agricultural Model?

As has been seen, the European Commission played a pivotal role in the various reforms, which took place between 1992 and the beginning of the twenty-first century. Above all, the achievements and failures of reform programs depended on the rejection or acceptance of measures by member States, farmer organisations, consumer associations and other parties involved. From a political point of view,

[53] Fouilleux (2004), pp. 235–255.

[54] European Commission (2003) A long term perspective for sustainable agriculture, COM (2003) 23 final.

[55] Hennis (2005).

[56] Steinberg and Josling (2003), pp. 369–417.

agricultural issues have, for a long time, represented the main battlefield on which the European Commission, with its supranational approach, and the intergovernmental Council confronted each other. While it may be easy to understand why some States—such as France, the Netherlands and Germany—defended the *status quo*, in spite of the clear deficit provoked by the CAP, it is much more difficult to examine why the European Commission failed to respond to the challenges of an economical system that had been showing its limits and weaknesses for such a long time. Adopting an unsophisticated point of view, it is easy to claim that the European Commission played a crucial role in establishing the CAP and, hence, holds full responsibility, together with the EEC institutional framework, for the negative trends witnessed in the field of European agriculture.[57] Alternatively, taking more of a political stance, the CAP allowed the European Commission to manage a considerable amount of financial funds, overrating its role in the European political arena and fostering favourable opinions among people across Europe, especially with regard to its successful fight against famine.[58]

However, the key question to ask is why the European Commission failed to change the disastrous CAP practises, which had stood out clearly since the very launch of the Mansholt Plan.[59] In order to address this problem, we need to extend our perspective to include some actors that are usually kept off the blacklist.

Firstly, we cannot ignore the electoral pressure enforced by farmers, unable to update their production systems; secondly, even if the mass exodus from rural areas greatly contributed, from a political point of view, to the loss in importance of farm workers, Agricultural Ministers in many European countries continued to foster agricultural lobbies. Moreover, broadening the scope of this analysis yet further, it inevitably becomes clear how the CAP was, for a long time, a key justification for the European integration process, as it demonstrated, without a shadow of a doubt, that the EEC was supporting one of the most important economical sectors. As a matter of fact, the CAP turned out to be a great loss for European consumers, who ended up sponsoring an expensive welfare system to support the agricultural sector.[60]

At this stage, another key question that cannot be ignored is why European consumers allowed themselves to be exploited by the privileges of European farmers for such a long time. A diachronic approach is required to answer this query, offering insights on people's views on agriculture over the last half-century: while initially Europeans maintained a sense of their agricultural roots, the 1970s brought about a sense of romanticising the countryside, due to both extensive urbanisation and pollution.[61]

---

[57] Skogstad (1998), pp. 463–490.

[58] Tallberg (2004), pp. 999–1022.

[59] Federico (2009), pp. 257–271.

[60] Daugbjerg (1999), pp. 407–428.

[61] Schmidt and Radaelli (2004), pp. 183–210.

During the 1980s, the increasing unemployment rate induced many people to think about agriculture as a possible solution to deindustrialisation. Meanwhile, the CAP was one of the main reasons that stopped the European Commission from implementing an ambitious plan to counter the European industrial crisis. It was only thanks to Jacques Delors that the EEC decided to adopt a strategy that would both reduce the overwhelming weight of the CAP on the European budget and promote environmental conservation through the engagement of farmers. The reforms proposed with the Single European Act presented a new structural policy to integrate price policy support with policies for rural development in order to transform rural societies into the main actors promoting new regional agendas.[62]

According to Jacques Delors, the European Commission was not only worried about environmental issues but also concerned about a reform program founded on the deep-set belief that a rural renaissance was necessary, seeing as rurality has always been an essential feature of European societies. In other words, once the fight against postwar famine was won, the CAP had to take concrete steps to preserve the European rural world and proceeded to do so by fostering a change of paradigm centred on a new multifunctional approach. The main feature of this new paradigm concerned farmers, now seen not only as good producers but also as key actors in the creation of a civilisation. In sum, this acknowledgement of the cultural and environmental importance of rural areas, together with the new role of farmers, represent the main features of the new multifunctional paradigm of European agriculture.[63]

It is now, moreover, crucial to examine the main reasons that led to a change of paradigm in European agriculture at the turn of the last century. With respect to this, widespread opinions focus on the unique combination of external factors—especially the final step of the Uruguay Round and the establishment of the WTO—and domestic ones, represented by the new European Commission chaired by Jacques Delors and the Treaty of Maastricht.[64] Clearly, the reasons driving the CAP reform not only represented an evolution of its ideological basis but were also the outcome of an incremental adaptation to an ever-changing framework. As a matter of fact, political paradigms are inevitably subject to change when they are unable to fulfil their goals. As regards the CAP, the key question was how to make the agricultural sector sustainable after it had absorbed up to 90 % of the EEC budget during the 1970s. In this sense, the CAP has undergone a shift from being a dependent model, or "State-assisted" paradigm, to becoming a highly competitive model in the case of some agricultural sectors, especially in Northern Europe.[65] As a result of the reforms implemented over the last 20 years, the European Union has come to embody a multifunctional paradigm. What is more, it seems that over the last few years European agriculture has been proceeding towards yet a fourth paradigm—namely,

[62] Slangen et al. (2004).

[63] Dobbs and Pretty (2004), pp. 220–237.

[64] Coleman (1998), pp. 632–651.

[65] Coleman et al. (1996), pp. 273–301.

a globalised production model—as the European Union has become the greatest world exporter of agricultural goods.[66]

All of these changes in paradigm have, over the years, allowed for a reduction of overall expenses and for the reallocation of funds to other item costs, thus transforming the CAP into a useful tool for promoting a wealthy Europe and for conserving unique rural landscapes.

In essence, the history of the CAP may be summarised as embodying "continuity in an ever-changing context": continuity represented by the ongoing need to support farmers' incomes, while ample changes have characterised the way in which financial resources have been transferred to farms. On the other hand, the complexity of the CAP has been so extensive that it is very hard to consider it as a systematic reform. Instead, it has resulted from the combination of many small reforms, taking place over the years and eventually leading to a highly transformed CAP format. Nowadays, European agriculture is not only devoted to production but also considered from an agroecological perspective, a source of common goods.[67] As a matter of fact, the CAP supports agricultural activities in Europe, ensures the stable provision of quality food, contributes to the safeguarding of the environment and constitutes an important means of countering hydrogeological instability.[68]

Finally, what lies ahead for European agriculture? All that can be said is that the CAP is now facing new major challenges, as reported in the "Agriculture 2020" report by the Commission. This highlights the instability of rural areas, the high agricultural diversity within the EU, the slowing down of growth rates on the world agricultural markets and the difficulties experienced with implementing processes of structural change. Inevitably, achievements and failures will depend on circumstances. However, in this never-ending story, much will depend on the ability of the European institutional framework to embody change, in order to foster sustainable development within a sector, which represents one of the most important features of an often evoked European identity.

References

Ackrill RW (2000) CAP reform 1999: a crisis in the making? J Common Mark Stud 38 (2):343–353

Brandow GE (1977) Policy for commercial agriculture, 1945–1971. In: Martin LR (ed) A survey of agricultural economic literature, vol 1. University of Minnesota Press, Minneapolis, pp 209–294

Bullock DS (1992) Redistributing income back to European community consumers and taxpayers through the common agricultural policy. Am J Agric Econ 74(1):59–67

[66] Van den Hoven (2004), pp. 256–283.

[67] Van Huylenbroek and Durand (2003).

[68] Glebe (2003).

Coclanis PA (2003) Back to the future: the globalisation of agriculture in historical context. SAIS Rev 23(1):71–84
Coleman WD (1998) From protected development to market liberalism: paradigm change in agriculture. J Eur Public Policy 5(4):632–651
Coleman WD, Tangerman S (1999) The 1992 CAP reform, the Uruguay Round and the commission: conceptualizing linked policy games. J Common Mark Stud 37(3):385–405
Coleman WD, Skogstad GD, Atkinson MM (1996) Paradigm shift and policy networks: cumulative change in agriculture. J Public Policy 16(3):273–301
Communauté européenne (ed) (1958) Résolution finale adoptée par les délégations des six États membres de la Communauté économique européenne (CEE) réunies pour la conférence agricole de Stresa, du 3 au 12 juillet. In: Recueil des documents de la Conférence agricole des Etats membres de la Communauté économique européenne à Stresa du 3 au 12 juillet 1958. Service des publications des Communautés européennes, Luxembourg, p 250
Daugbjerg C (1999) Reforming the CAP: policy networks and broader institutional structures. J Common Mark Stud 37(3):407–428
Daugbjerg C (2004) International trade negotiations and reforms of the CAP. Paper presented to the 5th SGIR pan-European conference 'Constructing World Orders', The Hague, Netherlands, 9–11 September 2004
Daugbjerg C, Swinbank A (2004) The CAP and EU enlargement: prospects for an alternative strategy to avoid the lock-in effect of CAP support. J Common Mark Stud 42(1):99–119
Dobbs TL, Pretty JN (2004) Agri-environmental stewardship schemes and "multifunctionality". Rev Agric Econ 26(2):220–237
European Commission (1960) Proposals for the working out and putting into effect of the common agricultural policy in application of article 43 of the treaty establishing the European economic community. DG VI/COM (60) 105, Brussels
European Commission (1968) Memorandum of the reform of agriculture in the European Community. DG VI/COM (68) 1000, Brussels
European Commission (1991) The development and future of the CAP: reflections paper of the Commission. COM (91)100 final
European Commission (1995) Study on alternative strategies for the development of relations in the field of agriculture between the EU and the associated countries with a view to future accession of these countries (Agricultural Strategy Paper), CSE (95) 607
European Commission (1997) Agenda 2000. COM(97) 2000 final
European Commission (2002) Mid-term review of the common agricultural policy. COM(2002) 394 final
European Commission (2003) A long term perspective for sustainable agriculture, COM(2003) 23 final
European Conference on Rural Development (1996) Cork Declaration 'A living countryside', 7–10 November 1996
Fearne A (1997) The history and development of the CAP 1945–1990. In: Ritson C, Harvey DR (eds) The common agricultural policy, 2nd edn. CAB International, Wallingford, pp 11–55
Federico G (2009) Was the CAP the worst agricultural policy of the 20th century? In: Patel KK (ed) Fertile ground for Europe? The history of European integration and the common agricultural policy since 1945. Nomos, Baden-Baden, pp 257–271
Foreman-Peck J (1983) A history of the world economy: international economic relations since 1850. Harvester Wheatsheaf, London
Fouilleux E (2004) CAP reforms and multilateral trade negotiations: another view on discourse efficiency. West European Politics 27(2):235–255
Galloway D (1999) Agenda 2000 – packaging the deal. J Common Mark Stud 37:9–35
Germond C (2010) The agricultural bone of contention: the Franco-German Tandem and the making of the CAP, 1963–1966. JEIH 16(2):25–44

Glebe TW (2003) Multifunctionality: How "Green" is the "European Model of Agriculture"? Discussion paper 01–2003, Technische Universität München, Environmental Economics, Resource Economics and Agricultural Policy Research Group
Hennis M (2005) Globalisation and European integration – the changing role of farmers in the common agricultural policy. Rowman & Littlefield Publishers Inc., Lanham, Boulder, New York, Toronto, Oxford
Hill BE (1984) The common agricultural policy: past, present and future. Methuen, London
Hofreither MF (2007) Origins and development of the common agricultural policy. In: Gehler M (ed) From common market to European Union building. 50 years of the Rome Treaties 1957–2007. Böhlau, Wien, pp 333–348
Jack B (2013) Agriculture and EU environmental law. Ashgate, Aldershot
Johnson DG (1947) Forward prices for agriculture. The University of Chicago Press, Chicago
Josling T (2009) Western Europe. In: Anderson K (ed) Distortions to agricultural incentives: a global perspective, 1955–2007. Palgrave, Macmillan – World Bank, London, Washington DC, pp 115–176
Josling T, Tangermann S (1999) Implementation of the WTO Agreement on agriculture and developments for the next round of negotiations. Eur Rev Agric Econ 26(3):371–388
Katranidis S, Vakrou A (2002) Member States positions and voting coalitions on the Agenda 2000 CAP reform. Cahiers d'économie et sociologie rurales 63:5–29
Kay A (1998) The reform of the common agricultural policy: the case of the MacSharry reforms. CAB International Publishing, Wallingford
Knudsen ACL (2009) Farmers on welfare: the making of Europe's common agricultural policy. Cornell University Press, Ithaca
Laqua D (2011) Internationalism reconfigured: transnational ideas and movements between the world wars. I. B. Tauris, London
Ludlow NP (2005) The making of the CAP: towards a historical analysis of the EU's first major policy. Contemp Eur Hist 14(2005):347–371
Ludlow NP (2006) The European community and the crises of the 1960s: negotiating the Gaullist challenge. Routledge, Abingdon
Mahé LP, Roe TL (1996) The political economy of reforming the 1992 CAP reform. Am J Agric Econ 78(5):1314–1323
Mansholt S (1963) Regional agreements for agricultural markets. In: Proceedings of the 11th international conference of agricultural economics. Oxford University Press, London
Meunier S (1998) Divided but united: European trade policy integration and EU–US agricultural negotiations in the Uruguay Round. In: Rhodes C (ed) The European Union and the world community. Lynne Rienner Publishers, Boulder, London, pp 193–211
Milward AS (1984) The reconstruction of postwar Europe, 1945–1951. University of California Press, Berkeley and Los Angeles
Milward AS (1992) The European rescue of the nation state. University of California Press, Berkeley and Los Angeles
Moyer W, Josling T (2002) Agricultural policy reform – politics and process in the EU and US in the 1990s. Ashgate, Aldershot
Neville-Rolfe E (1984) The politics of agriculture in the European community. Policy Studies Institute, London
O'Rourke KH (1997) The European grain invasion 1870–1915. J Econ Hist 57:775–801
OECD (1961) Trends in agricultural policies since 1955. 5th report on agricultural policies in Europe and North America. Organisation for economic co-operation and development, Paris
Schmidt VA, Radaelli CM (2004) Policy change and discourse in Europe: conceptual and methodological issues. West Eur Polit 27(2):183–210
Seidel K (2010) Taking farmers off welfare. The EEC commission's memorandum "Agriculture 1980" of 1968. JEIH 16(2):83–102
Sheingate AD (2001) The rise of the agricultural welfare state: institutions and interest group power in the United States, France, and Japan. Princeton University Press, Princeton

Skogstad G (1998) Ideas, paradigms and institutions: agricultural exceptionalism in the European Union and the United States. Governance 11(4):463–490

Slangen LHG, Jongeneel RA, Polman NBP (2004) Institutional analysis of multifunctional agriculture. Contribution presented to the 90th European association of agricultural economists seminar 'Multifunctional agriculture, policies and markets – Understanding the critical linkage', Rennes, October 27–29 2004

Slicher Van Bath BH (1963) The Agrarian history of Western Europe, A. D. 500–1850. St. Martin's Press, New York

Spoerer M (2010) "Fortress Europe" in long-term perspective: agricultural protection in the European community, 1957–2003. JEIH 16(2):143–162

Steinberg H, Josling T (2003) When the peace ends: the vulnerability of EC and US agricultural subsidies to WTO legal challenge. J Int Econ Law 6(2):369–417

Swinbank A (1993) CAP reform, 1992. J Common Mark Stud 31(3):359–372

Tallberg J (2004) The power of the presidency: brokerage, efficiency and distribution of EU negotiations. J Common Mark Stud 42(5):999–1022

Tangermann S (1998) An ex-post review of the 1992 MacSharry reform. In: Ingersent KA, Rayner AJ, Hine RC (eds) The reform of the common agricultural policy. Macmillan Press Ltd, London

Thiemeyer G (2009) The failure of the Green Pool and the success of the CAP: long-term structures in European agricultural integration in the 1950s and 1960s. In: Patel KK (ed) Fertile ground for Europe? The history of European integration and the common agricultural policy since 1945. Nomos, Baden-Baden, pp 47–59

Tracy M (1984) Issues of agricultural policy in a historical perspective. J Agric Econ 35:307–318

Tracy M (1989) Government and agriculture in western Europe, 1880–1988, 3rd edn. Harvester Wheatsheaf, New York

Tracy M (1994) The spirit of Stresa. Eur Rev Agric Econ 21:357–374

Van den Hoven A (2004) Assuming leadership in multilateral economic institutions: The EU's "Development Round" discourse and strategy. West Eur Polit 27(2):256–283

Van Huylenbroek G, Durand G (eds) (2003) Multifunctional agriculture: a new paradigm for European agriculture and rural development. Ashgate, Aldershot

Van Meijl H, Van Tongeren F (2000) The agenda 2000 CAP Reform, world prices and GATT-WTO export constraints. Eur Rev Agric Econ 29(4):445–470

Webber D (1999) Franco-German bilateralism and agricultural politics in the European Union: the neglected level. West Eur Politics 22(1):45–67

Environment, Landscape, Agriculture, and Food in the Framework of State and Regional Legislative Powers as per Art. 117 of the Italian Constitution

M. Troisi

Abstract In the Italian Constitution, subject matters as the environment, agriculture, food, and landscape have always been divided in terms of legislative jurisdiction and, hence, by sector.

Even after the constitutional reform of 2001, this compartmentalization remained, indeed it was institutionalized. As a result, the Regions, which have exclusive legislative powers about "agriculture", can only intervene in a narrow perimeter, due to the presence of other contiguous sectors that interfere with agriculture. Agriculture, indeed, is expressed in activities that, if carried out rationally, produce positive externalities for the environment and, if exercised irrationally, create negative externalities. Moreover, agricultural production is destined for the food market, which demands products that are healthy and genuine. It is clear that agriculture is a sector highly subject, by its very nature, to intervene with other material sectors.

So "which" agriculture is attributed exclusively to the Regions by the amended Constitution? The choice of the 2001 reform is symptomatic of a clearly sectoral and distinctly anachronistic view of agriculture. It does not take into account its intrinsic multifunctionality, the links and indivisible ties between agriculture and the environment and between agricultural production and agri-food products. These links broaden and expand the limits of agriculture and its regulation in the direction of "agroecology," which also appears to be more in tune with the Community framework.

From this perspective, the constitutional amendment is incoherent as regards the evolution of agriculture. In practice, rather than seeking to reduce it, it accentuates this hiatus among the various parts of the agri-food-environmental system, thus keeping separated (albeit formally) various equidistant subjects according to legislative competence.

M. Troisi (✉)
Department of Legal Sciences, University of Salento, Lecce, Italy
e-mail: michele.troisi@unisalento.it, http://www.giurisprudenza.unisalento.it/scheda_docente/-/people/michele.troisi

M. Monteduro et al. (eds.), *Law and Agroecology*,
DOI 10.1007/978-3-662-46617-9_10

Even after the constitutional amendment, the Constitutional Court itself continued to remain close to pre-reform jurisprudence: the basic core of agriculture remained the same, namely the "production of plants and animals for food" (Judgements 12/2004, 116/2006). Yet the presence of this "hard core," although it provides the subject matter "agriculture" with a relative stability, in order to prevent the risk of its being absorbed in other sectors, also sets up boundaries and greatly limits its potential.

Agriculture is indeed much more complex than a mere productive phenomenon.

Keywords Agriculture • Constitution • Multidisciplinarity • Multifunctionality

1 Introduction

An analysis of the protection provided by the Italian Constitution in relation to the modern concept of "sustainable agriculture" or "agroecology" is no simple task, for various reasons. Foremost, the Constitution represents the institutionalization of a sectoral and division-based approach to the subject. Sectors such as the environment, agriculture, food, and development, while being ontologically interdependent "worlds," are also realities that complement each other. Yet in the Italian Constitution and, as a consequence, in its *sub*-constitutional sources, they have always been divided in terms of legislative jurisdiction and, hence, by sector.

Furthermore, the problem is increased by a conflictual contingency among the Republic's various constituent bodies with regard to the exact scope of their respective legislative powers. The Italian Constitution of 1948 outlined a *regional* State in which the relations between the State and single Regions have over time been established, improved, and, in the last decade, overturned.[1]

While a certain tension is essential for any State that recognizes the powers of local autonomies, making competition between unity and differentiation quasi-normal, these two contrasting but inseparable principles are part of the same Constitutional system of values.[2] Having said this, we must recognize that State–Region relations have reached unprecedented levels of conflictuality in the wake of the (ill-fated)[3] reform of Title V, Part 2, of the Italian Constitution in 2001.

[1] On the new role of the Regions following the constitutional reform, see Caretti and Tarli Barbieri (2012); D'Atena (2013); Martines et al (2012); Mangiameli (2013).

[2] Benelli and Bin (2009), pp. 1185–1212, exp. p. 1188.

[3] Of the many voices expressing perplexity during the reform's period of approval, see the (prophetic) observations made by Bin (2001), pp. 122–153, exp. p. 129. According to Bin, the relation between the State, Regions, and Local Authorities, outlined in the reform, would soon become the idea that "the State act as arbitrator between the other two, mediating, distributing, removing and regulating functions, applying the divide et impera or, if you prefer, 'the three-card game' [...]. [...] an effective producer of future jurisprudential dispute" (all the translations are of the author of this paper).

The new Art. 117 of the Italian Constitution has in fact inverted the relation between the rule and the exception as far as the legislative powers of the State and the Regions are concerned. Prior to the 2001 amendment, the rule was the competence of the State and the exception was the competence of the Regions. With the reform, the rule has become the competence of the Regions and the exception is now the competence of the State.[4] In practice, however, the State's powers are anything but marginal. This is not only because it is vested with the lion's share of subject matters but also because some sectors attributed to the State have, in practice, tended to "invade" domains where the Regions should exercise exclusive legislative powers. This applies principally to the matters dealt with in this study. Assigning "environmental protection" to the State means giving the "center" a potentially unlimited sphere of competence, since a great many human activities have an impact on the environment.

This naturally means a continual dispute between the State and the Regions over jurisdictional rules, on the assumption that the competences of the one have been usurped by the other and *vice versa*.[5] In this scenario of confusion and conflict, it is superfluous to point out that the role of the Constitutional Court has expanded strikingly, from that of (mere) guardian of the Constitution to its "co-reformer" (or, according to some, "counter-reformer").[6]

In identifying matters of exclusive (or residual) legislative regional power, for example, in 2001 the Constitutional Court responded by expanding the exclusive and concurrent powers of the State, rather than by identifying residual powers. Yet even when it did identify residual powers, this occurred as a progressive limitation of the expression of other interests (and thus other "matters"). None of the matters granted constituted a sphere of exclusive legislative intervention reserved to the Regions.[7]

For example, agriculture is a central element of agroecology and has been explicitly identified by the Court (see below) as an exclusively regional competence. Despite this, it is not considered a *bona fide* matter with fixed and well-defined confines that can be described as fully delegated to the general and exclusive competence of the Regions. On the contrary, what we find are a number of "fragments," cut across by other competences, such as the State's exclusive jurisdiction in the "protection of competition," "protection of the environment," and the so-called concurrent competences, relating to "health protection," "government of the territory," and "food."

Hence, "the interference and inextricable interlocking with other interests and competences" (Constitutional Court, Judgement 407/2002) means that segments or areas of agriculture may be subject to a plurality of claims for legitimate State intervention. These claims satisfy the demands of the unitary State, but also affect

[4] Caranta (2004), pp. 990–1013.

[5] Lugaresi (2010), pp. 557–589.

[6] Benelli and Bin (2009), p. 1211.

[7] Benelli and Bin (2009), p. 1203.

the competences of the Regions, due to the presence of specific local requirements. The Regions are also entitled to regulate the primary economic activity of agriculture in the territorial area under their jurisdiction.[8] Any identification of constitutional protection for "agroecology" has to face the delicate phase of "agriculture's" transition from the "old" to the "new" version in the constitutional framework. That is, from an exceptional and, to some extent, justifiable sectoral regulation of agriculture to a "monofunctional" vision of agriculture, operated by the reform, and clearly in contrast with its actual evolution and cultural development.

2 The 1948 Constitution and the Concept of Agriculture

At the time of the Constituent Assembly, not only was the modern concept of "agroecology" a thing of the future, but also agriculture itself was perceived in a very limited way. It was still closely linked to the simple "production of goods," and there was certainly no idea of its positive externalities that went beyond the mere production of food. This formulation was consecrated in Art. 44 of the Constitution, which stipulates that "for the purpose of ensuring the rational exploitation of land and equitable social relationships, the law imposes obligations and constraints on the private ownership of land; it sets limitations to the size of holdings according to Region and agricultural area; it encourages and imposes land reclamation, the conversion of landed estates [*latifundia*] and the reorganization of farm units." The constitutional text makes it clear that the primary function of land, particularly in the postwar historical-political context in which the Constitution was drafted, was to obtain the greatest productive input from agricultural cultivation (through its "rational exploitation") in order to satisfy the massive national demand for food.

The content of Art. 117 on the division of legislative powers between the State and the Regions reflected this cultural model of agriculture and outlined concurrent State–Region legislative powers. Here the task of the State was to set the basic principles, whereas the Regions would legislate in detail for matters such as "agriculture and forestry," "water resources," "artisan skills," and "tourism and hotel industry." Matters not listed under Art. 117 (environment protection, health, etc.) were instead the exclusive competence of the State. This created a hiatus among the various parts of the agri-food-environmental system and was translated into administrative diversification, starting with a divergence of legislative competences. The concrete example is Decree 300/1999 (*Reform of the organization of Government*). This identified the responsibility of different government ministries

[8] Germanò (2003), pp. 117–194, exp. p. 194.

(agriculture, environment, productive activity)[9] and, in so doing, acritically opened the way to the subsequent fragmentation of agriculture.

Yet it is interesting to note how Law 382/1975 and then Presidential Decree 616/1977 transferred a series of functions relating to agricultural activity and production to the Regions, linking agriculture to a series of related and complementary instrumental interventions, beyond the mere production of agricultural goods. Indeed, Art. 50 transferred the administrative functions of the State for "fairs and markets," "tourism and hotel industry," and "agriculture and forestry" to the Regions "as matters," *nota bene*, "relating to the economic development of the respective populations."

The Constitutional Court reacted by accentuating the isolation of agriculture from other matters, giving it a merely objective meaning as a productive phenomenon. In so doing, it rejected the utility of instrumental or ends-oriented criteria and responded to *agriculture-production* with the concepts of *agriculture-environment* and *agriculture-market*, both of which withstood fragmentation or being localized on a territorial basis, all in the name of the national interest.[10]

Judgement 994/1988 is the best example of where the Court separated interventions in agricultural matters, understood as the "production of goods" and reserved to the Regions, from those "of national interest [...] characterized, objectively, as having a direct influence or effect on constituent terms of the market itself, such as supply and demand, prices, production costs etc., and from a functional perspective, for the tight correlation that links them to national planning and, in general, to the interests arising from the latter."[11] In the end, the Court blocked potential precipitous action, ruling that agriculture was (solely) a matter of the production of goods. Within this limited area, the Regions were (only) entitled to legislate (in line with basic principles set by the State).

3 The Evolution of Agriculture and the 2001 Constitutional Amendment

Towards the end of the twentieth century, the concept of agriculture was literally capsized by two phenomena, one "cultural" and the other legislative. The "cultural" evolution of agriculture was inevitable. It had become increasingly clear that agricultural activity was so broad, intercepted so many different demands, and

[9] This act institutionalized the various competences of government ministries. In particular, the following ministries were created: Productive Activities, Agricultural, [Food] and Forestry, Environment and Protection of the Territory, Infrastructure and Transport, and Cultural Heritage and Activities.

[10] Germanò (2003), p. 194.

[11] Constitutional Court Judgement 994/1988 (Author's translation). See Carrozza (1992), pp. 19 et seqq.

postulated such a series of obligations that it could no longer remain "limited" to the mere "production of goods." Agriculture's *multifunctionality* was a constituent given: food production was completed by the sustainability of productive processes, healthy food, animal well-being, and the protection of the landscape.

As a consequence, Art. 44 of the Constitution (which had *prima facie* a literal interpretation limiting agriculture) underwent a change that freed it from the problems of the historical period in which it was drafted. Thus, the "rational exploitation of land," mentioned in the article, is no longer interpreted solely in terms of production. On the contrary, by *land* we mean the comprehensive system taking in the territory as a whole, *in primis* the landscape, as per Art. 9 of the Constitution, which specifies the function of "safeguarding the natural landscape and the historical and artistic heritage of the nation." Agricultural production, thus, is not an end in itself but is compatible with the environmental protection of the territory and the production of resources for present generations and their conservation for those of the future.

The second phenomenon that helped overturn the concept of agriculture was legislative. The 2001 constitutional amendment was the most significant change to the Constitution since its inception in 1948. It revolutionized relations between the State, Regions, and Local Authorities, introducing the notion of subsidiarity as a guiding principle for redefining relations between the two bodies with legislative powers. This also created a new element insofar as "agriculture and forestry" disappeared from the amended Art. 117. Since, however, alongside the exclusive and concurrent legislative power of the State, Constitutional Law 3/2001 established an exclusive residual legislative power of the Regions, the first problem that the law posed was whether agriculture could be allocated to the regional domain.[12]

In fact, the positive response was not immediately forthcoming. For some time, the unlisted "old" subject matter agriculture, instead of passing under the legislative power of the Regions, appeared to have disintegrated into other matters listed in Art. 117.[13] The question was and remains very closely linked to the meaning of the term "agriculture" and the recognition of its "confines," contents, and objectives.[14]

There is little doubt that "agriculture" came fully within the pre-2001 Italian legal framework. In effect, in 2001, when the Court was about to identify various matters, it was also facing a series of national and EC laws dealing with agriculture as a matter in its own right. For this, the silence of the new Art. 117 on the issue did not mean that the old matter "agriculture and forestry" disintegrated into other matters, listed or otherwise. On the contrary, in applying the *historical-normative*

[12] On the negative consequences arising from the absence of a list of regional competences, see Mangiameli (2013).

[13] Jannarelli (2003), pp. 82 et seqq.

[14] Losavio (2012), p. 326.

criterion,[15] it meant and means, if anything, that this matter "existed" and "exists" in the system. As a consequence, the subject matter agriculture in its *continuum* with the past had made "a qualitative leap" from a concurrent matter to a matter reserved exclusively to the Regions.[16]

Moreover, the same Constitutional Court adopts this *continuum* interpretation in Judgement 12/2004. This is one of the rare cases[17] in which the Court singled out, expressly and without reserve, the existence of a subject matter emanating from the lists under Art 117, Paras. 2–3,[18] that catalogue the domains of concurrent State competence(s): "agriculture is a legislative competence entrusted to the Regions as a residual power and removed from the State's jurisdiction." Agriculture, then, is attributed to the Regions. But "which" agriculture? Is it the "monofunctional" agriculture typical of 1948 or the "multifunctional" agriculture that enhances the multiple profiles linked to it?

3.1 From Multifunctionality to Multidisciplinarity

Once an exclusive competence has been "conquered" in a key subject such as agriculture, the regional legislator soon realized that this field was destined to intersect with different sectors that intervene with agriculture and whose legislative competences are not univocal.[19] Agriculture, indeed, is expressed in activities that, if carried out rationally, produce positive externalities for the environment and, if exercised irrationally, create negative externalities. Moreover, agricultural production is destined for the food market, which demands products that are healthy and genuine. It is clear that agriculture is a sector highly subject, by its very nature, to intervene with other material sectors.

Let us try to analyze (albeit synthetically) those sectors outlined by the Italian Constitution:

I) *Protection of competition* (amended Art. 117, Para. 2, lett. e) is an exclusive State power. Precisely because agriculture goes beyond the mere production of goods, and includes their marketing and sale, there is the question of how agricultural products are distributed. "Commerce" is not listed in Art. 117, Paras. 2–3, so

---

[15] According to Mangiameli (2013), p. 122, this criterion represents "the interpretative technique that implies loyalty to the Constitution and that, in the case of listing matters, avoids the unlawful shifting of objects to the State and safeguards regional legislative powers."

[16] Germanò (2003), pp. 140 et seqq.

[17] Indeed, the Court's opinion on this point has always been oriented in the sense of the "impossibility of leading a particular object of regulatory action to the sphere of application attributed to the residual legislation of the Regions as per [. . ..] Art. 117, Para. 4, for the sole reason that this object is not immediately referable to one of the matters listed in the preceding sub-paragraphs" (Judgement 370/2007, author's translation).

[18] Tarchi (2006), pp. 65–86, exp. p. 80.

[19] On the relation between the matters listed in the amended Art. 117, see D'Atena (2007), pp. 59–66.

we can assume that it is an exclusively regional competence. Yet the question is more complex and is part of the delicate *protection of competition* and State competence. In a sector such as "the market," and steeped in Community regulations, the Court specified that subjective rights of an economic nature could not be regulated by regional laws, which are by nature territorially localized. On the contrary, this competence gives the State a very broad scope to "invade the other's turf."

Here too the Constitutional Court makes a distinction between the production of goods, and prices and the market. The latter by its nature needs to be regulated in order to maintain parity of access, competition rules, and transparency of information. In practice, the Court understands *protection of competition* in a "dynamic" sense that grants the State the power to make any interventions of "macroeconomic importance" (Sentence 14/2004).

II) *Protection of the environment and the ecosystem* is attributed by the 2001 reform to the exclusive competence of the State, as per Art. 117, Para. 2, lett. s).[20]

The environment is a value that transversally permeates the entire exercise of agricultural production[21] because, as already stated, only the "rational exploitation of land" (Art. 44) guarantees its protection against negative phenomena such as soil erosion. The "mode" in which agriculture is carried out can generate positive or negative externalities for the environment, understood as air, flora, fauna, land, and also "conservation of places." The environment is a subject matter that has greater intervening potential with agriculture, especially with regard to its intrinsic "transversality," which is dealt with below.

III) *Government of the territory*: Art. 117, Para. 3, attributes a concurrent legislative power to the Regions. The link between agricultural activity, also considered as productive activity, and the territory is *in re ipsa* insofar as "economic development is also an element which characterizes and transforms the territory, taken not as a mere *locus* of the life of a community, but as a distinguishing and identifying dimension of the community itself."[22] But the link between the two is also important because *government of the territory* and agriculture have the common mission of safeguarding the particular function of every "piece of rurality" on Italian territory: a conservation that is not just "material" but also applied to the demographic profile, rural tourism, and environmental protection.

The link between the government of the territory and agriculture is significant (also because it runs counter to Italian jurisprudence, which tends to favor State jurisdiction) and is dealt with in Judgement 96/2012 of the Constitutional Court on agro-tourism. Here, the Court rejected the State censure based on the violation of Art. 3 of the Region Law (Umbria) against that planned use for agro-touristic

---

[20] On the relation between agriculture and environment, especially after the 2001 constitutional reform, see CRISTIANI E in this book. More generally, see Caravita (2005); Galloni (1995), pp. 15 et seqq., exp. p. 27; Grassi (2012).

[21] Grassi (2003), pp. 39 et seqq.

[22] Nigro (1980) (author's translation).

activity of only those structures that existed before the regional law came into force. The Court validated the regional choice consistent with setting a certain chronological limit in order to allow and promote the use of Umbria's sizeable, albeit partly dilapidated or in ruins, structural heritage, whose renovation was promoted in various ways, for agro-touristic activity. In this way, the law became part of the "government of the territory," and this is not patently unreasonable "insofar as it responds to a legislative policy that is particularly rigid, yet compatible with the aims of a rational regulation of agricultural territory."

IV) *Food* is another area that the 2001 reform attributes to the concurrent legislative power. Yet, in this case, the intrusion into agriculture is very evident, as the latter is destined to end in food products.[23] Furthermore, this aspect is particularly significant if we take into account the widespread demand among the urban population targeted at quality food products that are closely linked to the material and immaterial (cultural, traditional, historical) values of the territories where they are produced.[24]

However, the Community definition of food is objectively all encompassing, covering the entire sequence of production, transformation, and marketing of agricultural products.[25] Regulation (EC) 178/2002 of the European Food Safety Authority (EFSA) stipulates that "in order to ensure the safety of food, it is necessary to consider all aspects of the food production chain as a continuum from primary production and the production of animal feed up to, and including, the sale or supply of food to the consumer because each element can have a potential impact on food safety" (Art. 12).

V) *Health protection* is also a matter of concurrent legislative powers. Since agricultural products are for human and animal consumption, we can hardly contest [State] involvement in the regulation of health protection and agriculture.[26]

In the light of this brief *excursus*, it is clear how agriculture, while a matter of exclusive Regional legislative power, is strongly conditioned by prescriptive State norms. Indeed, the regional discipline of agriculture *de qua* either submits when faced with the State discipline in cases of exclusive State competence (competition, environmental protection) or must respect the basic principles set by the State in cases within its concurrent competence (management of the territory, health, food). This is confirmed by constitutional jurisprudence, which repeatedly stresses how "in the jungle of competences there is nevertheless room for State intervention [...], even if the matter is presented as a so-called exclusive or full regional competence (Art. 117, Para. 4)."[27]

[23] For an interpretation of the interconnection of agriculture and food, see Benozzo (2007), pp 107–124; Cristiani (2004), p. 11.

[24] Marotta (2006), pp. 127–166, exp. p. 127.

[25] Germanò (2003), p. 169.

[26] From the extensive literature on the right to health, see Caravita (1984), pp. 522–545; Chieffi (2003); Cocconi (1988); Luciani (1980), pp. 769–787; Morana (2014); Nania (2006), pp. 121–132.

[27] Malo (2008), pp. 141–176, exp. pp. 152 et seqq.

Two Constitutional Court decisions apply to all the examples:

A) Judgement 116/2006 regulating genetically modified organisms (GMOs)[28]

We will not retrace the GMO debate here; it is enough to recall the dense web of Community provisions that allows local lawmakers room for maneuver on the basis of the precautionary principle. In Italy, Decree 279/2004, subsequently Law 5/2005 (the Alemanno Decree), immediately raised doubts as to whether the State had the competence to adopt the issue under the new framework of competences outlined by the 2001 reform. Indeed, it is difficult to contest that the entire Law dealt with a matter, namely agriculture, that was the exclusive competence of the Regions.

In response to the complaints made by some Regions, the Court, beyond reaffirming (see Judgement 12/2004) that "the production of plants and animals for food" constituted the "core" of agriculture, "distributes" this delicate sector, with its concurrent powers among different levels of competence. According to Court's judgement, the part of the Law that implicitly confirms the lawfulness of the use of GMO in agriculture is an expression of the State's exclusive competence in "environmental protection" and of its concurrent competence in "health protection"; the aspects dealing with (mere) cultivation are instead, to all effects, in the "production of plants and animals for food" and thus under the exclusive competence of the Regions in matters of agriculture.[29] Where the Court could have treated agriculture in a unitary manner, it instead confirmed its jurisdictional segmentation.

B) Judgement 339/2007 on agro-tourism.[30]

This decision does not reveal a great deal because the Court, responding to regional complaints, declared many provisions of Law 96/2006 unconstitutional, precisely because it undermined the residual competence of the Regions in agriculture and tourism. On the other hand, it is interesting because the Court took the opportunity to specify that the residual power was not *per se* sufficient to avoid legislative intervention by the State. Indeed, according to the Court, agro-tourisitic activity, "even if at first, [...] under the residual competence of the Regions for agricultural activity and tourism, nonetheless intervenes in other matters attributed to the exclusive or concurrent competence of the State." In the same judgement, the modes of production, preparation, and presentation of food consumed in agro-touristic structures are not judged illegitimate because the Court includes them under "health protection" and hence as an expression of basic principles, which are the exclusive competence of the State. On the contrary, some parts of the Law were declared unconstitutional, such as those authorizing the use of domestic kitchens and the condition that buildings destined to host a maximum of ten places must be fit for habitation. This was not because the matter came under agriculture and tourism but was due to the excessively detailed rules in the concurrent area of "health protection" (where only basic principles should have been established).

[28] See Borghi (2006), pp. 961–977.

[29] Milazzo (2005), pp. 225–264, exp. pp. 250 et seqq.

[30] See Lucifero (2008), pp. 834–839.

Finally, the Court declared constitutional the rule obliging the Regions to submit a succinct annual report to the Ministry for Agriculture, Food and Forestry on agro-tourism in the territory of their own competence, with sectoral data and any provisions made on the subject, because the Court considered it an expression of that exclusive State legislative power as per Art. 117, Para. 2, lett. r), in matters of "coordination of statistical information and data information of State, Regional and local administrations."

In the light of all this, let us turn to the question raised at the end of the previous section: "which" agriculture is attributed exclusively to the Regions by the amended Constitution? The Court's choice, which interferes in matters with different competences, is symptomatic of a clearly sectoral and distinctly anachronistic view of agriculture. It does not take into account its intrinsic multifunctionality, the links and indivisible ties between agriculture and the environment and between agricultural production and agri-food products. These links broaden and expand the limits of agriculture and its regulation in the direction of "agroecology," which also appears to be more in tune with the Community framework.

From this perspective, the constitutional amendment is incoherent as regards the evolution of agriculture. In practice, rather than seeking to reduce it, it accentuates this hiatus among the various parts of the agri-food-environmental system, thus keeping separated (albeit formally) various equidistant subjects according to legislative competence.

Even after the constitutional amendment, the Court itself continued to remain close to pre-reform jurisprudence: the basic core of agriculture remained the same, namely the "production of plants and animals for food" (Judgements 12/2004, 116/2006). Yet the presence of this "hard core," although it provides the subject matter "agriculture" with a relative stability, in order to prevent (as already stated) the risk of its being absorbed in other sectors, also sets up boundaries and greatly limits its potential.[31] Agriculture is indeed much more complex than a mere productive phenomenon.

With this definition, the Court seems to have completely neglected the progressive complexity and breadth of agriculture's economic functional aspects, in particular, the profound link with rural territory and its environmental and cultural resources. On the contrary, according to more aware treatment of the subject,[32] concepts such as "diversification, multifunctionality, pluriactivity, territoriality and pluri-subjectivity are the new paradigms of agricultural processes that remodel and reshape agriculture well beyond its traditional coordinates." Agriculture is no longer exclusively economic but finds itself alongside issues such as biodiversity, sustainable production, and the protection of the historical and cultural heritage of rural areas.

Yet the 2001 reform responds to the thorny issue of protecting the *multifunctionality* of agriculture by fragmentating agriculture into different institutional

[31] Losavio (2012), pp. 326–337, exp. p. 337.

[32] Adornato (2007), pp. 67–80, exp. p. 72.

levels (this is the rather narrow solution of *multidisciplinarity*). The consequent uncertainty regarding powers concretely downgraded the scope *pleno sensu* of agriculture, which was split into different ontologically related matters, leaving the sole phase of the production of goods to the Regions, exactly as before the 2001 reform.

3.2 The Contribution of the Constitutional Court Between Transversal Matters and Criterion of Predominance: A Further Downgrading of Regional Autonomy

The Constitutional Court has tried to resolve this complexity in its steadfast "dematerialization of subject matters."[33] Realizing the excessive rigidity of the lists in Art. 117, the Court tried to trigger attributive dynamics characterized by a degree of elasticity decidedly superior with respect to the mere literal datum, thus going beyond the formal concept of "subject matter." Judgement 407/2002 explicitly states that "not all material spheres specified in Art 117(2) can [. . .] constitute 'subject matters' in the strict sense."

This operation by the Court, however, has caused more confusion from a jurisdictional perspective and is often seen as a "centralist impulse"[34] that downgrades regional autonomy. After all, from the outset part of Italian jurisprudence, it had been affirmed that this sort of "dematerialization" would inevitably lead to "a sliding slope that facilitates the slipping of 'matters' from the Regions to the State, whilst making the transit from the State to the Regions an uphill exercise."[35] In this sense, the so-called transversal subject matters[36] are a typical instrument used by the Court to go beyond the schema of Art. 117. They represent those sectors of legislation where the identification of the competence depends not, technically speaking, on the search for a subject matter as per the lists of Art. 117 but on the pursuit of national interests that overlap with the interests entrusted to the Regions by the Constitution.[37] In the end, the distribution of competence in such matters set by Art. 117 conditions the legislation of other sectors belonging to the concurrent or residual competences of the Regions. This transition has a significant impact precisely in those sectors related to agroecology, whose boundaries have literally disintegrated, one Court judgement after another.

For example, Art. 117, Para. 2, lett. m), attributes to the exclusive competence of the State the following subject matter: the "determination of the essential levels of

[33] Benelli (2006).

[34] Mangiameli (2010), p. 8.

[35] Ruggeri (2004).

[36] In the vast literature for a first reconstruction of constitutional jurisprudence, see Arconzo (2005), pp 181–242.

[37] Benelli and Bin (2009), p. 1210.

well-being regarding civil and social rights that must be guaranteed throughout the entire national territory."[38] It is a criterion for the harmonization of the basic conditions of civil and social rights throughout Italy. Yet, precisely due to this particular function, according to the Court, it is not "a subject matter in the strict sense, but a competence of the State legislator which can be exerted in all subject matters" (Judgement 282/2002), hence a transversal competence able to invade other spheres not expressly reserved to the State.

But it is above all with reference to the complex theme of the environment that the Court has used this method most. Indeed, the Court holds that it cannot deal with a "subject matter in the technical sense that can be described the *protection of the environment*, insofar as it does not seem to be configurable as a rigorously circumscribed and delimited sphere of State competence. On the contrary, it involves and interlinks inextricably with other interests and powers. It is easy to treat the environment as a *constitutionally protected value*, and as such, as a sort of transversal subject matter, affecting different competences, that can be regional, while it is up to the State to make the decisions that meet the justifiable demand for uniform treatment across the entire national territory" (Judgement 407/2002, author's translation). The same interpretative reasoning is used by the Court for subject matters such as the (already cited) "protection of competition" (Judgement 14/2004) and the "protection of cultural heritage" (Judgement 232/2005).

Beyond transversal subject matters, we need to stress the multiple and different techniques used by the Court, regarding the special exceptions as per Art. 117, to delimit the discretion of the regional legislator and so that the national guarantee of rights prevails over the regional demand for autonomy.[39] This is the case of the "criterion of predominance," which, once established in a specific subject matter, forms the "basic core" of the disputed subject, or rather the prevailing legislative competence, and also "attracts" potentially related norms.

An example of the application of this criterion is Judgement 368/2008. This is particularly significant because it deals with the relation between the denomination of wines and agriculture. The object is Art. 1 of the regional Law of Friuli-Venezia Giulia 24/2007, which provided for "the denomination 'Tocai Friulano', a centuries-old heritage of the vine-growing region, continues to be used by producers wine-bearing vines of the Region Friuli–Venezia Giulia [....] to designate the wine, derived from the vine of the same name, that is marketed throughout Italy."[40]

---

[38] See Luciani (2002), pp. 345–360; Pinelli (2002), pp. 881–908; Ruggeri (2002).

[39] Belletti (2012), p 184; Ruggeri (2011), pp. 1461 et seqq.

[40] Among other reasons for complaint, the State noted that the Law violated Art. 117, Para. 2, lett. r) ("pesi, misure e determinazione del tempo: coordinamento informativo statistico e informatico dei dati dell'amministrazione statale, regionale e locale; opere dell'ingegno"). On the plaintiff's part, in practice, the discipline of the denomination did not concern agriculture but involved "opere dell'ingegno", an expression corresponding to "industrial property," which takes in the regulation of the distinctive branding of products, geographic indications, and denominations of origin. Moreover, the discipline of branding revealed on the level of "civil law," and the discipline set by the contested law was also applicable to the "protection of competition," both exclusive State competences.

In the Court's opinion, "the question of trademarks under which wine is marketed has an impact on a multiplicity of interests exceeding 'agriculture'". This conclusion is also consistent with the firm orientation of the European Court of Justice which has, in particular, stressed that "the objective of Community regulation in matters of the designation and presentation of wines is to reconcile the need to provide the consumer with exact and precise information on products [. . .] with that of protecting territorial producers against anti-competitive behavior" (Judgement 12 May 2005, C-347/05); according to the European Court of Justice, the "denominations of origin come within the field of industrial and commercial property rights" (Judgement 16 May 2000, C-388/95).

The problem is that "regardless of the exact configuration of the [trade]mark examined and the category to which it belongs, it is clear that the rule in question – bearing in mind its content and objectives – has a bearing on multiple interests: producers, consumers, the community with respect to the principle of truth, the correct exercise of competition, and thus intervenes in a multiplicity of matters. This intervention must apply the *criterion of predominance*. This is applicable since the core of the discipline belongs to different subject matters, apart from agriculture, such as the protection of competition, and the organization of civil society, none of which is attributed to the Regions, thus rendering the norm illegitimate."[41]

Using such criterion therefore, the Court made other matters of exclusive State competence prevail over agriculture (restrictively understood as simple production of goods) and, as a consequence, declared the illegitimacy of the regional regulation. Thus, one judgement after another, the *criterion of predominance* has become one of the basic criteria that orients the Court, while it is entirely without constitutional cover.

The problem is the concrete trend taken by this principle in recent years. In a first phase, it did not affirm the predominance of State subject matters over regional ones; taking the jungle of competences into account, it was intended to give preference to State ones, provided that the other competences involved could be expressed in terms of "fair participation." But in a second phase, the application of

[41] See again Constitutional Court, Judgement 368/2008 (author's translation). As regards so-called regional brands, Judgement 66/2013 dealt with the question raised by the President of the Council of Ministers on the constitutionality of Lazio Regional Law 1/2012 on "Provisions to support systems of quality and traceability of agricultural and agro-food products." The Region Lazio had introduced a "regional" brand of quality designed to apply to specific agricultural products and agro-food to promote the agriculture and gastronomic culture of the territory, on the basis of a ruling whose approval was requested by the Region. The Court contested that the Law directed consumer preferences to branded products rather than other homogeneous products of a different origin. This would have had a limiting impact, at least "indirectly" or "potentially," on the free movement of goods, precisely those restrictive effects that the regional legislator is prohibited to follow by Community law (in particular, Arts. 34–36 TFUE, which set the ban on quantitative limitations on imports and any measure having equivalent effect, the ban on quantitative limitation on exports and any measure of equivalent effect, and the repeal of such bans). The Court found that the censured norms violated the principle of Art. 117, Para. 1, on the limitations of the exercise of legislative power by the State and Regions derived from Community regulation.

the criterion changes in the sense of reinforcing State powers and downgrading regional powers. By law, the inevitable consequence is that the "criterion of predominance" has ended up representing a post-reform reedition of *national interests* that in the past had allowed the Court to affirm the full competence of the State, without too much delay in evaluating counterprotection in favor of the Regions.[42] In the last instance, if the reform of Title V of the Constitution was designed to invert powers, shifting the general competence to regional legislation, the "counter-reform" of constitutional jurisprudence appears to be moving in the opposite direction, consolidating State competences and marginalizing regional ones.[43]

4 The Role of Territory

This excessively "State-centered" phase is doubly contradictory, both with the spirit of the reform and with the *ratio* of agroecology itself. First, it runs counter to the "pre-legal" motivations of the 2001 reform: the impulses deriving from key changes in the economic context, also at the European level, and which led policymakers/jurists/legislators/the Court to believe that the growth of regional autonomy would become a key factor of economic development.

In this sphere, the philosophy underlying the reform was dictated precisely by the conviction that the Regions could function as centers of reference, coordination, and management to support activities and services, particularly in productive activity. This would give them the function of promoting the resources and qualities of the territory, within the framework of the competences of the global economy. In this sense, it is worth noting the *re-emergence of the territory* as a sphere of social and political relations of greater significance and with its own features and needs.

Indeed, there has been a series of causes such as the importance acquired by the territorial dimension in economic relations,[44] the collapse of social guarantees won under the service State, and the consequent widespread perception of [in]security in the community, as well as the appearance of an "identity question" as a response to the failure of some integration policies. These all favor "reactive phenomena with respect to a cosmopolitan vision of social ties,"[45] where the territory is understood as a *locus* of interests that are more restricted interests but of greater intensity and coherence.[46]

The concept of "territory" has always tended towards the homogenization of local specificity in its national form; but now it expresses the hope of having its own rules and traditions. Faced with this phenomenon, "States appear too uniform in

---

[42] Benelli and Bin (2009), p. 1210.

[43] Benelli and Bin (2009), p. 1211. On this point see Mangiameli (2013).

[44] Barbati (2009), pp. 243–260; Barbera (1999), pp. 1095–1107; Capello (2004).

[45] In this sense, O' Brien (2005), pp. 56 et seqq. (author's translation).

[46] Buratti (2010), p. 141.

their centralism to recognize the many diversities that operate across their territories and which are no longer reflected or reflectable in a representative political order. Their uniform territoriality appears to dominate the other, more uneven, territorialities, the territorialities that extend beneath it and that want to emerge and be heard."[47]

Thus, the Region becomes the *locus* for the protection of interests of a more restricted collectivity, which are promptly intercepted and interpreted by public territorial institutions. The latter thus increase their own protagonism even against the general interests of the national community.[48] The second contradictory point of State interference relates to the *ratio* of agroecology. We should emphasize that the sectors that make up this matter have a lowest common denominator: the *territorial vocation*, or a certain protection of the specificity of the field of action to be safeguarded. As regards this, we should emphasize a very interesting survey regarding the production of regulations, fruit of a study carried out 10 years after the constitutional reform.[49] The research divides the legislation into three broad categories of regional services: *the person*, *economic development and productive activity*, and *environment and infrastructure*.

In the macrosector *economic development and productive activity*, the regulation data reveal that 48–50 % of the laws deal with agriculture and rural areas, industry and trade, 7 % with commerce and tourism, and 14 % with nutrition and food (the so-called detailed laws).[50] This numeric data on sectors favored by the Regions in the first decade of reform should not be underestimated. They reveal how the regional legislator particularly cared about agriculture also as element to promote the specificity of each rural community; the "demand for rurality," in fact, has given rural areas a new "strategic centrality," in which the development passes through territories as vital points in a grid system made of interrelations, integrations, complementarity. Each of these new communities, each with its unrepeatable individuality, needs a special and specific diversified protection by the regional legislator.[51] It is clear that the lack of an adequate support function by the Regions would push back the macrosector as a whole, with serious repercussions for the economy and development, and the general conditions of social and relational life and the environment itself.[52]

The reform of Title V of the Constitution and agroecology share a common impetus to promote territorial differences. This is the response to a particularly complex system, where differentiation is inevitable, since it is the only way to

---

[47] Ferrarese (2006), p. 144.

[48] Ciarlo (2001), pp. 767–789.

[49] Arabia and Desideri (2010), pp. 949–965, exp. p. 959.

[50] We should stress that if we take a look at the geographical areas of Italy, there are a greater number of laws dealing with rural development and agriculture in the South in comparison with the North and Center. Yet if we look at artisan work-industry and commerce-tourism, the North and Center outdo the South.

[51] Marotta (2006), p. 127.

[52] Arabia and Desideri (2010), p. 963.

reestablish the basic equilibria to/that maintain the system.[53] Yet the path taken by the Court tends in the anachronistic direction of a *regionalism of uniformity*, typical of the 1948 Constitution, then justified by a (historical and social) need to eliminate the human and territorial inequalities of the time.

5 Conclusions: Some Proposals *De Jure Condendo*

This then is the doubly distressing picture for a positive jurist trying to identify the outlines of the protection of agroecology in the Italian Constitution: a fragmentation between the subject matters that constitute the sector/domain and a phase of clear State–Region conflict in the decade when the reform was made. Constitutional jurisprudence, for its part, tends to favor neo-centralist projects with the State as protagonist.[54] There have been a great many interventions by the Court to reaffirm the presence of unitary needs. This has been done to guarantee the principle of equality, through a series of competences, exclusive or transversal or criteria *sui generis*, such as the criterion of "predominance," which reined in attempts by the Regions to exercise their legislative power autonomously.[55]

On the other hand, we have stressed how "agroecology," in the same way as agriculture, can function as a motor for the territory when it interacts correctly with the environment and that the Regions were aware of this. This is revealed by the simple quantitative fact of the amount of regional laws specifically "dedicated" to agriculture and rural issues. We now need to ask what route to take for a legal protection that deals with the phenomena in a more systematic but also less conflictual way. Here the discourse inevitably slips into *de jure condendo*. At a time when the parliamentary debate on reform is also dealing with the (new) reform of Art. 117, in order to reduce the problems unleashed by the text currently in force, it would be beneficial (and timely) for agroecology to be explicitly dealt with in the Constitution. The question then is which scheme of distribution of legislative powers to back? We must reject any idea of exclusive regional powers. Indeed, many social rights (e.g., right to the environment or even a right to health) cannot be left entirely in the hands of the Regions without the risk of serious territorial disparities to the detriment of citizens. Social rights are, by definition, *aterritorial*.[56]

[53] Poggi (2005), pp. 599–622.

[54] Belletti (2012), p. 184; Mangiameli (2013).

[55] Violini and Rovagnati (2012), pp. 957–994, exp. p. 958.

[56] On the vast theme of social rights, and without claiming to be exhaustive, see Baldassarre (1989), pp. 1–34, exp. p. 28; Mazziotti (1964), pp. 802–807, exp. p. 804. The Italian Constitution clearly derives social rights from substantial equality, as social rights consist of "the body of norms through which the State carries out the balancing and moderating function of social inequality, in order to guarantee the equality of situations despite differences of circumstance"; Luciani (1995), pp. 97–134, exp. p. 121; Lombardi (1999); Caretti (2005); Bifulco (2003); Rimoli (2004). On the

We might consider making "agroecology" the exclusive competence of the State, but this would mean sacrificing the basic characteristics of this new domain—based on the territorial vocation and on promoting the specificity of all rural communities—on the altar of unitary needs. Eliminating the option of an (albeit minimal) discipline differentiated by the nearest legislator, in the name of unitary needs, would mean homogenizing matters that are intrinsically heterogeneous, because they are linked to the socio-cultural context of a particular territory.

It is not hard to understand that the question of the explicit and unitary protection of agroecology is set within a more general issue: rethinking the role of a central power that must not only contain but also concede space to regional autonomy (thus favoring it), rereading it in the light of the principle of responsibility of the political classes that adopt particular decisions. That is, it appears more than ever necessary to reflect on which tasks can be carried out at the central level in order to support, rather than undermine, regional autonomy, where this is virtuously interpreted by local actors.[57] The promotion of the material and immaterial externalities of agriculture should indeed be seen as a motor for any virtuously governed territory.

An explicit reference to agroecology in Art. 117 could be cited under the list of so-called concurrent powers. The State *qua* legislator would be left the task of setting the insuperable limits relating to fundamental principles in that domain. The Regions would instead be responsible for a detailed regulation, taking the specificity and particular features of their territories into account. A solution of this sort helps free constitutional control from the iron grip of the repressive function of regional intervention but considers it, on the contrary, as a unifying element helping to reorganize a system urgently in need of modernizing intervention.[58]

To conclude, the inclusion of "agroecology" with its many subjects (some relating to inviolable rights, others intrinsically linked to the territory) in the Constitution could mark the start of a new era in center-periphery relations.

References

Adornato F (2007) Agricoltura plurale e poteri regionali. In: D'Atena A (ed) Regioni e attività produttive, Rapporto sulla legislazione sulla spesa. V. Giuffrè, Milano, pp 67–80

Arabia AG, Desideri C (2010) Le materie dello sviluppo economico alla prova del federalismo fiscale. Rivista giuridica del Mezzogiorno 24(3):949–965

Arconzo G (2005) Le materie trasversali nella giurisprudenza costituzionale dopo la riforma del Titolo V. In: Zanon N, Concaro A (eds) L'incerto federalismo. Le competenze statali e regionali nella giurisprudenza costituzionale. Giuffrè, Milano, pp 181–242

Baldassarre A (1989) Diritti sociali. In: Enciclopedia giuridica, vol XI. Treccani, Roma, pp 1–34

delicate relationship between Regions and social rights, see Chieffi (1999); Longo (2007); Vivaldi (2013), pp. 545–560.

[57] Violini and Rovagnati (2012), p. 980.

[58] Violini and Rovagnati (2012), p. 980.

Barbati C (2009) Territori ed interessi economici: le politiche dei luoghi per lo sviluppo locale. Le istituzioni del federalismo 31(2):243–260
Barbera A (1999) Il governo locale nell'economia globale. Le istituzioni del federalismo 21 (1):1095–1107
Benelli F, Bin R (2009) Prevalenza e "rimaterializzazione delle materie": scacco matto alle Regioni. Le Regioni 37(6):1185–1212
Benelli F (2006) La "smaterializzazione" delle materie. Giuffrè, Milano
Belletti M (2012) Percorsi di ricentralizzazione del regionalismo italiano nella giurisprudenza costituzionale. Aracne, Roma
Benozzo M (2007) Alimentazione: una materia trasversale. In: D'Atena A (ed) Regioni e attività produttive, Rapporto sulla legislazione e sulla spesa. V. Giuffrè, Milano, pp 107–124
Bifulco D (2003) L'inviolabilità dei diritti sociali. ESI, Napoli
Bin R (2001) Le potestà legislative regionali, dalla Bassanini ad oggi. In: Ruggeri A, Silvestri G (eds) Le fonti di diritto regionale alla ricerca di una nuova identità, Atti del seminario di Messina. Giuffrè, Milano, pp 122–153
Borghi P (2006) Colture geneticamente modificate, ordinamenti e competenze: problemi di coesistenza. Le Regioni 34(5):961–977
Buratti A (2010) Rappresentanza e responsabilità politica. Jovene, Napoli
Capello R (2004) Economia regionale. Localizzazione, crescita regionale e sviluppo locale. Il Mulino, Bologna
Caranta R (2004) La tutela della concorrenza, le competenze legislative e la difficile applicazione del titolo V della Costituzione. Regioni 32(4):990–1013
Caravita B (2005) Il diritto dell'ambiente. Il Mulino, Bologna
Caravita B (1984) La disciplina costituzionale della salute. Diritto e società 12:522–545
Caretti P, Tarli Barbieri G (2012) Diritto regionale. Giappichelli, Torino
Caretti P (2005) I diritti fondamentali. Libertà e diritti sociali, Giappichelli, Torino
Carrozza P (1992) L'ambito costituzionale della "agricoltura e foreste" e la produzione di mangimi. Rivista diritto agrario 71(2):19 et seqq.
Chieffi L (ed) (1999) I Diritti sociali tra regionalismo e prospettive federali. Cedam, Padova
Chieffi L (ed) (2003) Il diritto alla salute alle soglie del terzo millennio. Profili di ordine etico, giuridico ed economico. Giappichelli, Torino
Ciarlo P (2001) Le nuove Regioni: vocazione all'economia territoriale, crisi del criterio di competenza e contrattualismo politico. Nuove Autonomie 10(6):767–789
Cocconi M (1988) Il diritto alla tutela della salute. Cedam, Padova
Cristiani E (2004) La disciplina dell'agricoltura biologica fra tutela dell'ambiente e sicurezza alimentare. Giappichelli, Torino
D'Atena A (2013) Diritto regionale. Giappichelli, Torino
D'Atena A (2007) Interferenze ed intrecci tra materie e competenze legislative. In: D'Atena A (ed) Regioni e attività produttive, Rapporto sulla legislazione sulla spesa. V. Giuffrè, Milano, pp 59–66
Ferrarese MR (2006) Diritto sconfinato. Inventiva giuridica e spazi nel mondo globale. Laterza, Bari-Roma
Galloni G (1995) L'incontro tra agricoltura e ambiente: fondamenti costituzionali. In: Galloni G, Tarolo C, Donnhauser C (eds) Agricoltura e ambiente, Il cambiamento di rotta della nuova PAC e la sua attuazione in Italia. Il Sole 24 ore, Milano, pp 15 et seqq.
Germanò A (2003) La "materia" agricoltura nel sistema definito dell'art. 117 Cost. Le Regioni 31 (1):117–194
Grassi S (2012) Problemi di diritto costituzionale dell'ambiente. Milano, Giuffrè
Grassi S (2003) Agricoltura e tutela dell'ambiente nel nuovo Titolo V della Costituzione. In: Germanò A (ed) Il governo dell'agricoltura nel nuovo Titolo V della Costituzione. Giuffrè, Milano, pp 39–66
Jannarelli A (2003) L'agricoltura tra materia e funzione: contributo all'analisi del nuovo art. 117 Cost. In: Germanò A (ed) Il governo dell'agricoltura nel nuovo Titolo V della Costituzione. Giuffrè, Milano, pp 67–124
Lombardi G (1999) Diritti di libertà e diritti sociali. Politica del diritto 30(1):7–16

Longo E (2007) Regioni e diritti. La tutela dei diritti nelle leggi e negli statuti regionali. EUM, Macerata

Losavio C (2012) La competenza legislativa residuale delle Regioni nella giurisprudenza costituzionale: il caso dell'agricoltura. In: Mangiameli S (ed) Il regionalismo italiano dall'Unità alla Costituzione e alla sua riforma, II. Giuffrè, Milano, pp 326–337

Luciani M (2002) I diritti costituzionali tra Stato e Regioni (a proposito dell'art. 117, comma 2, lett. m) della Costituzione. Politica del diritto 32(3):345–360

Luciani M (1995) Sui diritti sociali. In: Studi in onore di Manlio Mazziotti di Celso, II. Cedam, Padova, pp 97–134

Luciani M (1980) Il diritto costituzionale alla salute. Diritto e società 8:769–787

Lucifero N (2008) La legge statale sull'agriturismo a fronte del vaglio di legittimità della Corte costituzionale. Nota a sent. Cort. Cost. 12 ottobre 2007 n. 339. Giurisprudenza italiana 160 (4):834–839

Lugaresi N (2010) Il contenzioso costituzionale sul "Codice dell'ambiente", tra rapporti istituzionali complessi e confini sfuggenti tra materie. Le Regioni 38(3):557–589

Malo M (2008) Baricentro regionale in materia di turismo. In: D'Atena A (ed) Regioni e attività produttive, Rapporto sulla legislazione e sulla spesa. V. Giuffrè, Milano, pp 141–176

Mangiameli S (2013) Le Regioni italiane tra crisi globale e neocentralismo. Giuffrè, Milano

Mangiameli S (2010) 'Introduzione'. In: Osservatorio sulla legislazione, Rapporto 2010 sulla legislazione tra Stati, Regioni e Unione Europea, II. Camera dei Deputati, Roma, p 8

Marotta G (2006) L'agroalimentare del Mezzogiorno di fronte alle sfide dei mercati e della società. Rivista economica del Mezzogiorno 20(1–2):127–166

Martines T, Ruggeri A, Salazar C (2012) Lineamenti di diritto regionale. Giuffrè, Milano

Mazziotti M (1964) Diritti sociali. In: Enciclopedia del Diritto. Giuffrè, Milano, pp 802–807

Milazzo P (2005) Alcune questioni di interesse costituzionale in materia di organismi geneticamente modificati in agricoltura (a proposito del d.lg. n. 479/2004 sulla coesistenza fra le forme di agricoltura). Diritto pubblico 11(1):225–264

Morana D (2014) Diritto costituzionale alla salute. Lezioni. Giappichelli, Torino

Nania R (2006) Agricoltura e tutela della salute. In: Rook Basile E (ed.) Dopo la modifica dell'art. 117 Cost: problemi ed esperienze sulla competenza della materia agricoltura. Giuffrè, Milano, pp 121–132

Nigro M (1980) Il governo locale. Bulzoni, Roma

O' Brien M (2005) Cosmopolitismo e località. Pensa, Lecce

Pinelli C (2002) Sui "livelli essenziali delle prestazioni concernenti i diritti civili e sociali" (Art. 117 co. 2, lett m). Diritto pubblico 8(3):881–908

Poggi A (2005) Diritti sociali e differenziazioni territoriali. In: Balduzzi R (ed) I servizi sanitari regionali tra autonomia e coerenze di sistema. Giuffrè, Milano, pp 599–622

Ruggeri A (2011) Regioni e diritti fondamentali. Giurisprudenza italiana 163(6):1461 et seqq.

Ruggeri A (2004) Le Costituzioni passano, ma la giurisprudenza... resta. www.federalismi.it. Accessed 4 Jan 2004

Ruggeri A (2002) Neoregionalismo, dinamiche della formazione, diritti fondamentali. www.giurcost.org. Accessed 20 Sept 2002

Rimoli F (2004) Stato sociale. In: Enciclopedia giuridica, vol XIII. Treccani, Roma

Tarchi R (2006) Agricoltura e ambiente. In: Rook Basile E (ed) Dopo la modifica dell'art. 117 Cost: problemi ed esperienze sulla competenza della materia agricoltura. Giuffrè, Milano, pp 65–86

Violini L, Rovagnati A (2012) Rileggendo la giurisprudenza costituzionale sui rapporti tra Stato e Regioni del 2012. Le Regioni 40(5–6):957–994

Vivaldi E (2013) Dal criterio della prevalenza alla leale collaborazione: quale autonomia legislativa regionale in materia di politiche sociali? In: Cavasino E, Scala G, Verde G (eds) I diritti sociali dal riconoscimento alla garanzia. Il ruolo della giurisprudenza. Editoriale Scientifica, Napoli, pp 545–560

The Relationship Between Agricultural Law and Environmental Law in Italy

E. Cristiani

Abstract In Italy, the traditional view of agriculture has been that it is the most immediate, authentic and natural expression of the environment. This occurred even though the study of the emergence of environmental issues and their integration in agricultural law has only recently appeared as a problem. Furthermore, the affirmation of environmental law in Italy is a recent fact. Even today, there is debate as to whether, instead of building a "law *of* the environment" as an independent discipline, we should instead identify, in the various sectors of law, common principles as source of inspiration for a "law *for* the environment." Many books on agrarian law, published in the 1980s, deal with environmental issues only marginally, to define the borders of agrarian law *strictu sensu*. An important role in defining the boundaries and interactions between environmental law and agrarian law has been played by Italian journals of agrarian law, which starting from the beginning of the twentieth century opened to new subject matters, such as food, nutrition and environmental protection. This broadening of agricultural law towards environmental law had important repercussion on didactics, as proved by the inclusion in the sector of the scientific discipline of "agricultural law" of studies pertaining to the protection of the environment and to the commercialization of agricultural products.

Keywords Agriculture • Environment • Law

1 Introduction

This chapter develops along the following lines of analysis. First, we present the historical framing of the boundaries between agricultural law and environmental law. Second, we analyze the constitutional norms and actions of Italian constitutional jurisprudence in identifying the "subject matters" of agriculture and the environment. In conclusion, we give an account of how this issue has been dealt with in the journals, manuals and in didactics in general.

E. Cristiani (✉)
Institute of Law, Politics and Development (Dirpolis), Sant'Anna School of Advanced Studies, Pisa, Italy
e-mail: eloisa.cristiani@sssup.it, https://www.sssup.it/eloisa.cristiani

M. Monteduro et al. (eds.), *Law and Agroecology*,
DOI 10.1007/978-3-662-46617-9_11

The environmental side of Italian agricultural law may appear today quite obvious, from both a factual and a legislative point of view, though it was little known until a few decades ago. According to the traditional view, agriculture was the most immediate, authentic and natural expression of the environment. It was seen as a sector of intervention with a positive socio-environmental impact as opposed to industry, which conversely created a fracture in the relationship between a population and its environment. For many years, this has prompted Italian doctrine to take for granted a relationship of perfect symbiosis between agriculture and nature.[1] The study of the emergence of environmental issues and their integration in agricultural law is a very recent issue. Moreover, the affirmation of environmental law in Italy is a recent fact. Even today, there is debate as to whether, instead of building a "law *of* the environment" as an independent discipline, we should not identify, in the various sectors of law, common principles as a source of inspiration for a "law *for* the environment."[2] Indeed, environmental issues and interests overlap with many disciplines in this area of study, which are, however, called upon to ensure full protection in a substantially uniform manner, be it from a multiplicity of "perspectives." The formulation of the juridical concept of environment in Italy is generally said to date to a well-known essay by M. S. Giannini, published in the *Rivista trimestrale di diritto pubblico* in 1973.[3] However, we must wait until 1980 for the term "environment," subdivided into various subsectors, to appear in a legal encyclopaedia. Previously, it was merely a cross-reference. The first issue of the *Rivista giuridica dell'ambiente* only apperead in 1986, signing an important step for the establishment of environmental law in Italy.

The origin of the interest of agrarian scholars with respect to ecological issues must therefore be contextualized within the correct historical dimension, primarily by the evolution of socio-economic conditions in the agricultural sector in Italy.[4] The main objective of agricultural policy in the 1960s and early 1970s appears to have been maximum productivity in the name of competiveness. Public intervention in the agricultural sector was designed to improve the efficiency of the farm enterprise and to raise farmers' income to that of other professional categories. Concerns of a broadly speaking environmental character were generally seen as excessive, given a political and legislative schema focusing on farm modernization, particularly in terms of productive efficiency.

[1] Cristiani (2004), p. 11.

[2] Grassi (2012), p. 12; Bolognini (2007), pp. 723–737, 723.

[3] Giannini (1973), pp. 15–53, 15. There exist three distinct meanings of the term "environment": (1) the environment as referred to by the legislation and the movement of ideas relative to landscape; (2) the environment as referred to by the legislation and the movement of ideas relative to the defence of soil, air and water; (3) the environment as referred to by the legislation and in urban studies. For years, the doctrine continued to refer to this "tripartition" in order to describe the different profiles of environmental protection: Caravita (2004), p. 13, and, in greater detail, Caravita (2005), p. 16; Nespor (1986), pp. 1–8, 2.

[4] In this regard, it should be recalled the Proceedings of the "Giornate Camerti di Diritto agrario" (1–2 December 1989); see Pennisi and Calcinelli (1990).

The debate among agrarian scholars regarding agricultural law and protection of the environment emerges in the late 1970s, under the influence of EEC jurisdiction, initially through the Environmental Action Programmes.[5] At that time, in the face of EEC food self-sufficiency for basic products and faced with the problem of surplus agricultural products and their destruction, the myth of the gross national product, as the only parameter for assessing a country's progress, gave way to other more pressing needs for protection. A policy of coherent and economically sustainable development could no longer ignore the value of the environment. It became clear that agriculture is involved in and is a key participant in this process to reconcile protection and development: it must be oriented, in order to survive, towards forms of land use that respect the environment and natural resources. The legislator began to provide incentives for regulations to encourage agriculture with a lower environmental impact, which would involve the farmer who could in turn assume public functions in order to safeguard the territory.[6] An important moment in establishing a new concept of agriculture oriented towards interests that were not purely productive was the debate that developed in the 1970s. This involved moving the jurisdiction for agriculture to Italy's regional governments. Presidential Decree 616/1977 put into effect the transfer of functions from the Italian State to the Regions. This identified, as far as we are concerned, what is meant by agriculture and forests (Art. 66), town and country planning (Art. 80), environmental assets (Art. 82) and the protection of nature (Art. 83). This opened a debate on the boundaries of the subjects involved, their relationships and the transversal character of environmental issues, undoubtedly common to various disciplines. However, in the 1980s, the manuals of agricultural law dealt only marginally with environmental themes, in a sort of *actio finium regundorum* (action to regulate boundaries) in order to set appropriate "limits" for the boundaries of agricultural law in the strict sense. Besides, the ways in which agrarian scholars approached differed a great deal. There are those, from the pages of the *Rivista di diritto agrario*, who do not hesitate to theorize a redefinition of agricultural law as agri-environmental law: a science that is naturalistic-environmental in addition to being juridical, based on the study of the bio-society, its promotion and conservation.[7] On the other hand, there are those who defended the methodological coherence of the system of agricultural law, with a perspective of a "pure" conception of the discipline, and rejected the inherent loss of identity brought about by an assimilation of agricultural law into a law of nature or the environment. There is the hypothesis of a possible broadening of the object of agricultural law, given the dynamic nature of the discipline, but which also emphasizes limitations and dangers. In the light of this viewpoint where "the object of agricultural law remains specific and continues to regard essentially

[5] See Cordini et al. (2005); Costato (1987), pp. 512 et seqq; Manservisi (2008), p. 177; Costato and Manservisi (2012), p. 1.

[6] Costato (2003), pp. 75–85, 84.

[7] Capizzano (1987), pp. 433–451.

the organization of the juridical relations pertaining to agricultural exploitation of the land,"[8] the only way in which "ecological concern" could be expressed as part of the programme of agricultural law was in the form of "new limits to landed agricultural property and above all new constraints on the factory farm, but also through new parameters (those of an ecological character) for the assessment of general clauses such as 'good agricultural practice' in the obligatory relations between private entities and between public administration and private bodies."[9] Within the context of studies on the relationship between these two branches of law, Galloni has declared that we are not speaking about a genetic modification of the matter but that we need to highlight "the meeting point between agriculture and environment that can no longer be postponed and, consequently, [the meeting point] between agricultural law and environmental law [which] is represented by the local territory…"[10]

However, times and regulations change, and the study of agricultural law cannot disregard the social reality with which it is associated, isolating itself in the conceptual categories of tradition. Environmental concerns and emergencies have produced a vast legislative output that projects the agricultural sector into a broader dimension, which inevitably impacts on the boundaries of the discipline and on the correlated doctrinal positions. The new role for rural spaces and agricultural producers imposes a redefinition of the duties of agricultural law in modern society.[11] Together with its important productive function, agriculture carries out "a more complex social and environmental function" through land management and conservation of the environment, a function of public relevance from which the entire community benefits.[12]

On a legislative level, the agrarian scholars most sensitive to ecological issues, *ab origine*, focused on the notion of *rationality* in exploitation of land (Art. 44, Italian Constitution). It is precisely this concept of rationality, albeit combined with a blatantly economic term such as *exploitation*, that was able to tip the scales towards values of another sort, notably environmental values. The open-endedness of the constitutional wording has enabled Italian legislation to adapt to the evolution of agriculture. This prompts scholars to ratify the overcoming, in a developmental way, of the old opinion, associated with a merely productive aspect, according to which the economic interpretation of the word

[8] All the quotations in English between "" are author's translations from Italian, if not otherwise stated.

[9] Carrozza (1984), pp. 177–178, 177; Carrozza (1982), pp. 77 et seqq; Carrozza (1994); Carrozza (1988), pp. 93, 320. In this perspective see also Grossi (1986), pp. 427–438. The author affirms that, in comparison with many other disciplines, agricultural law is a future-oriented discipline; nonetheless, it needs to beware of the flattery of environmental law, for instance, because if "agrarian issues" becomes an opportunity, all is lost for agricultural jurists.

[10] Galloni (1995), pp. 17–48, 27; Galloni (2000), pp. 381–411, 402.

[11] Francario (1993), pp. 517–519; Galloni (1993), pp. 5–10.

[12] See Jannarelli (2002), pp. 729–753, 743.

exploitation prevailed.[13] The study of the rule of "rationality" as a guiding criterion of the productive process in agriculture therefore leads to results that are convincing from the perspective of safeguarding natural resources. Today, in the light of current national and EU legislation, there can be no doubt, in Italy, that *exploitation* can be defined as rational only insofar as it respects the equilibrium of the biosphere, by means of a sustainable management of nonrenewable natural resources. However, there is no doubt that it is in the field of EU agricultural law that development has occurred, and not only from a practical perspective, but also at a systematic level, of agricultural law substantially inspired by legislation intended to safeguard and improve the natural environment.[14]

Galloni himself, who identifies "protection of the environment as a meeting point of national agricultural law and EU legislation and as a basis for the launch of a modern system of agriculture," observes that the "sources" of "agricultural environmental law" are found mainly within the context of EU legislation. From this perspective, the environment becomes an interpretative criterion of current legislation and propositional criterion for the production of the new law. In this way we arrive at the most recent textbooks of agricultural law that do not hesitate to offer, regarding the issue at stake, complex definitions such as this: "Italian (and EU) agricultural law is a system of rules and regulations which no longer tends to only protect agriculture and agricultural entrepreneurs, but also and at the same time to conserve the environment, to ensure the healthy quality of agricultural products, also with the aim of safeguarding consumers, and maintaining a human presence in marginal geographical areas otherwise destined to be abandoned."[15]

2 Constitutional Norms

The problem of the boundaries between agriculture and environment and consequently between agricultural law and environmental law reemerges in Italy with the constitutional law of 18 October 2001/3, when the term "environment" makes its formal entrance into the Italian Constitution. Art. 117(2)*s* states that the State holds exclusive legislative power in issues of "protection of the environment, of the

[13] The literature is immense. See, for example, Irti (1972), pp. 391–401; Sandulli (1972), pp. 465–490; Carrozza (1985), pp. 4 et seqq; Ferrero (1979), p. 228; Capizzano (1979), pp. 23–88; Salaris (1981); Rodotà (1982), pp. 211–233; Comporti (1983), pp. 444–465; Carrozza (1983), pp. 466–473; Busnelli (1983), pp. 474–478; Costato (1983), pp. 479–488; Graziani (1985), pp. 309–316; Costato (1988), pp. 325–338; Francario (1986).

[14] This work is obviously organized within a typically Italian dimension in that the logic of the Convention at which it was presented was to identify the differences between the various national legal systems. It is clear that an analysis of the relations between agricultural law and environmental law within a European Union perspective would have been almost analogous among the various countries.

[15] Costato (2001), p. 34.

ecosystem and of cultural assets."[16] However, in accordance with Art. 117(3), to be considered concurrent legislation (State–Regions) are those issues such as the "protection of health," "government of the land" and "the exploitation of cultural and environmental assets," inherently associated with environmental protection. The subject matter of agriculture, which includes *ab antiquo* the so-called protective agriculture and which today presents itself, within the modern context of multifunctionality, as agriculture in respect of, and for the protection and exploitation of, the rural environment and local territory, is included within the disciplines reserved for regional management. Among the numerous judgements of the Italian Constitutional Court trying to identify the contents of the discipline of agriculture, we should recall the Judgement 12 of 13 January 2004, with which the Italian Regions challenged some regulations of the 2002 Financial Law. I will focus, in particular, on just two of the various issues brought to the notice of the Constitutional Court. The first is relative to the regulation that stipulates administrative sanctions in the case of the illegal planting of vines. In this case, the Court stated that the administrative competence regarding sanctions could not be presented as a subject matter *per se* but accedes to the underlying substantial matters. Thus, the question is valid because, the Court affirmed, "planting vines pertains to what could be defined as *the crux of the issue of agriculture, which relates to the production of plants and animals intended for consumption as food*. We are dealing, therefore, with legislative competence entrusted residually to the Regional governments and subtracted from state legislative competence."[17] There is no doubt that here the Court demarcates the essential nucleus of the matter in a restrictive way, referring only to the productive phenomenon (while from time immemorial, as we have already said, agriculture also carries out other functions, for example legislation regarding woodland, among the oldest legislation) and disregarding the transformation and use of the products.[18] The other question regards the provision of the Financial Law with which reference is made to a decree of the Ministry of Economics that specifies the operative regulations with which it is possible to concede incentives to horse breeders "for the development of hippotherapy and for the genetic improvement of trotters and gallopers." In this case, the Court distinguishes the two particular cases by affirming that hippotherapy is unrelated to agriculture but is ascribable to the issue of protection of health, of concurrent jurisdiction. Genetic improvement, on the other hand, is attributable to the issue of agriculture.

With regard to the issue in question, in order to identify the boundaries between agriculture and environment, of particular relevance is the Judgement 116/2006, with which the Constitutional Court declared the constitutional illegitimacy of Decree Law 279/2004 (Art. 3–5, pars. 3–4, 6, pars. 1, 2, 7–8) due to its invasiveness as regards the regional legislative competence in the matter of "agriculture."

---

[16] Doctrine has widely analyzed the sense in the introduction of the term "ecosystem," absolutely unprecedented and not legally defined, together with the concept of environment: Tarchi (2006), pp. 65–106, 71; Grassi (2003), pp. 39–65, 57. More generally, see Benozzo and Bruno (2003), p. 5.

[17] Emphasis added.

[18] Tarchi (2006), p. 79.

Let us recall that the Decree Law, converted—with modifications—into Law 5/2005, had as its objective the legislative discipline regarding the coexistence of transgenic, conventional and organic crops, implementing Recommendation 2003/556/CE of the European Commission. It deferred to a successive act of the Ministry of Agriculture and Forestry Policies to identify framework regulations, on the basis of which the Regions should have adopted their regional plans regarding the coexistence of these crops. The question of legitimacy had been raised first and foremost by the Region of Marche. The Court substantially agreed with this Region, prescribing clear rules for the allocation of competences between State and Regions, within the context of that very singular "issue" of GMOs. The Court recognizes the legitimacy of Arts. 1 and 2, which give the definitions of transgenic, organic and conventional crops and which affirm, in line with EU law, the principle of coexistence between these crops so as to safeguard their specific characteristics and productive peculiarities in that they constitute the practice of the exclusive state legislative competence regarding protection of the environment and competing in the matter of protection of health. On the other hand, the Court refers to the discipline of agriculture in that part of Decree Law 279/2004 (Arts. 3–4, 6, pars. 1, 7 and in addition, consequently, Art. 5(3–4), Art. 6(2) and Art. 8), whose object is cultivation for productive purposes, claiming its own jurisprudence in this matter (Judgement 12/2004) according to which "the crux of the issue of agriculture [. . .] relates to the production of plants and animals intended for consumption as food." The cultivation of GMOs, even though authorized, and the discipline of coexistence, has as an object, in fact, the economic aspect of the use of GMOs in agriculture and does not take into account the evaluation of the implications for health and the environment. On the other hand, it is the system of authorizations established by Directive 2001/18/EC (and successive modifications) that foresees, if necessary, specific measures in the matter of coexistence aimed at safeguarding human health and the environment, whose application is mandatory. Both as far as agriculture is concerned and with regard to the environment, we are, however, faced with heterogeneous and complex realities whose definition cannot easily be based on objective parameters but must be based on issues of a functional or finalistic nature.[19]

The many judgements published by the Constitutional Court (including some recent examples on various articles of the Environmental Code),[20] which constitute the protection of the environment as a combination of values to pursue and to be considered as an object of the action of public powers, illustrate how the different levels of government must contribute to its protection and share common responsibilities.[21] The Court, following a *jurisprudence constante* and with almost identical formulations, has highlighted how the issue of "environmental protection" has a content that at the same time is both objective, in that it refers to an asset, "the

[19] Tarchi (2006), p. 69.

[20] Reference is to Legislative Decree 3 April 2006, n. 152, containing "Norme in materia ambientale."

[21] Grassi (2003), p. 54.

environment," and finalistic, because it tends towards a better conservation of that asset.[22] For this reason, regarding the same asset "environment," there are different and concurrent competences, which remain distinct, pursuing their specific aims autonomously through the provision of different regulations. According to the scheme of the constitutional legislator, on the one hand, the State is entrusted with the protection and conservation of the environment, by setting levels of protection that are "adequate and not limitable." On the other hand, it is the competence of the Regions, with respect to the levels of protection established by the State, to exercise their jurisdiction in regulating the use of the environment and avoiding deterioration or alterations of the environment. In this sense, the State competence has been affirmed in that when it is an expression of protection of the environment, it constitutes a "limit" to the practice of regional competences.[23] The Regions cannot therefore violate the levels of environmental protection set by the State, but they can, in the practice of their own competences, fix higher levels of protection, impacting, also in this way, albeit indirectly, on the protection of the environment.[24]

The fact that Article 117 of the Italian Constitution entrusts the State to define direct regulations to safeguard ecological equilibria does not exclude the Regions from practicing their legislative functions, being able to adopt regulations that safeguard environmental needs within matters entrusted to their exclusive or concurrent jurisdiction. Moreover, in an analogous form, the principle of integration and transversality of environmental needs, at the EU level, excludes neither specific policy making in the sector of the environment nor the fact that member states, as is well known, can adopt provisions for even greater protection.[25]

In this way the Italian legal order ensures efficient protection of a value of primary relevance and of its multifaceted content. It is clear that there are various profiles of environmental protection on different subject matters, not always of

[22] See, among others, Judgements 367/2007 and 378/2007; 104/2008; 12/2009, 225/2009 and 315/2009. See the *notes* to some of the cited judgements Furno (2007), pp. 4119–4128; De Leonardis (2009), pp. 1455–1464.

[23] In this sense, see Judgements 62/2008, 180/2008, 214/2008 and 437/2008 and also 61/2009 and 164/2009. The reform of *Titolo V* of the Italian Constitution, if correctly interpreted, in the sense of ensuring a uniformity of direction in policies regarding protection of the environment, indicates "the reference framework within which there is coordinated and coherent commitment on the part of all the subjects (Municipalities, Provinces, Metropolitan councils, Regions and State) which, in the configuration foreseen by the reformed Article 114, constitute the Republic," in the terms efficiently expressed by Cordini (2009), pp. 611–634.

[24] Thus, see Judgements 104/2008, 12/2009, 30/2009, 61/2009, and 225/2009.

[25] On the one hand, Art. 37 of the Charter of Nice affirms that "a high level of environmental protection and the improvement of the quality of the environment must be integrated into the policies of the Union and ensured in accordance with the principle of sustainable development" (original text) and, on the other, that in Arts. 191–193 of the TFEU, which discipline protection of the environment, there is recognition of the right of Member States to maintain or make provisions for an even greater protection. Safeguarding the environment is, however, set out by Art. 4 of the TFEU as one among the concurrent competences of the EU. See Germanò and Rook Basile (2014), p. 81; Merusi (2007), pp. 495–501, 499.

secondary importance, within the State's exclusive or concurrent or even reserved for the Regions, as in, for example, the case of "our" (Italian) agriculture.

3 Conclusion

In conclusion, we will highlight the role of agricultural law journals in defining the boundaries and interactions between the two subject areas.[26] Clearly, it would be simplistic and acritical to argue that the these journals have always dealt with the environment insofar as they deal with an activity that takes place in the environment and uses its resources. Through an analysis of specialist literature, we highlight how the doctrinal approach has regarded the relation between agricultural law and environmental law. This allows us to see whether or not the environmental dimension is intrinsic to agriculture.

Editorial choices have led to the appearance of new journals or the renaming of older ones. These are sometimes accompanied by leading articles or lively debates among scholars who explain the essence or practical and conceptual implications of the change. To cite a few examples: starting from the first issue in 1993, *Giurisprudenza agraria italiana*, "in harmony with the evolution of European and national legislation and jurisprudence," changed its name to *Diritto e giurisprudenza agraria e dell'ambiente*. The first issue under the new title included articles treating the environment as a limit to the practice of agricultural activities (polluting and polluted agriculture), a form of agriculture (environmental restrictions on agricultural-forest land), a product of agriculture with reference to the new role that the Common Agricultural Policy (CAP) assigns to agri-business and the services it can carry out to safeguard the environment.[27] In 2006, the journal once again changed its name to *Diritto e giurisprudenza agraria, alimentare e dell'ambiente*, in order to formally include food law. To the question "What do we mean when we speak of agriculture?" Francesco Adornato, in the editorial opening of the first issue (2004) of the journal *Agricoltura Istituzioni Mercati, Rivista di Diritto Agroalimentare e dell'Ambiente*, responds: "We are speaking of 'plural' agriculture." In 2005, the historic *Rivista di diritto agrario*, founded in 1922, added the subtitle "Agriculture-Food-Environment" to its original title.

What emerges from the study of works published in 2013 in Italian scientific journals is what has been defined as "The third era of agricultural law facing us: the struggle to save the planet and to create a people-oriented habitat."[28] In this light, agricultural law can be seen as a front-line laboratory for formulating regulations and solutions for issues of more general importance. A recent essay illustrates what

---

[26] Galloni (1986), pp. 145–190; Galloni (2000), p. 381; Sirsi (1986), pp. 415–420; Cristiani (2008), pp. 464–479.

[27] In these terms, Francario (1993), p. 517.

[28] The reference is to the title of the fifth section of the work by Jannarelli (2013), pp. 11–35.

can be defined as the future pathway of agricultural law "faced with *product-oriented* agricultural law from the very beginning and *market-oriented* agricultural law, in which we are still partially immersed, the new agricultural law taking shape is *eco-oriented*."[29] By way of example, the author refers to the direct legislation to regulate the changes in land use in relation to land planning and protection of the environment, the provisions to protect agricultural activity, safeguarding productive potential in the presence of the incompatible use of land and the laws to preserve water and forestland, recently rediscovered as common assets to be protected for future generations. The opening up of the agricultural journals to new subject matters, or perhaps simply the clarification of the varied way in which Italian and EU legislation and jurisprudence deal with agricultural law, not only makes the Italian journals observant witnesses of how things are changing, and how juridical relations are evolving in this sector, but also makes them active protagonists in this transformation.

It is important to stress the didactic repercussions of this broadening of agricultural law towards environmental law. Appendix B of Ministerial Decree of 4 October 2000, which describes the content of the sector of the scientific discipline of "agricultural law," states: "The sector includes the studies relative to the organization and implementation of productive agricultural activities in their manifold juridical aspects, as identified by the discipline of the civil code, by special legislation and by European Union legislation, taking the issues of comparative law into consideration. *The studies pertain, moreover, to juridical issues relative to protection of the environment and to the commercialization of agricultural products*."[30] This consequently gives the teachers of agricultural law the opportunity to lecture on agricultural law, agri-environmental law and food law in the faculties of Law, Agriculture and Political Science. Thus, alongside textbooks on "forestry and environmental law" or dedicated to "profiles of environmental law in the European Union" and the "compendia of food law," we find, in the textbooks of agricultural law, chapters on the notion of environment, in relation to agricultural legislation. In these we can study the positive externalities deriving from the rational practice of agricultural activity, from the modern standpoint of multifunctionality and also agriculture-related soil pollution, agri-environmental contracts or the issues of legislative competence between State and Regions and the differentiation between what constitutes agriculture *stricto sensu* and what it is in relation to environmental protection.[31] In Germanò and Rook Basile's three-volume *Manuale di diritto agrario comunitario*,[32] the first volume deals with agricultural law and the circulation and protection of rights, the second agri-environmental law and the third agri-food law. Together with the wide range of themes dealt with, these volumes are a clear sign of the changing times and the now well-established broadening of the

[29] See Jannarelli (2013), p. 11.

[30] Emphasis added.

[31] See Carrozza et al. (1989), pp. 3–34; Costato (1992), pp. 72–77; Casadei (1992), pp. 78–88.

[32] See Germanò and Rook Basile (2014).

subject matter. Today there is no doubt that, in relation to the environmental needs intrinsic to a modern agricultural reality and associated with the activity of those working in the territory, we need to have a different politico-institutional regime to govern agriculture. This needs to be crystallized in the "environmental" modifications of the EU Treaties, at an operational (or hypothetically theoretical) level so that we can identify obligations or restrictions that establish a new content, a different "objectivity," for the definition of land areas as "assets to be used and circulated in a controlled manner," hopefully as common assets.[33] However, in my opinion, and in this new juridical context, the general principles of agricultural doctrine conserve their ontological autonomy. The environmental and nutritional dimensions of agricultural law simply testify to the extremely dynamic nature of the subject matter, particularly due to its sensitivity and permeability with regard to the emerging needs of civil society.

References

Benozzo M, Bruno F (2003) Legislazione ambientale (per uno sviluppo sostenibile del territorio). Giuffrè Editore, Milano

Bolognini S (2007) La definizione giuridica di ambiente: una chimera? Diritto e Giurisprudenza Agraria Alimentare e dell'Ambiente 12:723–737

Busnelli FD (1983) Dalla proprietà agraria alle proprietà agrarie, con ritorno alla proprietà. Riv Dirit Agrar 28:474–478

Capizzano E (1979) La razionalità dello sfruttamento del suolo nella disciplina costituzionale dei rapporti economici e nel "nuovo" quadro pubblico di riferimento dell'attività economica in agricoltura. Diritto dell'economia 25:23–88

Capizzano E (1987) Per un diritto agrario ambientale. Riv Dirit Agrar 32:433–451

Caravita B (2004) Profili costituzionali e oggetto del Diritto ambientale. In: Cristiani E, Massart A (eds) Diritto agrario e ambiente. Pacini Editore, Pisa

Caravita B (2005) Il diritto dell'ambiente. Il Mulino, Bologna

Carrozza A (1982) Agricoltura e tutela della natura (L'impatto ecologico sul diritto agrario). Giurisprudenza agraria italiana 29:77 et seqq.

Carrozza A (1983) La proprietà come istituto del diritto agrario. Riv Dirit Agrar 28:466–473

Carrozza A (1984) Due alternative alla concezione "pura" del diritto agrario: diritto ambientale e diritto agro-alimentare. Riv Dirit Agrar 29:177–178

Carrozza A (1985) Diritto agrario e Costituzione. Riv Dirit Agrar 30:4–31

Carrozza A (1988) Lezioni di diritto agrario. Giuffrè Editore, Milano

Carrozza A (1994) Lineamenti di un diritto agrario ambientale. I materiali possibili. I leganti disponibili. Riv Dirit Agrar 39:151–172

Carrozza A, Comporti M, Panuccio V (1989) La didattica del diritto agrario: i testi dell'ultima generazione, vol XXXV. Accademia Economico Agraria dei Georgofili, Stamperia Editoriale Parenti, Firenze

[33] See the recent EU regulations of 17 December 2013 adopted within the context of cross-compliance or greening. As effectively observed, the measures that provide for reinforced eco-conditionality can be interpreted as a "consideration to be respected by farmers for the support and help granted to them by the European Union," Germanò and Rook Basile (2014), p. 259, n. 119.

Casadei E (1992) L'insegnamento del diritto agrario nelle Facoltà di Agraria. Riv Dirit Agrar 37:78–88
Comporti M (1983) La proprietà agraria. Riv Dirit Agrar 28:444–465
Cordini G (2009) Principi costituzionali in tema di ambiente e giurisprudenza della Corte Costituzionale italiana. Rivista giuridica dell'ambiente 24:611–634
Cordini G, Fois P, Marchisio S (2005) Diritto ambientale, profili internazionali, europei e comparati. Giappichelli Editore, Torino
Costato L (1983) Proprietà terriera e potere politico. Riv Dirit Agrar 28:479–488
Costato L (1987) Alcune considerazioni sul diritto comunitario fra agricoltura e ambiente (anche a proposito dell'Atto unico europeo e dei regolamenti 1760 e 2242/87). Rivista di diritto agrario 32:512 et seqq.
Costato L (1988) Proprietà agraria. Enciclopedia del diritto XXXVII:325–338
Costato L (1992) L'insegnamento del diritto agrario nelle Facoltà giuridiche. Riv Dirit Agrar 37:72–77
Costato L (2001) Corso di Diritto agrario. Giuffrè Editore, Milano
Costato L (2003) Per una storia del diritto agrario. Riv Dirit Agrar 48:75–85
Costato L, Manservisi S (2012) Profili di diritto ambientale nell'Unione Europea. CEDAM, Padova
Cristiani E (2004) La disciplina dell'agricoltura biologica fra tutela dell'ambiente e sicurezza alimentare. Giappichelli Editore, Torino
Cristiani E (2008) Le riviste agraristiche italiane e i nuovi contenuti del diritto agrario. Riv Dirit Agrar 53:464–479
De Leonardis F (2009) La Corte Costituzionale sul Codice dell'ambiente tra moderazione e disinvoltura. Rivista giuridica dell'edilizia 4:1455–1464
Ferrero E (1979) La proprietà terriera. Giappichelli, Torino
Francario L (1986) Le destinazioni della proprietà a tutela del paesaggio. Jovene, Napoli
Francario L (1993) Agricoltura e ambiente: nuovi stimoli per l'approccio giuridico. Diritto e giurisprudenza agraria e dell'ambiente 10:517–519
Furno E (2007) La Corte Costituzionale "salva" la cogestione in materia paesaggistica ma non scioglie il nodo dei rapporti Stato-Regioni. Giurisprudenza costituzionale 6:4119–4128
Galloni G (1986) Il contributo culturale delle riviste agraristiche. In: Massart A (ed) Metodi e contenuti del Diritto agrario moderno. Giuffré Editore, Milano, pp 145–190
Galloni G (1993) Profili giuridici di un nuovo rapporto tra agricoltura ed ambiente. Diritto e Giurisprudenza Agraria e dell'Ambiente 1:5–10
Galloni G (1995) L'incontro tra agricoltura e ambiente: fondamenti costituzionali. In: Galloni G, Tarolo C, Donnhauser C (eds) Agricoltura e ambiente (Il cambiamento di rotta della nuova PAC e la sua attuazione in Italia). Pirola, Milano, pp 17–48
Galloni G (2000) Nuovi confini del Diritto agrario fra il diritto comunitario e il diritto ambientale. Riv Dirit Agrar 45:381–411
Germanò A, Rook Basile E (2014) Manuale di diritto agrario comunitario. Giappichelli Editore, Torino
Giannini MS (1973) Ambiente: saggio sui diversi suoi aspetti giuridici. Rivista trimestrale di diritto pubblico 23:15–53
Grassi S (2003) Agricoltura, ambiente ed ecosistema. In: Germanò A (ed) Il governo dell'agricoltura nel nuovo Titolo V della Costituzione. Giuffrè Editore, Milano, pp 39–65
Grassi S (2012) Problemi di diritto costituzionale dell'ambiente. Giuffrè Editore, Milano
Graziani CA (1985) Sull'attualità dell'art. 44 della Costituzione. Nuovo diritto agrario 12:309–316
Grossi P (1986) Relazione di sintesi. In: Massart A (ed) Metodi e contenuti del diritto agrario moderno. Giuffrè Editore, Milano, pp 427–438
Irti N (1972) Profili della programmazione agricola (o per una rilettura dell'art.44, I°comma Cost.). Riv Dirit Agrar 18:391–401
Jannarelli A (2002) Il diritto agrario tra profilo globale e profilo locale: spunti sul metodo e l'oggetto della ricerca. Rivista di diritto agrario 47:729–753

Jannarelli A (2013) Il divenire del diritto agrario italiano ed europeo tra sviluppi tecnologici e sostenibilità. Riv Dirit Agrar 58:11–35

Manservisi S (2008) L'interattività delle norme ambientali e di quelle agrarie internazionali nel contesto comunitario e nazionale. Aracne, Roma

Merusi F (2007) Il diritto ambientale. Riv Dirit Agrar 52:495–501

Nespor S (1986) Ambiente 1973–1986. Rivista giuridica dell'ambiente 1:1–8

Pennisi R, Calcinelli C (eds) (1990) Tutela ambientale e centralità dell'agricoltura. Centro interdipartimentale audiovisivi e stampa dell'Università, Camerino

Rodotà S (1982) Sub art. 44. In: Branca G (ed) Commentario della Costituzione. Zanichelli, Bologna, pp 211–233

Salaris F (1981) Considerazioni sulla proprietà agricola. Giuffrè Editore, Milano

Sandulli AM (1972) Profili costituzionali della proprietà privata. Riv Trimest Dirit Proced Civ 26:465–490

Sirsi E (1986) La "cultura" delle Riviste e la Rivista di Diritto agrario. In: Massart A (ed) Metodi e contenuti del diritto agrario moderno. Giuffrè Editore, Milano, pp 415–420

Tarchi R (2006) Agricoltura e ambiente. In: Rook Basile E (ed) Dopo la modifica dell'art. 117 Cost.: problemi ed esperienze sulla competenza della materia agricoltura. Giuffrè Editore, Milano, pp 65–106

The Relationship Between Agricultural Law and Environmental Law in France

C. Hermon

Abstract Agricultural law and environmental law are two distinct branches of law. However, there are common points between the two legal spheres. Agricultural law regulates environmental protection, and environmental law regulates agricultural pollution. This in itself, however, is not a sufficient basis for a coherent legal system around agroecology. Yet there are significant obstacles to the creation of a more coherent legal framework. These obstacles are the very existence of agricultural law, on the one hand, and of environmental law, on the other, according to their construction and development, and also existing rights that might possibly be violated by "agroecological law," and the lack of consensus for changing the dominant agricultural model. Nevertheless, agricultural law could structure its own legal framework to enable the development of agroecology, despite current opposing provisions. The current bill for the future of agriculture is pointing in this direction.

Keywords Agricultural law • Agroecology • Environmental law

1 Introduction

Even though they are closely related, agricultural law and environmental law are two distinct disciplines that rarely converge. This makes the concept of agroecology, developed by other disciplines,[1] harder to grasp in terms of agricultural production and environmental protection. This separation of legal spheres can be seen in several key areas, primarily from an institutional point of view. In France, the existence of both a Ministry of Agriculture and a Ministry of the Environment

[1] Agroecology comprises production systems based on the specific features of ecosystems. This may be a variety of methods as defined by law (or not), together with integrated pest management, organic farming or agroforestry.

C. Hermon (✉)
Institute of Legal Studies for Urban Planning, Construction, and the Environment (IEJUC), Toulouse 1 Capitole University, Toulouse, France
e-mail: Carole.Hermon@ut-capitole.fr, http://www.ut-capitole.fr/rubrique-outils/vous-etes/mlle-hermon-carole-4542.kjsp

M. Monteduro et al. (eds.), *Law and Agroecology*,
DOI 10.1007/978-3-662-46617-9_12

reflects this dichotomy and ensures its continuity. Merging the two ministries has never been a serious option and is probably not feasible in the current political context. Despite the 2010 general review of public policies,[2] and reform of public services, the two regional directorates were maintained. The regional Directorate for Industry merged with the Directorate for Nature and the Environment, but the Regional Directorate for Agriculture remained distinct from the new Regional Directorate for Environment, Housing and Planning.

On a more anecdotal, yet significant, note, this clear distinction is reinforced in higher education. In the private–public law division, agricultural law comes under private law,[3] whereas environmental law is historically associated with public law.[4] As a consequence, agricultural law is taught in the curriculum of private law and environmental law is generally reserved to students of public law.

Most importantly, in French law, the compartmentalization of agricultural law and environmental law is reinforced by the principle of the independence of the statutes. Thus, the authorization granted under one statute does not constitute an authorization under another statute.[5] As a result, the competent administrative authority issuing an authorization cannot base its decision on the legislation under which another authorization must be granted. For example, the Prefect (senior State official in a region), referred to for monitoring purposes defined by the *Code Rural* (Rural Code) (see Sect. 2.2.1), cannot refuse an agricultural farming permit on the grounds that it is contrary to a point of environmental law. Conversely, the administrative authority competent to issue licenses to operate a classified facility (see Sect. 2.1.2) cannot rely on the Rural Code to legally analyze an application before granting or denying it.[6] Agricultural law and environmental law are therefore two distinct branches of law.

However, this division sometimes appears artificial. For instance, the standards transposing the so-called nitrate directive (Directive 91/676/EC on the protection of water against pollution caused by nitrates from agricultural sources) are clearly governed by environmental law; they are designed to protect water quality and impose restrictions to this end. These restrictions have also been incorporated into the Environmental Code.[7] Since the directive and its transposing acts only cover

---

[2] See Circular of 7 July 2008 on the organization of the departmental administration of the State, with regional organization dealt with in the annex.

[3] Terré (2009), p. 95.

[4] Legal relations between private persons governed by environmental law are now sufficiently numerous that we can challenge the exclusive subordination of environmental law to public law; see Van Lang (2011), pp. 157 et seqq.

[5] See the Report of the Council of State (1992) on “L’urbanisme: pour un droit plus efficace”. In: La documentation française, pp. 37 et seqq.; Bétaille (2012), pp. 447–450. Delhoste (2001).

[6] The administrative authorities and judges are not authorized to rule on the issue of building permits by taking into account the definition of agricultural activities under the Rural Code, CE 14/02/2007, req. No. 282398.

[7] This information now appears in environmental law textbooks such as Prieur (2011), pp. 677 et seqq; Van Lang (2011), pp. 411 et seqq; Romi (2010), pp. 560 et seqq, but not in textbooks on agricultural law.

agricultural activities, it ought to fall within the realm of agricultural law and thus appear in the statutes on fertilizers contained in the Rural Code. Thus, the material criterion[8] does not always allow us to establish a clear distinction between agricultural law and environmental law. That is precisely why we use the formal and more functional criterion when analyzing the relationship between the two branches of law. We thus consider both agricultural law and environmental law as being limited to the elements they respectively codify. However, this approach does not mask the artifice; despite the apparent compartmentalization, common points between the two legal spheres will continue to surface. Agricultural law clearly regulates environmental protection, while environmental law regulates agricultural pollution,[9] a broad subject related to agriculture (Sect. 2). Their respective confines defined by legal codes are rather blurred. However, this is not a sufficient basis for a coherent legal system revolving around agroecology. On the other hand, there are considerable obstacles to the good functioning of such a system that are not solely due to the existence of two separate legal fields. Yet this does not mean that the legal field is unable to tackle agroecology (Sect. 3).

2 Agricultural Law and Environmental Law: Common Concerns

Environmental law and agricultural law overlap in several areas. Environmental law comprises certain standards in farming activity intended to prevent or limit their environmental impact, Similarly, agricultural law includes provisions designed to encourage or compel farmers to take environmental protection into account. Yet agroecology as such does not constitute a legal category defined either by environmental or agricultural law and therefore does not have its own statutory scheme. This research attempts to point out that environmental law and agricultural law can nonetheless converge and bring agriculture and environment closer together.

2.1 *Protection of the Environment in Agricultural Law*

Three major sets of rules stand out in the Rural Code; our research outlines them along with the gradual assimilation of environmental issues into agricultural law:

[8] According to which agricultural law regulates agricultural activities, whereas environmental law protects natural areas, species and resources and combats pollution and environmental harm.

[9] Environmental law does not take in agriculture as a whole; more specifically, the ecological services it can render are not integrated into environmental law. See Doussan (2009), pp. 125–141, 126–128; Langlais (2013).

for administrative policies, the section on the use of plant protection products; provisions for green farming leases; and agri-environmental measures supported under the second pillar of the CAP.

2.1.1 Regulating the Marketing and Use of Plant Protection Products

Given the impact of plant protection products on the environment and animal life,[10] the regulation of marketing has always been a key issue in environmental protection. Regulation was first introduced to ensure the protection of plants against "crop pests," and its stated public health objective was to guarantee the safety of consumers. This regulation is at the heart of the legal system intended to support the green revolution. From then on, legislation also sought to cover risk prevention and environmental damage, and thus illustrating the progressive interest in incorporating environmental protection into agricultural law.[11] The regulatory framework is now largely based on EU law so that French law is not allowed to differ from that of other member states. We can, however, still address the peculiar nature of this issue in France, which is the world's third largest consumer of pesticides and Europe's greatest consumer.[12]

The adoption of the new "pesticide package"[13] in EU law had several goals: to reduce the use of plant protection products, to phase out the most hazardous substances and to promote integrated pest management[14] and nonchemical methods of plant protection. This development is significant.[15] The provisions were more or less transposed into French law, when applicable, and became part of the Rural Code, under the section dealing with plant protection. Thus, the initial objectives are complemented by the goal of protecting the environment or of reducing environmental impact of pesticides. As such, and in accordance with EU law, the Rural Code calls for an authorization procedure prior to the marketing and use of products, in order to rule out environmentally hazardous effects; it also requires

[10] See Aubertot et al. (2005).

[11] See Hermon and Doussan (2012), pp. 95 et seqq.

[12] Aubertot et al. (2005).

[13] This system comprises three texts adopted on 21 October 2009: Regulation 1107/2009/EU of the European Parliament and the Council on the marketing of plant protection products, Directive 2009/128/EU establishing a framework for Community action for a sustainable use of pesticides, and Directive 2009/127/EU amending Directive 2006/42/EC on machinery for pesticide application.

[14] Defined by Directive 2009/128/EC as "the careful consideration of all available methods of protecting plants and, therefore, the integration of appropriate measures that discourage the development of pest populations and maintain the use of plant protection products and other types of interventions at levels justified on economic and environmental considerations and reduce or minimise the risk to human health and to the environment. IPM against pests emphasises the growth of a healthy crop with the least possible disruption to agro-ecosystems and encourages natural methods of pest control" (Arts. 3–6, author's translation).

[15] See Hermon and Doussan (2012).

farmers to stop using the most dangerous substances and their substitution with less harmful ones, regular monitoring of equipment and mandatory training for all professionals dispensing or using pesticides, including farmers.

In this context, they attend training courses on integrated pest management and alternatives to chemical pesticides under the aegis of the Ministry of Agriculture. Moreover, and in conjunction with the development of European legislation,[16] in 2008 the Ministry drew up Ecophyto,[17] an action plan to reduce the use of pesticides[18] and to develop alternatives. Public sector research and agricultural education were mobilized for this purpose, and a network of experimental farms was set up.[19]

The development of agricultural law, with respect to the environment, is thus worth noting and can lead to a striking shift in production systems. A study by the INRA (National Institute for Agricultural Research) shows that "halving the use of pesticides in farming by 2018 corresponds to a radical change in agricultural production, with a strong development of organic agriculture and the near generalization of integrated production."[20] The creation of green farming leases also falls within the scope of the environmental movement.

2.1.2 The Green Farming Lease

The green farming lease was introduced under the Framework Law on Agriculture of 5 January 2006 with the specific aim of introducing environmental clauses into farming leases. According to the bill's *rapporteur* in the Senate,[21] the farming lease at the time did not allow it. "The status of tenant farming consists of public policy regulation that may not be infringed upon and is intended to stimulate production and to ensure security for the farmer. The lessee thus remains free to organize farming practices as it best suits him [. . .]. It therefore seemed necessary to review the status of tenant farming in order to include the provisions of a green lease."[22]

[16] The "Ecophyto 2018 reduction of pesticide use from 2008 to 2018" was developed following the Grenelle Environnement Agreement of 2007 and assembled State representatives, local authorities, NGOs, employers and workers, divided into five colleges to discuss environmental issues.

[17] A national action plan pursuant to Directive 2009/128/EU.

[18] The goal was part of the National Strategy for Biodiversity and recalled since then, the last mention of it being through Law 2009/967 of 3 August 2009 relating to the implementation of the Grenelle Environment Agreement called Grenelle 1, which set the target of halving the use of plant protection products within 10 years (Art. 31).

[19] According to Ecophyto's 2012 annual report, the experimental farming network trying out crops with a lower pesticide intake (the DEPHY farms) includes 1,900 farms and 42 institutions of agricultural education.

[20] See Hermon and Doussan (2012), p. 98, note 12. All the quotations in English between "" are author's translations from French, if not otherwise stated.

[21] Contra Bodiguel (2011), according to which there is no guarantee that environmental provisions are prohibited by the public order status of tenant farming.

[22] César (2006), p. 185.

The green lease was designed to limit lessee's freedom to promote environmental principles,[23] whereas the freedom to operate was *per se* aimed at productivity.

The existence of this farming lease is particularly significant for our research insofar as it may qualify as a genuine agroecological project. Its contents are defined by regulation. Art. R. 411-9-11-1, created by the Decree of 8 March 2007 and as yet unamended, features an exhaustive list of 15 different cultural practices, ranging from the ban on ploughing up pasture, irrigation, the use of pesticides or fertilizers to the requirement to perform cover cropping, diverse crop rotation and organic farming, and the definition of tillage methods. However, as we have pointed out, the environmental lease was conceived as a dispensation scheme. It is therefore limited to certain lessors or certain plots; although the list of lessors and contractible spaces was nonetheless extended by law in 2010.[24]

The lessors referred to in Art. L. 411-27 (Rural Code) entitled to a green lease under the tenant farming status are legal persons governed by public law, qualified environmental protection organizations, registered socially responsible companies and public utility foundations or endowment funds.[25] As for the areas that meet the requirements for green farming leases, regardless of the lessor, they are either fragile landscapes and/or exceptional for environmental or health reasons and protected as such by the Environmental Code or by the Public Health Code, such as catchment areas supplying the beaches that are subject to green tides, drought-prone areas called water distribution zones, protection of areas surrounding water catchment points, coastal areas under the jurisdiction of the Conservatoire du Littoral (Coastal Protection Agency), national and regional parks, nature reserves, sites registered with Natura 2000, etc.[26] The list is long, and its close scrutiny reveals the fact that areas that are eligible for a green farming lease would most likely constitute a significant proportion of French farmland. From a legal standpoint, the green farming lease represents a breakthrough in relation to the common farming lease, although it is clear that in practice lessors have not exploited it fully. Last but not least, the support provided for agri-environmental measures also shows the gradual greening of agricultural law, especially since it appears that French

[23] See Bosse Platière et al. (2013), pp. 57 et seqq.

[24] As we will see later, Art. 4 of the bill on agriculture, food and forestry removes these provisions and opens the green lease to any person, anywhere and to any clause. See Sect. 2.2.2, bill passed by the National Assembly on 14 January 2014, doc. Senate, No. 279.

[25] For the presentation of talks for Terre de liens, see Cacciabue (2013), pp. 1 et seqq, 13.

[26] Art. L. 411-27 of the Rural Code states that the space for which the green lease was made must have been "subject to a formal management document," which is commonly the case, considering that these "management records" bear different names depending on the areas they apply to, and have different statutes and a separate scope, such as public utility easements within the perimeters of protection of water catchment points, regulation of activities through the decree creating national parks specified by its charter, contracts and programmes of regulatory action in areas subject to green tides, etc. The green lease must comply with the applicable "management document."

expenditure on this axis of the second pillar of the CAP[27] is one of the highest among all member states.

2.1.3 Agri-environmental Measures

Aid for agri-environmental practices under the second pillar of the CAP is enshrined in Title IV, Book III, and administered by the decentralized directorates of the Ministry of Agriculture. This is a major focus of the incentive to shift production systems towards more environmentally friendly practices and reflect the evolution of the CAP from unconditional support for production to the support of more sustainable forms of production. Agri-environmental measures do not have the same power as postwar income support, but their rationale is similar. It comes down to ensuring a guaranteed revenue for farmers who follow State guidelines in their productive activity.

In France, two national agri-environmental schemes were included in the programme for rural development. Their terms of access were defined by the Minister of Agriculture, and their provisions were to be formalized in different regions: rotational measures to encourage crop diversification and PHAE, and the agri-environmental grassland premium, supporting the maintenance of grassland and extensive livestock management. The former remained marginal, while the latter has been used extensively to strengthen livestock farming in areas where this type of farming is practiced. In addition, six schemes were decentralized; as a consequence, regional prefects could proceed to fully or partially contract them, depending on the practices and opportunities of their regions. Farmers were thus offered support with the reintroduction and conservation of breeds and plant species, aid for efficient forage-based "mixed farming" systems and specific aids for beekeepers. Most importantly, it was mainly in this context that support for organic agriculture was initially introduced with the target of 20 % organic production by 2020.[28] In other words, the financial assistance for converting to organic farming has been an important part of the budget for the second pillar of the CAP, and without this assistance, based solely on the support of the market, it is commonly agreed that the conversions would not have been as numerous.[29]

---

[27] An OECD assessment reports that European support to agro-environment represents 23 % of the total budget of the second pillar, but this share is much higher in France, namely around 43 %) ("Evaluation des réformes des politiques agricoles de l'Union européenne", OECD, 2011, p. 142 and Appendix C, p. 182).

[28] Art. 31, Law 2009-967 of 3 August 2009 relating to the implementation of the Grenelle Environment Agreement. Since 2010, aid to the conversion to organic farming and the preservation of organic farming in France fall under the first pillar on the basis of Art. 68 of Regulation 73/2009 of 19 January 2009 establishing common rules for direct support schemes to farmers under the CAP.

[29] Between 2007 and 2012, the number of organic farms increased by 104 %, bringing the total number of organic farmers to over 25,000 and the surfaces to 85 %, more than one million

Regional agri-environmental measures have been implemented in designated areas where environmental requirements were most pressing, namely, in sites registered with Natura 2000, which received assistance for hedgerow and pond restoration and delayed mowing, and in basins where the objectives for good water status set by Directive 2000/60/EC of 23 October 2000 establishing a framework for Community action in water policy cannot be met without aid. The reduction of the use of pesticides and fertilizers, the introduction of intermediate crops, and grass covers were also funded.

Agri-environmental measures have helped agroecology indeed become part of the Rural Code, and in a broader sense, the environment has acquired an important status. Alongside these developments, and indisputably so, environmental law has taken hold of agriculture.

2.2 *Agricultural Pollution Under Environmental Law*

Environmental law has had a long-standing interest in agricultural pollution. Firstly, there are special policies to regulate specific activities in order to prevent or minimize environmental risk and damage. Secondly, environmental taxation is designed to deter specific behavior. Finally, within the framework of energy policies and to control greenhouse gas emissions, farmers can now use the carbon market and energy efficiency schemes. Environmental law thus deploys a wide range of tools to regulate farming and to encourage farmers to take better care of the environment; it is therefore unlikely that many farmers in France are able avoid it.

2.2.1 Policies for the Prevention and Reduction of the Environmental Impact of Agriculture

In France, legislation on classified facilities (see Arts. L. 511-1 et seqq, R 511-9 et seqq of the Environment Code) has been a centrepiece of the pollution prevention plan, thanks to its broad scope and effective design. It covers all activities that may represent a hazard or harm to the environment and is designed to prevent any kind of risk to the environment as a whole. To achieve this, all potentially polluting activities are subject to authorization or prior registration or declaration according to their level of risk or harm, to mandatory monitoring and inspection and to a set of technical requirements meant to prevent or limit environmental damage. However, a significant part of agriculture has been incorporated into the law on classified facilities, livestock farming and also slaughterhouses, dairies, wineries, grain storage facilities and biogas plants. On the other hand, this policy does not cover crop

hectares, including plots under conversion; see BIO/OC Agency, on the website of the Ministry of Agriculture.

production, vineyards, tree crops and field crops. In addition, the production volumes required for the law on classified facilities are very low, e.g. 50 pigs, 50 dairy cows or 5,000 chickens. The thresholds requiring prior authorization are certainly higher: 2,000 pigs,[30] 200 cows or 30,000 chickens. On the one hand, the scheme encompasses many more activities and at lower thresholds than EU law. On the other hand, requirements for reporting and recording,[31] although more flexible, do not contain less binding environmental requirements for farmers. In livestock farming, for instance, these requirements cover manure storage and spreading and the mandatory distances related to the location of buildings.[32] Thus, the law on classified facilities firmly oversees and monitors a large part of the agricultural production in order to prevent or limit its impact on the environment. Moreover, in 1992, in order to incorporate aspects not covered by this policy, the legislature renewed a similar licensing and reporting system for facilities, structures, works and activities with a potential impact on water or the aquatic ecosystem (Arts. L. 214-1 *et seqq* and R. 214-1 *et seqq,* Environment Code). Therefore, fruit growers, market gardeners and grain farmers who are not subject to the law on classified facilities nevertheless have to apply for authorization under environmental law if they collect water, especially in drought-prone areas, if they intend to create a water reservoir or carry out waterproof or drain a wetland, etc. As such, they will have to comply with the legal conditions and will face sanctions if they do not. In the last instance, the bulk of agricultural production is subject to an environmental policy and to monitoring. In addition, farming may be subject to environmental taxation.

2.2.2 Environmental Taxation: Deterring Environmental Damage

In France, taxation is not the preferred tool to deter polluters. Environmental law regulations are policed and come with a few incentives. Moreover, taxation has two drawbacks: it is not sufficiently dissuasive and allows polluters to continue polluting in exchange for payment, and it is primarily intended to help consolidate the budget. However, environmental taxes exist and will be raised over time, and some of them apply to agriculture. For the time being, agricultural activities taxable under environmental damages come under charges levied by the water boards. These collect taxes from farmers who extract water or whose actions downgrade its quality and use tax revenue to improve water quality.[33] However, farmers can be

---

[30] According to the classification, amended by Decree 2013-1301 of 27 December 2013. The threshold was formerly set at 450.

[31] The smallest farms (up to 150 cows or 450 pigs) are subject to declaration, the largest to authorization and intermediate facilities (450–2,000 pigs or 151–200 cows) to registration.

[32] Finally, the orders of 27 December 2013, the first of which refers to the general requirements and to classified facilities subject to declaration under sections 2101 and 2102.

[33] Nicolazo and Redaud (2007).

subject to a tax on account of their water intake and based on the pollution they generate, basically caused by manure or the use of plant protection products.[34]

In France, the water extraction tax is paid by nearly all irrigating farmers. It is based on the amount of water extracted, measured or estimated at a flat rate set by each water board[35] in compliance with a ceiling set by law, which varies according to the irrigation technology used[36] and the quality of the water resources.[37] At the time of the last reform (Law 2006/1772 on water and aquatic environments of 30 December 2006), the tax was by and large increased, sometimes significantly. Although factored into farm operating costs, its deterrent action is still uncertain, assuming that this tax is indeed intended to limit the use of irrigation. Nevertheless, a clear message is sent to farmers: water is a scarce resource and can prove expensive.[38]

As for the fee to discourage widespread pollution, recently introduced to water boards, it taxes the use of plant protection products, thus reinforcing the goal of reducing their use under the Rural Code (Art. L. 253-6, Art. R. 253-44). This tax is based on the quantity of substances contained in plant protection products and classified by Regulation 1272/2008/EC on the classification, labeling and packaging of substances and mixtures as carcinogenic, mutagenic or toxic for reproduction or dangerous for the environment, and their rates are set by law. Once again, along with the reforms, legislation substantially increased the tax, so that it has become a burden for farmers using these products. This is especially so since, for transparency, retailers of plant protection products and processed seed who are liable to pay the tax and charge the end user must declare this explicitly on their bills.[39] This makes the price of pollution visible but not necessarily effective as a real deterrent.

The choice of the legislature with regard to the nitrogen taxation is, however, less convincing. Indeed, after all the official reports and parliamentary debates,[40] the bill on taxation was reduced to a tax imposed on stockbreeding; the use of mineral nitrogen is not subject to taxation. And the tax is based on the number of

---

[34] See Hermon and Doussan (2012), pp. 303 et seqq.

[35] Metropolitan France is divided into six drainage basins, each with its own agency. These are public institutions supervised by the Ministry of Ecology.

[36] Irrigation (other than surface irrigation) or surface irrigation, which allows agencies to limit the cost of surface irrigation, more water efficient compared to sprinkler irrigation.

[37] The maximum statutory fee is two times higher, depending on whether water extraction is performed in a water distribution zone, that is to say, in an area where the water deficit is structural, or outside such an area.

[38] Under the current legislation, the rate of the fee shall not exceed, for irrigation, €0.02 per m^3, €0.03 in case of water abstraction in water distribution zones (structural deficit areas), and, for surface irrigation, €0.10 per m^3 and €0.15 when collecting water in a water distribution zone.

[39] Since 1 January 2009, the failure to indicate this information on invoices makes the offender liable to a fine of approximately 750 euro for a fourth-class offense as defined by Art. R. 213-48-13-IV C. env.

[40] Tavernier (1999) and Flory (2003).

livestock units, which is a simple method but certainly not an incentive to a sound management of the nitrogen cycle.[41]

Overall, taxation for environmental purposes substitutes environmental policies in the sense that it puts a price, however small, on polluting behavior that is not subject to environmental law (i.e., the use of plant protection products) or not prohibited (i.e., intensive farming and irrigation). Conversely, environmental law is now trying to set a price for environmentally friendly practices by creating and organizing environmental securities markets. For the time being, these only cover energy efficiency policies and the reduction of greenhouse gas emissions, but they could be extended in order to pay farmers for environmental services.

2.2.3 Carbon Trading, Energy Efficiency and Agriculture: Promoting Ethical Practices

To meet the overall objectives of lowering greenhouse gas emissions and, more specifically, reducing the energy dependence of farms in general,[42] the Ministry of Environment has opened the carbon market and energy efficiency sector to the agricultural sector. Environmental law has an entirely different rationale from those analyzed above. It comes down to having third parties, compelled by law to reduce greenhouse gas emissions or to carry out energy savings, finance equipment or technical changes to enable farmers to cut their carbon emissions or energy consumption. Agriculture is thus incorporated into a collective effort where the contributions of some help meet the commitments of others, and in the end the environmental benefits of agriculture, rather than the damage caused to the environment, are recognized by the law. The approach is experimental, but its scope may become more ambitious.[43]

For the time being, two markets can help the agricultural sector enhance its practices.[44] First is the EU emissions trading market. Indeed, emission reduction units (ERUs) initiated under the Kyoto Protocol may be used, pursuant to Directive 2004/101/EC, for corporate compliance under the European system allowance trading. To this end, the Minister of the Environment has approved a number of

[41] Pursuant to Art. L. 213-10-2-IV of the Environmental Code, "the charge for a livestock farmer is based on the their number of livestock units and applies to a stock density exceeding 1.4 livestock units per hectare of utilized agricultural area (which excludes the most extensive types of farming). The charge rate is €3 per unit. The minimum amount of livestock for fee collection is set at 90 units and to 150 in alpine areas."

[42] The Law of 3 August 2009 ('Grenelle 1') set the objective to "increase energy efficiency in order to achieve a rate of 30 % of farms with low energy dependency by 2013," Art. 31. Yet according to Marion Guillou's report on le projet agro-écologique: vers des agricultures doublement performantes pour concilier compétitivité et respect de l'environnement (see Sect. 2.2.2), it is likely that the objective will not be achieved" (Guillou 2013).

[43] The bill on agriculture, food and forestry will apply this scheme to plant protection products (see Sect. 2.2.2).

[44] See Production agricole et droit de l'environnement, op. cit., pp. 371 et seqq.

agricultural projects that may lead to the issue of ERUs. In accordance with the procedure established by the Decree of 2 March 2007 (for the application of Decree 2006/622 of 2 May 2006 on the approval of project activities under the Kyoto Protocol), refer to three methods: the anaerobic digestion of livestock manure in 2007, feed changes for dairy cows (March 2011), and the introduction of leguminous crops in crop rotation (April 2011). In compliance with these methods, two projects have been approved; the first was geared towards reducing nitrous oxide emissions by inserting leguminous crops into crop rotations and involves 11 farming cooperatives and their members.[45] The second revolved around reducing methane emissions by introducing alpha-linolenic acid from natural sources in feed for dairy cows.[46] These projects could clearly not have been carried out without the funding provided by the resale of allotted emission reduction units.

The second market is the French one; it is more active and has not had the excess of carbon credits afflicting the European market. It was introduced by the Planning Act of 13 July 2005 setting out guidelines for the energy policy inspired by the carbon market scheme. Energy retailers are individually and collectively required to reduce national energy consumption. This legal requirement takes the form of transferable energy savings certificates. These are issued by the Ministry of the Environment to anyone, energy retailer or otherwise, conducting energy efficient operations or contributing to energy savings by a third party. In order to facilitate the implementation of these operations, standardized operations have been defined by ministerial decrees accompanied by fixed rates for energy savings. The agricultural sector has helped generate similar decrees for equipment such as biomass boilers, heat pumps, hot water tanks and equipment to monitor tractor engines.[47] So far, it has primarily been greenhouse operators, the large energy consumers, who have reaped the benefits of this system under the leadership of a professional trade union, *Légumes de France*. The system remains marginal, but it has nevertheless financed equipment and altered farming practices before they were rendered compulsory within the current context of climate change.

Our conclusion is that although environmental law is concerned with agriculture and agricultural law with the environment, these shared concerns have led to a patchwork of laws rather than a coherent legal framework. On the other hand, agroecology is not defined by law and is a legal field of its own. We would be hard

---

[45] Coordinated by the InVivo project, national union of agricultural cooperatives, approved 15 May 2012, for an expected service life of five crop years. The project will save approximately 546,403 tons of CO_2 per year.

[46] The project, led by the Blue White Heart Association, is to introduce this type of cattle feed in 1,000 farms averaging 61 dairy cows over 10 years, with an annual saving of 27,853 tons of CO_2. Projects financing the equipment of farming cooperatives meant to reduce energy consumption or the substitution of fossil fuel with renewable energy were also approved. See the website of the Ministry of Ecology (MEDD), under Energy, Air and Climate, greenhouse effect and climate change.

[47] See the website of the Ministry of Ecology (MEDDE), under Energy, Air and Climate, Energy savings.

pressed to identify agroecological law within agricultural law and environmental law. In other words, no current legal framework can define agroecology.

3 The Incoherent Handling of Agroecology at the Heart of Agricultural Law and Environmental Law

The obstacles to a more coherent legal framework for agricultural law and environmental law are neither negligible nor insurmountable. A report on agroecology submitted to the Minister of Agriculture on May 2013, followed by a bill, has opened up some perspectives on the matter.

3.1 Obstacles to a More Coherent System

The obstacles to developing a legal system better adapted to accommodate the model of agroecology are linked to several factors: the construction of agricultural law, environmental law, their respective development processes, and also the rights that they could encroach on. Yet the real root of the problem lies elsewhere; in France, the agroecology project does not have sufficient backing for it to be incorporated into the legal framework.

3.1.1 Internal Obstacles Present in Agricultural Law and Environmental Law

The structure of agricultural law and environmental law does not appear to be particularly well disposed towards the concept of agroecology. Environmental law does not seem to include many provisions that would affect farmers wishing to develop an agroecological approach in their industries.[48] More fundamentally, it is the method of structuring environmental law that is unsuited to the pursuit of agroecology. Environmental law is effectively adopted law, the main purpose of

[48] See, for example, the Decree of 19 December 2011 on action programmes to implement water protection against pollution by agricultural nitrates. In the context of prelitigation between France and the European Commission, on the application of the "nitrate" Directive of 12 December 1991, the Commission indicated that it considered the nitrogen equivalent standards produced annually by the French livestock sector to be underestimated (Opinion of 2 April 2003 under Art. 228 EC Treaty following the ruling of 8 March 2001 for failure to comply, aff. C-266/991, formal notice of 20 November 2009 stating that action programmes are insufficient in light of the objectives of the Directive); see Hermon and Doussan (2012), pp. 211–212. In response, France issued the decree of 19 December 2011, providing additional evidence as to the amount of nitrogen excreted by dairy cows, so that today, the longer the grazing period is, the higher is the production of estimated nitrogen.

which is to protect the environment, but it is based on a defensive model. Environmental law has repeatedly tried to address environmental risks and pollution, together with the need for technology and resources to protect natural areas or habitats. To date, the all-embracing ambition of the legislation on classified facilities to prevent any risk in any place by any activity has failed. Other policies have been established to meet the specific risks raised by certain products (e.g., the regulation of genetically modified organisms, GMOs) or linked to the particular sensitivity of a resource (e.g., water regulation). Calls for simplification and/or modernization of environmental law are recurrent and sometimes relayed by public authorities.[49] However, this basically becomes a question of assistance procedures, even leading to reduce the amount of environmental law, and not that of legal reform allowing environmental law to enhance its focus and implement a new techno-economic model that we would call ecological or sustainable. This is not the current state of affairs. On the contrary, such concepts are not even taken into account in current legislation. In other words, intensive farming built on an industrial model can conform to the norms of current legislation as much as a certified organic farm. However, it is our opinion that this is what awaits agroecology: the law setting the framework, and providing the support, for a new techno-economic model. Yet this kind of issue never arises in the calls for modernization.

In agricultural law, the issue is different. Some legal provisions may hinder the development of agroecology, particularly among the rules governing the control of facilities and farming leases. Nevertheless, structurally, agricultural law could forge its own legal regime to enable the development of agroecology. A "gear shift" may be just what is needed.

The process of monitoring farming facilities is specific to French agricultural law, and it applies to any kind of farming operation.[50] Monitoring is exercised for review purposes by a departmental committee made up overwhelmingly of agricultural professionals (Art. R. 313-2, Rural Code), and the authorization to operate falls under the jurisdiction of the Prefect. Decisions are taken on the basis of the departmental plan structure in accordance with Art. L. 331-1 setting priorities for policy on facilities. However, ever since the Agricultural Framework Law 99-574 adopted on 9 July 1999, the priority has been the arrival of new farmers and their specific projects, whatever they may be. Furthermore, the monitoring process is designed to prevent the disbanding of viable farms, to promote the expansion of small farms and, where appropriate, to encourage multipurpose farms insofar as "demographic change and economic perspectives allow it." Certainly, prefects may take into account a project's environmental interest (Framework Law on Agriculture 2006–2011 adopted on 5 January 2006) and, more specifically, the interest of maintaining a farm in organic production (since the Framework Law adopted on

[49] Barthélémy and Grimot (2006); de la Raudière (2010); Lambert and Boulard (2013).

[50] Pursuant to Art. L. 331-1 of the Rural Code, "the control of farms applies to the development of farmland or to landless production systems, regardless of the status or mode of legal organization thereof, and the capacity whereby the development is carried out."

9 July 1999). Yet this kind of action can take place just after the priorities set by the departmental agenda have been observed; only this agenda may provide for the setting up of new farms and farmers "on a full-time basis" as their top priority. From this point of view, between a multiactivity farmer gradually redesigning his farming with the end goal of organic production and a keen novice eager to start up an industrial factory farm, priority will be given to the latter.[51] As such, we can conclude that the monitoring process as it is now organized does not promote agroecology and may even hamper its development.

The same principle applies to the farming lease. It provides the tenant with complete freedom to carry out agricultural activity. This is irrespective of whether or not a farmer intends to run a conventional farm on land previously operated organically, to work on soil previously used for direct seed sewing or to implement off-land animal farming methods in an area where livestock had been previously farmed extensively; the lessor has no say in the matter (exceptions include green leases and/or clauses opposed to such methods seen above written in the lease). This could be likened to a degradation of the land as per Art. L. 411-29 or judged likely to impair the proper operation of the land as per Art. L. 411-31. In current jurisprudence, only certain matters such as animal abuse or serious lack of maintenance can qualify as an offense,[52] but this is certainly not the case insofar as a return to conventional practices is concerned. Thus, the Bourges Court of Appeal ruled that a landlord may not prohibit the use of sewage sludge from a wastewater treatment plant to a lessee if he intends to use it as a soil improvement; according to the regulatory framework, the use of such sludge does not qualify as land damage.[53] At the same time, however, the lessor cannot oppose a change in agricultural practices in the direction of agroecology. Under Art. 411-27(2), "The fact that the lessee applies, on land taken on lease, practices designed to preserve water resources, biodiversity, landscape, the quality of crops, soil and air, or to prevent natural hazards and erosion cannot be invoked in support of a request for cancellation made by the lessor." We can therefore agree that the landlord and tenant law does not guarantee environmental practices in agricultural production, and as such agricultural law is an obstacle to agroecology, yet we should also note that it does not go as far as to prohibit its development. This being as it may, there are clauses and implied provisions in the lease that would make more than one proponent of agroecology shudder. This refers to the fact that the removal of embankments, hedges, ditches and trees and separating and splitting up plots can be classified as land improvement for the leased plot (Art. L. 411-28, Rural Code).

All this reveals the history and the essence of agricultural law. It is not the monitoring of farming facilities or the freedom of management for the lessee that

[51] The prefect is required to reject the lower rank application: cf. CE 22/03/1999 Consorts Craquelin, RDR 2000, p. 54, CE 28/07/1999, Rec. 252.

[52] Ruling of the Third Civil Chamber of the Court of Cassation of 27 November 2007, No. 06-20.172. Ruling of the Third Civil Chamber of the Court of Cassation of 17 May 2011, No. 10-18.639.

[53] Bourges Court of Appeal, 2 May 2003 Bizouarne v. Cartier, RDR 2004, p. 40.

hampers the development of agroecology; it is rather the history and the essence of the law itself. As Professor Louis Lorvellec so aptly put it: agricultural law and the framework laws on agriculture from the 1960s onwards "support a policy of development based on intensive farming. Agricultural law organizes all aspects of farming life, based on the general objectives laid down in the framework laws, and they all contribute to this scheme." The law originating in the 1960 and 1962 Acts relating to the framework law is an "economic law emphasizing industrial farming." He added: "Agricultural law is expressly drafted for this purpose. Framework laws have defined specific economic goals with the goal of modernizing French agriculture through legal means. To this end, the legislature has developed this very specific project to create a business model and this model is used to encompass all agricultural production units."[54] However, current agricultural law still carries this original design, even more so than what we have so far described, and it is this orientation of agricultural law that hinders agroecology and needs to be radically reformed. Besides, drafting an "agroecology law" should respect basic human rights *and* attempt to build, if not a consensus, at least a large membership base.

3.1.2 Human Rights versus Agroecology?

We will proceed by recalling that in France the legislative package that was meant to protect farm operators and producers, if necessary, against their owners was adopted before the Constitutional Council actually extended its control to the Preamble to the Constitution;[55] in particular to the provisions of the Declaration of Human Rights and to the right to property. Nowadays, a proactive legislator could encourage measures instrumental to the development of agroecology and impose, to this end, restrictions on the freedom of enterprise, and the right to property would not be able to avoid the oversight of the Constitutional Council, either in its *a priori* or *a posteriori* form, assuming no one were to bring the matter before the Constitutional Council *ab initio* under Art. 61 of the Constitution. An ordinary farmer would be able to raise a priority issue of constitutionality on the occasion of routine litigation based on the fact that his choice of methods of production may be incompatible with agroecology and that his preferred method of production had been blocked.

In this case, the right to property or the freedom of enterprise are at odds with the right to a safe and stable environment, the duty to preserve it and the goal of sustainable development that have, since the adoption in France of the Environmental Charter of 1 March 2005, equal constitutional status,[56] and could be cited in support of the legislative provisions relating to agroecology, at least at the stage of

[54] Lorvellec (1988), pp. 2, 5, 11–12.

[55] Constitutional Council 16 July 1971 on freedom of association, AJDA 1971, p. 537.

[56] Constitutional Council, 19 June 2008, 2008-564 DC; Constitutional Council, 29 December 2009, 2009-599 DC.

an *a priori* oversight.[57] However, there is no guarantee that the Constitutional Council, whose role is to reconcile the various provisions of the Constitution, would rule in favor of agroecology. For this purpose, we can refer to the ruling of 24 May 2013, which can be enlightening for our research (No. 2013-317 QPC). In the present case, the Council was referred to by the Council of State for a priority issue of constitutionality raised by the French Cement Industry Trade Union and the Cement Industry Federation; the latter challenged an article of law that required new buildings to include a minimum amount of timber materials and referred to the regulatory authority for a clear definition of this minimum amount. The Council recalls that "legislators can apply limitations to the freedom of enterprise that arises from Art. 4 of the 1789 Declaration; these limitations may stem from constitutional requirements or be justified by the public interest, but their use must not result in disproportionate violations in relation to the aim pursued." In the present case, even though this provision was designed to fight air pollution, "to implement the right of individuals to breathe an air that is not harmful to their health" and "to make efficient use of energy," the Council found that it violated the freedom of enterprise insofar as the legislature mandated the regulatory authority to specify the minimum amount of timber to include, without limitation. However, beyond the referral to the regulatory authority, it was "the violation (of the freedom of enterprise) that was not justified by reasons of public interest directly linked to the objective pursued" that justifies the declaration of unconstitutionality and not the legislature's ignorance of the limits of its jurisdiction. In other words, the legislature that wanted to mandate the "unlimited" use of environmentally friendly materials in the construction industry, or the use of production systems and working methods covered by agroecology, would likewise, according to the Constitutional Council, infringe upon the freedom of enterprise and the right to property, whose violation is also assessed according to the principle of proportionality with regard to the objective pursued. The legislature should, at the very least, exercise restraint. In doing so, the judge merely alludes to the lack of agreement on the approaches that will help forge the future of agriculture.

3.1.3 Lack of Consensus

The postwar productivist regime was built through government efforts and reinforced by the social support it enjoyed. Its aims were to build an efficient farming system, to disseminate technological progress and to guarantee a

[57] In the context of the *a posteriori* verification, the applicant must report the violation of a right or freedom, and the Constitutional Council has already ruled that Art. 6 of the Charter on sustainable development "does not establish a right or freedom guaranteed by the Constitution; that this fact cannot, in itself, be relied upon to support a priority issue of constitutionality on the basis of Art. 61-1 of the Constitution" Constitutional Council, 23 November 2012, No. 2012-283, QPC. On the other hand, the violation of Arts. 1 (right to a safe and healthy environment) and 2 (the duty to preserve it) could be relied upon: see Constitutional Council, 8 April 2011, No. 2011-116, QPC, AJDA 2011 p. 1158; Constitutional Council, November 23rd, 2012, No. 2012-282, QPC.

satisfactory standard of living for farmers and food at affordable prices for all. The project was supported by the trade unions and the Catholic Agricultural Youth (JAC) movement, whose strong presence in rural areas became a vehicle for the dissemination of support for the new "modern farming." It did not meet with resistance in society, quite the contrary.

The situation is different for agroecology. It does not enjoy a consensus within mainstream unionism, as seen by the resistance, albeit insignificant, to the CAP reform and greening. There is currently no opinion leadership that could promote this project to young people such as the JAC in the past. Furthermore, it is not certain that the majority of consumers actually support it. In this respect, agroecology does not enjoy the sort of progress that we witness for industrial agriculture.

This does not imply that the law is powerless. Agroecology is a complex domain still lacking consolidated data and full technical references;[58] it also requires behavioral changes that may constitute an economic risk and calls for top-level agronomic skills. Moreover, agroecology requires flexibility and adaptability. It is therefore understandable that, from a legal standpoint, the situation is not straightforward. Yet the current bill on the future of agriculture, food and forestry opens up several leads (passed by the National Assembly 14 January 2014).

3.2 *Agroecology and the Law*

The current debate about the future of agriculture in France is extremely thought provoking. In a mission letter of 5 September 2012, the Minister of Agriculture instructed Marion Guillou, former director of the INRA (National Institute for Agronomic Research), to examine the development of the agricultural model towards a system that "combines competitiveness with environmental protection," in preparation for the country's next major farm bill. The report entitled "The agroecological project: for highly efficient farming that reconciles competitiveness with respect for the environment" was released on May 2013. It provides an inventory of experience and knowledge and analyzes over 200 practices with respect to their performance in terms of production and conservation of natural resources and the environment and in economic and social terms. Based on this study, the report promotes six production systems[59] and suggests that public policies provide the financial assistance required to ensure the transition toward these systems. In order to achieve this transition, the future CAP and aids listed under its second pillar would provide agri-environmental measures (AEMs) labeled "system measures" that would partially replace the current AEMs. However, the report insists that there

[58] The report cited, drafted under the direction of Marion Guillou, also calls for the establishment of a catalogue of accurate, reliable and readily available data and the introduction of a shared-information system (Guillou 2013, pp. 46–49).

[59] Two of them in livestock farming: mixed farming-dairy cows and pig production on straw-bedded systems, and four in crop production: low-input extensive crops, field crops with reduced tillage, perennial crops using IPM and agro-forestry.

can be no "one size fits all" approach; the methods must adapt to the pedo-climatic and agroecological environment. A useful agroecological project must therefore adapt to an area, comply with its local conditions and be collectively implemented to ensure that the environment benefits from it. We are therefore moving away from the scheme traditionally taken by agricultural law and environmental law whereby a person is requested or coerced to carry out or to execute an obligation or to refrain from an action. The report tries to establish a new structure called Economic and Environmental Interest Grouping (EEIG) with the aim of creating agroecological projects based on the membership of farmers within one small agricultural area and/or stakeholders in an industry, which would thus attract the bulk of the funding and particularly that of aids granted by the reformed CAP and therefore serve as an advocacy tool for consumers. It is not so much the legal innovation of the EEIGs or AEMs[60] that is interesting; for that matter, nothing is said about the status of these new EEIGs, but it is the more general purpose of the device that deserves attention.

No previous Minister of Agriculture has ever been as supportive of agroecology as Stéphane Le Foll is today, and the call for collective regional action in favor of the environment has rarely been stronger.[61] The analysis of the current outlook on the environment/agriculture ratio is equally new and relevant; it supports our research on agricultural law: the environmental aspect ought to be integrated into the operating model from its very design. Hastily applying environmental restrictions to a production model that had not originally included them has proved unsuitable and often ineffective.[62] EEIG schemes must therefore satisfy these requirements; they have to enable the creation of collective and regional agroecological production systems that are originally conceived as such.[63]

On 12 September 2013, a bill submitted to the CSO (High Council for the Guidance and Coordination of the Agricultural and Food Economy) stipulated that the State shall promote the "development of the production and processing industries combining economic and environmental performance." Agricultural law will also be designed to serve this purpose. In line with the recommendations of the Guillou Report, the bill provides for the creation of an Economic and Environmental Interest Grouping involving farmers and, if applicable, "other interested parties

[60] Some former agro-environmental measures are indeed close to what the report coins as system AEMs such as the agro-environmental grassland premium or the AEM called "low-input forage-based mixed farming and stockbreeding system."

[61] We should note that these undertakings have already been supported as part of the former AEMs called "regional" measures; see Hermon and Doussan (2012), pp. 291 et seqq.

[62] Keynote address at the delivery of the report, available on the website of the Ministry of Agriculture.

[63] In the name of collective regional measures, the report calls for the implementation of exchanges between crop farms and stockbreeding farms in the same agricultural area. Crop farms would provide grain and straw in exchange for manure and slurry, so that the use of mineral nitrogen would be greatly reduced, or, better still, the report suggests a collectivization of the objectives of ecological focus areas: a certain surface could be used by each farm, and an area greater than the sum of all individual farms put together would be collectively managed; report quoted above, pp. 39 et seqq.

[...] the members of which have to collectively commit to implementing a multi-annual project for sustainable changes in the practices used to manage their production systems, aiming for both economic and environmental performance" (Art. 3). Individual farming units or interest groups that bring them together as part of the EEIGs will be targeted "by priorities and increases in the allotment of state aid." This is the central feature of the bill and the main catalyst for agroecology; if it becomes law, a new chapter will be added to the Rural Code to address EEIGs, which will introduce the Third Book on the farming unit. As regards its legal status, the legislature refers to an implementing order, but the scheme provides for a great deal of flexibility. There is no requirement that the EEIG be a legal person, and its membership is not limited to farmers. Its scope is also broadly defined;[64] it allows for a multitude of initiatives. However, since it attempts to achieve an objective of public interest and to facilitate the granting of state aid, a simple contract is not enough to set up an EEIG, as was the case for the EIG, Economic Interest Grouping (Art. L. 251-1, Commercial Code). The bill mentions "submitting the grouping's multi-annual project to the administrative authority" and "a procedure for certifying the grouping's economic and environmental interest." We can therefore infer that the EEIGs will be subject to an approval procedure or something similar.

Moreover, several amendments were made to the Rural Code that promote the development of agroecology. Under Art. 4, entitled "Means for an environmentally friendly agriculture," the text does away with the various conditions cited above, referred to at the end of a green farming lease. The new Art. L. 411-27(3) reads as follows: "Clauses designed to secure the lessee's compliance with cultivation practices mentioned in the second paragraph (i.e. practices aimed at preserving natural resources), including requirements to maintain a minimum amount of ecological infrastructure, can be included in the lease upon its termination or renewal." Lessors everywhere may now opt for any environmental clause, if they can find a lessee. In order to acquire a greater understanding of the nitrogen cycle and to control it more effectively, Art. 4 also provides for a requirement applying to any person distributing or disposing of nitrogen fertilizers to report his activity to the administration.[65] Furthermore, the bill includes a new tool to reduce the use of plant protection products, in addition to the current Ecophyto plan and license fee (see Sect. 2.1.2).[66] As proposed by the Guillou Report, the text sets "restrictions on plant protection products" (Art. 23) applicable to suppliers. The observance of these restrictions can be direct or indirect through the purchase of plant protection product savings certificates similar to energy savings certificates (see Sect. 2.2.3); these certificates can be purchased from any professional who has conducted an

[64] Perhaps too broadly—in the sense that the concepts underlying the EIG, "sustainable change in practice" and "dual economic and environmental performance," are themselves not defined. It is up to the administrative authority that validates the project to show rigour and purpose.

[65] Art. L. 255-2-1. "The administrative authority may make it compulsory for individuals or legal persons who distribute or dispose of nitrogen fertilizers for agricultural use in a specific area- at no cost or for a fee, to provide an annual activity report."

[66] The creation of this market for savings certificates for plant protection products is presented by the Guillou Report as "an alternative or complementary to a fee increase" (Guillou 2013, p. 142).

operation meant to reduce the use of these products. If this market opens up, it will involve most suppliers, as only 5 % of sales will not be subject to savings. However, there is no mention of the extent of the savings that should be achieved through this scheme.[67]

Finally, references to the environment are highlighted in the SAFERs' mission statement (French Real Estate and Rural Planning Agency) and especially in policies on installing and monitoring facilities/equipment. SAFERs' current missions are "to improve land tenure with the installation or preservation of farms or timber contractors, to increase the size of certain farming and forestry enterprises, to improve soil quality and to plan and reorganize land plots." Moreover, since the Act of 23 February 2005 on the development of rural areas, "they contribute to landscape diversity, to the protection of natural resources and to the conservation of biodiversity" (Art. L. 141-1, Rural Code). But this environmental mission is not part of the SAFERs' agricultural mission; it is most frequently conducted in conjunction with public entities and can lead to a shift towards a nonagricultural use of the preempted land. The reform has brought about a change in the outlook on the issues, as environmental objectives are now embedded in the mission of agriculture. In the future, SAFERs could thus become instrumental in developing agroecology; their primary goal should be to contribute to "the protection of agricultural and natural areas." Their actions should target the creation and preservation of viable farms and "the diversity of production systems within the territories, including systems that combine economic and environmental performance and organic production systems" (the new Art. L. 141-1).

The shift is also significant with respect to the policy on setting up farms. Currently, there is no reference to the environment in the preliminary chapter to the section on setting up farms and monitoring policy. If the text is approved in its current state, the installation policy will probably try to "create, adapt and pass on farming enterprises" and also "to promote diverse production systems across the country, particularly those liable to generate employment and value added and to combine economic and environmental performance, and especially those pertaining to agroecology" (Art. 330.1, Rural Code). As for the master plans that provide the basis for monitoring facilities, their approach and priorities take into account economic issues as well as social and environmental ones. Their primary objective is to support farmers to set up agricultural activities; other objectives include "1. controlling the excessive expansion and concentration of farms under the management of a single individual or legal entity in order to maintain a biodiverse agriculture that can generate employment and value added; 2. promoting the diversity of production systems across the country, including systems that combine

[67] Knowing that voluntary measures organized under the Ecophyto plan will not meet the established goals, this scheme should be ambitious enough to overcome "the first inconclusive results." The target of reducing the use of pesticides—seen earlier and established by Grenelle 1—is thus considered out of reach: Guillou (2013), p. 73.

both economic and environmental performance and those under organic production."

The governing principle underlying the bill on the future of agriculture is straightforward: France needs to reformulate its agricultural model, shifting from the industrial agriculture of the 1960s to agroecology. All the necessary means are deployed for agroecology to thrive, and the Rural Code has undergone significant changes as a result of the reform. It is now up to the stakeholders to put the proposed legal framework into practice.[68] Only then can we safely declare, "agricultural law is dead; long live the law of agroecology!"

References

Aubertot JN, Barbier JM, Carpentier A, Gril JJ, Guichard L, Lucas P, Savary S, Savini I, Voltz M (eds) (2005) Pesticides, agriculture et environnement. Réduire l'utilisation des pesticides et limiter leurs impacts environnementaux. Expertise scientifique collective, synthèse du rapport. INRA – Cemagref, Paris

Barthélémy F, Grimot M (2006) Rapport sur la simplification de la réglementation des installations classées

Bétaille J (2012) Les conditions juridiques de l'effectivité de la norme en droit public interne: illustrations en droit de l'urbanisme et en droit de l'environnement. Dissertation, University of Limoges

Bodiguel L (2011) Les clauses environnementales dans le statut du fermage. Droit Rural 398:study n°16

Bosse Platière H, Collard F, Travely B (2013) Bail rural environnemental. Juris-Classeur périodique, édition Notariale 7:57 et seqq

Cacciabue (2013) Le bail rural à clauses environnementales: retours d'expériences et besoins. Sécuriser des engagements environnementaux. Séminaire d'échange sur les outils fonciers complémentaires à l'acquisition. In: Commissariat Général au Developpement Durable – Etudes et documents 82:1 et seqq

César G (2006) Rapport sur le projet de loi d'orientation agricole, vol 1. Doc. Senate, 2005–2006, no. 45

de la Raudière L (2010) Simplification de la réglementation et amélioration de la compétitivité industrielle

Delhoste M-F (2001) Les polices administratives spéciales et le principe d'indépendance des legislations. LGDJ, Paris

Doussan I (2009) Les services écologiques: un nouveau concept pour le droit de l'environnement? In: Cans C (ed) La responsabilité environnementale. Dalloz, Paris, pp 125–141

Flory J-C (2003) Les redevances des agences de l'eau: enjeux, objectifs et proposition d'évolution dans la perspective de la réforme de la politique de l'eau. Ministère de l'écologie et du développement durable, Paris

Guillou M (2013) Le projet agro-écologique: vers des agricultures doublement performantes pour concilier compétitivité et respect de l'environnement, Rapport au Premier ministre. Agreenium, Paris

Hermon C, Doussan I (2012) Production agricole et droit de l'environnement. LexisNexis, Paris

[68] At a conference organized by INRA on 17 October 2013, the Minister of Agriculture, Le Foll announced the goal of moving towards better economic and environmental performance for 50 % of the farms by 2025.

Lambert A, Boulard J-C (2013) Rapport de la mission de lutte contre l'inflation normative, followed by the opening of the Etats généraux de la modernisation du droit de l'environnement in June 2013

Langlais A (2013) Les paiements pour services environnementaux, une nouvelle forme d'équité environnementale pour les agriculteurs? – Réflexions juridiques. Environnement 1:study n°4.

Lorvellec L (1988) Droit rural. Masson, Paris

Nicolazo J-L, Redaud J-L (2007) Les agences de l'eau. Quarante ans de politique de l'eau. Johannet, Paris

Prieur M (2011) Droit de l'environnement, 6th edn. Dalloz, Paris

Romi R (2010) Droit de l'environnement, 7th edn. Montchrestien, Paris

Tavernier Y (1999) La fiscalité au secours de l'eau. In: Assemblée Nationale, Rapport d'information No. 1807, 22 Sept 1999

Terré F (2009) Introduction générale au droit, 8th edn. Dalloz, Paris

Van Lang A (2011) Droit de l'environnement, 3rd edn. PUF, Paris

The Relationship Between Agricultural Law and Environmental Law in Hungary

J.E. Szilágyi

Abstract The present paper provides a review on the relationship between agricultural law and environmental law from the point of view of a Hungarian lawyer. Taking this relationship into consideration, the Hungarian legal system applies similar solutions and provides legal institutions to the other Member States of the European Union (especially to Germany and Austria) in order to arrange differences between the interests of agriculture and environmental protection. The cause of this similarity primarily rests on the significant role of the European Union (EU) to regulate the relationship of these two laws. However, there are some challenging features of the Hungarian law that can be regarded as specialities, such as the specific theoretical approaches of the Hungarian lawyers or the constitutional basis of agri-environmental law (e.g., on GMOs) in the Hungarian constitution (the so-called Fundamental Law). Finally, this paper is a commitment to agroecology, as a scientific discipline and philosophical paradigm, which may play a remarkable role in bridging agriculture and environmental protection.

Keywords Agricultural law • Agri-environmental law • Environmental law

The Hungarian legal order has a close relation to "German-Austrian legal systems," and accordingly Hungarian "agricultural law" and "environmental law" also have numerous similarities to the agricultural and environmental law of Germany and Austria. Nevertheless, other national laws, international law and—mainly in the last 20 years—the EU law have equally had a great effect on the development of the Hungarian law. The relationships between the Hungarian law and international, EU and other national laws are not assessed in detail in the present paper.

As far as the topic of this article is concerned, the approach of the author of this chapter cannot reflect the approaches of every Hungarian expert in the field of the affected laws; however, the author endeavored to invoke all relevant scholars'

This research was supported by the European Union and the State of Hungary, cofinanced by the European Social Fund in the framework of TÁMOP 4.2.4. A/2-11-1-2012-0001 "National Excellence Program."

J.E. Szilágyi (✉)
Department of Agricultural and Labour Law, University of Miskolc, Miskolc, Hungary
e-mail: civdrede@uni-miskolc.hu, http://www.uni-miskolc.hu/~agrarjog/edeonel.htm

M. Monteduro et al. (eds.), *Law and Agroecology*,
DOI 10.1007/978-3-662-46617-9_13

sources of both agricultural law[1] and environmental[2] law. Current Hungarian legal studies consider the relationship between agricultural law and environmental law as a topical issue. Indeed, in the last few years, numerous PhD theses were written on topics connected to the relationship of these laws, namely, on the agro-aspects of nature conservation law,[3] on the environmental aspects of agricultural supports law,[4] on the legislation of GMOs.[5]

1 The Horizontal Classification of Legal Branches and Fields in the Hungarian Law

As regards the Hungarian law, similarly to the other national legal orders, both agricultural law and environmental law significantly differ from the other legal fields. To determine the features and specialities of agricultural law and environmental law, debates among scholars were renewed, especially at the time of Hungary's accession to the EU (in 2004). Such debates typically focused on the localization of these laws in the legal system and also concentrated on the determination of agricultural law and environmental law. Therefore, before the assessment of the relationship between agricultural law and environmental law, the present paper examines the issues of these debates, namely, the localization and determination of the definitions and contents of agricultural law and environmental law.

According to the classical Hungarian jurisprudence,[6] there are two aspects to be considered for the localization of a certain group of legal provisions in the Hungarian legal order: on the one hand, the position of a certain group of legal provisions between public law and civil law and, on the other hand, the problem of a separate "legal branch" (or only "legal field").

As regards the localization of a certain group of legal provisions between *public and civil law*, in the opinion of the author, many of the legal branches include both civil and public laws, if all of them are analyzed with an adequate critique. Therefore, the consistent delimitation of civil and public laws may only be relative and often may not support the localization of certain groups of legal provisions in the legal order.

[1] Bobvos and Hegyes (2011), pp. 9–12; Csák (2006), pp. 75–90 and 83–87; Fodor (2005), pp. 17–54; Kurucz (2007), pp. 41–86; Novotni (1991), pp. 275–280; Olajos (2008b), pp. 14–18; Prugberger (1999), pp. 3–22; Tanka (2007), pp. 371–394 and 391–394; Veres (1993), pp. 519–528.

[2] Bándi (2011), pp. 11–26; Csák (2008), pp. 9–13; Fodor (2012c), pp. 9–31; Fodor (2014), pp. 9–38; Bobvos (2011), pp. 7–36 and 13–16; Bakács (1992).

[3] Horváth (2010).

[4] Farkas Csamangó (2012).

[5] Tahyné Kovács (2013).

[6] See Szilágyi (1998), pp. 307–324; Szilágyi et al (2001), pp. 75–94; Prugberger (1975), pp. 602–611.

According to the abovementioned classical approaches to the *criterion of a separate legal branch* (in Hungarian: *jogág*),[7] a definite group of provisions can be classified as a separate legal branch if (a) the *direct subject* (i.e., *human attitudes*) of the provisions includes a homogeneous circle of social relations and (b) the *method* of the provisions is also specific (e.g., specific method might be the cogent or dispositive way of the legislation).[8] The *homogeneous subject* alone (for example, the human behavior connected to the environment) is not adequate for the fulfilment of the requirements of a separate legal branch; the criterion of a separate legal branch also requires a *specific method*.[9] If a group of certain legal provisions fulfils only the requirement of the homogeneous subject (but not that of a specific method), this group may (only) be considered a legal field (in Hungarian: *jogterület*).

As far as the theoretical definition of these laws is concerned, according to the main stream of the Hungarian jurisprudence, *agricultural law* can be regarded as an independent branch of our (Hungarian) legal system, but not in a traditional way. As opposed to this approach, *environmental law* is rather an integrated and integrating legal field. Because of these features of agricultural law and environmental law, their distinction is considerably problematic.

1.1 The Localization of Agricultural Law in the Hungarian Legal Order

According to the approach widely accepted in the Hungarian legal studies, the antecedents of the "modern agricultural law"[10] may date as far back as the early period of the formation of law(s) and state(s) (namely, to the antiquity). Modern Hungarian agricultural law dates back to the so-called change of regime[11] (1989/1990), when (after a 40-year-period of totalitarian socialism) *agricultural land law*

[7] The theory of Mihály Kurucz is based on a similar classical approach. He does not designate agricultural law as a separate legal branch, only as a nascent legal branch. As to the classical approach of a separate legal branch, Kurucz noted "that agricultural law is developing, and is annoyingly fragile, especially for those who work with it." Kurucz (2007), pp. 69–70.

[8] According to András Jakab, this approach of a separate legal branch was spread during the socialist period of Hungary; see Jakab (2005), pp. 52–54. Numerous authors (e.g., Miklós Szabó) contradict this theory; for him, the criterion of a separate legal branch descends from the nineteenth century. The author of the present paper shares the opinion of these latter authors.

[9] See Szabó (2002), p. 37.

[10] The word "modern" refers to the historical determinations of the different legal categories.

[11] The name of the "change of regime" refers to a bloodless, economic and political inner reform to a democracy in which the political leaders of the previous totalitarian regime were not called to account (or impeached), and due to this they were allowed to follow their activity in the new system, e.g., as political or economic leaders (i.e., it was not a revolution).

and, in a certain sense, *agricultural cooperative law* have been transformed into a capitalist agricultural law.

According to the definition (which is relatively widely accepted by scholars[12] of agricultural law), modern agricultural law means "the totality of the rules adopted for the implementation of the agricultural policy."[13] This definition is rather general; therefore, some notes are necessary, especially in connection with the localization of agricultural law in the Hungarian legal order. There are numerous authors who have claimed that agricultural law is a *separate legal branch* that compounds the provisions of both public and private laws.[14] A minority of agricultural scholars states that agricultural law is merely a legal field.[15] Finally, there is the opinion of András Jakab, who criticizes the criteria set out by the theory of separate legal branches. He noted that no correct criterion can be found on how to distinguish one legal branch from another. Therefore, the separate-legal-branch feature of agricultural law could not be confirmed by the legal theory. At the same time, taking the legal practice into consideration, agricultural law is a separate legal branch. The components of the abovementioned legal practice include the following: the tradition of the Hungarian universities (according to which agricultural law is a separate discipline), the huge number of agricultural provisions, the importance of the agricultural legislation in the EU and the tradition of the separate existence of agricultural law in the Hungarian jurisprudence.[16] However, in 2005, at the National Conference of Agricultural Lawyers, the discipline of agricultural law and the fields of agricultural law have been determined (five out of the six then existing departments teaching agricultural law signed the resolution[17]). According to the concept in force today, agricultural law is a "mixed vocational law" that includes in a specific way the institutions of both private and public laws, and that is also particular and original in its subject.[18]

Besides the definition of agricultural law, the contents of agricultural law have also been a questionable issue at the time of Hungary's accession to the EU. The most widespread method among Hungarian scholars for determining subjects (and

[12] See, for example, Bezdán (2012), p. 227; Bobvos and Hegyes (2011), pp. 10–11; Csák (2006), p. 89; Horváth (2010), pp. 14 and 62; Tanka (2007), p. 390.

[13] Szilágyi (2007), pp. 112–121.

[14] E.g., agricultural law is considered a "private-law-based mixed legal branch" by Pál Bobvos, which definition, inter alia, means that the basis of agricultural law is private law; see Bobvos and Hegyes (2011), p. 10. As opposed to this, István Olajos considered agricultural law as a "primal legal branch" in which the public and private law elements have not been separated; see Olajos (2008b), p. 14.

[15] E.g., according to the thesis of Tamás Prugberger, agricultural law is a legal field with sources of different legal branches systematized by a special view of agriculture; Prugberger (1999), pp. 3–22.

[16] Jakab (2005), p. 54.

[17] See the Resolution of the National Conference of Agricultural Lawyers on the Standard Conception concerning Agricultural Law as a Discipline of Legal Education (Miskolc, 14.1.2005).

[18] About the resolution, see Raisz and Szilágyi (2012), pp. 107–148 and 107–108.

therefore fields) of agricultural law is through determining its "regulated objects." The objects are typically the following: agricultural holding, agricultural producer, agricultural activity, agricultural product, foodstuff, rural area.[19] In the author's opinion, the relationship between agricultural law and environmental law may also be analyzed on the basis of the regulated objects of the agricultural law (see below).

In connection with the Hungarian determination of agricultural law, it is worth noting that, in other Member States of the EU, this legal field can be called in different ways. Thus, taking the French, English and German systems into consideration, besides the names of *droit agraire*, agricultural law, *Agrarrecht*, there are other names as well, namely, *rural law*, *droit rural*, *Landwirtschaftsrecht*, *Agrarrecht und das Recht des Ländlichen Raumes*. Of course, these different denominations may include different contents even in the same country. However, it is important to mention that the notion "agricultural law" in the frame of the Hungarian regulation basically includes rural development, forestry as well as fisheries.

1.2 The Localization of Environmental Law in the Hungarian Legal Order

According to a generally accepted approach of the Hungarian lawyers,[20] the beginning of environmental law dates back to the period after World War II (hereinafter referred to as WWII). Before WWII, merely the antecedents of the environmental law existed. The law after WWII began to protect the environment as a system incorporating all environmental components (i.e., land, air, water, the biosphere as well as the artificial environment created by humans) and the processes and structure thereof.[21] Hungarian lawyers distinguish between "environmental protection law" (i.e., before the 1980s) and "environmental law" (namely, after the 1980s).[22]

The author of the present paper shares the opinion of *László Fodor* about the definition of environmental law. According to his opinion, environmental law is the totality of provisions adopted for the implementation of the environmental policies, namely for the protection, conservation, maintenance and improvement of the environment.[23] However, in a later book, Fodor also called environmental law "the legal framework of the rational management with natural resources" and "the law of ecological sustainability."[24]

---

[19] Raisz and Szilágyi (2012), pp. 108–109.

[20] Cf. *inter alia* Bándi (2011), p. 13; Bobvos (2011), p. 11; Csák (2008), p. 10; Fodor (2001), pp. 16–17.

[21] Namely, before WWII, the legislation regulated merely the direct environment of human beings.

[22] Fodor (2001), pp. 17–18.

[23] Fodor (2001), p. 12. A similar definition can be found in Horváth (2007), pp. 333–355 and 343.

[24] Fodor (2014), p. 34.

According to the approach widely accepted among the Hungarian scholars,[25] environmental law is not a separate legal branch but merely a legal field. The reasons of this approach are as follows: on the one hand, environmental law does not have a homogeneous subject because its *indirect subject* (i.e., the environment) and likewise its *direct subject* (i.e., environment related human attitudes) are subjects of other legal fields and/or legal branches as well. As regards the *specific method*, though, environmental law has some special ways to regulate its field; nevertheless, it also applies the regulating methods of other legal branches/fields (inter alia, administrative law, civil law, criminal law). Otherwise, the exposition (i.e., detailed reasoning of the law) of the Hungarian environmental act[26] notes: "the feature of the new environmental protection legislation shall be the environmental friendly legal order ... The legal field of environmental protection cannot coexist parallel to separate, 'non- environmental protection' legal fields."

As for the contents of environmental law, the latter includes two main parts:[27] the general part and the sectoral (or specific) part. The *general part* comprises (a) the right to a healthy environment; (b) the legal principles of environmental law; (c) civil, criminal, administrative and other liability regimes in connection with environmental protection; (d) the licensing of the use of the environment; (e) different types of environmental assessment; (f) the economic instruments of environmental protection; (g) the eco-management and audit scheme; (h) eco-labels. The *specific part* of environmental law is made up of three groups. The first group (*environmental law of elements*) focuses on the protection of environmental components (e.g., land contamination and soil quality, climate change, air pollution and air quality, water pollution and water quality, conservation of nature, protection of animals, GMOs). The second group (the so called *causal environmental law*) concentrates on hazardous impacts (e.g., waste management, hazardous substances and technologies, noise and vibration, radiation). The third group includes the rules concerning different *economic branches* (e.g., agri-environmental law, energy-environmental law). Nevertheless, it is worth stressing that the different sectors of environmental law may overlap in numerous ways (e.g., agri-environmental law and (a) economic instruments of environmental protection, (b) environmental law of elements, (c) causal environmental law).

[25] Bándi (2011), pp. 11–23; Bobvos (2011), pp. 14–15; Csák (2008), p. 10; Fodor (2012c), p. 19–21.

[26] Act LIII of 1995 on the general rules of environmental protection.

[27] Cf. Bándi (2011); Csák (2008); Fodor (2012c).

2 The Relationship Between Agricultural Law and Environmental Law

The relationship of these two laws was the topic of numerous academic events and reports.[28] The relationship between agricultural law and environmental law is based on an obvious fact, that is, agriculture is one of the biggest environment-using economic branches. In connection with this fact, agriculture is not only the biggest possible polluter of the environment but also one of the biggest sufferers of environmental damage. Based on these consequences, the relationship of the two laws might be examined from numerous aspects. From the point of view of the author, it is nearly impossible to exactly categorize all aspects, but some important links and aspects may be highlighted.

1. *Historical link.* Because the formation of environmental law dates back to merely after WWII, it is from this period that the relationship between environmental law and agricultural law should be analyzed. Nevertheless, the establishment of nature conservation considered as a systematized antecedent of environmental protection dates back to the nineteenth century.[29] Nature conservation and its legal background may be regarded as a reaction to the negative effects of modern agriculture.[30] Therefore, it can be established that agriculture had intense effect on the formation of environmental law. Subsequently, environmental protection and environmental law have also formed the legislation of agriculture.

2. *Theoretical link.* Another link between agricultural law and environmental law is their theoretical grounds, namely the concept (or model or principle) of *sustainable development*,[31] the EU's model on *multifunctional agriculture*,[32] the theory of *ecological services*, etc. These theoretical grounds strengthen the role of public law in the legislation of both agricultural law and environmental law.

The interpretation of these natural-science-based theoretical grounds is not easy for lawyers. Therefore, it can be established that *agroecology*[33] could properly support the interpretation of these theoretical grounds and the conversion of these into the legislation and in the legal practice.

3. *International link.* There are numerous links between agriculture and environmental protection at the international level and in connection with international law.[34] The comprehensive classification of these links is not the subject of this

---

[28] See, e.g., the Hungarian report of the Commission II of the 27th CEDR congress held in Lucerne, Switzerland, from 11th to 14th September 2013; Csák and Raisz (2011) Accessed 15 June 2014; Csák and Jakab (2013). Accessed 15 June 2014.

[29] Bell et al. (2013), pp. 721–722; Kubasek and Silverman (2014), pp. 351–353.

[30] Fodor (2001), pp. 17–18.

[31] The Hungarian jurisprudence also enters into details in connection with the determination of sustainable development; see Bándi (2013); Bányai (2012), pp. 15–61; Szabó (2012), pp. 161–174.

[32] See Fodor (2012b), pp. 128–137.

[33] On determination of agroecology, see Monteduro (2013), pp. 2–11 and 3–4.

[34] See, for example, Raisz (2010), pp. 241–253.

article. Nevertheless, it is worth noting the role of the World Trade Organization, because the WTO has a great effect on both agricultural and environmental laws. For example, the recruitment of *direct income supports* in the EU can also be attributed to the significant influence of the WTO law.[35] These direct supports (e.g., Single Farm Payments, Single Area Payments), which are acceptable forms of domestic supports according to the WTO law (i.e., they are the so-called green box subsidies), can be gained by farmers on the condition of fulfilling certain environmental requirements. Similarly, the individual dispute cases of the WTO dispute settlement system also have a remarkable effect on the EU legislation. For instance, the *EC-Biotech case*[36] determined the feature of the EU GMO legislation in the long run.[37]

At the end of this point, it is worth noting that *Multilateral Environmental Agreements* also play a significant role in the legislation of agriculture.[38]

4. *European integration link.* Since the Treaty of Rome, the European integration (i.e., the European Economic Community, the European Community and the European Union) has had its own *agricultural policy*, i.e. the Common Agricultural Policy, CAP (see Chapter "The Common Agricultural Policy (CAP): Achievements and Future Prospects"), while the European integration established the contractual base of the *environmental policy* only with the Single European Act. The *European Court of Justice* played a significant role in determining the relationship of these policies.[39]

The integration of environmental protection requirements into EU policies and activities is determined in Article 11 of the *Treaty on the Functioning of the EU.* Similarly, the *Charter of Fundamental Rights of the EU* also stipulates the compulsory requirement of integration. Namely, according to Article 37 of this Charter, "A high level of environmental protection and the improvement of the quality of the environment must be integrated into the policies of the Union and ensured in accordance with the principle of sustainable development." The *EU Strategy for Sustainable Development* also prescribes "the process of integration of environmental concerns in sectoral policies,"[40] and previously the fifth EC *environment programme* determined the requirement concerning the integration of "environment into other policy areas."

On the abovementioned political and legal basis, environmental protection gathers more and more ground in the CAP, especially in connection with *direct supports*, *rural development*, the provisions of *cross compliance* and the new

---

[35] Cf. Nagy (2013), p. 83.

[36] WT/DS291/R, WT/DS292/R, WT/DS293/R.

[37] See McMahon (2007), pp. 322–330.

[38] See Norer (2012), pp. 3–20 and 15–16.

[39] See European Court of Justice, Joined Cases C-164/97 and C-165/97, European Parliament vs. Council of the European Union, Judgment of 25 February 1999. On the details, see Fodor (2014), p. 187.

[40] See COM(2001) 264 final.

requirements of *greening* (see Chapter "Ecosystem Services: European Agricultural Law and the Rural Development").[41]

5. *Constitutional link.* The Hungarian constitution, the socalled Fundamental Law of Hungary (entered into force in 2012), also contains direct provisions concerning both agriculture and environment. The environmental provisions of the Fundamental Law are assessed in detail by the Hungarian jurisprudence[42] since the previous Hungarian constitution also included provisions[43] concerning environmental protection (namely, the right to a healthy environment) and these provisions also became part of the Fundamental Law complementing further, new provisions to the old ones. Contrary to environmental protection, the provisions concerning agriculture are absolutely new elements of the Fundamental Law.[44]

Taking the topic of this article into consideration, it is worth drawing the attention to two provisions of the Fundamental Law. According to Article P of the Fundamental Law, "all natural resources, especially agricultural land, forests and drinking water supplies, biodiversity—in particular native plant and animal species—and cultural assets shall form part of the nation's common heritage, and the State and every person shall be obliged to protect, sustain and preserve them for future generations."[45] Another important provision is Article XX concerning the right to physical and mental health. According to Article XX of the Fundamental Law, Hungary shall promote the exercise of the right to physical and mental health by ensuring that its "*agriculture remains free from any genetically modified organism*, by providing access to healthy food and drinking water. . ., and by ensuring environmental protection." As for the GMO-free agriculture[46] mentioned in Article XX, it is worth emphasizing that this provision does not concern GM food. Therefore, Hungary does not interfere with the free movement of goods (i.e., GM food) by virtue of Article XX. Essentially, Article XX concerns merely the growing of GM plants in Hungary. Otherwise, it is also worth stressing that at this moment growing GM plants is forbidden in Hungary. It seems that while, at the time of writing of this article, the authorization process of the *Pioneer 1507* GM maize is coming to its end, Hungary has prohibited cultivation through safeguard measures[47]

[41] On the details, see Jack (2012), pp. 258–273. See, furthermore, Csák (2012), pp. 423–433; Csák and Olajos (2008), pp. 31–42; Farkas Csamangó (2009), pp. 151–170; Olajos (2012), pp. 41–92.

[42] See, for example, Bándi (2012), pp. 6–15; Fodor (2006), pp. 41–70 and 102–163; Fülöp (2012), pp. 76–87; Raisz (2012), pp. 37–70.

[43] Otherwise, the previous Hungarian constitution and the jurisdiction of the Constitutional Court of Hungary included utterly progressive provisions in connection with the "non derogation" principle; see Fodor (2014), p. 109.

[44] On the agricultural aspects of the jurisprudence of the Constitutional Court of Hungary before the Fundamental Law, see Téglási (2009), pp. 18–29.

[45] As to the interpretation of the concept of the "nation's common heritage" in the Hungarian Fundamental Law, see Raisz (2013), pp. 84–96 and 88–96.

[46] However, this topic is regarded, *inter alia*, as both an agricultural and environmental issue by the Hungarian jurisprudence; see, for example, Tahyné Kovács (2013), p. 14.

[47] See Art. 23 of Directive 2001/18/EC of the European Parliament and of the Council of 12 March 2001.

on individually authorized GMOs, such as the *MON810* GM maize and the *Amflora* GM potato. Furthermore, Hungary has even successfully sued the EU Commission at the General Court in connection with the Amflora GM potato.[48] The Hungarian government has already expressed its intention to obstruct the Pioneer-maize cultivation in Hungary as well. Consequently, it is worth emphasizing that hitherto, Hungary has prohibited the cultivation of GM plants applying EU law (e.g., by the safeguard clause of Dir. 2001/18/EC)[49] without citing Article XX of the Fundamental Law. Nevertheless, although Hungary adopted national coexistence rules,[50] these have never been applied because of the safeguards. Anyway, the Hungarian law scholars have differing interpretation and assessment of Article XX.[51] In the opinion of the author of the present paper, according to Article XX, the Hungarian state is obliged merely to "promote" a GM-free agriculture and not to achieve it.

6. *Links on the basis of the different levels of agricultural activities.* As above mentioned, the Hungarian agricultural lawyers may systematize the agricultural law on the basis of its regulated objects. As regards these objects, there have been significant changes and developments both in the number of regulated objects and in the contents thereof. *Inter alia*, having in mind the tendencies of Western Europe, these objects can be described as a multilayer idea,[52] according to four levels.

The first level is the core of the concept and includes the growing of crops and the keeping of animals. Typically, the rules of *cross compliance* of the Common Agricultural Policy are connected to this level of agricultural activity. The second level is really close to the first level and means the processing and sale of the agricultural products in the primary form. The classification of the next two levels as agricultural activity is not so clear, and there are some differences in the legal orders of the Member States. The third level corresponds to the secondary activities in the frame of an agricultural holding, e.g. agro-tourism in the rooms of a farm building. The fourth level means the secondary activities outside of an agricultural holding. *Inter alia*, the secondary activities may include activities connecting to waste management, and, in addition, this level has a direct relationship with another regulated object, i.e. the rural area. Namely, several EU supports relate to the regulated object of rural areas (i.e., the supports of the second pillar of the Common Agricultural Policy; e.g., NATURA 2000). These developments in connection with the change of the regulated objects of agricultural law have created many links between agricultural law and environmental law.

---

[48] See General Court, Case T-240/10, Hungary vs. European Commission, Judgment of 13 December 2013.

[49] On an individual Hungarian case, see Raisz (2012), pp. 111–112.

[50] Article 21/B-21/F of Act XXVII of 1998 on genetic modification; see Olajos (2008a), pp. 73–88 and 82–87.

[51] See Fodor (2012a), pp. 65–75 and 74; Raisz (2015); Tahyné Kovács (2013), pp. 72–77.

[52] The determination of agricultural activity is also difficult in other Member States. The German Handbook includes 40 different definitions of agriculture; see Käb (2003), p. 3 and Grimm (2001), pp. 1–4.

7. *Links on the basis of different kinds of integration. The agri-environmental law.* First of all, it is worth distinguishing between integrating and integrated rules of environmental protection. *Integrated environmental protection* means the situation where environmental protection and its rules become part of another legal field, different from environmental law. In contrast, *integrating environmental protection* means the situation where the core of environmental law, which can be determined only with difficulty, is extended to a new part. A typical example of integrated environmental provisions is the cross-compliance rules of agricultural law. Integrating provisions are the characteristic rules of environmental law, which are also applicable in the agricultural sector. However, this can mean *agri-environmental law in a wider sense*; i.e., this is the situation when both the general rules and the specific part of the Hungarian environmental law are applied on agriculture as well.[53] In a more precise or narrower sense, agri-environmental law merely includes rules concerning specifically the users of the environment in the agricultural sector.[54] Taking this narrower category of agri-environmental law into consideration, according to the Hungarian jurisprudence,[55] the main components of this are the following: water protection, land and soil protection, nature conservation and green[56] genetic modification. Nevertheless, air protection, noise protection, animal protection, waste management and light pollution are also relevant in connection with agri-environmental law. Generally speaking, the objects of agri-environmental law are the environment, the effects and the natural resources of agricultural activities.[57]

The Hungarian agri-environmental law's features summarized by *László Fodor* are as follows:[58] (a) although agricultural activity is the cause of environmental effects, at the same time, the direct environment and the natural resources of agriculture suffer the negative effects of environment use as well; (b) both the quantitative and qualitative sides of environmental law are significant (e.g., in connection with land or water protection); (c) it is nearly impossible to separate the provisions of use and management from the provisions of quantitative protection of the environment and natural resources (e.g., the unity of land use and land protection in the Hungarian law); (d) environmental rules are in a close relationship with the norms concerning consumer protection and public health, as the primary service of agriculture is food production; (e) the defensive rules of environmental protection typically became integrated parts of management acts (e.g., in Act LVII of 1995 on water management or in Act LV of 1996 on hunting); (f) the number of agri-environmental provisions (acts, decrees, etc.) is high; (g) typically, EU law

---

[53] Horváth (2009), pp. 80–101 and 80.

[54] Horváth (2009), p. 80.

[55] Horváth (2009), pp. 83–85.

[56] On the difference between green and red genetic modification, see Erbguth and Schlacke (2008), p. 336.

[57] Fodor (2014), p. 182.

[58] Fodor (2014), pp. 184–185.

determines the frame of agri-environmental law (except for the qualitative protection of lands), and the Hungarian law often follows even the structure of the EU provisions; (h) the economic and financial legislative method has a significant role in connection with agri-environmental law, as both the first and the second pillars of the CAP confirm.

3 Conclusions

Both agricultural law and environmental law are quickly changing parts of the legal orders of the EU and its Member States. Therefore, the analysis of their relationship is utterly difficult, and other scientific disciplines, such as agroecology, have to participate in order to help the work of lawyers. However, according to the best belief of the author, the jurisprudence can, *vice versa*, help to tackle the twenty-first century's environment-related problems as well.

Consequently, the Hungarian legal system applies similar solutions and legal institutions to the other Member States of the European Union (especially, to Germany and Austria) in order to arrange differences between the interests of agriculture and environmental protection. The cause of this similarity primarily rests on the significant role of the European Union in regulating the relationship of these two laws. However, there are some challenging features of the Hungarian law and jurisprudence that can be regarded as specialities. For example, the specific theoretical approaches of the Hungarian lawyers or the constitutional basis of agri-environmental law (e.g., on GMOs) in the Hungarian constitution (the so-called Fundamental Law) may present such specialities.

References

Bakács T (1992) Magyar környezetjog. Springer Hungarica, Budapest

Bándi G (2011) Környezetjog. Szent István Társulat, Budapest

Bándi G (2012) Gondolatok a környezethez való jogról. In: Raisz A (ed) A nemzetközi környezetjog aktuális kihívásai. University of Miskolc, Miskolc, pp 6–15

Bándi G (2013) A fenntarthatóság értelmezésének egyes jogi szempontjai. Dissertation, Hungarian Academy of Sciences

Bányai O (2012) Az energiafelhasználás csökkentésére és a megújuló energiaforrásokra irányuló szabályozás az ökológiai fenntartjhatóság nézőpontjából. Dissertation, University of Debrecen

Bell S, McGillivray D, Pedersen OW (2013) Environmental law, 8th edn. Oxford University Press, Oxford

Bezdán A (2012) A vadászati jog, annak gyakorlása és a vadásztársaságok. Dissertation, University of Szeged

Bobvos P (2011) Környezetjogi alapvetések. In: Miklós L (ed) A környezetjog alapjai. SZTE-ÁJK – JATEPress, Szeged, pp 7–36

Bobvos P, Hegyes P (2011) Agrárjog. JATEPress, Szeged

Csák C (2006) Az agrárjog rendszerbeli sajátosságai és fejlődési tendenciái. In: Miskolczi-Bodnár P (ed) A Civilisztika fejlődéstörténete. Bíbor Publisher, Miskolc, pp 75–90

Csák C (2008) Környezetjog. Novotni Publisher, Miskolc

Csák C (2012) Cross compliance, avagy környezetvédelem a támogatás feltételeként. Publicationes Universitatis Miskolcinensis Sectio Juridica et Politica 30(2):423–433

Csák C, Jakab N (2013) Agriculture and the requirements of a sustainable development: national report for Hungary. Paper presented at the Commission II of the 27th CEDR congress, Lucerne 11–14 September 2013. http://www.cedr.org/congresses/luzern/luzerncom2.php

Csák C, Olajos I (2008) The application of the single payment by national administrations and national courts. J Agric Environ Law 3(5):31–42

Csák C, Raisz A (2011) Legal framework of environmental law for agricultural production: national report for Hungary. Paper presented at the Commission I of the 26th CEDR congress, Bucharest 21–24 September 2011. http://www.cedr.org/congresses/bucarest/bucharestcom1.php

Erbguth W, Schlacke S (2008) Umweltrecht. Nomos Verlagsgesellschaft, Baden-Baden

Farkas Csamangó E (2009) Kölcsönös megfeleltetés teljesülésének ellenőrzése és a szankciók. Acta Universitatis Szegediensis de Attila József Nominatae Sectio Juridica et Politica 72 (5):151–170

Farkas Csamangó E (2012) A kölcsönös megfeleltetés környezetjogi aspektusai. Dissertation, University of Szeged

Fodor L (2001) Környezetjog előadások. Bíbor Publisher, Miskolc

Fodor L (2005) Agrárjog. Kossuth Academic Publisher of the University of Debrecen, Debrecen

Fodor L (2006) Környezetvédelem az Alkotmányban. Gondolat Publisher, Budapest-Debrecen

Fodor L (2012a) A GMO szabályozással kapcsolatos európai bírósági gyakorlat tanulságai. In: Csák C (ed) Jogtudományi tanulmányok a fenntartható természeti erőforrások tükrében. University of Miskolc, Miskolc, pp 65–75

Fodor L (2012b) A multifinkcionális és fenntartható mezőgazdaság európai modellje. Pro Futuro 2 (2):128–137

Fodor L (2012c) Környezetvédelmi jog és igazgatás. Debrecen University Press, Debrecen

Fodor L (2014) Környezetjog. Debrecen University Press, Debrecen

Fülöp S (2012) Az egészséges környezethez való jog és a jövő nemzedékek érdekeinek védelme az Alaptörvényben. In: Csák C (ed) Jogtudományi tanulmányok a fenntartható természeti erőforrások tükrében. University of Miskolc, Miskolc, pp 76–87

Grimm C (2001) Von der Landwirtschaft zur Wirtschaft auf dem Lande?: Gedanken zum Begriff der Landwirtschaft. Agrarrecht 31(1):1–4

Horváth G (2007) A környezetjog és az agrárjog közeledése, találkozása és metszete a magyar jogrendszerben. Állam- és Jogtudomány 48(2):333–355

Horváth G (2009) Az agrár-környezetvédelem speciális részterületei. Jog – Állam – Politika 1 (2):80–101

Horváth G (2010) Az agrár-környezetvédelmi jog elméleti alapvetése és hazai szabályozásának rendszerezése, különös tekintettel agrár-természetvédelmi részterületére. Dissertation, Eötvös Loránd University

Jack B (2012) Ecosystem services and the rural environment. Eur Energy Environ Law Rev 21 (6):258–273

Jakab A (2005) A jogági felosztás problematikája, különös tekintettel az agrárjogra. Collega 9 (4):52–54

Käb P (2003) Multifunktionale Landwirtschaft als agrarrechtliches Problem? Agrar- und Umweltrecht 33(3):Annex I

Kubasek NK, Silverman GS (2014) Environmental law, 8th edn. Pearson Education, Upper Saddle River

Kurucz M (2007) Agricultural law's subject, concept, axioms and system. J Agric Environ Law 2:41–86

McMahon JA (2007) EU agricultural law. Oxford University Press, Oxford

Monteduro M (2013) Environmental law and agroecology. Eur Energy Environ Law Rev 22 (1):2–11

Nagy Z (2013) Környezeti adózás szabályozása a környezetpolitika rendszerében. University of Miskolc, Miskolc

Norer R (2012) Agrarrecht – eine Einführung. In: Norer R (ed) Handbuch des Agrarrechts. Verlag Österreich, Wien, pp 3–20

Novotni Z (1991) A magyar agrárjog fejlődésének vázlata. Magyar Közigazgatás 41(3):275–280

Olajos I (2008a) A géntechnológiai tevékenység szabályozása Magyarországon. In: Szilágyi JE (ed) Környezetjog. Novotni Publisher, Miskolc, pp 73–88

Olajos I (2008b) A vidékfejlesztési jog. Novotni Publisher, Miskolc

Olajos I (2012) Támogatási rendszereink és a megújuló energiák. University of Miskolc, Miskolc

Prugberger T (1975) A jogrendszer tagozódásának egyes időszerű kérdései. Jogtudományi Közlöny 30(10):602–611

Prugberger T (1999) Az agrárjog tárgya, rendszere, forrásai és jogágisága. In: Fodor L, Mikó Z, Prugberger T (eds) Agrárjog I. Bíbor Publisher, Miskolc, pp 3–22

Raisz A (2010) Földtulajdoni és földhasználati kérdések az emberi jogi bíróságok gyakorlatában. In: Csák C (ed) Az európai földszabályozás aktuális kihívásai. Novotni Foundation, Miskolc, pp 241–253

Raisz A (2012) A constitution's environment, environment in the constitution: process and background of the new Hungarian constitution. Est Europa, Special edn (1):37–70

Raisz A (2013) Water as the nation's common heritage in the frame of the common heritage of mankind. In: Greksza V, Szabó M (eds) Right to water and the protection of fundamental rights in Hungary. University of Pécs, Pécs, pp 84–96

Raisz A (2015) GMO as a weapon – a.k.a. a new form of aggression? In: Szabó Marcel et al (eds) Hungarian Yearbook of International Law and European Law 2014. Eleven International Publishing, The Hague, pp 230–239

Raisz A, Szilágyi JE (2012) Development of agricultural law and related fields (environmental law, water law, social law, tax law) in the EU, in countries and in the WTO. J Agric Environ Law 7(12):107–148

Szabó M (2002) A jogi alapfogalmak. Bíbor Publisher, Miskolc

Szabó M (2012) A fenntartható fejlődés: nemzetközi jogi elmélet és szerződéses gyakorlat. In: Raisz A (ed) A nemzetközi környezetjog aktuális kihívásai. University of Miskolc, Miskolc, pp 161–174

Szilágyi P (1998) Jogi alaptan. Osiris Publisher, Budapest

Szilágyi JE (2007) Az agrárjog dogmatikájának új alapjai – útban a természeti erőforrások joga felé? Jogtudományi Közlöny 62(3):112–121

Szilágyi H, Olajos I, Loss S (2001) A jogrendszer. In: Szabó M (ed) Bevezetés a jog- és államtudományokba. Bíbor Publisher, Miskolc, pp 75–94

Tahyné Kovács A (2013) A genetikailag módosított szervezetekre vonatkozó európai és magyar jogi szabályozásról egyes környezetjogi alapelvek, különösen a fenntartható fejlődés tükrében. Dissertation, Pázmány Péter Catholic University

Tanka E (2007) Transformation of Hungarian agricultural law (1985/1990–2005). In: Allan FT, Jakab A, Takács P (eds) The transformation of the Hungarian legal order 1985–2005. Kluwer Law International, The Netherlands, pp 371–394

Téglási A (2009) How is property ownership guaranteed constitutionally in the field of agriculture? J Agric Environ Law 4(7):18–29

Veres J (1993) Egy lehetséges agrárjogi koncepció vázlata. In: Tóth K (ed) Emlékkönyv dr. Kemenes Béla egyetemi tanár 65, Születésnapjára. Comissio Scientiae Studiorum Facultatis Scientiarum Politicarum et Juridicarum Universitatis Szegediensis de Attila József nominatae, Szeged, pp 519–528

The Relationship Between Agricultural Law and Environmental Law in the United States of America

E. Dooley

Abstract Agricultural law and environmental law in the United States are individually vast areas of legislation, regulation, and jurisprudence involving a wide array of objects, actors, rights, and duties. There is also significant overlap between the two areas due to agriculture's dependence on the natural environment for production of food, feed, fiber, and fuel. In relation to the concept of agroecology, this chapter explores the legal and regulatory framework to identify how it influences agricultural practice in relation to environmental protection and the design and management of agroecosystems within the US. Laws and regulations implemented by different actors at both the federal and state/local levels are explored according to various agroecosystem components and environmental media (e.g., land, water, air). In general, the US framework for agricultural law and environmental law was found to not promote a holistic agroecology approach. Instead, the system aims to ensure a basic level of environmental protection in the design and management of agroecosystems through a fractionated approach, including multiple different regulatory schemes administered by agencies at varying levels that apply differently to various types and sizes of actors.

Keywords Agricultural law • Agroecosystem • Conservation • Environmental law • Pollution

1 Introduction

Agricultural law and environmental law in the United States are individually vast areas of legislation, regulation, and jurisprudence involving a wide array of objects, actors, rights, and duties. With regard to agriculture, the production of food, feed, fiber, and fuel is carried out in many different ways, ranging from organic to varying degrees of sustainable agriculture to intensive agriculture. Regardless, agricultural practice depends on the natural environment and can have a large impact on its quality; thus, there are significant overlaps in the public and private

E. Dooley (✉)
Ecologic Institute, Berlin, Germany
e-mail: elizabeth.dooley@ecologic.eu

M. Monteduro et al. (eds.), *Law and Agroecology*,
DOI 10.1007/978-3-662-46617-9_14

regulation of agricultural conduct in relation to the environment within the US legal system.

The concept of agroecology presents a holistic approach to the relationship between agriculture and the environment. Discussed extensively in the contribution of CAPORALI, agroecology can be defined as "the science of applying ecological concepts and principles to the design and management of sustainable agroecosystems."[1] Applying this concept to agricultural practice would necessitate a more balanced approach to promotion of ecosystem services in addition to the production of food. Moreover, the way in which the regulatory framework is structured to either promote or discourage certain types of behavior can have a significant impact on the design and management of agroecosystems. This article aims to highlight the structure of the US agricultural and environmental legal and regulatory systems with regard to various agroecosystem components and to demonstrate the varying ways in which agricultural practice is encouraged to protect or exploit the environment through that structure.

2 Background on US Agricultural and Environmental Laws

2.1 US Legal System

The United States is a federalist system with a Constitution that is the supreme law of the land.[2] It enumerates certain powers to a central federal government, and the remaining powers are within the jurisdiction of the fifty states.[3] The Constitution establishes the three branches of the federal government: executive, legislative, and judicial.[4] The branches are granted specific powers, and the principle of "separation of powers" allows each branch to act as a check and balance on the constitutionality of the other branches' actions.[5] This is intended to prevent one branch from becoming too powerful, an approach rooted in the American rejection of a monarchical system.[6]

Additionally, administrative agencies are an important part of the United States' government, having even been referred to as the "fourth branch."[7] These agencies

[1] International Assessment of Agricultural Knowledge, Science and Technology for Development (2006).

[2] Art. VI, U.S. Constitution.

[3] Calabresi (1995), pp. 752–831; Pryor (2001), pp. 1167–1182.

[4] Arts. I-III, U.S. Constitution.

[5] Levi (1976), pp. 371–391; Pryor (2001), see note 21, quoting "The Federalist No. 51, at 67 (James Madison) (Lester DeKoster ed., 1976)".

[6] Younger (1958), pp. 755–84; Ackerman (2000), pp. 633–792.

[7] Strauss (1984), pp. 573–669.

play a significant role in implementing the laws by issuing detailed regulations after "notice and comment," enforcing the regulations (e.g., compliance orders), and adjudicating regulatory violations.[8] The regulations flesh out the laws by detailing the "how" and "to whom," which lawmakers leave to the expert judgment of the administrators specializing in that area. For instance, the Clean Air Act delegates power to the Environmental Protection Agency (EPA) to create specific rules regarding the acceptable emission levels of pollutants and to design and implement a permitting scheme for installations.[9] However, the agency must proceed within the limits of the delegated power; attempting to regulate something that is beyond the scope is considered overreaching, against which injured parties can bring suit.[10]

The US judicial system is a common law system wherein court rulings set the precedent to be followed by subsequent courts under the doctrine of *stare decisis.*[11] Additionally, the courts interpret the laws passed by the legislature under the doctrine of judicial review.[12] In general, the US system has both public law, governing the conduct of individuals by the state such as through constitutional, criminal, and regulatory laws, and private law, which governs the conduct between individuals or legal entities.[13] Private law includes contracts, property, torts, and commercial law.[14] Both types of law are important within agricultural law and environmental law.

2.2 *Concept of Agricultural Law*

The most important agricultural statute in the United States is the Farm Bill, which is a comprehensive piece of legislation passed every 5–7 years with various forms of agricultural support over the years (e.g., price support, direct payments, crop insurance), conservation programs, and food assistance programs.[15] The first Farm

[8] Rosenbloom (1983), pp. 219–227 (discussing the multifaceted role of administrative agencies in terms of "managerial," "legal," and "political" functions following the separation of powers divide).

[9] Clean Air Act of 1970 (CAA), 42 U.S.C. § 7401 et seqq.

[10] Shapiro (1982), pp. 1487–1522; Seidenfeld (1999), pp. 429–495. For an interesting discussion regarding the interaction between the judicial branch's and administrative agencies' carrying out of legislative mandates, see Mikva (1986), pp. 1–9.

[11] Rehnquist (1986), pp. 345–376.

[12] Marbury v. Madison, 5 U.S. 137 (1803); Corwin (1914), pp. 538–572; Van Alstyne and Marshall (1969), pp. 1–47; Nelson (2000).

[13] Horwitz (1982), pp. 1423–1428.

[14] Horwitz (1982).

[15] Agricultural Act of 2014, H.R. 2642, Pub. L. 113–79; see Dimitri et al. (2005) The 20th Century Transformation of U.S. Agriculture and Farm Policy. Economic Research Service, United States Department of Agriculture, Economic Information Bulletin No. 3:i–14. http://www.ers.usda.gov/media/259572/eib3_1_.pdf.

Bill was passed in 1933 in response to the Dust Bowl crisis.[16] As the western part of the US was settled, the farmers plowed the sod and planted wheat, which grew abundantly with adequate rainfall. Extensive livestock production was also practiced, leading to devegetation. When a decade-long drought hit the Great Plains in 1931 (including the central western states of Kansas, Oklahoma, Texas, New Mexico, Colorado, and Nebraska), the soil was literally blown away in huge dust storms, along with farmers' crops.[17] Thus, the Agricultural Adjustment Act was passed to provide price support for commodities (based on historical prices from 1910 to 1914) and establish supply control through set-asides of land.[18]

It became clear, however, that the way the soil had been managed by planting "soil depleting" crops had contributed to the Dust Bowl.[19] Thus, in the Farm Bill passed 3 years later, payments were authorized for farmers to plant "soil conserving" crops.[20] The inclusion of this type of conservation provision gradually developed into a conservation title with multiple programs that now provide additional payments to incentivize farmers to manage their production in an environmentally sustainable way.

Many different areas of private law fall under the umbrella of agricultural law. For example, cash-rent agreements fall under contract law, where a tenant farmer rents a plot of land from the landowner. These are often oral agreements, so the conditions of the contract are based on common law, e.g., length of the contract, stewardship standards, etc.[21] Property law is highly important within agricultural law (e.g., for land transfers), incorporation of farming businesses uses commercial law, and intellectual property law governs the protection of genetically modified seeds from patent infringement. Public regulation of agriculture instead tends to focus on the intersection of agricultural production and use of natural resources (discussed below in Sect. 3.1).

2.3 *Concept of Environmental Law*

Modern environmental law stems from worsening environmental conditions and the responsive environmental movement during the 1960s. In particular, the Cuyahoga River in Ohio catching on fire due to high industrial pollution levels became

[16] Worster (1982).

[17] Worster (1982).

[18] For a historical look at the development of the Farm Bill, see McGranahan et al. (2013), pp. 67A–73A.

[19] McGranahan et al. (2013).

[20] McGranahan et al. (2013).

[21] Benschoter v. Hakes, 8 N.W.2d 481 (Iowa 1943); Pollack v. Pollack, 72 N.W.2d 483 (Iowa 1955); Denton v. Moser, 241 N.W.2d 28 (Iowa 1976); Morling v. Schmidt, 299 N.W.2d 480 (Iowa 1980); Ganzer v. Pfab, 360 N.W.2d 754 (Iowa 1985).

the symbol for the need to adopt the modern Clean Water Act of 1972 (CWA).[22] During a 4-year period from 1969 to 1972, additional major environmental statutes were adopted, including the National Environmental Policy Act (NEPA),[23] the Clean Air Act of 1970 (CAA),[24] the Endangered Species Act of 1973,[25] and the Resource Conservation and Recovery Act (regulating toxic waste disposal).[26]

Environmental law tends to be mostly public law through governmental regulation of natural and legal persons' use of and impact on the environment. Within these public law statutes, agriculture is actually often exempted or given safe harbors of operation.[27] For instance, the Clean Water Act focuses primarily on industrial point-source pollution rather than agricultural pollution that is diffuse, smaller in scale, and not similarly released from a fixed pipe (see below Sect. 3.2 for discussion of nonpoint source regulation).[28] However, in terms of private law, environmental effects from agriculture could be challenged as nuisance and trespass, e.g., private suits brought by neighbors claiming that the smell from a pig operation restricts their use and enjoyment of their property. Recently, however, several states have adopted "right-to-farm" statutes, which offer farmers a statutory defense in nuisance suits challenging their agricultural activities as long as they are in compliance with applicable legislation and regulations.[29] Some states have also adopted fee-shifting provisions to deter people from filing nuisance suits, which require the plaintiff to pay the farmer's attorney fees and costs if the farmer successfully defends the suit.[30]

3 Regulatory Framework for Various Agroecosystem Components

3.1 *Land/Soil Management*

As mentioned above (Sect. 2.1), the federal government was granted enumerated powers under the Constitution to regulate issues affecting interstate commerce,

[22] Clean Water Act of 1972 (CWA), 33 U.S.C. § 1251 et seqq.; Meiners and Morriss (2000).

[23] National Environmental Policy Act of 1969 (NEPA), 42 U.S.C. § 4321 et seqq.

[24] See note 8.

[25] Endangered Species Act of 1973 (ESA), 16 U.S.C. § 1531 et seqq.

[26] Resource Conservation and Recovery Act of 1976 (RCRA), 42 U.S.C. § 6901 et seqq.

[27] Ruhl (2000), pp. 263–349.

[28] House (2006).

[29] Rumley (2014) States' Right-To-Farm Statutes. National AgLaw Center Research Publication. http://nationalaglawcenter.org/state-compilations/right-to-farm/. Accessed 20 Oct 2014.

[30] New York Right-To-Farm statute, N.Y. Agric. & Mkts. §§ 300–310; Bormann v. Board of Supervisors in and for Kossuth County, 584 N.W.2d 309 (Iowa 1998); Farm Foundation, Right-To-Farm Laws (2014) History & Future. https://www.farmfoundation.org/news/articlefiles/129-hipp.pdf. Accessed 20 Oct 2014.

taxes and spending, immigration, military, etc.[31] Issues outside of the federal enumerated powers are governed by the states, which then may be delegated even further to the local level. Land use is generally an issue within the states' power, and often local governments are delegated control over cities' and/or counties' comprehensive planning of different types of uses (e.g., industrial, residential, commercial) through planning and zoning commissions.[32] At the same time, environmental issues are generally under federal jurisdiction due to the interstate nature, though the EPA and a state may sign a Memorandum of Understanding delegating federal authority to the state to implement the regulation.[33]

Thus, direct regulation of agricultural activities is often determined at the local level in terms of best management practices, which are highly contextual. For example, the Iowa Code establishes the Soil Conservation Division under the Iowa Department of Agriculture and Land Stewardship, which is mandated to assist the 100 Soil and Water Conservation Districts throughout the state with developing district soil and water resource conservation plans.[34] The Natural Resources Conservation Service (NRCS) under the federal Department of Agriculture offers assistance to private landowners (not covered by federal mandatory conservation compliance provisions) through Conservation Technical Assistance, but as it is not within the federal jurisdiction, the measures are voluntary.[35]

However, the federal Farm Bill does include both mandatory restrictions for production on certain types of land and voluntary conservation programs. The Sodbuster and Swampbuster conservation compliance measures require landowners who have highly erodable land (HEL) and/or wetlands to create and abide by a conservation management plan developed with the NRCS and to not convert wetlands.[36] Compliance is now tied to crop insurance premium subsidies since direct payments were eliminated by the 2014 Farm Bill, so farmers who do not come into compliance lose their eligibility for federally subsidized crop insurance.[37] Voluntary programs include the

[31] Art. I, Section 8, U.S. Constitution.

[32] Mandelker (1976), pp. 899–973; Nolan (2002), pp. 365–416.

[33] Houck and Rolland (1995), pp. 1242–1314 (discussing various delegations of authority); Gutherz (2011), pp. 289–320 (presenting an interesting discussion of the recent court findings leading to uncertainty about the constitutionality of environmental statutes and how they could affect adoption of climate change legislation).

[34] Iowa Code § 161A.4 ("measures including but not limited to the control of floods, the control of erosion by water or by wind, the preservation of the quality of water for its optimum use for agricultural, irrigation, recreational, industrial, and domestic purposes, all of which shall be presumed to be conducive to the public health, convenience, and welfare, both present and future").

[35] See Natural Resources Conservation Service, Conservation Planning (2014). http://www.nrcs.usda.gov/wps/portal/nrcs/main/national/technical/cp/. Accessed 20 Oct 2014.

[36] Food Security Act of 1985, Public Law No. 99–198, 99 Stat. 1354; Hamilton (1989), pp. 637–674; Malone (1988).

[37] Agricultural Act of 2014, H.R. 2642, Pub. L. 113–79, § 1118.

- Conservation Reserve Program: the landowner contracts with the government to take HEL or environmentally sensitive land out of production and convert it to vegetative cover for ca. 15 years.
- Conservation Stewardship Program: the landowner takes on extra conservation measures under 5-year contracts with the NRCS.
- Environmental Quality Incentives Program: the landowner must demonstrate conservation practices and activities from an EQIP plan of operations under a contract up to 10 years.
- Agricultural Conservation Easement Program: financial assistance up to 50 % of the market value of the agricultural land may be contributed to help protect productive farmland from conversion to other uses.[38]

3.2 *Water Management*

Agriculture is a major consumer of water through irrigation. Some of the leading agriculturally producing states in the country (e.g., California and the Great Plains, the latter notably being where the Dust Bowl occurred (see above Sect. 2.2) are highly dependent on irrigation.[39] In terms of extraction from local water bodies for irrigation purposes, water rights vary between the states and indeed by the part of the country. In the eastern part of the country, states follow the riparian doctrine, which recognizes anyone whose property borders a watercourse as having riparian rights to use water from it.[40] The western part of the US follows the prior appropriation doctrine, which grants water rights to the user who first put the water to beneficial use.[41]

Agriculture also impacts water quality through run-off of soil particles and nutrients into surface waters and leaching of nutrients into groundwater. These contributions from agriculture are classified as nonpoint source pollution under the CWA as opposed to point source pollution, e.g., from industrial actors discharging pollutants into water bodies.[42] Generally, point sources are regulated under the statute by the National Pollutant Discharge Elimination System (NPDES), requiring certain emitters to obtain a permit to discharge pollutants.[43] Nonpoint sources are regulated separately as diffuse, typically small-scale contributors, and each state has the authority under the federal statute to implement a nonpoint source pollution program.[44] These programs cover the identified water bodies that cannot be

---

[38] *Id.* §§ 2001–2508.

[39] See Moore et al. (1996), pp. 319–357.

[40] United States v. Gerlach Live Stock Co., 339 U.S. 725 (1950); Beuscher (1960), pp. 448–458; Ausness (1982), pp. 547–590.

[41] Bean v. Morris, 221 U.S. 485 (1911).

[42] Clean Water Act of 1972 (CWA), 33 U.S.C. § 1329.

[43] *Id.* § 1342.

[44] *Id.* § 1329(a).

reasonably expected to meet the state's water quality standards for the total maximum daily load of pollutants without addressing nonpoint sources' contribution.[45] Best management practices and measures are identified within the programs to help reduce nonpoint source pollution, which acts as a public form of regulation over agricultural activities.[46]

One exception to agriculture's general treatment as a nonpoint source is the inclusion of confined animal feeding operations (CAFOs) under the CWA's NPDES permitting scheme due to the high amount of manure generated and their discharge of pollutants to water bodies.[47] CAFOs are classified by the EPA as small, medium, or large, depending on how many animals are housed by the operation, which changes the requirements for method of discharge.[48] The permitting system generally aims to reduce environmental impacts from pollution discharges into water, but particularly for CAFOs, the requirements for adequate manure storage (e.g., lagoons) are important in case heavy rainfall events occur in order to prevent overflow and contamination.[49]

Finally, agriculture has also historically manipulated waterways, e.g., straightening streams to allow for more arable production. The CWA regulates the act of dredging and filling applicable wetlands and surface waters by requiring a permit to "discharge" dredge or fill material into "navigable waters."[50] If a wetland is authorized for conversion to another use, the permit holder may be required to complete some form of compensatory mitigation for unavoidable impacts.[51] "Wetland mitigation banking" is one example whereby the permit holder either creates or restores a wetland elsewhere or contributes funds to another third-party wetland mitigation bank.[52] An argument for mitigation banking is that larger, maintained wetlands are able to be developed in order to provide an offset for multiple converted wetlands.[53] An argument against this concept though is that the converted wetland was a habitat for flora and fauna, and simply because another wetland is created elsewhere, it does not mean that the same biodiversity can or will be shifted to another location.[54]

---

[45] *Id.*

[46] *Id.* § 1329(b).

[47] *Id.* § 1362(14).

[48] 40 C.F.R. 122.23; EPA (2003).

[49] Warrick (1995).

[50] See note 21, §1344 ("discharge" in this case refers to some of the dredged earth falling back onto the wetland as it is being excavated and/or fill material being put onto the land to alter the wetland).

[51] 33 C.F.R. 320.4(r); 40 C.F.R. 230.

[52] See EPA, Mitigation Banking Factsheet (2014) Compensating for Impacts to Wetlands and Streams. http://water.epa.gov/lawsregs/guidance/wetlands/mitbanking.cfm. Accessed 20 Oct 2014.

[53] See Silverstein (1994), pp. 129–161.

[54] See Steinhoff (2008), pp. 1–11. But see Spieles (2005), pp. 51–63 (finding through assessment of multiple nonnatural wetlands that they have undergone processes that suggest they may become vegetation equivalent with natural wetlands).

Administered by the Army Corps of Engineers, the dredge and fill permitting scheme specifically exempts normal farming activities "such as plowing, seeding, cultivating, minor drainage, harvesting for the production of food, fiber, and forest products, or upland soil and water conservation practices" from coverage.[55] Additionally, the scope of waters within the scheme's jurisdiction has been the subject of various court interpretations, yet uncertainty remains (e.g., whether intermittent streams that only have flowing water after heavy rainfall are included).[56] In general, the term "navigable waters" is understood to encompass "interstate waters, plus waters that are navigable, wetlands adjacent to navigable waters and other waters with a substantial connection to navigable waters."[57] However, the EPA has currently proposed a rule to clarify which waters are covered as navigable waters that need dredge and fill permits based on a recent study of the connectivity between water bodies.[58]

This is a contested issue between agricultural practice and environmental law. The permitting system is intended to prevent unauthorized conversion of wetlands and protect the environment. Regulatory uncertainty as to which bodies of water will be covered (especially if the water only appears rarely during the year) makes it difficult for farmers to know whether they will receive a compliance order from the EPA to restore a piece of their land if they dredge or fill it.[59] However, aversion to regulatory control of small waterways on farms can be seen in the current debates over the proposed "Waters of the United States" rule since the process to obtain a Section 404 permit for every small stream on the farm would be timely and expensive and does not demonstrate an obvious connection to interstate navigable waters.[60] During the notice and comment period of administrative rulemaking, hundreds of thousands of comments have been submitted, expressing concerns that the jurisdiction would be extended to more waters than are already covered and that agriculture activities would no longer be exempt.[61] The EPA insists, however, that the rule will actually benefit farmers and will not encompass any

---

[55] Clean Water Act of 1972 (CWA), 33 U.S.C. § 1344(f).

[56] Most recently, two cases were decided that caused ambiguity in terms of whether the Army Corps of Engineers' jurisdiction would extend to certain water bodies. Solid Waste Agency of Northern Cook County v. United States Army Corps of Engineers, 531 U.S. 159 (2001); Rapanos v. United States, 126 S. Ct. 2208 (2006).

[57] Gruenhagen (2014), pp. 14–15 (on file with author).

[58] Gruenhagen (2014).

[59] However, see http://water.epa.gov/lawsregs/guidance/wetlands/upload/2008_12_3_wetlands_CWA_Jurisdiction_Following_Rapanos120208.pdf for guidance issued by EPA following Rapanos to provide clarity about the agency's jurisdiction over waters.

[60] See, e.g., American Farm Bureau (2014), It's Time to Ditch the Rule. http://ditchtherule.fb.org/ (advocating for members to submit comments to the EPA advising against approval of the proposed rule). Accessed 20 Oct 2014.

[61] See, e.g., Fatka (2014) (discussing the legislative act proposed in the House of Representatives aimed at blocking the EPA from adopting the Waters of the US rule); The Hagstrom Report (2014) EPA responds to SBA Advocacy office on WOTUS. http://www.hagstromreport.com/2014news_files/2014_1002_epa-responds-sba-advocacy-office-wotus.html; Traxler (2014) Proposed EPA

new waters, though there is conflicting information suggesting that there would be a (potentially large) increase in coverage.[62] This example demonstrates the tension that can exist between agricultural practice and environmental protection within the US system, complicating legislative and regulatory coverage of some issues.

3.3 Air/Particulate Matter

The Clean Air Act generally establishes a national framework for state-run programs that help to reduce emissions of specified pollutants in order to achieve the National Ambient Air Quality Standards (NAAQs).[63] Agricultural operations are mostly small scale enough that they are outside the scope of the statute, but the EPA adopted the Air Quality Compliance Agreement for CAFOs due to the large-scale amount of emissions released from manure storage and the potential impact on air quality.[64] In addition, in 2009 EPA adopted a Mandatory Reporting of Greenhouse Gases (GHGs) rule for all major emitters of at least 25,000 metric tons of greenhouse gases (CO_2-equivalent) per year.[65] The application of that rule to agriculture was also controversial due to concerns that the regulation would overreach and cover operations' fossil fuel emissions from tractor use or similar equipment.

3.4 Genetically Modified Organisms

Genetically modified organisms (GMOs) in the form of seeds used for crop production occupy an interesting place in the US agricultural and environmental regulatory scheme. In general, once the products are harvested, e.g., maize, soybeans, tomatoes, etc., the Food and Drug Administration (FDA) considers them to be substantially equivalent to products from traditionally bred plants.[66] Thus, the FDA requires no special labelling for products containing GMOs. However, the process of approving new varieties of GMOs for planting does trigger NEPA.

water rules worry farmers, Prairie Business. http://www.prairiebizmag.com/event/article/id/21216/.

[62] See note 53 (highlighting the extensive interpretation that can be drawn from the EPA's Connectivity Study that all waters are connected and thus all waters could potentially be covered); EPA, Waters of the US. (2014). http://www2.epa.gov/uswaters. Accessed 20 Oct 2014.

[63] See note 8, §§ 7409–7410.

[64] See Hoover (2013), pp. 1–29 (outlining the way in which the CAA and NAAQs in particular apply to CAFOs).

[65] Federal Register 74(209):56481, Friday, Oct 30, 2009, Subpart JJ, § 98.360 et seqq.

[66] Federal Register 57(104):22984, Friday, 29 May 1992, Section VII Guidance to Industry for Foods Derived from New Plant Varieties. http://www.fda.gov/Food/GuidanceRegulation/GuidanceDocumentsRegulatoryInformation/Biotechnology/ucm096095.htm.

NEPA is the environmental statute that governs federal actions that may have a significant impact on the environment by requiring integration of this consideration into the government agency's planning and decision-making process.[67]

First, the Plant Protection Act of 2000 mandates that the Secretary of Agriculture, who delegated the authority to the Animal and Plant Health Inspection Service (APHIS), create regulations to prevent the "introduction of plant pests into the United States or the dissemination of plant pests within the United States."[68] Thus, APHIS issued regulations regarding GMOs believed to be plant pests, which require the company or the individual who seeks approval of the GMO to petition APHIS to determine whether it presents a plant pest risk or whether it is not covered by the regulations.[69] If APHIS gives the GMO a nonregulated status, then NEPA is triggered to provide a procedural check. An environmental assessment is prepared by APHIS, and if a finding of no significant impact is determined, then a full environmental impact statement does not need to be completed.[70] *Monsanto Co. v. Geertson Seed Farms* is an example of a challenge to this procedure, finding that the environmental assessment was inadequate and an environment impact statement should have been done instead of just deregulating the new GMO.[71]

Despite this public regulatory procedure, GMO approval ends up being somewhat of a formality as suggested by the fact that GMOs are approved and planted on a very wide scale in the United States.[72] One private law aspect of GMOs that has potentially significant environment implications as well as agricultural effects on other farmers' fields is the patent for genetic manipulation. Not unique to agricultural or environmental law, intellectual property law protects the GMO developer from (even unintentional) patent infringement. Thus, farmers are not allowed to save seeds to plant in the following year pursuant to the licensing agreement with the company to use their patented seeds, but also crops on an organic farm close to a GMO farm could become contaminated with the GMO genes through cross-pollination.[73] In the latter case, the developer will win the patent infringement case against the organic farmer under a strict standard of liability, but Monsanto Company, for example, does not sue farmers whose fields have accidentally become contaminated.[74]

[67] See note 23.

[68] 7 U.S.C. § 7711.

[69] 7 C.F.R. 340.

[70] 7 C.F.R. 372.

[71] Monsanto Co. et al. v. Geertson Seed Farms et al., 130 S. Ct. 2743 (2010).

[72] The US Department of Agriculture (USDA) National Agricultural Statistics Service (NASS) in 2012 found that 88 % of all corn, 94 % of cotton, and 93 % of the soybeans planted were biotechnology crops. USDA, Biotechnology Frequently Asked Questions (FAQs). http://www.usda.gov/wps/portal/usda/usdahome?navid=AGRICULTURE&contentid=BiotechnologyFAQs.xml. Accessed 20 Oct 2014.

[73] Bowman v. Monsanto Co., 133 S. Ct. 1761 (2013).

[74] Reuters, Top U.S. court refuses to hear appeal of Monsanto see case, Monday, 13 Jan 2014. http://www.reuters.com/article/2014/01/13/usa-court-monsanto-idUSL2N0KN1CA20140113. Accessed 20 Oct 2014.

This natural movement of plant genes to other areas and plants is a widespread concern of critics of GMOs due to the development of "superweeds" and reduction of genetic diversity in crop species.[75] However, it is important to recognize in the US context that there are also strong proponents of GMOs. They argue that environmental benefits are gained through reduced trips across the field in machinery, translating into fewer greenhouse gas emissions and better soil structure due to avoided compaction; decreased spraying of pesticides and herbicides due to the plant's resistance to pests, weeds, and diseases; and adoption of conservation tillage for less soil erosion and increased soil quality due to the ability to combat weeds with herbicide since the seed is resistant.[76]

3.5 *Pesticides*

Regulation of pesticides in US agriculture is another area where environmental law is highly relevant to avoid "unreasonable adverse effects on people and the environment" from their production, sale, and use.[77] The Federal Insecticide, Fungicide, and Rodenticide Act (FIFRA) requires registration and labelling before a pesticide can be sold, and the assessment as to whether a pesticide should be approved for certain uses rests on a cost-benefit analysis of the scientific data about the potential environmental, social, and economic impacts.[78] Pesticide application levels must then conform to the labelling amounts and methods, which for most farmers applying on a noncommercial basis do not require a special license.[79] Additionally, the EPA started the Pesticide Environmental Stewardship Program in 1994 to try to encourage users to reduce their pesticide application and employ more sustainable techniques such as integrated pest management through soft law measures.[80]

4 Concluding Remarks

The discussion above points to key topics, legislation, and regulations within the US that integrate both agricultural law and environmental law. Environmental law is heavily focused on public regulation of conduct in order to protect natural resources, and agricultural law encompasses public law regulations regarding treatment of wetlands, environmentally sensitive land, pollutant discharges into water and

[75] Qiu (2013) and Entine and Lim (2014).

[76] See Hails (2000), pp. 14–18.

[77] Federal Insecticide, Fungicide, and Rodenticide Act of 1947 (FIFRA), 7 U.S.C. §§ 136–136y.

[78] *Id.*

[79] See, e.g., 21 Iowa Code § 45.29.

[80] EPA, PestWise: An EPA Partnership Program (2014). http://www.epa.gov/pestwise/pesp/. Accessed 20 Oct 2014.

emissions into the air, approval processes for GMOs and pesticides, etc. However, private claims of action are also available and influence agricultural practice, e.g., for breach of a rental contract, nuisance suits, patent infringement, etc.

As seen from the assessment of the regulatory framework above, one may conclude that it generally only aims to ensure a basic level of environmental protection in the design and management of agroecosystems. Actions aimed at a higher level of protection often depend on an incentive system (e.g., conservation programs under the Farm Bill), which are also susceptible to change based on the political climate or availability of funding. Additionally, the regulatory structure contributes to a nonholistic approach through the separate regulation of various agroecosystem components, not only by different agencies but also between different levels of government and applied differently to various actors. For instance, the EPA issues rules regarding pollution standards for water and air that are binding from the federal level for some agricultural actors but for others are based on best practice at the local level. At the same time, other rules for treatment of water bodies are issued by the Army Corps of Engineers (i.e., Section 404 permits for dredging and filling wetlands) and may affect local actions to varying degrees based on the interpretations applied to statutory terms (e.g., the "navigable waters" controversy and pending "Waters of the US" rule from EPA). The USDA also enforces mandatory restrictions on wetland conversion under the Swampbuster program. Additionally, US state courts may intervene with opinions that change the way water rights are upheld, causing variations between different parts of the country as well as federal courts declaring agency actions beyond the scope of delegated authority or issuing a statutory interpretation that alters agency action.

Thus, the US legal and regulatory framework for agriculture and the environment does not promote a holistic approach according to the agroecology concept but rather takes a fractionated approach to minimizing environmental effects from agricultural production. Integrating regulatory actors' processes and approaches to try to make the framework more comprehensive would likely not be easy due to institutional structures and potentially coveted responsibilities, but increasing levels of cooperation must be considered and adopted in order to effectively address multifaceted issues within broader agroecosystems.

Acknowledgment The author would like to thank John Baker, J.D. of the Beginning Farmer Center, Iowa State University Extension for his generous assistance and advice in writing this chapter.

References

Ackerman B (2000) The new separation of powers. Harv Law Rev 113(3):633–792

American Farm Bureau (2014) It's time to ditch the rule. http://ditchtherule.fb.org/. Accessed 20 Oct 2014

Ausness R (1982) Water rights legislation in the east: a program for reform. William Mary Law Rev 24:547–590

Beuscher JH (1960) Appropriation water law elements in Riparian doctrine States. Buffalo Law Rev 10:448–458
Calabresi SG (1995) A government of limited and enumerated powers: in defense of United States v. Lopez. Mich Law Rev 94(3):752–831
Corwin ES (1914) Marbury v. Madison and the doctrine of judicial review. Mich Law Rev 12 (7):538–572
Dimitri C et al (2005) The 20th century transformation of U.S. agriculture and farm policy. Economic Research Service, United States Department of Agriculture, Economic Information Bulletin No. 3:i-14. http://www.ers.usda.gov/media/259572/eib3_1_.pdf
Entine J, Lim X (2014) Union of concerned scientists blames GMOs for 'superweeds' but issue more complex. Genetic Literacy Project. http://www.geneticliteracyproject.org/2014/05/06/video-union-of-concerned-scientists-blames-gmos-for-superweeds-but-issue-more-complex/
EPA, Mitigation Banking Factsheet (2014) Compensating for impacts to wetlands and streams. http://water.epa.gov/lawsregs/guidance/wetlands/mitbanking.cfm. Accessed 20 Oct 2014
EPA, PestWise: an EPA partnership program (2014) http://www.epa.gov/pestwise/pesp/. Accessed 20 Oct 2014
EPA (2003) Producers' compliance guide for CAFOs: Revised Clean Water Act Regulations for Concentrated Animal Feeding Operations (CAFOs), EPA 821-R-03-010. http://www.epa.gov/rfa/documents/Compliance-CAFOs.pdf. Accessed 20 Oct 2014
EPA, Waters of the US (2014) http://www2.epa.gov/uswaters. Accessed 20 Oct 2014
Farm Foundation (2014) Right-to-farm laws: history & future. https://www.farmfoundation.org/news/articlefiles/129-hipp.pdf. Accessed 20 Oct 2014
Fatka J (2014) House sends strong message against EPA's water rule. Feedstuffs. http://feedstuffs.com/story-house-sends-strong-message-against-epas-water-rule-45-117479
Gruenhagen C (2014) Environmental law conference regulatory update. Iowa State Bar Association
Gutherz IW (2011) Comment, cap and trade meets the interstate commerce clause: are greenhouse gas regulations constitutional after Lopez and Morrison? Pace Environ Law Rev 29:289–320
Hails RS (2000) Genetically modified plants – the debate continues. TREE 15:14–18
Hamilton ND (1989) Legal issues in enforcing federal soil conservation programs: an introduction and preliminary review. U C Davis Law Rev 23:637–674
Hoover JN (2013) Can't you smell that smell? Clean Air Act fixes for factory farm air pollution. Stanford J Anim Law Policy 6:1–29
Horwitz MJ (1982) Comment, the history of the public/private distinction. Univ Pa Law Rev 130:1423–1428
Houck OA, Rolland M (1995) Federalism in wetlands regulation: a consideration of delegation of Clean Water Act Section 404 and related programs to the States. Maryland Law Rev 54:1242–1314
House FJ (2006) Agricultural programs, terms and laws. Nova Publishers, New York
International Assessment of Agricultural Knowledge, Science and Technology for Development (IAASTD) (2006) Agriculture at a Crossroads. Global report of the international assessment of agriculture knowledge, Science and technology for development. Island Press, Washington
Levi EH (1976) Some aspects of separation of powers. Columbia Law Rev 76(3):371–391
Malone LA (1988) Swampbuster, Sodbuster, and Conservation Compliance Programs. Popular media, Agricultural law update, Paper 103. http://scholarship.law.wm.edu/cgi/viewcontent.cgi?article=1105&context=popular_media
Mandelker DR (1976) The role of the local comprehensive plan in land use regulation. Mich Law Rev 74:899–973
McGranahan DA et al (2013) A historical primer on the US farm bill: supply management and conservation policy. J Soil Water Conserv 68(3):67A–73A
Meiners RE, Morriss AP (eds) (2000) The common law and the environment: rethinking the statutory basis for modern environmental law. Rowman & Littlefield, Lanham

Mikva AJ (1986) How should the courts treat administrative agencies? Am Univ Law Rev 36 (1):1–9
Moore MR et al (1996) Water allocation in the American west: endangered fish versus irrigated agriculture. Nat Resour J 36:319–357
Natural Resources Conservation Service, Conservation Planning (2014) http://www.nrcs.usda.gov/wps/portal/nrcs/main/national/technical/cp/. Accessed 20 Oct 2014
Nelson WE (2000) Marbury v. Madison: the origins of judicial review. University press of Kansas, Lawrence
Nolan JR (2002) In Praise of Parochialism: the advent of local environmental law. Harv Environ Law Rev 26:365–41
Pryor WH Jr (2001) Madison's double security: in defense of federalism, the separation of powers, and the Rehnquist court. Ala Law Rev 53(4):1167–1182
Qiu J (2013) Genetically modified crops pass benefits to weeds: herbicide resistance could confer an advantage on plants in the wild. Nature. doi:10.1038/nature.2013.13517
Rehnquist JC (1986) The power that shall be vested in a precedent: stare decisis, the constitution and the supreme court. Boston Univ Law Rev 66:345–376
Rosenbloom DH (1983) Public administrative theory and the separation of powers. Public Adm Rev 43(3):219–227
Ruhl JB (2000) Farms, their environmental harms, and environmental law. Ecol Law Q 27 (2):263–349
Rumley ER (2014) States' right-to-farm statutes. National AgLaw Center Research Publication. http://nationalaglawcenter.org/state-compilations/right-to-farm/. Accessed 20 Oct 2014
Seidenfeld M (1999) Bending the rules: flexible regulation and constraints on agency discretion. Adm Law Rev 51(2):429–495
Shapiro M (1982) Administrative discretion: the next stage. Yale Law J 92:1487–1522
Silverstein J (1994) Taking wetlands to the bank: the role of wetland mitigation banking in a comprehensive approach to wetlands protection. Boston Coll Environ Aff Law Rev 22:129–161
Spieles DJ (2005) Vegetation development in created, restored, and enhanced mitigation wetland banks of the United States. Wetlands 25:51–63
Steinhoff G (2008) Wetlands mitigation banking and the problem of consolidation. Electron Green J 27:1–11
Strauss PL (1984) The place of agencies in government: separation of powers and the fourth branch. Columbia Law Rev 84(3):573–669
The Hagstrom Report (2014) EPA responds to SBA advocacy office on WOTUS. http://www.hagstromreport.com/2014news_files/2014_1002_epa-responds-sba-advocacy-office-wotus.html
Traxler M (2014) Proposed EPA water rules worry farmers. Prairie Business. http://www.prairiebizmag.com/event/article/id/21216/
Van Alstyne WV, Marshall J (1969) A critical guide to Marbury v. Madison. Duke Law J 1969 (1):1–47
Warrick J (1995) Hog-waste spill fouls land, river in Onslow. The News & Observer of Raleigh, North Carolina. http://www.learnnc.org/lp/editions/nchist-recent/6172
Worster D (1982) Dust bowl: the Southern Plains in the 1930s. Oxford University Press, New York
Younger I (1958) Congressional investigations and executive secrecy: a study in the separation of powers. Univ Pittsbg Law Rev 20:755–784

Crossing the Boundaries Between Agricultural Law and Landscape Law: The Rural Landscape

M. Brocca

Abstract This chapter aims to highlight some juridical links between "landscape" and "agriculture". The analysis of the legislation allows us to show the various features these close links have known during the time, moving from indifference or even opposition between the two areas of interest and coming to achieve a gradual harmonisation and integration. The first Italian regulations concerning landscapes used to neglect agricultural issues, mainly because they were inspired by aesthetic concerns, aiming to protect the Kant's "natural beauty". The contemporary concept of landscape (established by the European Landscape Convention in 2000 and, on the domestic side, by the Italian Code on Cultural Heritage and Landscape in 2004) is inclined to highlight both interaction between man and the environment and the importance of landscapes in building cultural identity, also as a consequence of the impact of agricultural activities on the land.

In this line, the analysis of various regulatory frameworks (about town planning, landscape, agriculture, agri-tourism, conservation of nature, etc.) will allow us to point out the legal concept of "rural landscape" in the light of its characteristic legislation.

Keywords Agricultural law • Integration • Landscape law • Rural landscape

1 The Terms of the Relationship

The rural landscape is one of the most eloquent expressions of landscape. It sums up and highlights landscape's different components: historical, cultural, environmental, aesthetic and economic. According to the famous definition by Sereni E., it is the "shape that man, during agricultural production activities, and in order to fulfill them, gives to the natural landscape consciously and systematically".[1] It is, therefore, the expression of an indissoluble combination whose elements feed and

[1] Sereni (1961), p. 29 (author's translation).

M. Brocca (✉)
Department of Legal Sciences, University of Salento, Lecce, Italy
e-mail: marco.brocca@unisalento.it, http://www.scienzegiuridiche.unisalento.it/scheda_personale/-/people/marco.brocca

M. Monteduro et al. (eds.), *Law and Agroecology*,
DOI 10.1007/978-3-662-46617-9_15

condition each other: agriculture transforms and shapes the land (and this transformation reflects the economic and social needs and the production techniques of the time), and the land affects agricultural activities, encouraging some productions and discouraging other ones, by virtue of its invariant morphology and also of the *cultural* and *identity* vision that local populations have of land.

The rural landscape is a result and a representation of mixing and co-evolving of various factors (determined by the technological, economic, social situation and also by the natural, climatic, soil condition of a certain place at a certain age), and, thanks to this wealth of interdisciplinary information, it also becomes a privileged observatory of the legal world, in particular about "the evolution of legal relationships that the law has built in the constant search for some sort of reconciliation of the labour and property interests".[2] From the legal point of view, the relations between *landscape* and *agriculture* have changed considerably over the time, from an initial indifference (or even opposition) to an increasing convergence and integration. So the rural landscape has become a synthesis and an expression of a conceptual rather than a prescriptive evolution of the reference matter (agricultural law and landscape law). Since agricultural law is based on the biological processes for the production of living plants or animals,[3] it needs to intercept new perspectives and subject matters corresponding to the evolution of agriculture. Multifunctional agriculture is, principally, "an agriculture with a leading part in a model of sustainable development, with an active role in landscape and environmental protection".[4] It is, therefore, an agriculture that does not give up its primary objective (that is to acquire food supplies) but that is oriented towards forms of environmental sustainability, involving the selection of agricultural land use and the preservation of certain land resources.

The rural landscape is a significant evidence that the law is no longer anchored to a purely aesthetic canon, but it reflects a historical and an anthropological principle, that is to say, "a land expressing identity whose characters derive from the natural and human factors and their reciprocal interrelations"[5] according to the current

[2] Ferrucci (2007b), pp. 451–456, 453 (author's translation), who reminds the remarks made by Sereni (1961), p. 293, about the impact of sharecropping on landscape; the features and effects of the agrarian reform policy, thanks to land reclamation (pp. 423 et seqq.); and the crushing of latifundism (pp. 439 et seqq.). See also Predieri (1981), p. 513, who emphasises the mutual influence between landscape and regulation so that the former can also be shaped as a result of standardization introduced by the latter.

[3] Carrozza (1994), pp. 151 et seqq.

[4] Adornato (2004), pp. 5–9, 5. As for the multifunctionality of agricultural law, which also includes the environmental safeguard, see Galloni (2001), pp. 5 et seqq. About the expansive *vis* of agricultural law, which reflects and is affected by the change of economic and social situation (with the *caveat*, however, that the essence of the matter cannot be transfigured), see Jannarelli (2006), pp. 183–203; Jannarelli (2008), pp. 3–14; Costato (2001), pp. 3–17, 14; Costato (2008), pp. 15–24. As in Carrozza (2007), pp. 495 et seqq., 497, states that "agricultural law is not overturned by environmental law, it is not absorbed by nor to be confused with the environmental law. The environmental requirements will have to integrate with farming" (author's translation). See also Cristiani (2008), pp. 464–479.

[5] Author's translation.

legal definition in Italy (Art. 131, Para. 1, Legislative Decree 42/2004). There is no doubt that the conformation of territory as a result of certain techniques of agricultural cultivation and production and the so-called rural architecture (the building of facilities functional to farm practices) are today perceived and recognised as typical and fundamental elements of the landscape.

2 Models in a Changing Relation

The first Italian organic law on the protection of the landscape (Law 1497/1939, so-called Bottai Law) totally ignored the rural component of landscape, protecting only the "natural beauties", namely areas of great value and rarity from an aesthetic and a cultural point of view, such as geological peculiarities, villas, gardens and parks of uncommon beauty, scenic beauties. It reflected and consolidated the approach that had inspired the first regulatory action,[6] according to which the natural environments rising to public interest are the ones able to provide aesthetic enjoyment and whose "beauty has been recognised and consecrated by literature, painting and other forms of art or which have been scenery of historical events".[7]

Such an approach inevitably left in the shade the agricultural conformation of land or, at most, tended to identify and play down the rural landscape in places such as gardens, parks and villas important from an aesthetic and contemplative point of view, often idealised by travellers in the Grand Tour as perfect products of the interaction between man and nature.

---

[6] The reference is, in particular, to Law 778/1922 (so-called Croce Law), which submits to a special protection "properties whose preservation has a significant public interest because of their natural beauty and their special relationship with the civil and literary history" and the "scenic beauty"—author's translation—(Art. 1). According to the proponent minister, Benedetto Croce—quoted by Cartei (2003), pp. 2110 et seqq.,—the defence of natural beauty meets the need to protect the public interest, "which is identified with the interest underpinned to the laws which protecting monuments and literary and artistic property" (author's translation). The legal protection of the landscape as an "extension" of the historical and artistic heritage, because of substantial similarities between *art* and *nature*, appears in the draft of Law 364/1909 on the inalienability of antiquities and fine arts (the so-called Rava-Rosadi Law), whose bill provided for the inclusion of "gardens, forests, landscapes, waters and all those places and natural objects which have the above-mentioned interest" (author's translation) in the list of assets to be protected according to Art. 1. The category of "natural monuments" will be expunged from the final text during the debate in the Senate for the incomplete assimilability of certain natural environments to the artistic monuments (Acts of Parliament, doc. no. 760-A, the Senate of the Kingdom, legisl. XII, first session, 1904–1907). In relation to Law 1497/1939, it should be noted—as underlined by Cartei (2003), p. 2111—that the criterion informing the categories of protected goods, of aesthetic and cultural kind, is not the only one, as also a scientific criterion emerges (as for the reference to properties with substantial geological peculiarities) and a socio-historical one (for the reference to properties having a distinctive aesthetic and traditional value).

[7] Ventura (2001), pp. 555–576 (author's translation).

It was really ignored the ability of agriculture to "create" landscapes; on the contrary, the attention (and fear) to possible deleterious effects of farming practices on consolidated landscapes appeared. In this sense, we can read the legislative initiative of the early twentieth century aiming to protect the pine forest in Ravenna (Law 411/1905, so-called Rava Law) as a place of Italian literary and artistic memory.[8] The need to preserve the cultural importance of the pine forest in Ravenna reveals the will to protect that portion of territory against the agrarian reforms of the time that, through land reclamation and deforestation, tried to satisfy the hunger for land and labour of the farm workers.[9]

Law 431/1985 (so-called Galasso Law), which marked an important evolution in the landscape law, still left the rural landscape in the shade. While protecting areas very important in terms of agricultural activities (land allocated to "agrarian universities", areas subject to civic uses, highlands, woods and forests), the Galasso Law enhanced the environmental dimension of the landscape by typifying the so-called areas of environmental concern, vast portions of the national territory—subjected to constraints established *ex lege*—relevant in terms of environmental protection for their morphological and geographical characteristics (e.g., coastlines, rivers, glaciers, wetlands, and so on). As stated by the Italian Constitutional Court, Law 431/1985 has moved "attention from the natural beauties, designed as an (only) aesthetic dimension of the land, to the environmental goods".[10]

Only the latest regulatory actions, both international and domestic, have given a more solid and significant recognition to the agricultural component of landscape.[11] The reference is, primarily, to the European Landscape Convention, adopted on 19 July 2000 and signed in Florence on 20 October 2000. It receives the instances made by the scientific world long time ago for a new definition of landscape, whose distinctive features are essentially two: first, the shape of the territory resulting from interactions between man and nature; second, the result of the perception of people living in that landscape, as expression of the diversity of their shared cultural and natural heritage underlying their identity.[12] The rural landscape

[8] It is the site mentioned by several authors, including Boccaccio in his Decameron (the tale of Nastagio degli Onesti) and Dante's Divine Comedy ("the divine thick and alive forest" XXVIII Canto of Purgatory), and painters such as Botticelli.

[9] It is remarked by Malfitano (2002), pp. 1–18.

[10] Corte Costituzionale 27 June 1986 no. 151.

[11] In this regard, we cannot ignore the attempts of law reform proposed by the Franceschini Committee (Committee of Inquiry for the protection and enhancement of historic, archaeological, artistic heritage and landscape established by Law 310/1964). Considering the cultural value of the landscape, the Franceschini Committee theorised the category of "cultural heritage", which included goods "that present singularity of agrarian culture"—author's translation—(statement XXXIX).

[12] We can define landscape as "an area, as perceived by people, whose character is the result of the action and interaction of natural and/or human factors" (European Landscape Convention, Art. 1, Lett. a).

seems to embody at best the landscape model proposed by the European Convention, and, similarly, its protection derives significant confirmations by the principles of landscape policy as defined by the international act.[13] The references to the landscape of the "countryside" (Convention's preamble) and "rural areas" (Art. 2) and to the dependence of the formation and transformation of the landscape on the "developments in agriculture production techniques" (preamble) are emblematic. We can consider also very interesting the general statements according to which the landscape policies and the definition of landscape quality targets must be modulated and conformed to the needs and particular types of landscape (Art. 1, Lett. b and c) and cannot assume an absolute immutability irrespective of places, having primarily "to guide and harmonise changes which are brought about by social, economic and environmental processes" (Art. 1, Lett. e).

The Italian Code on Cultural Heritage and Landscape (Legislative Decree 42/2004) recognises the requests of the European Landscape Convention related to landscape and to the protection and development of the latter.[14] The above-mentioned definition of landscape is well suited to the reality of the rural landscape because of the element of interaction between man and nature and considering cultural and social identity that agriculture gives to the territory. Moreover, the analysis of the Italian Code on Cultural Heritage and Landscape leads to additional useful references for the protection of the rural landscape. We think about, for example

- the reference to "safeguard of rural landscapes" as an element of the compatibility advice given by the Regional Landscape Plan when it identifies the lines of urban and building development (Art. 135, Para. 4, Lett. d),
- the reference to the "creation of new landscape values" consistent with the "dynamic" definition of landscape (Art. 131, Para. 5),
- the ability of the Regions to recognise "further local contexts" in the landscape planning to submit to specific safeguard and utilisation measures (Art. 143, Para. 1, Lett. e).[15]

There is, finally, the recognition that a characterisation of the landscape may also result from agricultural uses and that the legal instruments for the safeguard of

[13] Ferrucci (2011a), pp. 240–244.

[14] On the relations between the concept of landscape in Legislative Decree 42/2004 and that of the European Convention, see Sciullo (2008).

[15] We may consider the experience of Regione Puglia, whose Landscape Plan (PPTR) mentions, among the "further contexts", the "rural landscapes" (Art. 38, Section 3.1, Lett. t), NTA of PPTR adopted by resolutions of Regional Council 2 August 2013 no. 1435 and 29 October 2013 no. 2022). Agricultural areas are described also in the Code on Cultural Heritage and Landscape, in Art. 136, Lett. b) (villas, gardens and parks that are distinguished by their uncommon beauty) and Lett. c) (the complex of properties that has an aesthetic and traditional value). Among the categories of goods listed in Art. 142, we may find lands covered by woods and forests, areas allocated to the "agrarian universities" and areas burdened with civic uses, parks and national and regional reserves, territories under external protection related to parks, highlands: see Picozza (2006), pp. 82–100. See also Ferrucci (2011b), pp. 202 et seqq.; Canfora (2011), pp. 304–329.

landscape (legal constraints and planning) are also functional to the preservation and continuity of agricultural activities, and the promotion of well-established and typical agricultural practices of certain territories, to the extent that they contribute to "the creation of new landscape values".

The Italian Code of Cultural Heritage and Landscape (referred to from hereon simply as the Code) configures the relationship between landscape and agriculture in terms of integration; the same option emerges from the varied regulatory framework that affects the rural landscape. We can derive some guidelines from it: the first concerns the intersection between the rural landscape and the general discipline of land uses; the second looks at the individual elements that make up the rural landscape—different in kinds but united by cultural value—and requires a detailed analysis of the legislation related to traditional crops, tree heritage and rural architecture.

2.1 The Planning Law

The planning law, in regulating land use, looks at the rural landscape not as a legal asset in itself (differently from the way the Code does, according to which landscape incorporates cultural and environmental values). In planning law's perspective, the relevance of rural landscape depends on the different activities that take place in it and varies as a function of the uses and consumption of other parts of territory.[16]

In Italy, the traditional classification of agricultural areas in homogeneous zones (called *E*) of the township, according to the zoning method of the town planning scheme (Ministerial Decree 1444/1968, putting into effect Art. 17 of Law 765/1967), originally did not respond to the need to identify and exploit the agricultural vocation of the territory, but it was essentially designed to prevent, or at least defer, the building transformability of the area.[17] The basic idea was that, among the various typified destinations, the agricultural one was more proper to impose restrictions to buildability in order to ensure a reasonable balance between built-up areas and free areas.

In the logic of town planning scheme, the agricultural interest ended up becoming a "weak" interest,[18] designed for an indirect safeguard, which depends conversely on building perspectives of the affected area. This core of protecting rules has laid itself open, on one hand, to a broad interpretation by the courts and, on the other, to a legislative development, mainly by the Italian Regions.

From the first point of view, we point out the case law according to which

[16] Urbani (2010), pp. 29–48.

[17] Urbani (2010), p. 30.

[18] Urbani (2006), pp. 117–124.

- the agricultural purpose imprinted in a specific area by town planning does not pose constraints to farm practice and is not used to recognising the utilisation of the area itself for agricultural purposes;
- in contrast, it typically expresses a discretionary choice on behalf of administration about the conservation of existing structure, and this option (which is of conservative concern) reflects the interest in preserving the landscape or natural values, that is, it is needed to relieve the congestion or to control building expansion.[19]

This law reflects the "conservative" peculiarity of zone "E"; by recognising a close continuity between agriculture and land conservation, it connects the target of preserving the existing structure to the protection of landscape, natural and environmental values.

Only later, the regional legislation has imposed the awareness that agricultural areas are an essential and non-fungible component of the territory and that, as such, they express a "public differentiated interest"[20] relevant and conditioning for urban discipline. This option is implemented by the regional planning regulations through different solutions: for example, admitting the building activity only if it is functional to farming activities and performed by certain entitled persons;[21] distinguishing the agricultural areas according to the soil, climatic, agronomic characteristics and to the presence of crops and agricultural facilities in order to identify agricultural areas of great value subjected to a regime of "reinforced" protection;[22] introducing a regulative principle of primacy of agricultural interest

[19] Consiglio di Stato IV 30 December 2008 no. 6600; Consiglio di Stato IV 20 September 2005 no. 4828; Consiglio di Stato IV 31 January 2005 no. 256; TAR Trentino-Alto Adige—Trento I 6 April 2011 no. 105; TAR Veneto—Venezia I 31 March 2010 no. 1118; TAR Campania—Napoli VII 3 November 2009 no. 6825; TAR Lombardia—Brescia I 24 June 2009 no. 1318. The criminal law too states in this sense: for example, Corte di Cassazione III 13 July 2009 no. 39078. About the theoretical reasons in support of the administrative practice of "urban" environmental protection and the jurisprudential endorsement, see Portaluri (2011), pp. 241–255.

[20] Urbani (2006), p. 120.

[21] Among the first ones, Law 64/1995 of Regione Toscana (today Law 1/2005). See Salvia (2012); Urbani (2010), pp. 33–36, who recalls also the adhesive jurisprudence.

[22] This is the case of Trento, Provincial Law 29/1987, which distinguishes between agricultural areas of primary interest (where only agricultural production is possible, with the exception of the industrial activities and the industrial breeding) and agricultural areas of secondary interest, where it is also possible to find new urban perimeters for new development areas. We can find another example in Toscana, Regional Law 1/2005, whose Art. 40 provides for the identification, in the context of urban planning, of areas with exclusively or predominantly agricultural function: in the former, which correspond to areas of great value for the purpose of agricultural production, we can use soil solely for purposes connected with conservation or development of agriculture and related activities. The areas of renewable energy sources are legally regulated in order to limit the intensive use of agricultural land for production facilities. See, for example, Puglia, Regional Law 31/2008, whose Art. 2 forbids the construction of photovoltaic plants in agricultural areas of particular value, as qualified by the planning regulations or by the landscape plan, as well as in areas where monumental olive trees are cultivated: the same Art. 2, however, has been declared invalid by Corte Costituzionale 22 March 2010 no. 119 and Corte Costituzionale 11 June 2014

and prevalence of agriculture on the related town planning regulations by including them in spatial plans (e.g., the Provincial Coordination Territorial Plan)[23] or sector-based ones (e.g., the Regional Landscape Plan).[24]

2.2 The Protection of Traditional Crops

The shape given to the land by farm activities reflects the choices made by men relating to production techniques and the type of cultivated products, certainly influenced by morphological and climatic characteristics of the available areas. Yet the capacity to resist and adapt of the soil has led to the testing of agricultural practices of various kinds, often brought about by economic reasons, aiming to achieve higher productivity, sometimes supported by the legislator himself. The evolution of Common Agricultural Policy (CAP) is significant. The first EEC measures for the agriculture (since the 1960s)—that is to say, policies based on economic incentives and price supports for market stabilisation and support to farmers' incomes—have triggered an increase in domestic production and an improvement in the economic well-being of agricultural workers, but it has also had distortive effects of different nature: apart from problems of financial sustainability in the long term, as well as unequal distribution of benefits in favour of the larger firms (able to develop production processes on an industrial scale, at the expense of small businesses),[25] the EU agricultural policies have favoured processes of intensification/specialisation and simplification/homologation of production that led to the phenomena of marginalisation of certain areas, such as mountain and pasture, and certain traditional crops, although typical of the place, and with it a progressive loss of biodiversity.

The correctives, initially inspired by economic reasons (to reduce the "load" of surplus) and introduced on a voluntary basis, were based on the idea of discouraging the destination of arable land in agricultural production directing them to different uses, such as fallow, reforestation, the use for non-agricultural purposes, the creation of pastures. Progressively, the solutions have explicitly been influenced

no. 166 because it violates Art. 117 of the Constitution as for the distribution of legislative competences between State and Regions. The current state legislation, that is Ministerial Decree 47987/2010, permits the location of photovoltaic plants in areas classified as agricultural by urban plans (Art. 15, Para. 3), but, at the same time, it gives the regions the opportunity to qualify as sites non-suitable for the building of photovoltaic plants some areas, in particular agricultural zones affected by quality agro-industrial productions and/or of particular value from the landscape-cultural point of view.

[23] Trento, Provincial Law 22/1991; Umbria, Regional Law 27/2000.

[24] Marche, Regional Law 34/1992. To have an excursus of the different regional options, see Picozza (2006), pp. 89 et seqq.

[25] Rissolio (2010), pp. 197 et seqq.

by the ecological and landscape role of agriculture.[26] A key step in this evolution was represented by "Agenda 2000"[27] and the Regulations adopted in 1999 [(EC) 1257–(EC) 1260] aiming to reform the financing of the common agricultural policy and the system of direct support to farmers. In the name of the development of the rural world, the new rules recognised the importance that agriculture can play in this direction, favouring a rediscovery of its vocation as guardian of the land, keeper of natural resources and preservation of the landscape, as well as guarantee of food quality and safety.[28] In relation to support for farmers, the new rules made the granting of subsidies subordinate to "the environmental measures" introduced by Member States "in view of the situation of the agricultural land used or the production concerned and which reflect the potential environmental effects" (Art. 3, Regulation (EC) 1259/1999).

The rule of conditionality has been developed by later Regulations (Regulation (EC) 1782/2003 and Regulation (EC) 1698/2005). The strategic objective of preserving the rural landscape, based on the awareness that "in Europe, much of the valued rural environment is the product of agriculture" (Art. 3.2, Lett. ii), Decision 2006/144/EC), stands out. The financial aids are contingent upon the bond of maintaining the land in "good agricultural and environmental condition" and to respect "statutory management requirements", as defined in Annexes III–IV (Arts. 3–5, Regulation (EC) 1782/2003; Art. 51, Regulation (EC) 1698/2005), among which we find the observance of the specific local conditions, crop rotation, protection of permanent pasture, the maintenance of the terraces, the maintenance of landscape features.

The new set of regulations on rural development,[29] which is the outcome of the debate on the revision of the Common Agricultural Policy triggered by a specific

[26] See Regulation (EEC) 2328/1991 (which enables Member States to grant aid in sensitive areas in terms of environmental protection, conservation of nature and landscape) and Regulation (EEC) 2078/1992 (focusing on the promotion of agricultural productive methods able to reduce the polluting effects of agriculture, the promotion of land management consistently with environmental protection, the promotion of farming for abandoned agricultural and forest lands or lands at risk of ecological damage).

[27] European Commission, Agenda 2000: For a stronger and larger Union, COM (1997) 2000 final.

[28] Ragionieri (2003), pp. 201 et seqq.

[29] Regulation (EU) 1303/2013 laying down common provisions on the European Fund for Regional Development, on the European Social Fund, on the Cohesion Fund, on the European Agricultural Fund for Rural Development and on the European Maritime and Fisheries and general provisions on the European Regional Development, the European Social Fund, the Cohesion Fund and the European Fund for maritime Affairs and fisheries, and repealing Regulation (EC) 1083/2006; Regulation (EU) 1305/2013, on support for rural development by the European Agricultural Fund for Rural Development (EAFRD) and repealing Regulation (EC) 1698/2005; Regulation (EU) 1306/2013 on the financing, management and monitoring of the common agricultural policy and repealing Council Regulations (EEC) 352/1978, (EC) 165/1994, (EC) 2799/1998, (EC) 814/2000, (EC) 1290/2005 and (EC) 485/2008; Regulation (EU) 1307/2013, establishing rules for direct payments to farmers under support schemes provided by the common agricultural policy and repealing Regulation (EC) 637/2008 of Council and Regulation (EC) 73/2009; Regulation (EU) 1308/2013 establishing a common organization of markets for agricultural products and repealing Council Regulations (EEC) 922/1972, (EC) 234/1979, (EC) 1037/2001 and (EC) 1234/2007; Regulation (EU) 1310/2013 laying down certain transitional provisions on support for rural

Communication of EU Commission in 2010[30] and tries to incorporate the objectives of the "Europe 2020 Strategy", shows further awareness of the role of rural landscapes in the rural development policies. References are manifold: foremost, among the "priorities" defined by Art. 5 of Regulation (EU) 1305/2013 (to replace the previous "axes"), which in turn make explicit the thematic objectives of the Common Strategic Framework as in Art. 10 of Regulation (EU) 1303/2013, there is the "improving competitiveness of primary producers by better integrating them into the agri-food chain through quality schemes, adding value to agricultural products, promotion in local markets and short supply circuits, producer groups and organisations and inter-branch organisations" (Art. 5, no. 3, Lett. a), as well as the priority of "restoring, preserving and enhancing biodiversity, including in Natura 2000 areas, and in areas facing natural or other specific constraints, and high nature value farming, as well as the state of European landscapes" (Art. 5, no. 4, Lett. a). About support measures, we can see different innovative profiles, including the provision of additional measures for the promotion of small farms and short supply chains (Art. 7, no. 3, Regulation (EU) 1305/2013), favouring both "shortening" distances between farmers and markets, both consumers in terms of better prices and greater benefits for the quality of production; recognition of economic support measures on behalf of farmers who participate in quality schemes for agricultural products and foodstuffs (Art. 16, Regulation (EU) 1305/2013); the introduction of stricter conditions for provision of grants, as the support is provided only to farmers who undertake, on a voluntary basis, to carry out operations consisting of one or more agri-environment-climate commitments on agricultural land to be defined by Member States, able to go beyond the relevant mandatory standards and the minimum activities established pursuant to EU legislation (Art. 28, no. 2–3, Regulation (EU) 1305/2013); the specification of the statutory management requirements that characterise the rule of conditionality, including specifically the "minimum level of maintenance of the landscape" (Annex II, Regulation (EU) 1306/2013);[31] the dependence of the payments (the so-called ecological fee and progressive reduction of the basic payment) on agricultural practices beneficial for the climate and the environment, consisting in crop diversification, maintenance of existing permanent grassland, creation of an area of ecological interest[32] or

development by the European Agricultural Fund for Rural Development (EAFRD), amending Regulation (EU) 1305/2013 of the European Parliament and of the Council as regards resources and their distribution in relation to the year 2014 and amending Regulation (EC) 73/2009 of the Council and Commission Regulations (EU) 1307/2013, (EU) 1306/2013 and (EU) 1308/2013 of European Parliament and of Council as regards their application in the year 2014. About the evolution of direct payments to farmers, as for the "greening", see Ferrucci (2014), pp. 327 et seqq.

[30] European Commission, The CAP towards 2020: Meeting the food, natural resources and territorial challenges of the future, COM (2010) 672 final.

[31] Under the following conditions and standards: retention of landscape features, including where appropriate, hedges; ponds; ditches, trees in line, in group or isolated; field margins and terraces; and also a ban on cutting hedges and trees during the bird breeding and rearing season and, as an option, measures for avoiding invasive plant species.

[32] The Regulation provides a list of the possible areas of ecological importance, among which: land lying fallow; terraces, landscape features; buffer strips (Art. 46).

equivalent practices (Art. 43, Regulation (EU) 1307/2013), according to the idea that we must award firms oriented towards the improvement of the general environmental performance of the holding, in particular as regards biodiversity, the improvement of soil and water quality, the preservation of landscape and meeting the climate change mitigation and adaptation objectives (*Whereas* no. 45, Regulation (EU) 1307/2013).

We can find the focus on the maintenance and reclamation of traditional crops as typical elements of the rural landscape, also in national law. One example is Law 394/1991, the framework law on protected areas: adopting a new "dynamic" and "promotional" vision about the natural parks,[33] the law allows and encourages a variety of activities considered "sustainable", among which there are the agricultural ones managed according to traditional uses (Arts. 4–12). Furthermore, the Italian legislator has formalised the landscape value of protected natural areas: according to Art. 142 of Legislative Decree 42/2004, protected natural areas are landscape assets that have legal constraints and are guarded by law.

Traditional crops and products are also subject to incentives as they may be attractive resources inside the economic system of an area. In this way, the preservation of the landscape, which is guaranteed by maintaining the shape given to a territory by traditional agricultural practices, combines with the other goals typical of the landscape's policies, such as increasing knowledge and enjoyment of landscape goods, and tourism becomes the best vehicle for the pursuit of these goals and at the same time for the start of economic development processes. The combination of protection and improvement of the landscape and economic development through the promotion of typical agricultural resources inspires that normative trend aiming to recognise and appreciate "roads", "paths", "routes" focused on the promotion of typical local products like wine, oil, milk. The territorial marketing,[34] particularly interesting because it is a cultural and economic one, results in rules that acknowledge facilities and financial support for various projects: specific traffic signs for cultural routes, information and documentation centres, promotion through advertising channels, promotion of cultural activities, training of skilled operators. There is a first season of local and regional initiatives, then implemented and formalised by the State Legislator with Law 268/1999, "Discipline of wine routes" (and Ministerial Decree 781/2000, concerning the definition of the requirements and minimum standards for the recognition), whose specimen was taken up and adapted to the local needs of the regions.[35]

---

[33] About the new conception of "park" expressed by Law 394/1991, see, among others, Salvia (2012), p. 271; Crosetti (2008), pp. 511 et seqq.; Cartei (1993), pp. 601 et seqq.

[34] Ferrucci (2011a), p. 241.

[35] Among the regional initiatives, see Emilia-Romagna, Regional Law 23/2000, "Disciplina degli itinerari enogastronomici dell'Emilia-Romagna"; Trento, Provincial Law 19/2001, "Disciplina dell'agriturismo, delle strade del vino e delle strade dei sapori"; Toscana, Regional Law 45/2003, "Disciplina delle strade del vino, dell'olio extravergine di oliva e dei prodotti agricoli e agroalimentari di qualità", modified by Regional Law 4/2013; Molise, Regional Law 50/2005, "Istituzione delle strade del vino, dell'olio, del tartufo, del latte e dei sapori del Molise"; Lazio,

The promotion of local agricultural products from a touristic point of view characterises also the legislation on agri-tourism (Law 96/2006). The Law aims to emphasise the role of the farmer, on the one hand, and to bring back the agri-touristic organisation to the traditional pattern of agricultural farm, on the other hand. The agri-tourism activities cover a wide range of interventions: not only hospitality and supply of food and drinks but also the organisation of tastings of farm products, including wine shop, and the organisation (even outside of the farm property) of recreational, cultural, educational, sporting activities, as well as hiking excursions and horse-riding tourism.

Law 96/2006 recognises the close connection between the agri-tourism activities and the agricultural ones (Art. 2, Para. 1) and formalises the prevalence of the latter over the former, with particular reference to the labour time necessary to perform the same activities (Art. 4, Para. 2); it expresses a preference for using agricultural products of its own and the adoption of traditional methods of processing, as indicated by the obligation to use, above all, products of its own agricultural production, and those of local farms, for the supply of meals and drinks (Art. 2, Para. 3, Lett. b; Art. 4, Para. 4, Letts. a and b).[36]

In the model outlined by the legislator, the landscape "essence" of agri-tourism is clear:[37] it combines the support of agriculture and rural economy with the promotion of sustainable kinds of tourism and the development of territory.

Moreover, the multifunctionality of the agricultural firm has been stressed by Legislative Decree 228/2001, entitled "Guidance and modernization of the agricultural sector", which defines the "activities related" to properly agricultural ones: among them, we find the ones "intended for goods or services supplying through the

Regional Law 21/2001, "Disciplina delle strade del vino, dell'olio d'oliva e dei prodotti agroalimentari tipici e tradizionali"; Friuli Venezia Giulia, Regional Law 21/2000, "Disciplina per il contrassegno dei prodotti agricoli del Friuli Venezia Giulia non modificati geneticamente, per la promozione dei prodotti agroalimentari tradizionali e per la realizzazione delle strade del vino"; Veneto, Regional Law 17/2000, "Istituzione delle strade del vino e di altri prodotti tipici del Veneto"; Abruzzo, Regional Law 101/2000, "Disciplina delle strade del vino in Abruzzo"; Umbria, Regional Law 38/1999, "Disciplina delle strade del vino dell'Umbria"; Piemonte, Regional Law 29/2008, "Individuazione, istituzione e disciplina dei distretti rurali e dei distretti agroalimentari di qualità e modifiche della legge regionale 12 maggio 1980, n. 37 (Le enoteche regionali, le botteghe del vino o cantine comunali, i musei etnografico-enologici, le strade del vino)". For further information on national and regional regulations, see, in particular, Strambi (2006), pp. 204–238; La Torre (2010), pp. 573 et seqq. At international level, some areas relevant for their agricultural character have been included in the UNESCO list of world heritage areas: recently, the "wine landscape of Piemonte: Langhe-Roero and Monferrato" recognised by UNESCO with the decision of 22 June 2014, with the following justification: "the site encompasses the whole range of technical and economic processes relating to the winegrowing and wine making that has characterised the region for centuries". On this experience, see Lombardi (2009), pp. 389–397.

[36] The expressed preference for the local products is the priority given in the supply of products characterised by DOP, IGP, IGT, DOC and DOCG trademarks or included in the national list of traditional food products (Art. 2, Para. 3, Lett. b).

[37] Ferrucci (2007a), pp. 679–683.

prevailing use of equipment or resources on the farm usually employed in the agricultural activity, including the activities which promote the area and the rural and forest heritage, or rather which are addressed to accommodation and hospitality as defined by law" (Art. 1, Para. 1). The Legislative Decree 228/2001 marks also an evolutionary approach about management measures of the rural landscape, starting a consensual model that provides for two types of agreements: Collaboration Contracts (Art. 14) and Conventions (Art. 15) between farmers and public authorities. Collaboration Contracts are functional to promoting production activities on the territory, ensuring the quality of products and protecting local food traditions: they may also include specific measures to provide appropriate information to consumers about the origin of raw materials and the characteristics of products. It is worth noting that Collaboration Contracts have to establish the commitment for the entrepreneur to "ensure the protection of natural resources, biodiversity, cultural heritage and agricultural and forest landscape". Conventions are specifically designed to promote an "active" role for farmers in provision and safeguard of the territory, preservation of the agricultural and forest landscape, care and maintenance of the hydrogeological aspects, and so on: the public authorities grant farmers considerable advantage in return (like funds, public licenses, rate reductions or public works).[38] In both cases, it is clear the intent of the legislator to introduce measures able to safeguard and promote traditional agricultural methods and products because protection and improvement of rural landscape could also depend on the survival of those very typical productions.[39]

2.3 *Rural Architecture*

Farmhouses, country houses, farms, pens, *masi*, *masserie*, bakeries, mills, oil mills, stone walls, terraces: these are examples of man-made facilities for operational agricultural needs, on which in recent times the legislator—and, long before, scholars—has polarised the attention as significant evidence of the history of rural communities and their economies, as characteristic and tangible landscape's features. They are expressive elements of rural architecture, even minor, but

[38] Think about the typifying of the rural districts and the districts of high-quality food (Art. 13): the former are local production systems characterised by a homogeneous historical and territorial identity arising from the integration between agricultural activities and other local activities, as well as from the production of typical goods or services, consistent with the traditions and the natural and territorial vocations; the latter are characterised by a significant economic presence and by interrelation and interdependence of agricultural and agri-food productions, as well as by one or more productions, certified and safeguarded in accordance with the current EU or domestic legislation or by traditional or typical productions. About the "rural district", see, among others, Albisinni (2002), pp. 459 et seqq.; Albisinni (2011), pp. 103–148; Varotto (2005), pp. 2–20; Galluzzo (2010).

[39] Ferrucci (2010), pp. 296–301; Lombardi (2009), p. 395.

extremely widespread and layered in rural areas, whose landscape importance is unquestionable and denotes at best the conceptual evolution of the landscape, no longer anchored to purely aesthetic criteria (which would lead merely to the emergence of the architectural heritage particularly important from a historical and an artistic point of view), nor only to natural and environmental elements of the area, but, above all, characterised by a collective interpretation of perceived identity of the places. Both structural features of buildings, which refer to specific styles and techniques of rural architecture, both their original function, which reflects the location of the property in a specific production system, are significant elements of this interpretation. The regulatory framework regarding rural landscape reflects this dual vision. We can find provisions designed to safeguard the structural elements of rural buildings and other provisions designed to maintain and reclaim their original purposes. Law 378/2003, entitled "Provisions for the protection and improvement of the rural architecture", is emblematic: it addresses "types of rural architecture, such as agricultural settlements and rural buildings or premises on the national territory, built between the thirteenth and nineteenth century, which are evidences of the traditional rural economy" (Art. 1, Para. 1), and the incentive measures are directed, on one hand, to the "preservation of the traditional elements and historical, architectural and environmental features of agricultural settlements, traditional rural buildings and premises" (Art. 2, Para. 1, Lett. a) and, on the other hand, to "the preservation of the original use of rural settlements or buildings or premises, the protection of surrounding areas, of traditional methods and sort of cultivation, and the establishment of activities that are compatible with typical cultural traditions" (Art. 2, Para. 1, Lett. b). The law outlines an administrative model shared between State and Regions. The State has to identify the types of rural architecture that can access the benefits provided by law, to establish technical/scientific criteria that the restoration and conservative reclamation[40] must adapt to, as well as to create and to manage the National Fund for the protection and improvement of the rural architecture. The Regions are responsible for the recognition of buildings related to the categories typified at the State level and for the arrangement of programs aiming to point out the individual measures of intervention, as well as the management of shares of national fund allocated to them. Financial assistance may cover up to 50 % of expenditure recognised in the financial plan; grants are

[40] Ministerial Decree 238/2005. The importance of the relationship between these buildings and agricultural activities emerges from the description of the types of rural architecture provided by the decree: they are examples of rural architecture relevant according to Law 378/2003 (among others: spaces and buildings used for agricultural activities; material evidence contributing to the definition of the recognisable historical-anthropological units, with particular reference to the link between productive settlement and space and, in this context, between agricultural property and land; fences and paving of open productive or residential spaces; historical rural roads; canalization, irrigation, water supply and terraces' control systems; temporary shelters (Art. 1, Para. 2–3, Ministerial Decree 238/2005). The conservative works getting public subsidies must take into account the "needs of technological renovation of agricultural holdings" (Art. 2, Para. 1, Lett. a), Law 378/2003).

dependent on a special agreement that establishes, inter alia, the non-transferability of the property for at least a decade.

The above-mentioned Law on agri-tourism 96/2006 follows the same approach. It considers the regeneration of rural building heritage as one of the essential purposes of agri-tourism (Art. 1, Lett. e) and pursues it by means of incentives rules and a rigorous system of conditions. The assimilation of the premises used for agri-tourism purposes to the rural buildings (Art. 3, Para. 3) leads in the same category both the farmer's premises used for accommodation purposes and the other buildings on the farm subordinated to agri-tourism.[41] The preferential treatment towards the farm buildings recycled for agri-tourism emerges also from sanitary measures: when they define the fitness for use, from a sanitary point of view, Regions are required to take due account of the "special architectural and rural characteristics of the buildings, in particular as regards to the height and volume of the premises in relation to window/floor areas, as well as the limited size of the activity concerned" (Art. 5, Para. 1). On the other hand, the building changes designed to have a rural premise functional to agri-touristic activities (above all, for the accommodation) are possible only if they are "in compliance with the specific typology of architectural features, as well as the landscape and environmental characteristics of places" (Art. 3, Para. 2). The option that emerges is the "conservative transformation",[42] where change interventions are oriented in the delicate pursuing of a twofold aim: to adapt buildings from a technological point of view (also in order to comply with changing needs of the owners) and, at the same time, to strengthen or to restore the original shapes and locations of the properties. At the same time, the techniques of rural architecture, developed in a context of scarce resources and oriented to the functionality of the buildings with respect to the agricultural needs, offer a model for sustainable building, which is suitable (or even inspires) the modern bio-ecological architecture (the so-called bio-building).[43] The same approach is present in Legislative Decree 99/2004, whose Art. 12 provides a significant fiscal incentive: the income deriving from the leasing of farm buildings (for at least 5 years and not more than nine) is included in the cadastral and agricultural income of the land they stand on, so becoming tax-free income. The objective of the development of rural building heritage and its close connection with the agricultural sector emerge from the conditions to which this fiscal incentive is made subject: the owner of the property to be leased must be a farmer, and the property must have acquired the habitability condition by means of a regular restructuring.[44]

We can find an expression of the greater sensitivity about the cultural value of rural architecture even in the most often mentioned Code on Cultural Heritage and

---

[41] Ferrucci (2010), p. 298; Sciaudone (2007), pp. 146 et seqq. About the legislation on municipal property tax-ICI and single tax-IMU, please refer to Bagnoli (2012), pp. 1826 et seqq.

[42] See Ferrucci (2008), pp. 562–605 (author's translation).

[43] Ferrucci (2010), p. 298; Lucifero (2010), pp. 159–279.

[44] Valletta (2004), pp. 341–344; Del Mastro (2005).

Landscape (Legislative Decree 42/2004) in the section that refers to cultural heritage. Among the typified categories of cultural heritage, it is significant the reference to "rural architectures having historical or ethno-anthropological evidence as a witness of traditional rural economy" (Art. 10, Para. 4, Lett. l), where the awareness that they are unique assets stands out, because the cultural value derives from the instrumentality of them to the economic and productive function. The rule allows to apply the regulatory constraints typical of cultural heritage—where the conservation is combined with the aim of public enjoyment and appreciation—to rural architectural heritage, of course to the most significant evidence susceptible of declaration of particularly important interest.

The focus on rural architecture as part of the landscape is also present at Regional level. Here, indeed, there is an effort to preserve above all the architectural examples reflecting the construction techniques and productive system typical of that region, of which, in the unanimous opinion, they are distinctive and identity elements: think about the *masi* in South Tyrol, the *tratturi* in Puglia, the *cascine* in the Po Valley, the *trabucchi* in Abruzzo and Molise.

The two guidelines in the state legislation emerge even in the regional one: the preservation of the integrity of handmade buildings and facilities that are evidence of traditional economic systems, outdated and no more repeatable, because they are elements of the history and culture of an area and need, therefore, knowledge and fruition and the promotion and reclamation of their original agricultural destination, also in order to improve and enhance rural development.

Expression of the first approach, for example, is Law 29/2003 of Regione Puglia aiming to protect the *tratturi* (sheep tracks, old transhumance routes), in which it is clear the goal to recognise and protect the tracks that retain their original characteristics or that can be taken back to it in order to promote forms of cultural-touristic enjoyment.[45] In the same direction, there is Law 9/1997 of Regione Molise, as amended by Regional Law 17/2003, on the protection and improvement of *tratturi*, *tratturelli*, *bracci* and *riposi*, as well as other regional laws aiming at the protection of traditional paths (mule tracks, cartways, and so on), through the establishment of suitable excursion networks.[46]

In this sense, e.g., Law 11/1996 of Regione Sicilia establishes rules for the protection and improvement of the ancient windmills and promotes the traditional

[45] The law has been transfused into Regional Law 4/2013, whose Art. 8 establishes the "Park of Puglia cattle tracks" (*Parco dei tratturi di Puglia*). About Regional Law 29/2003, the Constitutional Court has recognised the presence of a balanced structure of powers between state, regions and municipalities: see the judgment 14 October 2005, no. 388, examined by De Giorgi Cezzi (2006) and by Angiuli (2007), pp. 580–586.

[46] Toscana, Regional Law 17/1998, "Excursion network in Tuscany and Regulation of excursion activities"; Marche, Regional Law 2/2010, "Establishment of excursion network in Marche"; Emilia-Romagna, Regional Law 14/2013, "Excursion network in Emilia-Romagna and enhancement of excursion activities"; Piemonte, Regional Law 12/2010, "Restoration and enhancement of the excursion heritage in Piemonte"; Puglia, Regional Law 21/2003, "Discipline of tourism and excursion networks in Puglia".

production of sea salt: the goal of "maintaining the traditional characteristics of the handmade articles" (Art. 2, Para. 2) is pursued by granting contributions (Art. 2, Para. 3) and even by resorting to expropriation proceedings against the owners who do not implement conservation and development interventions required by the Law (Art. 3, Para. 1).

The conservation of formal and structural characteristics of rural buildings is pursued also by permitting (and providing incentives for) their reuse and requalification, even for destinations different from the original ones: e.g., the residential use, with the related purpose of containment of soil consumption and energy consumption, is favoured by Law 40/2003 of Regione Veneto, which recognises contributions to those who intend to recover the property as their own home; the economic-productive use, different from the agricultural original destination, is permitted by Law 20/1998 of Regione Puglia on rural tourism, which admits the restoration and renovation of rural buildings, farmhouses, *trulli*, towers, fortifications, and, in general, ancient artefacts assessed in urban agricultural land register, in order to change the building into accommodation, but only on the condition that the owner maintains "unchanged the existing volume above ground and with the safeguard of the original prospectuses and architectural and artistic features of the property" (Art. 1, Para. 2). The same approach is followed by Law 1/2009 of Regione Lazio, which promotes the establishment of agricultural parks.

A focus of the Italian Regions towards the goal of protection and restoration of the former destination of rural buildings emerges from different laws establishing "incentive rules" on the basis of two options (which can also be combined): the provision of economic and/or administrative incentives and the permissibility of supplementary economic or socio-cultural activities, in addition to the agricultural ones.

The Law of Bolzano Province on closed *masi*[47] (typical self-sufficient farms characterised by a legal constraint of indivisibility) is a significant example of the first option: the Law establishes in favour of *masi* some financial contributions, measures for credit assistance and special derogations to town planning and building rules.

Specific "intervention programs for the protection and improvement of the rural architecture", functional to the "preservation of the original use, the protection of the surrounding areas, traditional cultivation types, methods and activities compatible with typical cultural traditions", characterise the regulation of Regione Lazio. Art. 31-*bis* of Regional Law 24/1998 (introduced by Art. 71 of Regional Law 4/2006) focuses on the historical and landscape features of the region and includes even "the artifacts related to farming and rural manufacturing activities and services on the territory, as examples of post-unification rural landscape characterised by colonisation of the territory through holdings, reclamations, and land subdivisions".

[47] Provincial Law 17/2001; Decree of the President of the Province 19/2006; Provincial Law 13/1997; Provincial Law 3/2007.

In the same direction, we can find also Law 12/2005 of Regione Umbria on the protection of historic water mills, which promotes measures to ensure the restoration of functionality of these goods, aiming to a quality milling production.[48]

The regional legislation about agri-tourism is one of the expressions of the second option. Regional laws repeat the state model in defining relationship between agricultural and accommodation activities, as well as in establishing incentive rules, even derogating to planning and building regulation, on the condition that the structural and architectural characteristics and the types of use of rural buildings are maintained (for example, Puglia, Regional Law 42/2013; Marche, Regional Law 21/2011; Bolzano, Provincial Law 21/2011; Bolzano, Provincial Law 13/1997).

Other rules reflect the typical rural and distinctive architecture of different Italian Regions. This is the case of Regione Lombardia: Regional Law 17/2007, on the establishment of parks, devotes a section to the system of farmsteads. In order to enhance their cultural and landscape relevance, the Regional Law permits a range of complementary activities to agriculture: e.g., use of buildings for agri-tourism activities; specialised accommodation activities in support of wine activities, such as wine tasting, wine cellars visiting, sale of products from the farm; additional educational activities, such as the "didactic farms" and specialised agricultural schools.

Another prime example concerns the *trabucchi*, ancient wooden constructions useful for fishing activities along the Adriatic coast. The current policy of Regione Abruzzo provides incentives for renovation and restoration of those ancient structures, on the condition that they retain the original instrumentality for the purpose of fishing, and permits also additional catering activities that use fish products of the same structure or local fish products or from surrounding areas or, anyway, from Adriatic Sea (Arts. 1, 3-*bis*, 3-*ter*, Regional Law 13/2009, introduced by Art. 15, Regional Law 38/2010).

Another experience is that of the educational and didactic farms (*masserie didattiche*). Law 2/2008 of Regione Puglia accredits and promotes[49] agro-industrial and agri-tourism firms that, as a supplement to traditional production activities, undertake initiatives of teaching and training about rurality in addition to accommodation. The intent of the legislator to recognise the multifunctionality of the agricultural entrepreneur is clear, by creating additional sources of income that are linked to cultural and economic background of the territory.

[48] The law also promotes projects with educational and entertaining aims as for the knowledge and development of water mills as a "witness of the ancient rural civilization" (Art. 1, Para. 2).

[49] There are different promotional measures: the development of a logo able to distinguish the educational farms, the establishment of a regional register, the provision of economic incentives and measures useful to spread the knowledge of educational farms.

2.4 *The Monumental Trees*

Trees are traditionally an important part of the agricultural sector for their productive function (production of fruits, wood, energy, etc.), as well as for the protective one (soil protection, water regulation, barriers to the wind, conservation of biodiversity, etc.). As well as this consolidated awareness, trees are today considered also as "actors" of landscape because of their multidimensionality—agricultural, aesthetic, botanical, historical and anthropological—that is comprehensively represented into the landscape interest.

This multidimensionality is recognised in the most recent legislation, which results in the classification of "monumentality" of trees.[50]

The Legislative Decree 63/2008 (which amends the Code on Cultural Heritage and Landscape) has supplemented the list of landscape goods by adding the "monumental trees". The new category is formulated as an example of the natural beauty in Art. 131, Lett. a), Legislative Decree 42/2004: the monumental trees can be declared as landscape goods of significant public interest and, in this way, be subjected to regulatory constraints typical of the landscape law.[51]

More recently, Law 10/2013 ("Standards for the development of public parks and gardens") has established a specific rule dedicated to the protection of monumental trees, which represents an important step forward. The effort of the Law in giving a more detailed legal definition is significant: it considers "monumental" the trees, individual or as part of woodlands, including rows of trees, which are characterised by their grandeur and longevity, their botanical rarity, their connection to events or memories relevant from a historical, cultural, documentary or local tradition point of view, as well as by their location in architectural complexes of historical and cultural importance (Art. 7, Para. 1). This definition is large and reflects the many facets—scenic, natural, historical, cultural—of the status of monumental trees. The protection system outlined in the rule is also very interesting: it provides for the establishment of an official list of monumental trees in Italy, managed by the State Forestry Corps; the census is ensured by the individual municipalities and Regions on the basis of criteria and guiding principles laid down by the Ministry of Agriculture, Food and Forestry; the inclusion on the national list of monumental trees involves specific restrictions on their use, in particular the prohibition of cutting them down and damaging (except in duly justified cases of extreme urgency, upon municipal authorisation and mandatory

---

50 In jurisprudence, it was pointed out the sense of the new formulation that, on one hand, evokes the grandeur and longevity of trees and, on the other hand, emphasises the action of nature as creator of masterpieces similar to those of art: so De Giorgi Cezzi (2005), pp. 2955 et seqq. (comment on the decision of TAR Puglia—Lecce, I, 5 July 2005, no. 3611).

51 Art. 137 regulates the composition of the regional committees entrusted with proposals for declaration of public interest. Paragraph 2 provides that the Committee is integrated by the representative of the competent Regional Headquarter of the State Forestry Corps when the proposal relates to rows of trees and monumental trees.

advice of the State Forestry Corps); the transgression of these prohibitions is to be punished by pecuniary sanctions ranging from 5,000 to 100,000 euros.[52]

The regulatory framework is completed by the regional legislation, which is substantial even about this topic.

Making use of the legislative power in the area of "woods and forests" (until the constitutional reform of 2001, it was a subject matter of concurrent legislation that is shared between State and Regions; today, it is a matter of exclusive legislative competence of the Regions, except for the limit of "environmental protection" that remains under the exclusive competence of the State),[53] the Regions have adopted specific rules aiming to protect and enhance the monumental trees. They are individual provisions included into regional forest laws,[54] but there are cases of laws on the monumental trees passed *ad hoc*.[55] Generally, the regional rules consider the entire category of monumental trees; in some cases, regulatory interventions relate to individual varieties of trees (e.g., as for the olive trees, subjected to specific regional laws).[56]

Regional laws have a number of common characteristics. In terms of definition, the classification as "monumental tree" is based on the following indicators: botanical rarity, aesthetic and morphological interest because of the tree's grandeur; longevity or peculiar form; link with historical-cultural events or memories or local traditions.[57] Some laws provide, as an additional indicator of the monumentality, for the reference to the territorial context of the tree, and according to this, trees are considered as being monumental when "they are an essential part of a traditional

---

[52] See Ferrucci (2013); Manservisi (2013a); Manservisi (2013b), pp. 357 et seqq.

[53] About the distribution of legislative powers between the state and regions in the area of woods and forests according to the constitutional principles (Art. 117 of the Constitution), see, among others, Amirante (2012), pp. 233–275; Maddalena (2009), pp. 635–647; Abrami (2008), pp. 529–536.

[54] Among the regional forestry laws: Art. 17, Basilicata, Regional Law 42/1998; Art. 12, Liguria, Regional Law 4/1999; Art. 12, Umbria, Regional Law 28/2001; Arts. 31–34, Lazio, Regional Law 39/2002. In other cases, the provisions about monumental trees are included in laws relating to the regional flora (Art. 6, Emilia-Romagna, Regional Law 2/1977; Art. 12, Lombardia, Regional Law 10/2008) or in planning laws (Art. 69, Trento, Provincial Law 1/2008).

[55] Valle d'Aosta, Regional Law 50/1990, "Protection of monumental plants"; Piemonte, Regional Law 50/1995, "Protection and enhancement of monumental trees of outstanding natural and historic value in Piemonte"; Toscana, Regional Law 60/1998, "Protection and enhancement of monumental trees and amendment of Art. 3 of Law 11 April 1995, n. 49"; Veneto, Regional Law 20/2002, "Protection and enhancement of monumental trees"; Molise, Regional Law 48/2005, "Protection and enhancement of monumental trees"; Calabria, Regional Law 47/2009, "Protection and enhancement of monumental trees and wild flowers from Calabria".

[56] Puglia, Regional Law 14/2007, "Protection and enhancement of the monumental landscape of olive trees in Puglia"; Abruzzo, Regional Law 6/2008, "Provisions for the protection of adult olive trees for their classification, recovery and disposal. Discipline on the cutting down and explants of olive trees"; Veneto, Regional Law 6/2011, "Rules concerning the cutting down of olive trees"; Calabria, Regional Law 48/2012, "Protection and enhancement of the olive legacy in Calabria".

[57] The additional criterion provided by the law of Trento Province, which describes the connection of the trees with "agricultural activities fallen into oblivion", is significant.

complex and peculiar landscape" or "they are in public and private areas such as urban areas, places of worship and areas used for recreational, touristic and leisure purposes" (Umbria) or they are located "in the surroundings of goods of historical-artistic, architectural, archaeological interest" (Puglia) or they are "specimens placed in special architectural complexes such as villas, monasteries, churches, botanical gardens and private historic homes" (Lombardia). Sometimes, the recognition of monumentality is given to a set of trees, as in the case of the particularly valuable tree rows (Piemonte, Lombardia, Emilia-Romagna, Umbria, Puglia), monumental forests (Lazio) and monumental olive trees (Puglia).

The survey of monumental trees, which leads to the compilation of an official list, is provided. The survey work is entrusted to the State Forestry Corps or to the local authorities and the operators of natural protected areas or technical bodies (such as the Committee for the Assessment of Plant Requirements in Valle d'Aosta or the Technical Committee for the Protection of Monumental Trees in Puglia), and voluntary reporting (from environmental associations and individual citizens) is usually allowed. The list is updated and approved by the Region (generally by the Regional Council; a different case is that of Regione Veneto, where the competence is given to Veneto Regional Office for Agriculture), except in the case of Regione Calabria, which has conferred administrative power to the provinces about this subject matter.

As for the forms of protection, the regional laws follow two alternative lines of policy: reiterating landscape constraints established by Legislative Decree 42/2004 or providing for a special and additional protection regime.

The first solution has been adopted by Regione Piemonte, the second by the other Regions. Calabria and Puglia (the latter only as for the monumental olive groves) have followed both the approaches.

The second line of policy is characterised by special measures prohibiting to cut down, to explant or to damage monumental trees. Some laws prohibit constructing or changing the intended use of buildings inside the areas where the trees stand (regional laws of Toscana, Calabria, Lazio prohibit it for 20 years, Molise for 50 years); they impose also to replant tree species similar to the ones cut down (Molise, Toscana and Veneto give to municipalities the faculty to apply the rule).

The prohibition admits some derogations (upon a specific authorisation granted by the Region or the municipalities) related to plant health, public safety and (in some laws) public interest and subjected to a set of conditions, that is, the impossibility to adopt alternative solutions, the acquisition of technical advice, the replanting of the same trees or of specimens belonging to the same species.

The regime of monumental trees affects obviously planning regulations, and it prevails on them in case of contrast (conflicts are solved by the specialty—*lex specialis*—criterion). Some regional laws (Toscana, Veneto, Molise) burden municipalities to fit planning regulations in order to provide for a special protection. At local level, the regulation on urban green spaces can introduce specific—and more strict rules than the regional and state ones—provisions on monumental trees.[58]

---

[58] See the regulations of the municipalities of Torino, Bologna and Genova. On this point, see also Graziosi (2012), pp. 189–204.

Control tasks are generally entrusted to the State Forestry Corps, as well as provincial and municipal police, hunting and ecological guards.

Regional laws focus also on monumental trees' enhancement. Several laws (Lazio, Veneto, Molise and Piemonte) set up initiatives for promotion of monumental trees in order to make known the meaning of their protection, as well as to improve the surrounding local and environmental context. For this purpose, the Regions

- grant funds for ordinary and extraordinary care of monumental trees (Piemonte);
- award the best student papers about the relationship between the regional history and the trees and offer also thesis scholarships (Molise);
- establish special mentions for oil produced by the monumental olive trees and grant funds for projects aiming to develop tourism, maintain the growing of monumental olive trees, improve quality of products, recovery and maintenance of countryside (Puglia).

3 Concluding Remarks

Rural landscape corresponds to the modern conception of the landscape, free from the cult of the "beauty of nature" (which marked the first legislation on the subject) and anchored to a historicist criterion, which enhances the land's identity whose characters derive from the natural and human factors and their reciprocal interrelations over time.

It is clear that agricultural practices, especially the typical and well-established ones, shape the territory by giving it visual, chromatic and perceptive aspects recognised by people as distinctive and identity elements of the places they belong to. The "forging" of the territory by the hand of man is made for productive and economic reasons, and it reveals, in legal terms, a further feature of the rural landscape: to be a positive expression of the principle of sustainable development and of the principle (functionally connected) of integration.[59] In rural landscape, farming faces countryside, and this comparison is not only possible but also necessary as farming is a constitutive element, indeed a source of "production" of the landscape; therefore, such as agriculture integrates into the modern conception of the landscape, the landscape's protection requires the exercise of farming activities in an environmentally sustainable way.

A regulatory reference to this integrated approach is to be found, from the point of view of landscape law, in the aforementioned European Landscape Convention,

[59] On the relationship between the principle of sustainable development and the integration one, as defined by EU Law, see Ferrara (2005), pp. 509 et seqq.; Cafagno (2007); Renna (2012), pp. 62–84; Rota (2012), pp. 169 et seqq.

where it reminds the need "to integrate landscape into its regional and town planning policies and in its cultural, environmental, agricultural, social and economic policies, as well as in any other policies with possible direct or indirect impact on landscape" (Art. 5, Lett. d);[60] this option has been incorporated in the Code on Cultural Heritage and Landscape, particularly in the aforementioned Art. 135, Para. 4, Lett. d), where the "preservation of rural landscapes" becomes the limit and, at the same time, the aim of landscape planning. From the point of view of agricultural law, the different rules (previously discussed), especially from the EU, stand out and promote the maintenance or recovery of traditional and typical crops of the area and discourage the allochthonous ones.

The principle of integration faces and addresses conflicting interests underlying the rural landscape: the immaterial cultural one, related to enjoyment and fruition of the landscape by the community, and the material, economic and productive one, typical of the entrepreneurs and the owners. The first interest invokes the administrative functions of protection and enhancement of the rural landscape and clearly shows the intimate connection between the two functions: the landscape incorporates a strong identity, but it can be shaped and can flourish only by the use and, therefore, by the development of the area.[61]

The second interest evokes the classic question of the limits to private property and feeds the ongoing debate on the legal regime of "land" or, in other terms, of "soil".[62] Land is a finite resource, not extendable or renewable and, at the same time, a source of utility not only for its owner but moreso for the community (think, for example, of ecosystem services). Land has an existence value that cannot be captured by the market, not only an exchange and use value. For this reason, the legal doctrine is reconsidering the problem of ownership's rights over the land. Agricultural property supplies both benefits that are an exclusive property of the owner (linked to agricultural production) and non-excludable ecosystem and landscape services for the community.[63]

---

[60] About the importance of the principle of integration in the landscape framework, cf. Cartei (2013), pp. 703–743.

[61] The link between protection and enhancement of the landscape is even closer than the one occurring for the cultural heritage because "if the landscape is perception of the area, its protection implies in itself an element of enjoyment, and thus enhancement" (author's translation): so Casini (2014), pp. 385–396.

[62] Graziani (2005), pp. 45–72; Graziani (2013); Boscolo (2014), pp. 129–146. These scholars focus on the problem of property with reference to "common goods", a category well known to economic sciences and taken from legal doctrine in order to affirm the existence of a *tertium genus* compared to the traditional dichotomy between public goods and private goods. Returning to the definition suggested by the so-called Rodotà Commission (commission established in 2007 at the Ministry of Justice for the amendment of the rules of the Civil Code in matters of public goods), they are goods "which are suffering from a highly critical situation, problems of scarcity and depletion and from the absolute inadequacy of legal safeguards [. . .] defined as things that express functional utility to the exercise of fundamental rights and the free development of the person, and are aware of the principle of intergenerational safeguarding" (author's translation).

[63] Boscolo (2014), p. 135.

The landscape interest and the agricultural one, once conflicting, can integrate, and this mutual influence reveals (at least) two postulates: the functions of protection and enhancement of the landscape, which nourish the interest to public use, must also encompass support measures to agriculture in order to ensure the maintenance or recovery of agricultural practices; the agrarian property may expand only if identity characteristics of the traditional landscape, produced by the agriculture itself, are safeguarded.

References

Abrami A (2008) I nuovi approdi dell'ordinamento giuridico forestale. Diritto e giurisprudenza agraria, alimentare e dell'ambiente 16(9):529–536

Adornato F (2004) Di cosa parliamo quando parliamo di agricoltura. Agricoltura, Istituzioni, Mercati 1(1):5–9

Albisinni F (2002) Commento all'Art. 13 del D.Lgs. 18 maggio 2001, n. 228. Rivista di diritto agrario 81(2–3):459 et seqq.

Albisinni F (2011) Prodotti agro-alimentari e distretti in agricoltura. In: Goldoni M, Sirsi E (eds) Il ruolo del diritto nella valorizzazione e nella promozione dei prodotti agro-alimentari. Atti del Convegno Pisa, 1–2 July 2011. Giuffrè, Milano, pp 103–148

Amirante D (2012) Profili di diritto costituzionale dell'ambiente. In: Dell'Anno P, Picozza E (eds) Trattato di diritto dell'ambiente, vol I. Cedam, Padova, pp 233–275

Angiuli A (2007) Attori e competenze nella tutela e valorizzazione dei beni paesaggistici: il caso dei tratturi di Puglia. Giurisprudenza italiana 146(3):580–586

Bagnoli M (2012) L'IMU rurale fra terreni e fabbricati agricoli. Corriere Tributario 35(24):1826 et seqq.

Boscolo E (2014) Oltre il territorio: il suolo quale matrice ambientale e bene comune. Urbanistica e Appalti 18(2):129–146

Cafagno M (2007) Principi e strumenti di tutela dell'ambiente come sistema complesso, adattativo, comune. Giappichelli Editore, Torino

Canfora I (2011) Agricoltura, tutela del paesaggio e sviluppo delle energie alternative. Rivista di diritto agrario 90(3):304–329

Carrozza A (1994) Lineamenti di un diritto agrario ambientale. I materiali possibili. I leganti disponibili. Rivista di diritto agrario 73(2):151 et seqq.

Carrozza MF (2007) Il diritto ambientale. Rivista di diritto agrario 86(4):495 et seqq.

Cartei GF (1993) Tutela dei parchi naturali e nozione costituzionale del paesaggio. Rivista trimestrale di diritto pubblico 43(3):601 et seqq.

Cartei GF (2003) Il paesaggio. In: Cassese S (ed) Trattato di diritto amministrativo – Diritto Amministrativo speciale, vol II. Giuffrè, Milano, pp 2110 et seqq.

Cartei GF (2013) Autonomia locale e pianificazione del paesaggio. Rivista Trimestrale di Diritto Pubblico 63(3):703–743

Casini L (2014) La valorizzazione del paesaggio. Rivista Trimestrale di Diritto Pubblico 64 (2):385–396

Costato L (2001) L'evoluzione del diritto agrario dal codice civile del 1942 ad oggi. Rivista di diritto agrario 80(1):3–17

Costato L (2008) La lezione di Antonio Carrozza (e qualche possibile fraintendimento). Rivista di diritto agrario 87(1):15–24

Cristiani E (2008) Le riviste agraristiche italiane e i nuovi contenuti del diritto agrario. Rivista di diritto agrario 87(4):464–479

Crosetti A (2008) Le tutele differenziate. In: Crosetti A, Ferrara R, Fracchia F, Olivetti Rason N (eds) Diritto dell'ambiente, 7th edn. Laterza, Roma-Bari, pp 511 et seqq.

De Giorgi Cezzi G (2005) Tutela dell'identità dei luoghi e rilevanza "monumentale" di un ulivo secolare. Foro Amministrativo – TAR 4(9):2955 et seqq. (comment on the decision of TAR Puglia – Lecce, I, 5 July 2005, no. 3611)

De Giorgi Cezzi G (2006) Le "lunghe strade verdi" degli armenti. Gli antichi tratturi tra competenza statale e regionale. Aedon 9(1). http://www.aedon.mulino.it/archivio/2006/1/degiorgi.htm. Accessed 22 Oct 2014

Del Mastro E (2005) La tutela del paesaggio rurale: tendenze evolutive a livello nazionale e comunitario. Aedon 8(2). http://www.aedon.mulino.it/archivio/2005/2/delmastro.htm. Accessed 22 Oct 2014

Ferrara R (2005) I principi comunitari della tutela dell'ambiente. Diritto Amministrativo 13 (3):509 et seqq.

Ferrucci N (2007a) L'agriturismo: icona della trasversalità dell'agricoltura tra impresa, alimentazione, ambiente e paesaggio. Rivista di diritto agrario 86(4):679–683

Ferrucci N (2007b) Riflessioni di una giurista sul tema del paesaggio agrario. Diritto e giurisprudenza agraria, alimentare e dell'ambiente 16(7–8):451–456

Ferrucci N (2008) Il bene forestale come bene ambientale e paesaggistico. In: Crosetti A, Ferrucci N (eds) Manuale di diritto forestale e ambientale, 1st edn. Giuffrè, Milano, pp 562–605

Ferrucci N (2010) Architettura e paesaggio rurale tra permanenze, recuperi e trasformazioni. Diritto e giurisprudenza agraria, alimentare e dell'ambiente 18(5):296–301

Ferrucci N (2011a) Il paesaggio agrario tra Convenzione europea del paesaggio, Codice Urbani e normativa agraristica. Diritto e giurisprudenza agraria, alimentare e dell'ambiente 19 (4):240–244

Ferrucci N (2011b) La tutela del paesaggio e il paesaggio agrario. In: Costato L, Germanò A, Rook Basile E (eds) Trattato di diritto agrario, vol II. Utet Giuridica, Torino, pp 202 et seqq.

Ferrucci N (2013) Un altro passo in avanti verso la tutela giuridica degli alberi monumentali. In: Georgofili.it. http://www.georgofili.info/detail.aspx?id=1212. Accessed 23 Oct 2014

Ferrucci N (2014) Agricultura e Ambiente. Rivista Giuridica dell'ambiente 29(3–4):327 et seqq.

Galloni G (2001) Da una recente ricerca su agricoltura e ambiente. Diritto e giurisprudenza agraria, alimentare e dell'ambiente 10(1):5 et seqq.

Galluzzo N (2010) Ruolo dei distretti e impatto della politica agricola comunitaria sulle aree rurali. Aracne, Roma

Graziani CA (2005) Proprietà della terra e sviluppo rurale. Studi Parlamentari e di Politica Costituzionale 38(3–4):45–72

Graziani CA (2013) Terra. Oltre la proprietà. In: Territorialmente.it. https://www.google.it/search?q=+Graziani+CA+Terra+Oltre+la+propriet%C3%A0%2C+in+www.territorialmente.it&ie=utf8&oe=utf8&aq=t&rls=org.mozilla:it:official&client=firefoxa&channel=sb&gfe_rd=cr&ei=7qhIVP7PBcyO8QeJrIDYBw#; http://www.territorialmente.it/wordpress/wp-content/uploads/2013/06/Oltre-la-propriet%C3%A0-Carlo-Alberto-Graziani.pdf. Accessed 23 Oct 2014

Graziosi B (2012) I nuovi regolamenti comunali per il verde urbano e la pubblicizzazione del "verde privato". Rivista giuridica dell'edilizia 55(6):189–204

Jannarelli A (2006) Pluralismo definitorio dell'attività agricola e pluralismo degli scopi legislativi: verso un diritto post-moderno. Rivista di diritto agrario 85(2):183–203

Jannarelli A (2008) Antonio Carrozza e le nuove sfide per il diritto agrario: brevi meditazioni dopo i convegni pisani in sua memoria. Rivista di diritto agrario 87(1):3–14

La Torre ME (2010) Attività turistica e categorie contrattuali. In: Degrassi L, Franceschelli V (eds) Turismo. Diritto e diritti, 1st edn. Giuffrè, Milano, pp 573 et seqq.

Lombardi R (2009) La candidatura dei paesaggi vitivinicoli del basso Piemonte a "territorio patrimonio dell'Umanità": brevi note a margine della tutela dei paesaggi rurali. Nuove autonomie 18(2–3):389–397

Lucifero N (2010) Paesaggio, agricoltura e territorio. Nuovi modelli di tutela. In: Rook Basile E, Carmignani S, Lucifero N (eds) Strutture agrarie e metamorfosi del paesaggio. Dalla natura delle cose alla natura dei fatti, 1st edn. Giuffrè, Milano, pp 159–279
Maddalena P (2009) Il bosco e l'ambiente. Rivista giuridica dell'ambiente 24(5):635–647
Malfitano A (2002) Alle origini della politica di tutela ambientale in Italia. Luigi Rava e la nuova Pineta "storica" di Ravenna. Storia e futuro 1(1):1–18
Manservisi S (2013a) Alberi monumentali e infrastrutture verdi. Tutela, valorizzazione e incentivi nella recente normativa nazionale e dell'Unione europea. Aracne, Roma
Manservisi S (2013b) Tutela e valorizzazione degli alberi monumentali nella recente normativa nazionale e regionale nel contesto delle linee evolutive della politica ambientale dell'Unione Europea. Rivista di Diritto Agrario 92(2):357 et seqq.
Picozza E (2006) La tutela del paesaggio nelle zone agricole tradizionali. In: Cugurra G, Ferrari E, Pagliari G (eds) Urbanistica e paesaggio: atti dell'8° Convegno nazionale AIDU, Parma, 18–19 November 2005. Editoriale Scientifica, Napoli, pp 82–100
Portaluri PL (2011) L'ambiente e i piani urbanistici. In: Rossi G (ed) Diritto dell'ambiente, 2nd edn. Giappichelli Editore, Torino, pp 241–255
Predieri A (1981) Landscape. In: Enciclopedia del diritto, vol XXXI. Giuffrè, Milano, p 513
Ragionieri MP (2003) L'impresa agricola multifunzionale: l'evoluzione della PAC verso un nuovo modello agricolo. In: Costato L (ed) Trattato breve di diritto agrario italiano e comunitario, 3rd edn. Cedam, Padova, pp 201 et seqq.
Renna M (2012) I principi in materia di tutela dell'ambiente. Rivista Quadrimestrale di Diritto dell'Ambiente 3(1–2):62–84
Rissolio A (2010) Il paesaggio creato dall'uomo: profili di evoluzione normativa. Teoria e storia del diritto privato 3(3):197 et seqq.
Rota R (2012) Profili di diritto comunitario dell'ambiente. In: Dell'Anno P, Picozza E (eds) Trattato di diritto dell'ambiente, vol I. Cedam, Padova, pp 151–231
Salvia F (2012) Manuale di diritto urbanistico. Cedam, Padova
Sciaudone S (2007) Commento dell'Art. 3 della L. 20 febbraio 2006, n. 96. Le nuove leggi civili commentate 30(1–2):146 et seqq.
Sciullo G (2008) Il paesaggio fra la Convenzione e il Codice. Aedon 11(3). http://www.aedon.mulino.it/archivio/2008/3/sciullo2.htm. Accessed 22 Oct 2014
Sereni E (1961) Storia del paesaggio agrario italiano. Laterza, Roma-Bari
Strambi G (2006) Le strade del vino, dell'olio e dei sapori: il quadro giuridico di riferimento. Rivista di diritto agrario 85(2):204–238
Urbani P (2006) Governo del territorio e agricoltura. I rapporti. Rivista giuridica dell'edilizia 49 (3):117–124
Urbani P (2010) Le aree agricole tra disciplina urbanistica e regolamentazione dell'attività economica. Rivista giuridica dell'edilizia 53(1):29–48
Valletta M (2004) Valorizzazione del patrimonio abitativo rurale. Diritto e giurisprudenza agraria e dell'ambiente 12(5):341–344
Varotto M (2005) Nuovi modelli di "*governance*" dello sviluppo locale: i distretti rurali. Azienditalia 20(5):2–20
Ventura F (2001) Alle origini della tutela delle "bellezze naturali" in Italia. In: Cazzato V (ed) Istituzioni e politiche culturali in Italia negli anni Trenta, vol II. Istituto poligrafico e Zecca dello Stato, Roma, pp 555–576

The Agri-Food Market and Eco-Oriented Consumer Law: Towards a New Model

S. Tommasi

Abstract The agri-food sector requires a comprehensive and cross-cutting policy that also meets the needs of environmental protection. Environmental issues are always dealt with in different ways, depending on the socio-cultural context at any given time. One salient element that emerges, however, is the overexploitation of land, and permanent environmental damage is closely linked to an approach that considers agriculture exclusively as a question of producing goods. The increasing levels of welfare in industrialized countries and the ability to produce a surplus over and above domestic demand have led policymakers and practitioners to neglect the basic function of agricultural activities: providing food for people. Taking this function as a starting point, we can demonstrate the need to regulate agricultural activity in a way that goes beyond the demands of market logic. We must also recognize the importance of standards in agri-food law that enable consumers to make informed choices. In contemporary society, consumers often choose products not only for their quality or price but on the basis of other intangible values. A growing share of consumption choices take into account the environmental impact of production processes or, more generally, their compliance with ethical rules also linked to energy consumption and the proximity of the market to the area of production. In this sense, we can detect a clear spread of aggregation methods that go beyond the usual patterns [of consumer behavior] and the [growing] role of the consumer in the shift from market-oriented to eco-oriented choice.

Keywords Agroecology • Consumer law • Sustainable agriculture

S. Tommasi (✉)
Department of Legal Sciences, University of Salento, Lecce, Italy
e-mail: sara.tommasi@unisalento.it, https://www.unisalento.it/web/guest/scheda_personale/-/people/sara.tommasi

M. Monteduro et al. (eds.), *Law and Agroecology*,
DOI 10.1007/978-3-662-46617-9_16

1 Introduction

The relationship between law, agriculture and the environment is affected by changes in society and by technological development.[1] Two examples may be sufficient to explain this. First, there are the changes related to the collapse of the feudal system and the consequent shift from an extremely hierarchical social order to a society based on the formal equality of subjects.[2] Second, there are the even stronger transformations associated with business in the form of agricultural enterprise. Agricultural activity is no longer exclusively oriented to meet the food needs of the territory in which it operates, but it is influenced by the signals coming from the market.[3] At this point, the overexploitation of the territory begins with the use of products that poison the land and groundwater, endangering their future use.[4] While this certainly does not justify nostalgic calls for a return to the past, it may help to rediscover the basic role of agriculture for human and animal nutrition or the awareness that agrarian law needs to deal with systemic problems because it is at this level that the recent transformations have occurred.[5]

Another element that merits examination is the circular relationship between agriculture and the environment in the sense that agricultural activity without rules to safeguard the environment also takes place in an environment that cannot safeguard agriculture, its products and those who consume them.[6] What is good for nature is good for man, and that is why the agricultural policy of the future—as emerges clearly from COM(2011) 631—should go beyond strictly economic aspects and develop as a strategic policy taking into consideration food security, health and the environment.[7] The link, then, is not only between agriculture and the environment but also between agriculture, the environment and consumer health and safety.

This is confirmed by a great deal of data and is in line with the provisions of Art. 169 of the Treaty on the Functioning of the European Union (TFEU), according to which the purpose of ensuring a high level of consumer protection is closely linked to the protection of health.[8] This linkage is confirmed, for instance, by the European programs on public health and COM(2011) 709 final, on the establishment of the third program of Community action in the field of health for the period 2014–2020,[9]

[1] On this point, see Jannarelli (2013a), pp. 11–35; Bologna (2010), pp. 359–362; Maddalena (2007), pp. 477–482.

[2] Jannarelli (2013b), pp. 405–438.

[3] Jannarelli (2013a), p. 22.

[4] Jannarelli (2013a), p. 24.

[5] Jannarelli (2013b), p. 405.

[6] Leccese (2011), p. 45.

[7] COM (2011) 631 final.

[8] Tamponi (2011a), pp. 579–616.

[9] Cf. http://eur-lex.europa.eu. Accessed 2 Oct 2014.

where the role of the factors in the prevention of environmental health risks is recognized.[10] Decision 2008/721/EC of 5 September 2008 is already significant since it sets up an advisory structure of Scientific Committees and experts in the field of consumer safety, public health and the environment.[11] Consumers should be aware of the close relationship between problems connected to food consumption, as emerges significantly in the drawing up of European food laws that are based on the need to provide the consumer with information useful to ensure the deliberateness of his choices on health issues.[12] On the other hand, the need for an effective system of prevention and control of food-related health risks are monitored by the White Paper on Food Safety, COM(1999) 719 final, and Regulation 178/2002/EC, which constitute a key moment in the vast and varied program of Community action in this field[13] and which form the basis for ensuring a high level of protection of human health and consumer interests in relation to food.[14] Regulation 1924/2006/EC[15] on nutritional information and Regulation 1169/2011/EC,[16] dictated in terms of food information to consumers, amending Regulation 1924/2006/EC, have also had an important role as they affirm that the free movement of safe and wholesome food is an essential aspect of internal market and contributes significantly to the health and well-being of the citizens. This regulation is aimed also at integrating the general principles on Unfair Commercial Practices Directive 29/2005,[17] through specific rules concerning the provision of food information to consumers. Consumer choices can be influenced by, *inter alia*, health consideration,[18] and it is a general principle of food law to enable consumers to assume informed choices in relation to food they consume and to prevent any practices that may mislead them.[19]

[10] For the previous programs, see White Paper Together for Health: A Strategic Approach for the EU 2008–2013 COM(2007) 630 final and to Decision 1350/2007/EC of the European Parliament and of the Council of 23 October 2007 establishing a second programme of Community action in the field of health (2008–2013). http://europa.eu/legislation_summaries/consumers/consumer_safety/I28153_it.htm. On this point see Sirsi (2011), pp. 496–523.

[11] Cf. http://eur lex.europa.eu. Accessed 2 Oct 2014.

[12] Jannarelli (2012), pp. 38–46; Canfora (2012), pp. 114–138; Costato (2011a), pp. 1–18; Ragusa (2011), pp. 457–487.

[13] On this point, see Jannarelli (2011), p. 144.

[14] Article 1 of Regulation 178/2002/EC of the European Parliament and of Council, of 28 January 2002, stipulating general principles and requirements of food law, establishing the European Food Safety Authority and stipulating procedures in matters of food safety, in http://eur-lex.europa.eu.

[15] On this point, see Costato (2008), pp. 299–316.

[16] Jannarelli (2012). Regulation 1169/2011 is read in the more general framework of regulation in order to give direct orders to information processes that are the basis of market relations. Albisinni (2012), pp. 66–78; Canfora (2012); Giuffrida (2012), pp. 79–93; Forti (2012), pp. 94–113; Ragusa (2011), p. 480, which points out that the regulation concerns allowing consumers to obtain relevant information about the products they're interested in, apart from labeling, which, although a necessary tool, is not the only one possible or necessarily desirable because it is not usable for nonprepacked products; Russo (2012), pp. 47–65.

[17] Jannarelli (2012), p. 41.

[18] Whereas 3) Regulation 1169/2011/EC.

[19] Whereas 4) Regulation 1169/2011/EC.

2 Negative Externalities Related to the Weakness of the Agriculture–Environment Linkage

Exploring the relationship between agriculture and the environment is required not only in terms of deontic logic or value but also legally. Agriculture should not be seen as a mere recipient of prohibitions or obligations made with a view to protecting environmental assets but as functionalized activities to protect the environment as a whole.[20] In particular, agriculture is a privileged perspective that allows consumers to understand the intrinsic transversal dimension of the environment[21] and its essential function of allowing any primary human expression.[22] Agricultural activities are, by definition, both productive *and* conservative. Alongside the soil, agriculture needs to preserve natural resources, even if only to maintain its productive capacity.[23]

The lack of attention, particularly at a legal level, on the link between agriculture and the environment is also exposed to numerous irreversible risks and negative externalities. Let us consider the use of agricultural land on the basis of more convenient production locations, even if completely independent of the link between agricultural produce and foodstuffs. Indeed, the food outlet of agricultural crops has become a simple alternative compared to other possible outlets. This is the case, for instance, of the installation of wind farms or solar panels to produce electricity. The use of land for non-agricultural purposes has a more innovative range than in the past and is likely to alter the previous frame by requiring the introduction of legal measures. Unfortunately, these have often been written hastily and do not always cope with the complexity of the matter.[24] The key issue is the possible conflict between opposing interests of collective importance.[25] Such a conflict requires a proper balance, which cannot certainly be left to casual and not sufficiently meditated solutions.[26] It is not only the suggestively proposed

[20] Carmignani (2012), p. 98.

[21] Bigliazzi Geri (1987), pp. 495–507.

[22] Carmignani (2012), p. 155.

[23] Carmignani (2012), p. 91.

[24] Cf. Tamponi (2011b), pp. 481–495.

[25] Canfora (2011), pp. 304–329.

[26] Brambilla (2013), pp. 406–421. The significant point is on the Communication from the Commission to the European Parliament, the Council, the Economic and Social Committee and the Committee of the Regions—Flagship initiative under Europe 2020 Strategy, COM(2011) 21 final, where we read that in order to operate appropriate choices, even over longer distances, we need to examine the entire cycle of how we use resources, including the value chain and trade-offs among different priorities. To have the necessary information in order to evaluate different choices will help policymakers to decide where to focus their action. Among many examples, reference is made to the fact that the use of the land to produce foodstuffs may compete with the use of the land for energy, and both uses may be an obstacle to the competitive use of soil that encourages biodiversity or acts as an ecosystem, for example by absorbing carbon from the atmosphere.

alternative between flour or gasoline[27] but also the possible friction between the development of economic activities—meant to reduce the emissions of pollutants and greenhouse gases—and the protection of the landscape and the environment in its physical appearance.[28] We should also distinguish between the use of agricultural soils for forms of non-polluted energy production and the use of such soils as mere land surface on which to install structures that produce alternative energy. Only the former utilization should be considered as an agricultural activity since it relies on natural resources and involves activities based on the life cycle of organisms.

Insufficient attention paid to the link between agriculture and the environment can have serious consequences for the environment and society. For example, the relationship between society and the market has been reversed: it is not that the individual and companies conform to the market relationships of agricultural produce but that a purely economic logic outlines the overall structure of social relationships without the law succeeding in its primary function as guarantor of balanced coexistence and in simultaneously satisfying the economic and non-economic interests involved.[29] We should not forget, moreover, that large areas are now removed, long term, from their natural fruition and utilization at the local level. The seriousness of this fact is demonstrated by a simple observation. On the basis of the information handed down from father to son, some populations grow herbs and plants that are part of their diet and rich in substances that have therapeutic function in treating many diseases. The experience of men, even in countries in the developing world, has done much more than multinational companies, which are only ready to patent the therapeutic substance isolated through complex chemical procedures.[30]

Agricultural activity also focuses on the more fertile soil of the plains, so the hilly and mountainous areas are increasingly neglected with serious consequences for the overall condition of the territory.

The damage is also significant from the point of view of food as mountainous areas are often those where there is the production of a diverse range of quality

[27] Adornato (2008), pp. 5–8.

[28] On these issues, and the case law of the Constitutional Court on the point, see Canfora (2011), p. 311; De Leonardis (2005), pp. 889–914, he shows that if, in fact, on the one hand, those that, with an effective metaphor, have been called "windmills" of our century can contribute greatly to the reduction of greenhouse gases, on the other hand, it can have a negative effect on the landscape: it is well known that the areas of greatest wind potential are those of the ridges, hills and mountains, all extremely scenically relevant.

[29] Jannarelli (2013b), p. 405.

[30] Germanò (2011), pp. 589–603. A particular attention to safeguard production methods and traditional recipes is found in Art. 17, Regulation (EU) 1151/2012 of the European Parliament and of the Council on quality schemes for agricultural products and foodstuffs, where it is stated that a scheme for traditional specialties guaranteed is established to safeguard traditional methods of production and recipes by helping producers of traditional product in marketing and communicating the value-adding attributes of their traditional recipes and products to consumers. On this point, see Costato (2012), pp. 648–667.

products. The problem is not overlooked at Community level. On the point we should note Regulation (EU) 1151/2012 of the European Parliament and the Council on quality schemes for agricultural products and foodstuffs.

The reference is particularly in whereas 4 of the Preamble, where we read that operating quality schemes for producers that reward them for producing a diverse range of quality products can benefit the rural economy. This is particularly the case in less-favored areas, in mountain areas and in the most remote regions, where the farming sector accounts for a significant part of the economy and production costs are high. Of particular interest is also Art. 31 of the said Regulation, where it is stated that "mountain product" is established as an optional quality term.[31]

This regulation is also central to the provisions referring to "labelling for local agriculture and direct sales." The new labeling can indeed help protect the environment by reducing the environmental costs of long-distance transport, reducing waste on the assumption, for example, of packaging reduction.[32] Therefore, labeling takes on a central role for the protection of the environment, even setting aside the reference to the quality of the product and only for the mere fact that it certifies the origin of the same produce from a geographical neighbor.[33]

Commission's Green Paper on promotion measures and information provision for agricultural products: a reinforced value-added European strategy for promoting the tastes of Europe, COM(2011) 436 final is particularly sensitive on the promotion of local food products. It states that regional and local farming has hidden potential that is not currently fully exploited. Regional and local markets are an essential meeting place for producers and consumers. They enable the former to receive the rewards for their labors more efficiently and the latter to contribute to the development of their local areas, reduce the environmental impact of their consumption habits and access a wide variety of products rooted in local traditions and ways of life.[34]

Protecting local communities and the maintenance of hill and mountain areas also mean trying to go beyond the current model of market enterprise, to the benefit of the plurality of existing agricultural models. The resources of the Common Agricultural Policy (CAP) have been oriented towards the type of intensive, specialized, capitalized enterprise in a substantially unidirectional way. Yet diversity is a great asset for Italy, and maintaining and disseminating a variety of agricultural models are crucial if we are to address the current crisis adequately. The economic crisis increasingly demonstrates the failure of the current industrial model of agriculture.[35]

---

[31] On the important profiles regarding the "option quality terms" (according to the terms of Regulation (EU) 1151/2012), as related to the Commission's resistance to any initiative by the Member States for further indications on labeling, see Costato (2012), p. 660.

[32] Canfora (2013), pp. 149–161.

[33] Canfora (2013), p. 151.

[34] Canfora (2013), p. 150.

[35] Koo and Jiang (2014), pp. 279–302.

According to a recent decision of the Italian Constitutional Court, the products that are grown on a limited portion of territory may be deemed necessary for the protection of the environment. For this reason, the Court has justified those public tenders for catering services that establish as a preferential requirement the use of food products originating from near territories. The Court specifies that unlike the priority given to individuals who use goods whose transport leads to a reduced amount of harmful emissions—this priority is justified by the benefits that the limitation of emissions gives in terms of environmental protection—the priority, accorded to those subjects who use the products transported exclusively within a region, regardless of their level of emissions, constitutes a measure that has an equivalent effect to the one prohibited by Article 34 TFEU—which includes all trading rules that may hinder, directly or indirectly, actually or potentially, intra-Community trade—and cannot be justified under Article 36 TFEU. In other words, the preference for a product made within a limited territory can be given for environmental requirements, such as those expressed by reference contained in the regional appealed law, referring to the level of carbon dioxide emissions during transport, and not for the mere regional origin of the goods; this law by itself does not guarantee that the goods are actually kilometre zero and that their transport has a minor negative impact on the environment.[36]

3 New Challenges of Agricultural Law with Particular Focus on the Role of Consumers

The importance of promoting local food products has clearly been seen. The theme is closely linked to the need, even in the context of agri-food law, which recognizes the importance of standards that enable the consumer to make an informed choice.[37] It is not uncommon for consumers to choose products not just for their taste or price but also because of their origin or of certain intangible qualities of the same product.[38] A growing share of consumption choices take into account the environmental impact of production processes or, more generally, compliance with ethical rules related to them, for example, in relation to energy consumption.[39]

A recent survey, initiated at Community level, reveals that more than three-quarters of consumers would be willing to pay more for environmentally friendly products if they had the certainty of the effective protection of the environment. At

[36] Italian Constitutional Court, 6 December 2013, n. 292.

[37] On the need to strengthen the position of consumers with clear information, see Rubino (2012), pp. 668–679; Jannarelli (2012), p. 38. On the notion of consumer, see Costato (2011b), pp. 19–52; Lucifero (2011), pp. 321–422; Saija and Tommasini (2011), pp. 493–532.

[38] Li et al. (2014), pp. 69–87; Costantino (2013), pp. 166–189.

[39] Carmignani (2012), p. 155.

the same time, only slightly over half of EU citizens consider themselves well informed about the environmental impact of the products they buy and use.[40]

The EU's current focus on consumers is a known datum, and it would be out of place here to retrace the normative data that confirm what has already been said.[41] Instead, we must look at the future objectives of European consumer policy, a policy that—at least according to the *Proposal for a Regulation of the European Parliament and of the Council on a consumer programme 2014–2020*, COM(2011) 707 final—puts the consumer at the center of the Single Market.[42] The proposal, which follows the program of *Community action in the field of consumer policy 2007–2013*,[43] states that it has become increasingly apparent that at a time when Europe needs new sources of growth, consumer policy is one area that can make a significant contribution in meeting Europe 2020 objectives, and the more consumers are able to make informed decisions, the greater the impact they can have on strengthening the Single Market and stimulating growth. Empowered consumers who are well protected and able to benefit from the Single Market can thus drive innovation and growth by demanding value, quality and service.[44]

The awareness that consumer choices can contribute to sustainable growth, based on an efficient use of resources, appears obvious even from European Consumer Agenda[45] and "Horizon 2020" initiatives to deepen the scientific study of consumer behavior and the implications for health, safety and sustainability of consumer choices.[46] It is no coincidence that the European Consumer Agenda pays particular attention to sustainable consumption intended as a model of behavior that cares about the environmental impact of the products purchased,[47] all the more because rising consumption around the globe has increased pressure on the environment, including climate change, and created greater competition for resources.[48]

---

[40] In http://www.synergy-net.info/default.cfm?fuseaction=link&id=3,39160, 9. Accessed 10 Oct 2014.

[41] On the point, see Rubino (2012), p. 668; Ragusa (2011), p. 457; Jannarelli (2012), p. 38; Korn (2012), pp. 663–709.

[42] In http://ec.europa.eu/consumers, 2. Accessed 10 Oct 2014.

[43] Decision 1926/2006/EC of the European Parliament and of the Council of 18 December 2006 establishing a programme of Community action in the field of consumer policy (2007–2013).

[44] COM(2011) 707 def. 2; COM(2012) 225 final.

[45] COM(2012) 225 final.

[46] Proposal for a Regulation establishing a framework program for research and innovation, COM (2011) 809. On this point, see COM(2012) 225 final, 1.

[47] Cohen (1998).

[48] COM(2012) 225 final, 4, stipulates that consumers should be empowered, assisted and encouraged to make sustainable and healthy choices that will lead to cost saving for themselves and for the society as a whole. Consumers have the right to be informed on the environmental impact throughout the life cycle of the products (goods and services) they intend to buy. They should be supported in identifying truly sustainable choices. Effective tools are needed to protect them against misleading and unfounded environmental and health claims.

The connection between the issues related to agri-food products and consumer protection has been made clear in the Italian Consumer Code since its earliest general provisions, which show how, in accordance with the principles contained in the Treaties, to harmonize and reorder legislation on purchasing and consumption processes in order to ensure a high level of consumer and end-user protection. In the same direction, Article 2(2) of the Italian Consumer Code recognizes as fundamental, among other things, the rights to protection of health, safety and product quality; to appropriate information; to the exercise of trade practices according to the principles of good faith, fairness and [brand] loyalty and consumer education.[49] However, Article 6 of the Italian Consumer Code relative to the minimum content of information for the consumer is not applied since food is the subject of specific provisions in the field of information.

Legislation on unfair commercial practices also makes significant references to food issues by thus confirming the suitability of the rules protecting the consumer in order to have a positive impact on the organization of the food market. In this sense, Directive 2005/29/EC qualifies in absolute terms as unfair commercial practices those "misleading commercial practices falsely claiming that a product is able to cure illnesses, dysfunction or malformations."[50] This approach has been reaffirmed by the *Green Paper on unfair trading practices in the Business to Business food and non-food supply chain in Europe* COM(2013) 37 final.[51] Advertising may constitute an unfair commercial practice:[52] to determine misleading advertising indications of geographical origin of the goods should also be taken into account. Comparative advertising shall be permitted for products with designation of origin; in each case, it relates to products with the same designation and it does not take unfair advantage of the reputation of a trademark, product name or other distinguishing marks of a competitor or of the designation of origin of competing products.[53]

All this shows that consumer protection is closely linked to procedures regulating the production of agricultural food as well as to the adoption of measures protecting the environment and the consumers' right to health.

[49] Tamponi (2011a), p. 593.

[50] On this point, see Costantino (2011), pp. 230–247.

[51] In http://eur-lex.europa.eu. Accessed 5 Oct 2014.

[52] On the impact of the Directive in the food sector, see Di Lauro (2011), pp. 547–578.

[53] Tamponi (2011a), p. 593.

4 Agricultural Products as Mere Commodities: An Unsuitable Approach

The primary function of agricultural products is to satisfy man's basic food needs. Taking this function as a starting point, a regulation of agricultural activities responding exclusively to a market logic would be manifestly inadequate.[54]

The relationship between agriculture and the environment cannot be entrusted to a discipline based solely on incentives and sanctions. A producer who has the ability to pay the penalty, or who cannot benefit from ads, may not want to take environmental rules on board to the detriment of those who respect them. Therefore, we need to change the methods of carrying out agricultural activities. Environmental protection should be adopted as a basic rule of conduct for operators in the industry.

Monetization, moreover, cannot be used as a long-term strategy, which would irreparably lead to a loss of resources, hence the need for efficient legal techniques of enforcement. The value of the environment safeguard seems invisible to the economy that guides worldwide choices of politics. Yet the natural capital is the basis for the well-being of society and individuals. What happens is that man destroys the basis of this natural capital, even before recognizing the value of what is being lost. The continuing depletion of and damage to soil, water and other biological resources impact negatively not only on our health, food safety and consumer choices but also on opportunities for agricultural enterprises.[55] Sustainability must be reconceived, therefore, as a business model, which is normatively relevant and able to impact on agricultural law.[56] Only in this way sustainability can be realized.

It is no wonder that it is the primary sector of the economy, namely agriculture, that plays the role of a sentinel to give warning of failures resulting from an economic policy and law projected exclusively to the globalization of the economy and the free market. The objective is to have a territorially balanced EU agriculture and environment actions within an open economic environment. This requires a strong public policy because the agricultural goods cannot be adequately remunerated and regulated through the normal market functioning.[57]

In addition to the approaches highlighted above, the unsustainability of an approach that treats agricultural products as mere commodities points towards even more dramatic developments.

The vital function of agricultural products means thinking of their constitutional value as in the case, for example, of work. Work is formally recognized as a constitutional right precisely because it cannot be treated as a commodity. On the

---

[54] For the necessity of following new legal perspectives, see Adornato (2012), pp. 405–414; Borghi (2012), pp. 3–36.

[55] Bologna (2010), p. 359.

[56] Jannarelli (2013a), p. 12.

[57] COM(2010) 672.

other hand, the unique bond between agricultural law and labor law has not escaped the doctrine that, in fact, shows that bond precisely because of the crucial—or rather vital—role of agriculture and work for human beings, for their health and dignity.[58] Agriculture, which has a functional complementarity with the environment, satisfies not only man's basic need for food but also his need, as well as primary, to live in a healthy environment.

References

Adornato F (2008) Farina o benzina? Il contributo dell'agricoltura ad un nuovo modello di sviluppo. Agricoltura Istituzioni Mercati 5:5–8

Adornato F (2012) La sicurezza alimentare tra primato della tecnica e crisi del diritto. Rivista di Diritto Agrario 91:405–414

Albisinni F (2012) La comunicazione al consumatore di alimenti, le disposizioni nutrizionali e l'origine dei prodotti. Rivista di Diritto Agrario 91:66–78

Ascarelli T (1959) Ordinamento giuridico e processo economico. Giuffrè, Milano

Bigliazzi Geri L (1987) Divagazioni su tutela dell'ambiente ed uso della proprietà. Rivista critica di diritto privato 5:495–507

Bologna G (2010) Biodiversità e scienza della sostenibilità. Ital J Forest Mountain Environ 65:359–362

Borghi P (2012) Diritto alimentare italiano e integrazione europea. Rivista di Diritto Agrario 91:3–36

Brambilla P (2013) Produzione di energia da fonti rinnovabili e tutela delle aree protette: le nuove frontiere del bilanciamento degli interessi. Rivista Giuridica dell'Ambiente 28:406–421

Canfora I (2011) Agricoltura, tutela del paesaggio e sviluppo delle energie alternative. Rivista di Diritto Agrario 90:304–329

Canfora I (2012) La responsabilità degli operatori alimentari per le informazioni sugli alimenti. Rivista di Diritto Agrario 91:114–138

Canfora I (2013) I marchi regionali di qualità e la correttezza dell'informazione dei consumatori: libera circolazione delle merci vs. tutela dell'agricoltura locale? Rivista di Diritto Agrario 92:149–161

Carmignani S (2012) Agricoltura e ambiente (Le reciproche implicazioni). Giappichelli, Torino

Cohen J (1998) Quante persone possono vivere sulla terra. Il Mulino, Bologna

Costantino L (2011) Il ruolo dell'Autorità Garante della Concorrenza e del Mercato nel settore agroalimentare. In: Costato L, Germanò A, Rook Basile E (eds) Trattato di diritto agrario, vol 3. Utet, Torino, pp 230–247

Costantino L (2013) La tutela del contraente debole nelle relazioni negoziali lungo la filiera agroalimentare nelle più recenti esperienze giuridiche europee e statunitensi. Rivista di Diritto Agrario 92:166–189

Costato L (2008) Le indicazioni nutrizionali nel reg. n. 1924/06. Rivista di Diritto Agrario 87:299–316

Costato L (2011a) La politica agricola comune nel settore dell'alimentazione. In: Costato L, Germanò A, Rook Basile E (eds) Trattato di diritto agrario, vol 3. Utet, Torino, pp 1–18

Costato L (2011b) Principi e requisiti generali della legislazione alimentare. In: Costato L, Germanò A, Rook Basile E (eds) Trattato di diritto agrario, vol 3. Utet, Torino, pp 19–52

[58] Ascarelli (1959), p. 39. On the point, see Jannarelli (2013b), p. 405.

Costato L (2012) Il regolamento n. 1151/2012 del Parlamento Europeo e del Consiglio sui regimi di qualità dei prodotti agricoli e alimentari. Rivista di Diritto Agrario 91:648–667

De Leonardis F (2005) Criteri di bilanciamento tra paesaggio ed energia eolica. Diritto Amministrativo 13:889–914

Di Lauro A (2011) La comunicazione e la disciplina della pubblicità dei prodotti alimentari. In: Costato L, Germanò A, Rook Basile E (eds) Trattato di diritto agrario, vol 3. Utet, Torino, pp 547–578

Forti A (2012) Determinazioni nutrizionali. Rivista di Diritto Agrario 91:94–113

Germanò A (2011) Le nuove sfide del diritto dell'agricoltura (nuovi cibi e nuove norme). Rivista di Diritto Agrario 90:589–603

Giuffrida M (2012) Pratiche leali di informazione e informazioni volontarie. Rivista di Diritto Agrario 91:79–93

Jannarelli A (2011) Profili giuridici del sistema agro-alimentare tra ascesa e crisi della globalizzazione. Cacucci, Bari

Jannarelli A (2012) La fornitura di informazioni sugli alimenti ai consumatori nel nuovo reg. n. 1169/2011 tra l'onnicomprensività dell'approccio e l'articolazione delle tecniche performative. Rivista di Diritto Agrario 91:38–46

Jannarelli A (2013a) Il divenire del diritto agrario italiano ed europeo tra sviluppi tecnologici e sostenibilità. Rivista di Diritto Agrario 92:11–35

Jannarelli A (2013b) I prodotti agricoli tra alimenti e merci: alle radici moderne dell''eccezionalità' agricola. Rivista di Diritto Agrario 92:405–438

Koo WW, Jiang Y (2014) Estimating the local effect of weather on field crop production with unobserved producer behavior: a bioeconomic modeling framework. Environ Econ Policy Stud 16:279–302

Korn R (2012) Tutela dell'ambiente, consumatori e responsabilità sociale d'impresa: i nuovi strumenti della sostenibilità aziendale. Contratto e impresa Europa 17:663–709

Leccese E (2011) Danno all'ambiente e danno alla persona. Angeli, Milano

Li X, Clark CD, Jensen KL, Yen ST (2014) Will consumers follow climate leaders? The effect of manufacturer participation in a voluntary environmental program on consumer preferences. Environ Econ Policy Stud 16:69–87

Lucifero N (2011) La comunicazione simbolica nel mercato alimentare: marchi e segni del territorio. In: Costato L, Germanò A, Rook Basile E (eds) Trattato di diritto agrario, vol 3. Utet, Torino, pp 321–422

Maddalena P (2007) L'ambiente: riflessioni introduttive per una sua tutela giuridica. Ambiente e sviluppo 21:477–482

Ragusa VE (2011) I prodotti alimentari 'preincartati' e gli alimenti 'preimballati per la vendita diretta', alla luce del regolamento n. 1169/2011 relativo alla fornitura di informazioni sugli alimenti ai consumatori. Rivista di Diritto Agrario 91:457–487

Rubino V (2012) La responsabilità degli operatori del settore alimentare per violazione degli obblighi informativi del consumatore dopo il regolamento (UE) n. 1169/2011. Rivista di Diritto Agrario 91:668–679

Russo L (2012) Deleghe normative e atti di esecuzione nel reg. UE n. 1169/2011. Rivista di Diritto Agrario 91:47–65

Saija R, Tommasini A (2011) La disciplina giuridica dell'etichettatura degli alimenti. In: Costato L, Germanò A, Rook Basile E (eds) Trattato di diritto agrario, vol 3. Utet, Torino, pp 493–532

Sirsi E (2011) Il diritto all'educazione del consumatore di alimenti. Rivista di Diritto Agrario 90:496–523

Tamponi M (2011a) La tutela del consumatore di alimenti nel momento contrattuale: valore delle indicazioni obbligatorie e volontarie nella formazione del contratto. In: Costato L, Germanò A, Rook Basile E (eds) Trattato di diritto agrario, vol 3. Utet, Torino, pp 579–616

Tamponi M (2011b) I diritti della terra. Rivista di Diritto Agrario 90:481–495

Emerging Land-Use Cross-Scale Patterns and the Pirsig's Monkey Trap

G. Zurlini, T. Semeraro, R. Aretano, M.R. Pasimeni, A. De Marco, I. Castorini, N. Zaccarelli, and I. Petrosillo

Abstract We want to draw the attention to some emerging land-use cross-scale patterns resulting from social-economic factors and associated with an historical characteristic sequence of different land-use regimes that could indicate overregulation in social-ecological landscapes (SELs). We postulate that these emerging patterns with clearly defined spatial areas with fixed rules and increasing merging and enlargements of specific functions in some SEL locations are early warning signal of regime shifts and can be typical in many different human-dominated parts of the world. This current overall tendency could make in fact land administration inflexible, and planning may reinforce rigidity, erode resilience, and promote regime shifts and collapse in SELs instead of the adaptability required to counter surprises due, for instance, to climate change. The problem we presently face is how a "static" and "ordered" landscape condition in SELs, provided by the cross-scale intersections of land use, plans, and norms can be made sustainable in face of unpredictable disturbance and change. If we don't have proper mechanisms to monitor and predict changes and if we are not able to adapt through feedback mechanisms to changes in the environment, we might get stuck in a rigidity trap like the Pirsig's monkey and we are at high risk for failing. We show that a potential way to address such issues is to look at recent trends of different land-use regimes, along with a simple framework to interpret resulting spatial patterns across scales. We provide examples of this approach and discuss what a cross-scale land-use pattern could mean, what it tells about the condition of SELs, and what the effects could be of changing observed conditions in SELs because of, for instance, climate change. We exercise the approach for the Apulia region in southern Italy taking advantage of recent historical trends observed in main drivers and of the rich information provided by cross-scale pattern analysis in the pattern transition space provided by classic neutral landscape models. We suggest that the degree to which the observed pattern departs from a particular neutral model can indicate whether major constraints or organizing structure has been placed on the landscape

G. Zurlini (✉) • T. Semeraro • R. Aretano • M.R. Pasimeni • A. De Marco • I. Castorini • N. Zaccarelli • I. Petrosillo
Landscape Ecology Laboratory, Department of Biological and Environmental Sciences and Technologies (DISTEBA), University of Salento, Lecce, Italy
e-mail: giovanni.zurlini@unisalento.it, https://www.scienzemfn.unisalento.it/scheda_docente/-/people/giovanni.zurlini

M. Monteduro et al. (eds.), *Law and Agroecology*,
DOI 10.1007/978-3-662-46617-9_17

and how those landscapes might evolve/react to additional variation due to land use and climate change. The degree of overregulation provided by cross-scale patterns of land use is a warning to planners and managers that the problem is becoming widespread and can no longer be addressed simply with short-term and local-scale solutions. To manage a transition toward more environmentally efficient and, therefore, more sustainable land use, we should design and manage landscape elements and structure to create less contagious and more heterogeneous landscapes. Nevertheless, we have to change societal values at the root of overregulation and rigidity. We have to be aware that we might get stuck in a rigidity trap to appreciate the similarity of our common condition and to start real cooperation.

Keywords Cross-scale patterns • Landscape overregulation • Neutral landscape models • Rigidity traps • Social-ecological landscapes

1 Introduction

By tradition, land planning addresses the spatial arrangement given by the composition and configuration of landscape elements (e.g., land uses and habitats) contained in a heterogeneous landscape[1] and represents an embodiment of hypotheses about the effects of landscape change.[2] Planning is a typical decision-making process under uncertainty[3] resulting from social-political, economic, and technological changes; climate and natural calamities (e.g., earthquakes and floods); and rapid changes in demand and prices.[4]

Some of the uncertainties are intrinsic in the process of planning itself, like disturbance regimes that are no longer thought as rare, external events, but rather as intrinsic and inherent features of complex adaptive system dynamics.[5] Others arise from the shortcoming of the techniques of planning and in the planners themselves.[6] Any kind of planning, for example, is based on forecasts that are mere estimates about the future, and any change in the anticipated situation may render plans ineffective for complex adaptive systems like social-ecological landscapes (SELs).[7] Furthermore, there are growing demands on space from residential and economic activities, which leads to uncertainty among them, in order to define who has to manage and control these processes to improve economic, social, and ecological performances of SELs.[8]

[1] Marsh (2005).

[2] Nassauer and Corry (2004), pp. 343–356.

[3] Kato and Ahern (2008), pp. 543–559.

[4] Owens and Cowell (2011).

[5] Levin (1999) and Gunderson and Holling (2002).

[6] Kato and Ahern (2008).

[7] Berkes and Folke (1998); Zaccarelli et al. (2008), p. 26.

[8] Dijst and Schenkel (2002), pp. 1–18.

In addition, the last 20 years have shown a growing emphasis on a rather nonhierarchical role for local authorities, and an important role has been created for various stakeholders, organized at different levels (e.g., from household to global) in a panarchy of SELs.[9] Such stakeholders have often differing views as to which system states are desirable or which ecosystem services are to be exploited. This has introduced a new type of uncertainty for planners as top-down, centralized, and hierarchical management of policies has been often converted into a more decentralized, reticular, and interactive process where communication processes, such as bargaining, negotiation, and arguing, are seen as essential elements in a policy process.[10]

SELs necessitate increasing predictability and stability (order) in order to ensure uninterrupted provisions of resources for human use because dependent industries require predictability and desire maximization in productivity.[11] In the face of those uncertainties, planners usually keep on shaping the future spatial configuration of SELs based on local fine-scale knowledge and with a rather static, short-term perspective. Zoning ordinances, for instance, generate fixed rules and clearly defined spatial areas, usually apply to all developments, and hold for long periods of time without change.[12] Social-economic factors are imposed on SEL biophysical components to generate change in landscape pattern as households and businesses need more space for residential, recreational, and economic activities.[13]

In general, SELs appear to follow a historical characteristic sequence of different land-use pattern regimes, e.g., from presettlement natural vegetation to frontier clearing, then to subsistence agriculture and small-scale farms, and finally to intensive agricultural and urban areas and confined recreational areas.[14] This general trend is characterized by expansion of global croplands, pastures, plantations, and urban areas, with large increases in the consumption of energy, water, and fertilizer, along with the drainage of wetlands and floodplain embankments, conflicts between housing and economic land use, the loss of biodiversity following habitat fragmentation by urbanization and transport infrastructure.[15] This trend results in a pattern transition leading to clearly defined spatial areas with fixed rules in many parts of the world with increasing merging and enlarging of specific functions in some SEL locations like intensive agriculture, urban, and recreational areas.[16] This can make land administration inflexible, and planning may reinforce rigidity, erode resilience, and promote collapse in SELs[17] instead of the adaptability required to counter uncertainty and surprises due, for instance, to climate change.

---

[9] Gunderson and Holling (2002); Zaccarelli et al. (2008); Petrosillo et al. (2010), pp. 359–367.

[10] Slezak (1999), pp. 3–22.

[11] Zurlini et al. (2013), pp. 1161–1173.

[12] Booth (2003).

[13] Black et al. (2003), pp. 51–67; Millennium Ecosystem Assessment (MEA) (2005).

[14] Foley et al. (2005), pp. 570–574.

[15] Lambin et al. (2001), pp. 261–269; MEA (2005).

[16] Foley et al. (2005).

[17] Holling and Meffe (1996), pp. 328–337; Allison and Hobbs (2004), p. 3; Anderies et al. (2006), pp. 865–878.

The current land-use planning system is insufficiently equipped to stimulate sustainable development,[18] where sustainability has been incorrectly envisioned as a durable, stable condition that, once achieved, could persist for generations,[19] and the emergence of land overregulation may not be evident through traditional single-scale approaches.

As land-use transformation is becoming a main global driver given the world-wide changes to forests, farmlands, waterways, and air,[20] there is a need to go beyond such static and short-term conceptions in landscape planning and management, and global analysis is required to determine the net effect of local land-use decisions and assess global implications for greenhouse gas concentrations and climate. This is because regulations to protect natural ecosystems and trade policies may merely shift land uses from one country to another, by increasing imports, and mitigating climate change by the use of biofuels in one place may increase global greenhouse gas emissions due to the parallel response of land-use changes in remote locations.[21]

The problem we presently face is how a "static" and an "ordered" landscape condition in SELs, provided by the cross-scale intersections of land uses, plans, and norms (order) can be made sustainable in face of unpredictable disturbance and change (disorder).[22] However, we are still underestimating the tendency to lock into certain patterns that come at the cost of the ability to adjust to new situations and occur on levels varying from the cell and the mind to societies.[23] This resulting rigidity limits the ability of persons, groups, and companies to respond to new problems, and some may have contributed to the collapse of ancient societies.

This is well illustrated by the example in Pirsig's (1974) book, where the South Indian monkey trapper drills a hole in a coconut, puts a ball of rice inside, and chains the coconut to a stake. The monkey smells the rice and inserts his hand to grasp the rice. But now he is trapped since his fist with the ball of rice is now too big to pass through the hole and he does not let go of the rice. Pirsig calls this trap "value rigidity." The usually high value the monkey places on rice needs reevaluation in this life-threatening situation. In this metaphor, value rigidity skews the value we attach to facts, and because of value rigidity we might get stuck in a "rigidity trap."[24]

If we don't have proper mechanisms to monitor and predict changes and if we are not able to adapt through feedback mechanisms to changes in the environment, we might get stuck in a rigidity trap like the Pirsig's monkey and we are at high risk

[18] Diamond (1995), pp. 131–138.

[19] Ahern (1999), pp. 175–201.

[20] Turner et al. (2007), pp. 20666–20671; MEA (2005).

[21] Lambin and Meyfroidt (2011), pp. 3465–3472.

[22] Zurlini et al. (2013).

[23] Scheffer and Westley (2007), p. 36.

[24] Carpenter and Brock (2008), p. 40.

for failing. From theoretical studies, an increase in spatial connectivity may be a leading indicator as an early warning signal for an impending critical transition of regime shifts.[25] Most of the proposed indicators have been developed in simple ecological models and, even if potentially useful for managing transitions of real SELs, still remain elusive in their application and have not yet been tested in the field.[26]

In this paper, we aim at drawing the attention to some emerging cross-scale patterns of land-use resulting from social-economic factors and related to historical characteristic sequence of different land-use regimes that could be signals of overregulation and rigidity in SELs. A potential way to address such issue can be based on the recent historical trends of different land-use regimes and a simple framework to interpret current spatial patterns of land use across multiple scales also with the aid of simulated landscape patterns. We provide examples of this approach and discuss what a cross-scale land-use pattern could mean, what it tells about the condition of SELs, and what the effects could be of changing observed conditions in SELs. In this attempt, we exemplify concepts and methods that have received extensive treatment elsewhere,[27] taking the Apulia region in southern Italy as an example. We first illustrate how different patterns can be defined in the pattern transition space of composition and configuration, allowing us to determine the cross-scale nature of land-use pattern, which is very central to understanding the resulting trend and the kinds of management and/or policy actions to take at different scales. We then exercise the framework with real and simulated maps. In this respect, classical null models[28] are applied as baselines for comparison to the real landscapes on the same pattern transition space. We show that a potentially useful way to look for early warning signals for an impending critical transition in SELs is to look at how different cover types are patterned at multiple scales to postulate how those landscapes might evolve or react to variation of land use in the face of, for example, changes in climate conditions. Finally, we argue that to manage a transition toward more environmentally efficient and, therefore, more sustainable land use, we have to change societal values at the root of overregulation and rigidity.

---

25 Dakos et al. (2010), pp. 163–174.

26 Scheffer et al. (2009), pp. 53–59.

27 Zurlini et al. (2006), pp. 119–128; Zurlini et al. (2007), pp. 705–721; Zaccarelli et al. (2008); Petrosillo et al. (2010).

28 Gardner et al. (1987), pp. 19–28; Gardner and Urban (2007), pp. 15–29.

2 Data and Methods

2.1 Study Area: Apulia Region and Its Recent Historical Trends

The Apulia region (southern Italy) (Fig. 1) is considered here as an example of panarchy of nested SELs made up of people and nature.[29] Overall, more than 82 % of the Apulia region contains agroecosystems. The northern and somewhat the central part of the region include arable lands (39.8 %), producing cereals and vegetables, while extensive century-old as well as intensive olive groves (22.6 %), fruit orchards and vineyards (6.4 %), and heterogeneous agricultural areas (13.3 %) dominate the central and southern parts of the region,[30] which are mainly karstic with no surface water bodies. Major towns and small urban settlements account only for 3.8 % of the entire region, while natural habitats are unevenly distributed with major forested areas (7.3 %) concentrated in the Gargano peninsula.

Agriculture is still the primary economic resource as shown by the recent historical trends of productive and unproductive land use and of main employment sectors (Fig. 1). Trends are characterized by a contraction of arable lands, vineyards, and expansion of olive groves, plantations, and urban areas, with large increases in energy, water, and fertilizer consumption, along with extensive biodiversity loss.[31]

Oil production is one of the most important economic drivers in Apulia region followed by tourism,[32] and the recent historical intensification of olive groves in Salento peninsula is evident (Fig. 2), along with olive yield, as a result of the European Common Agricultural Policy (CAP) reform of 2003. CAP decoupled production and direct subsidies to olive growers and introduced the "single payment scheme" according to which farmers receive payments provided they meet certain standards concerning plant health and the environment and keep their land in good agricultural and environmental condition (cross-compliance). A regional law[33] preserves ancient olives for their support for biodiversity, hydrogeological protection, local climate regulation, and their value for the culture and the beauty of landscapes of the region.

Agricultural intensification—defined as higher levels of inputs and increased output (in quantity or value) of cultivated or reared products per unit area and time—has been a worldwide phenomenon since 1961 that doubled the world's food production with only a 10 % increase in the area of arable land globally,[34] but

[29] Zaccarelli et al. (2008).

[30] Zaccarelli et al. (2008).

[31] Zaccarelli et al. (2008) and Petrosillo et al. (2010).

[32] Istituto Nazionale di Statistica (ISTAT) (2012).

[33] Puglia, Regional Law 14/2007; Regione Puglia (2012).

[34] Tilman (1999), pp. 5995–6000.

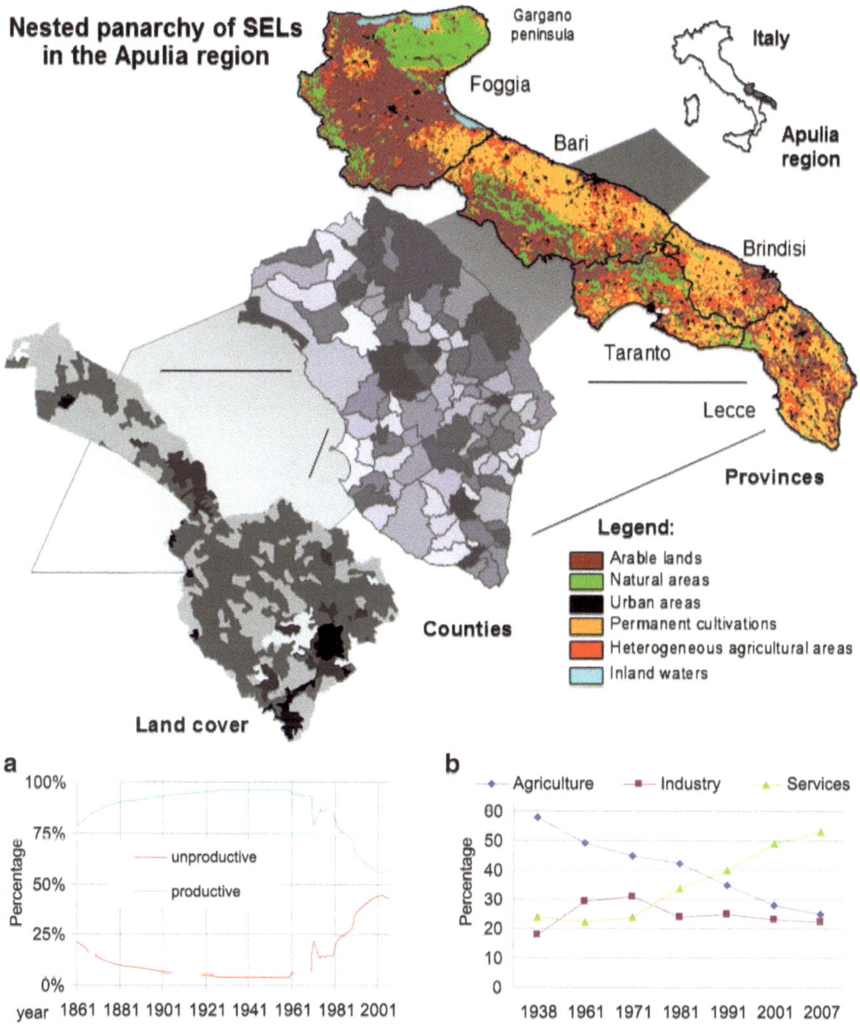

Fig. 1 Nested panarchy of SELs in Apulia (southern Italy) with main land-use composition (*top*) and recent historical trends of productive and unproductive land use (**a**), and of main employment sectors (**b**) (source of census data: ISTAT 2011)

increasing the risk of desertification. Highly intensive production increased in the Apulia region significantly because of mechanization with a decrease in the relative employment sector (Fig. 1). Olive oil, wheat, and wine production (the three main agricultural products) are critically dependent on sufficient and cheap water availability.[35] As a result, vulnerabilities for the region are mostly related to its persistent

[35] Kapur et al. (2010), pp. 1470–1478.

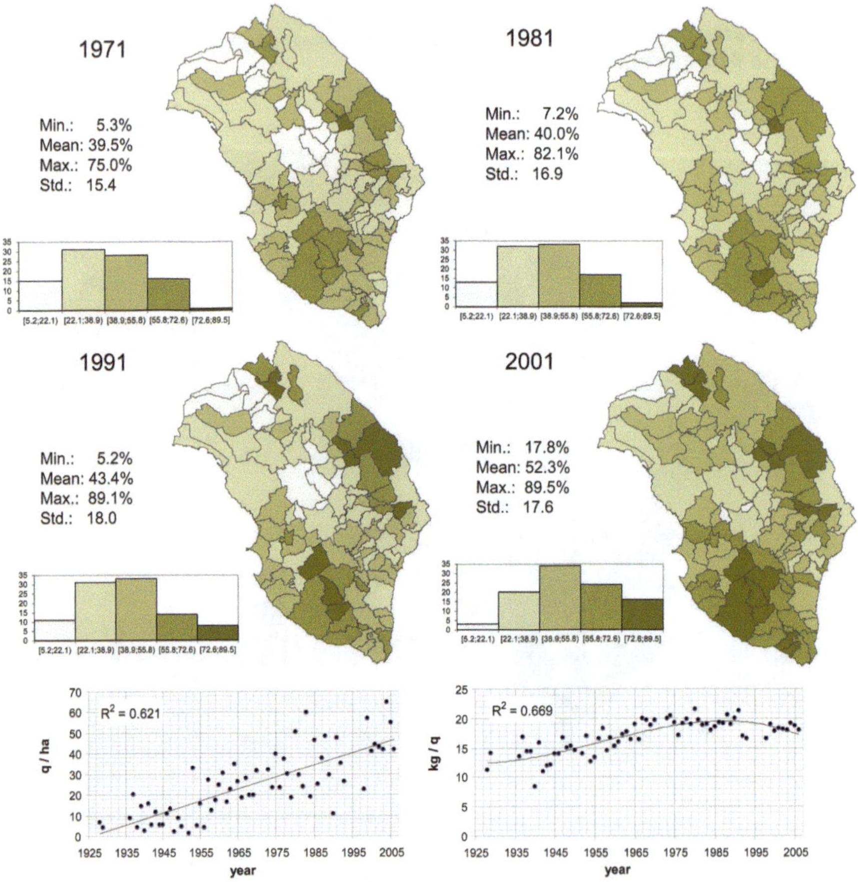

Fig. 2 Map of olive grove intensification for the municipalities of the Salento peninsula. Trends in olive yield and olive oil production efficiency are presented below; a *darker color* corresponds to a greater intensification of olive groves (source of census data: ISTAT 2011)

water budget deficit (about 350 mm/year), which requires constant water imports from nearby regions and heavy exploitation of aquifers.[36]

In the last decade, a number of plans has been developed in the Apulia region at different jurisdictional levels in the panarchy of SELs,[37] such as for hydrogeological risk and water management; for landscape and coastal area management; for cultural heritage, biodiversity, and parks protection; for regulating energy development and production; and for the development of rural environments (Fig. 3). Plans generated clearly defined spatial areas with fixed rules. Due to the very complicated intersection of all these plans, it is often very difficult, if not

[36] Gualdi et al. (2011); Salvati et al. (2011), pp. 1216–1227.

[37] Regione Puglia (2012).

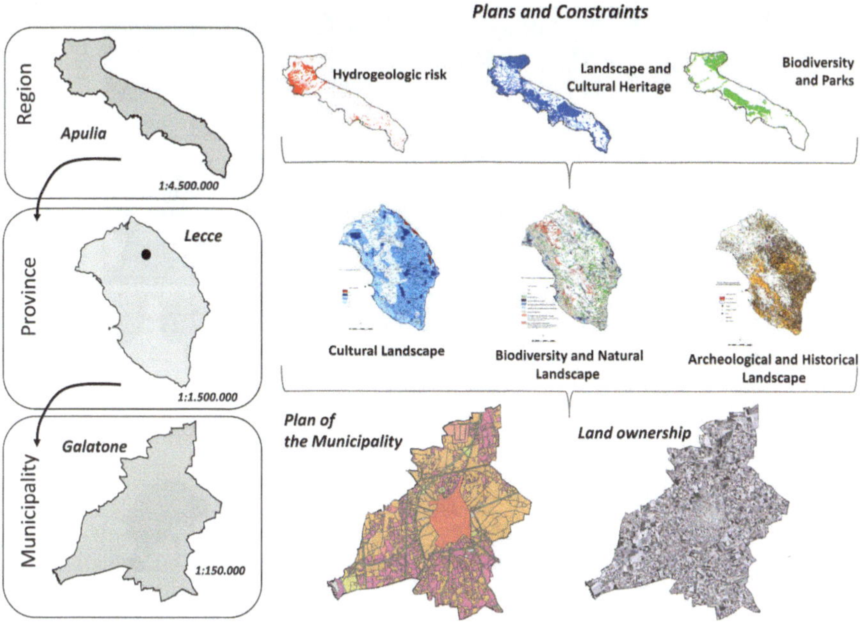

Fig. 3 Example of overlay of plans and constraints at different jurisdictional levels in the Apulia region (from Regione Puglia 2012)

impossible, to determine where a new particular activity can establish consistently with all the rules (Fig. 3), and this is always at the root of many conflicts.

2.2 *The Pattern Transition Space*

The transition space is where changes in landscape pattern across scales occur. Many authors[38] have suggested focusing on the two most fundamental measures of pattern that are landscape composition (what and how much there is) and landscape configuration (spatial arrangement).

In this respect, even simple binary maps generated by neutral landscape models (NLMs)[39] can produce a surprisingly rich array of spatial patterns to illustrate how different combinations of composition and configuration can lead to different kinds of land-use patterns. As an example, a set of 25 neutral landscapes is given in Fig. 4, where the focal land-cover area (Pc) is related to connectivity (H) as the degree of spatial autocorrelation among adjacent cells.[40] There are a number of

[38] See, e.g., Li and Reynolds (1994), pp. 2446–2455; Riitters et al. (2000), pp. 27–56; Neel et al. (2004), pp. 435–455; Zurlini et al. (2006, 2007); Proulx and Fahrig (2010), pp. 1479–1487.

[39] Gardner et al. (1987).

[40] Neel et al. (2004).

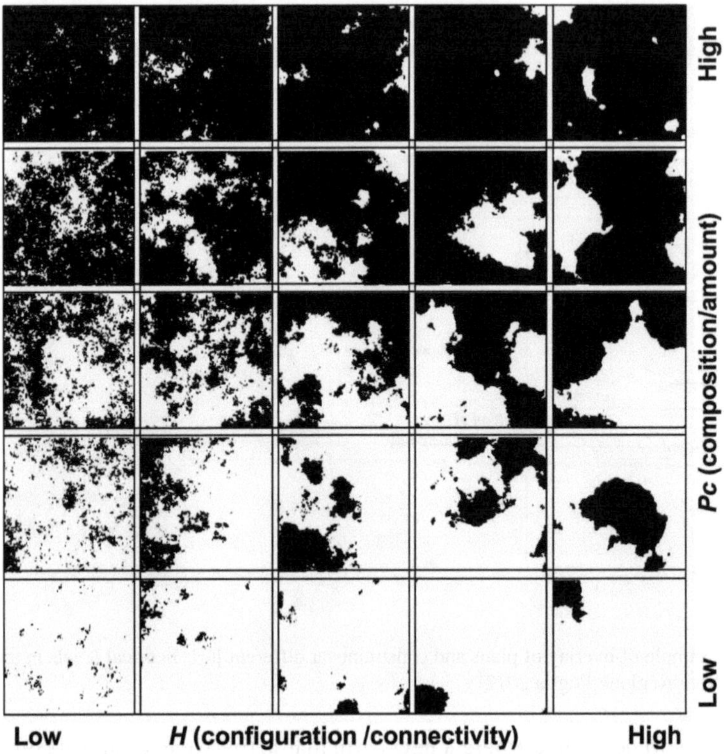

Fig. 4 Pattern transitions illustrated by 25 binary multi-fractal neutral landscape maps (256 × 256 cells) generated by the computer program RULE (Gardner 1999) and ordered by connectivity as the degree of spatial autocorrelation among adjacent cells (H) and the amount of the focal land cover type (Pc) (modified after Neel et al. 2004)

circumstances where quite a few transitions of focal area (black) pattern occur. There is a transition from a background matrix (top) to foreground patches (down), while, at the top, there is a transition of perforated matrix from small or diffuse holes (left) to large or distinct holes (right). Below, there is a transition from smaller patches (left) to fewer, larger patches (right). In general, from left to right, at similar composition values, there is a gradient from more to less fragmented landscapes and from less to sharper contrast with the nonfocal cover type. These transitions can occur across spatiotemporal scales.

NLMs do not adequately represent linear features such as rivers and hub-and-spoke landscape patterns within agriculture and urban landscapes.[41] However, the focus on focal class aggregation of neutral models is justified by the fact that changes in class aggregation are a major, if not the dominant, component of the fragmentation process at the landscape level.[42]

[41] See, e.g., Jones et al. (2013), pp. 1175–1192.

[42] Neel et al. (2004); Li et al. (2004), pp. 137–148.

Connectivity and fragmentation can be estimated for real landscapes, for example, using the proportional adjacency of a focal cover type pixels, i.e., the conditional probability that a focal cover pixel is adjacent to another focal cover pixel of the same type (Pcc).[43] One advantage of using it as a structural connectivity measure is that it can be easily calculated for both real and simulated binary landscapes and, therefore, can be used for generating the same pattern transition space for either real or simulated landscape patterns. Another advantage is that it represents image "texture," which is one of the fundamental aspects of pattern measured by popular pattern metrics.[44]

As a result, we characterize a pattern transition space [Pc, Pcc] by the amount (composition Pc) and spatial arrangement (configuration or connectivity Pcc) of a focal cover type (Fig. 4) at different scales resulting from different-sized spatial windows. Taken together, they can describe wide-ranging spatial patterns that are encountered on real maps for different focal land surface features, including habitats, land use/cover types, disturbance regimes, and any other focal feature.[45]

2.3 Real and Simulated Cross-Scale Patterns

One way to describe land-use patterns across scales is by using an overlapping moving window device to measure map composition, i.e. the proportion of focal cover type (Pc) for different window sizes over the entire region. We use the adjacency within a window (Pcc) both for simulated and real maps with the same set of window sizes.

As to real landscapes, we set few broad focal land-use/land-cover (LULC) categories from the 2006 CORINE land cover of Apulia region (www.sit.puglia.it) as the thematic base level for the whole panarchy of SELs, that is, forests, olive groves, and arable lands summing up to 93 % of the entire available land. We also generated 1,000 random, multifractal, and two-level hierarchical landscape pattern maps of size 1,024 × 1,024 cells using the RULE model.[46] Then we measured the connectivity of simulated maps by proportional adjacency to see how we might find real landscapes to differ from neutral landscapes.

A critical component of this approach is the "convergence point" (CP), which represents the global [Pc, Pcc] value that is exactly equal to the extent of the entire map. For any smaller window, the value of [Pc, Pcc] will necessarily depart from the CP of reference if the local pattern at the scale of the window size is different from the global pattern. With decreasing window size at a given geographic

[43] Riitters et al. (1995), pp. 23–39.

[44] Riitters et al. (1995).

[45] Riitters et al. (2000), Zurlini et al. (2006), Zaccarelli et al. (2008), and Petrosillo et al. (2010).

[46] Gardner (1999), pp. 43–62.

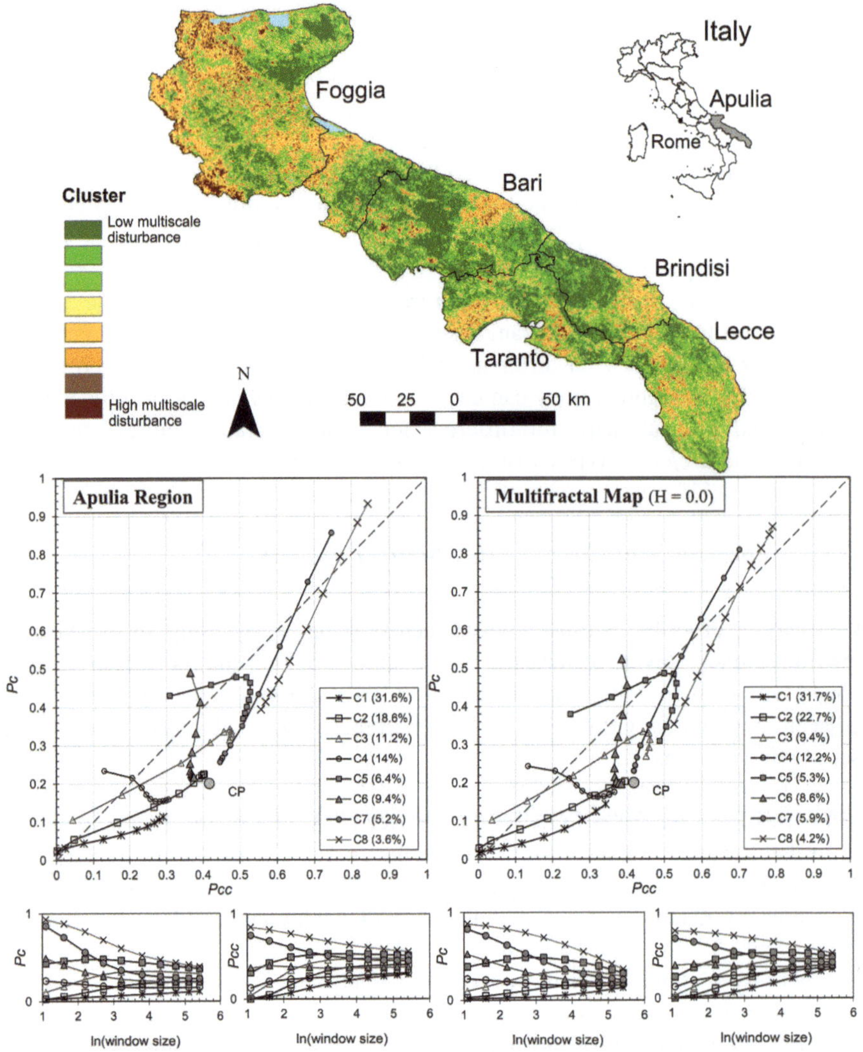

Fig. 5 Example of cross-scale cluster profiles in the Apulia region and their geographical representation (Pc and Pcc refer to disturbance; modified after Zurlini et al. 2007; Zaccarelli et al. 2008). Real (*left*) and simulated multifractal (*right*) profiles. Percentages refer to the number of pixels for each cluster profile with respect to all pixels of the entire region

location, the profile away from the CP in [Pc, Pcc] space describes the cross-scale "profile" of pattern surrounding that location at different window sizes[47] (Fig. 5).

Such pixel-level profiles are clustered to show the main profiles of eight groupings of cross-scale pattern sizes.[48] The CP can shift in the pattern transition space

[47] Zurlini et al. (2006, 2007).

[48] Zurlini et al. (2007).

according to corresponding changes in global [Pc, Pcc] values.[49] An idea of the cross-scale path of CP is given by the two extreme cluster profiles, that is, from the lowest and from the highest composition to the CP (e.g., C1, C8 in Fig. 5). Any two geographic locations with the same cross-scale cluster profile experience the same pattern at different neighborhood sizes.

3 Results

3.1 Simulated Cross-Scale Patterns

A random map, by definition, has no local domains at any scale in either pattern metric space or in real geographic space so every location on the random map experiences the same pattern. Simulated CPs at various disturbance compositions are always located above the main diagonal in the transition space.[50]

Multifractal maps do not exhibit convergence, and none of the profiles reaches the CP (Fig. 5). By definition, a multifractal is constructed to have the higher moments grow increasingly with scale, making for nonstationary parameters. This implies that cluster profiles will not converge to CP except asymptotically, and this occurs both for the real and simulated landscapes (Fig. 5, profiles for Pc and Pcc, below). Multifractal patterns refer to the alternation of various elements scattered in the landscape across a range of scales and are very common and similar to real patterns we can observe in the real geographic world[51] (Fig. 5).

As to hierarchical patterns, they do exhibit a convergence at intermediate window sizes with a typical fish-bone scaling pattern for structural connectivity (Fig. 6). Cluster profiles of hierarchical maps look like strings of a frayed rope starting at local scales from different regions and then quickly aggregate along scale to form a common "rope" with variations in composition but with contagion that does not vary much across scales.[52] In general, hierarchical patterns are characterized by fewer, larger patches with clearly defined boundaries and a sharp contrast with the nonfocal cover type (Fig. 4, lower right-hand corner).

[49] Zurlini et al. (2006).

[50] Zaccarelli et al. (2008).

[51] See, e.g., Milne (1991), pp. 199–235; Li (2000), pp. 33–50; Halley et al. (2004), pp. 254–271; Zurlini et al. (2007).

[52] Zurlini et al. (2007).

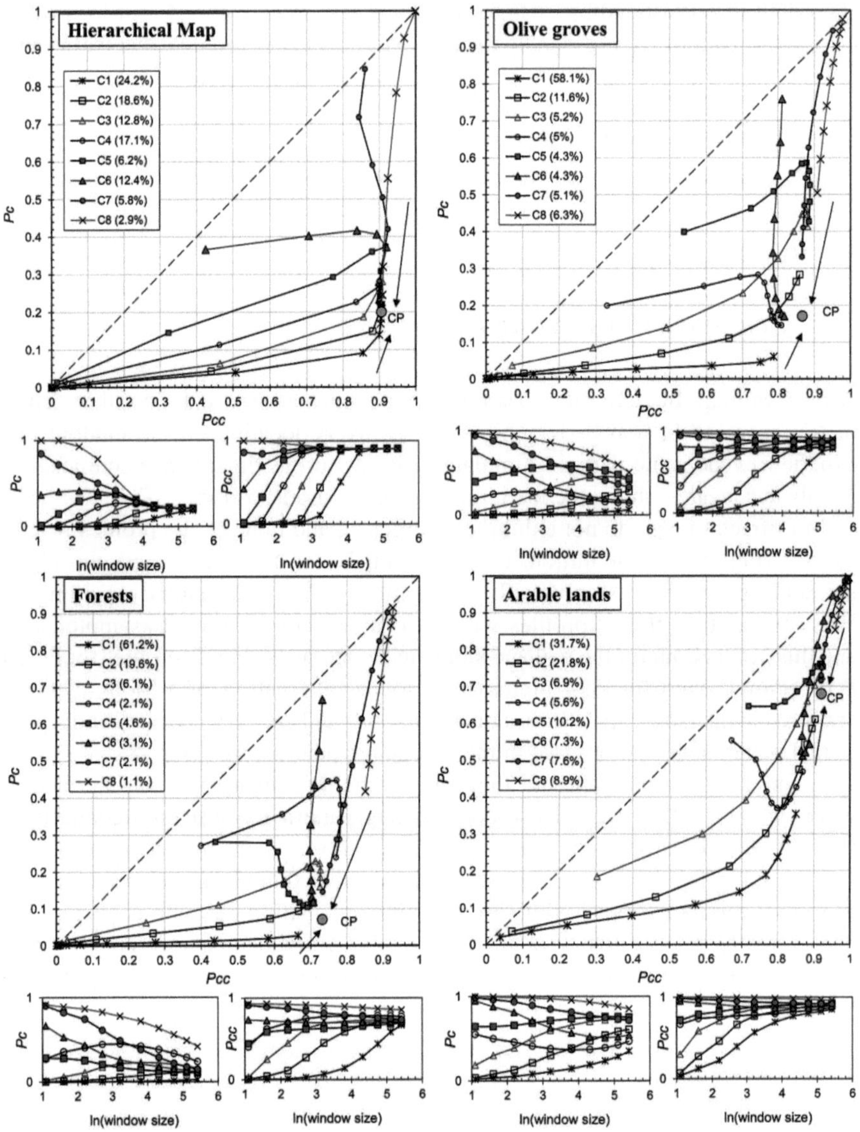

Fig. 6 Simulated hierarchical patterns (*top left*) and cross-scale patterns for Olive groves, Forests, and Arable lands in Apulia. Land-use patterns are clearly additive and sum up to 93 % of the entire available regional land

3.2 Real Patterns

Cross-scale profiles obtained for broad land use classes like forests, olive groves, and arable lands of the study area appear to follow a hierarchical-like pattern

(Fig. 6). Those patterns are characterized by fewer larger patches with clearly defined boundaries (Fig. 4), with the typical fish-bone scaling pattern for structural connectivity (Fig. 6; down: Pcc). Hierarchical-like patterns arise because some land uses (large fields, industrial and urban areas) and land covers (conservation areas, natural parks) are typically constrained within the bounds established by planning actions and/or economic necessities (Fig. 3). Economies of scale seem largely responsible in southern Europe for the current, observed trend in merging and increasing farm sizes and thus fewer farms and farmers[53] (Fig. 1). Not only fields have been merged and enlarged to enhance farming efficiency resulting in rather homogeneously farmed landscapes (e.g., arable lands, Figs. 1 and 6), but in general, also the expansion of industrial and residential areas tends to be confined in fewer, larger patches in the landscape for economic reasons. Also, planning for conservation both on land and sea often implies identification and confinement of land and coastal areas in fewer, larger portions to be protected to make management easier and to favor habitat clumping for mitigating the negative effects of fragmentation and habitat loss on species survival (e.g., forests, Fig. 6).

As to olive groves, the olive tree is well adapted to the Mediterranean climate and karstic conditions and requires little water in its natural state. Whereas a traditional olive farm is made of large, ancient, widely spaced trees providing cover for grass and grazing animals, the trees in new plantations are tightly packed, scrubbier, and usually grow on shallower soil. Such intensive cultivation has been merged and enlarged (Figs. 2 and 6), producing up to 20 times as many olives as a traditional grove, but it needs much more irrigation.

3.3 *Composition and Structural Connectivity Across Scales*

We can identify three main types of relationship between composition and structural connectivity with some invariant properties at particular scale ranges in pattern transition space for random, hierarchical, and multifractal patterns (Fig. 7). To get an idea of underlying landscape patterns behind, one can refer to simulated maps (Fig. 4).

At higher composition values, connectivity in multifractal patterns increases almost proportionally to composition (Fig. 7, top right). On the other hand, in highly fragmented maps, a rise of focal cover, for example, of 20 % (from 0.0 to 0.2) results in a 45 % increase in connectivity (Fig. 7, bottom left). Within these two opposite situations, transitions between multifractal patterns entail a corresponding smaller increase in connectivity for each unit of percentage increase. Interestingly, 45 % of the overall structural connectivity resides in the first 20 % of landscape composition.

[53] Metzger et al. (2006), pp. 69–85.

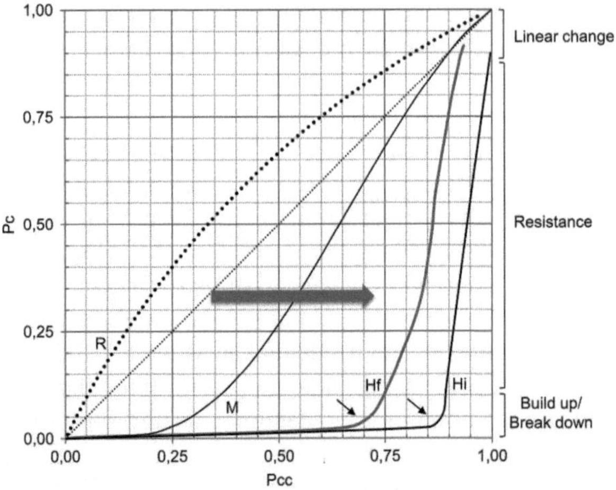

Fig. 7 Summary of results for random, multifractal and hierarchical patterns in the pattern transition space defined by composition (Pc) and configuration or connectivity (Pcc): R = pattern of simulated random maps; M = general pattern of a simulated and observed multifractal maps (Fig. 5); Hf = observed hierarchical-like pattern of Forests (Fig. 6), and Hi = simulated hierarchical maps. The *grey arrow* indicates the postulated current direction of pattern transition. *Black arrows* indicate thresholds

In hierarchical patterns, connectivity at higher composition values can vary much less than proportionally to composition (Fig. 7, top right). On the other hand, at lower composition values (Fig. 7, bottom right), a reduction of 20 % (from 0.2 to 0.0) results in a remarkable 90 % decrease in structural connectivity. Similarly, a loss from 0.01 to 0.05 of focal cover type composition results in nearly 85 % increase in fragmentation. These two opposite situations are distinct in simulated patterns by a sharp shift at very low composition (about 2.5 %) and at 88 % connectivity that is determined by the early 2.5 % of focal cover type in the landscape. Landscape pattern of natural areas (e.g., forests, Fig. 6) could be very vulnerable to habitat loss at very low composition since a small amount of habitat reduction (e.g., less than 7 %) may disrupt structural connectivity up to 73 %. These two opposite situations are distinct at very low composition (about 7 %) by a sharp shift at 73 % connectivity that can be deemed as a critical threshold (Fig. 6).

We can then identify three main types of relationship between composition and structural connectivity with some invariant properties at particular scale ranges in pattern transition space of CPs (Fig. 7): *Build-up (or break-down)*, where the first 10 % of composition of focal land cover builds up almost 35 % of total connectivity for multifractal patterns; in hierarchical-like patterns, 10 %, for example, of forest and olive grove composition provides more than 70–80 %, respectively, of relative total structural connectivity; *Resistance*, where a unit of percentage increase in composition determines a parallel much lower increase in connectivity (about 50 % for multifractal and very much less for hierarchical patterns); *Linear change* between composition and connectivity only for multifractal patterns at higher composition values.

4 Discussion and Conclusions

4.1 Overregulation and Rigidity Traps

The arrow in Fig. 7 indicates the postulated direction of pattern transition due to historical trends, from a multifractal-like to hierarchical-like patterns, as well as the potential increase of connectivity imposed, for instance, by land-use change to adjust to climate change. This is due to increasing spatial aggregation of intensive agriculture, urban areas, and recreational areas[54] with higher levels of connectivity at the same amount of the focal land-cover type (Fig. 4). As a consequence of the interplay between climate change and general agricultural intensification, negative impacts are foreseen in southern Europe for major sectors relying on ecosystem services, like agriculture, because of their substantially lower adaptive capacity.[55] Such intensification and the introduction of new farm machinery, new strains of cereals and tree crops, and extensive application of fertilizers are the main social-economic drivers in charge for contemporary desertification in Mediterranean Europe, along with the overexploitation of water resources for greater demand for irrigated agriculture.[56] However, other causes are coresponsible of negative impacts like urban development and tourism, the extensive rural migrations and the abandonment of nonprofitable and marginal agricultural lands (cf. Fig. 2), the concentration of economic activity in coastal areas,[57] and the low level of perception of the authorities and the public.[58]

Agricultural intensification is presently the main cause of desertification and soil erosion in the Apulia region, which is the second most vulnerable Italian region for degradation of soil,[59] which is already less naturally productive.[60]

Besides, evidence from *in situ* time series highlights that during the second half of the twentieth century, the regional climate has become moderately warmer and drier[61] and responsible for a significant proportion of the interannual production variability for olive oil (15 %) and wheat (16 %).[62] The net irrigation requirements for major crops in the Apulia region are expected to increase in the next 100 years with a maximum of 65 % for intensive olive groves along with further overpumping

[54] Foley et al. (2005).
[55] Metzger et al. (2006).
[56] Briassoulis (2003).
[57] Briassoulis (2003).
[58] Petrosillo et al. (2013), pp. 609–620.
[59] Perini et al. (2009), pp. 45–55.
[60] Parise and Pascali (2003), pp. 247–256; Briassoulis (2003).
[61] Gualdi et al. (2011).
[62] Gualdi et al. (2011).

of aquifers.[63] This will increase water stress by the intensity, extent, timing, and duration of changes in normal water resource availability.[64]

As a result, an overregulated structure is expected to emerge across scales with hierarchical-like patterns and potential local increase of soil and water stress, less capacity to dissipate it, and dramatic breakdowns of connectivity at low habitat composition below certain thresholds (Fig. 7). This aggregation can be detrimental to the necessary adaptability of natural landscape elements and biodiversity, can increase vulnerability;[65] and can alter, in turn, the patterns of regional temperatures, precipitation, vegetation, and other climate variables[66] with adverse effects on the ability of ecosystems to provide goods and services.[67]

A "rigidity trap" in SELs is formally characterized by low heterogeneity and higher aggregation and connectivity of entities (e.g., land uses, land covers), great capacity to focus on a singular approach, and low capacity to explore alternatives; there is little capacity to dissipate stress, and stress may accumulate to high levels through the panarchy.[68]

An overregulated planning of land use along with intensive agricultural systems like olive groves and arable lands can lead in the Apulia region to a rigidity trap. This can occur through a pathological cycle of resource degradation and stress (soil and water), followed by social-economical response aimed at reestablishing or maintaining productivity of the resource-degrading activity, with consequent further degradation and erosion of system adaptive capacity to cope with shocks and surprises.

We argue that this circumstance not only applies to the current conditions of the Apulia region, but it can also become typical in many different human-dominated parts of the world. It can also trigger at a global scale the displacement, rebound, cascade, and remittance effects, amplified by economic globalization, that accelerate land conversion in developing countries.[69]

We are still underestimating the tendency to lock into certain patterns that come at the cost of the ability to adjust to new situations.[70] Value rigidity is at the root of the rigidity trap produced by the pathological cycle of resource degradation and stress and can make it hard to learn new facts and to recognize important facts as we preselect facts as important, or not, in line with our established values.[71] This can likely result in hierarchical-like patterns of overregulated planning and management of land use as it has been shown for the Apulia region.

---

[63] Kapur et al. (2010).

[64] Acosta-Michlik et al. (2008), pp. 151–160.

[65] Adger (2006), pp. 268–281.

[66] Pielke (2005), pp. 1625–1626.

[67] Tilman et al. (2002), pp. 671–677.

[68] Holling (2001), pp. 390–405; Carpenter and Brock (2008).

[69] Lambin and Meyfroidt (2011).

[70] Scheffer and Westley (2007).

[71] Armson (2011).

We have still to change beliefs, analyses, or hunches that can immobilize us far more effectively than preparedness to live with uncertainty and surprises, and, foremost, we must be fully aware that we might get stuck in a rigidity trap. Self-sealing beliefs can be self-correcting when extreme events such as wildfires or hurricanes foster change in long-established rules and practices like planning and management.[72] However, extreme events may also provoke other feedback processes working to maintain the status quo, such as the financial and/or political support that accompanies continued crisis management. We might nevertheless get stuck in a trap through the pathological cycle of resource degradation even when resource managers recognize that things would improve if they approached their work differently.[73]

4.2 *Perspectives in Adaptive Design and Management*

An adaptive approach still provides a fundamental framework for the implementation and adaptation of management and polices over time as more information is collected.[74] A crucial issue then could be developing landscape planning (e.g., restoration) that might accommodate for surprises[75] and for variation of CP (Fig. 7) as humans will change land use, and especially land management, to adjust to climate change. In this respect, new conceptual frameworks for the design of SEL sustainability are emerging to establish how landscape condition can be made sustainable in face of unpredictable disturbance and change.[76]

Managing a transition toward more environmentally efficient and, thus, more sustainable land use implies better information on consequences of land-use decisions at local to global scales, the creation of proper incentives for agents, and a greater capacity to adopt new land-use patterns and practices.[77] This can be crucial when facing the challenge of increasing provisioning services such as food production by 70 % for 2050[78] for an expanding population while, simultaneously, conserving or enhancing other ecosystem services (regulating and cultural services) in agricultural systems.[79]

[72] Schusler et al. (2003), pp. 309–326.

[73] Repetto and Allen (2006), pp. 110–136.

[74] Walters (1986); Vernier et al. (2009), pp. 3–14; Cushman and McKelvey (2010), pp. 111–130.

[75] Scheffer et al. (2001), pp. 591–596.

[76] See, e.g., Olsson et al. (2004), pp. 75–90; Folke et al. (2005), pp. 441–473; Musacchio (2009), pp. 993–1013; Opdam et al. (2009), pp. 715–721; Ostrom (2009), pp. 419–422; Benayas and Bullock (2012), pp. 883–889; Zurlini et al. (2013); Jones et al. (2013).

[77] Lambin and Meyfroidt (2011).

[78] FAO (2009).

[79] Kiers et al. (2008), pp. 320–321.

Strategies to this end could involve the design and management of landscape elements and structure to shift land-use pattern in the opposite direction to that of the arrow in Fig. 7 to create less contagious and more heterogeneous rural landscapes. This can imply the strategic placement of managed and seminatural ecosystems in SELs to reduce stress intensity so the services of natural ecosystems (e.g., commodities, water availability, pollination, reduced land erosion, soil formation) can be even enhanced.[80] Land separation and land sharing are examples of such strategies.[81] The first involves restoring or creating nonfarmland habitat in agricultural landscapes through, for example, woodland, natural grassland, hedgerows, wetland, and meadow on arable land,[82] or riparian habitats[83] to benefit wildlife and specific services. Land sharing involves the adoption of biodiversity-based agricultural practices, learning from traditional farming practices, transformation of conventional agriculture into organic agriculture and of "simple" crops and pastures into agro-forestry systems. Some existing small-scale farming systems have high water-, nutrient-, and energy-use efficiencies and conserve resources and biodiversity without losing yield.[84]

A key aspect is to implement monitoring programs to evolve iteratively as new information emerges and research and managing questions change.[85] This helps evaluate how environmental targets and ecosystem services respond to specific landscape pattern designs[86] and whether or not certain landscape patterns at multiple scales result in synergies and trade-offs among different types of ecosystem services.[87]

The degree of impending rigidity that we start to face is a warning to planners and managers that the problem is becoming widespread and can no longer be addressed simply with narrow-minded and local-scale solutions. If we don't have proper mechanisms to monitor and predict changes and if we are not able to adapt to changes in the environment through feedback mechanisms, we might get stuck in a rigidity trap.

Our simple regional example suggests that the degree to which the observed pattern departs from a particular neutral model can be an indicator of whether major constraints or organizing structure has been placed on the landscape and how those landscapes might evolve/react to additional variation of land use. This is because we can reasonably guess where the CP is in the pattern transition space for the

[80] Jones et al. (2013).

[81] Benayas and Bullock (2012).

[82] Benayas and Bullock (2012).

[83] Jones et al. (2010), pp. 1261–1275.

[84] Kiers et al. (2008).

[85] Lindenmayer and Likens (2009), pp. 482–486.

[86] Ahern (1999) and Jones et al. (2013).

[87] Wu and Hobbs (2002), pp. 355–365; Naidoo et al. (2008), pp. 9495–9500.

region of interest, how the pattern could be across multiple scales and its environmental consequences, and whether a particular landscape scenario will shift the CP (Fig. 7). The goal could be then to build in a safety factor as the CP is expected to shift, to help identify where one can solve problems at multiscaled patterns where a more coordinated, cross-scale approach is needed. Multiscaled land-use patterns will not be easier to manage, but having such knowledge will be necessary for multiple stakeholders in the panarchy of SELs to cooperate in social networks within and between organizational levels for managing SEL resilience[88] under uncertainty and change.

Cross-scale collaborative planning networks such as the U.S. Fire Learning Network[89] can facilitate overcoming the rigidity traps that prevent resource management agencies from responding to complex cross-scalar problems. Changing circumstances demand to reappraise values like in the case of Pirsig's monkey and his rice. The intentional induction of cooperation could be promoted across the panarchy of SELs through the establishment of social initiatives that increase the perception of similarity within and among stakeholders to reach a minimal level that makes cooperation advisable.[90] In other words, we must be fully aware that we might get stuck in a rigidity trap to appreciate the similarity of our common condition and to start real cooperation.

References

Acosta-Michlik L, Kavi Kumar KS, Klein RJT, Campe S (2008) Application of fuzzy models to assess susceptibility to droughts from a socio-economic perspective. Reg Environ Chang 8:151–160

Adger WN (2006) Vulnerability. Glob Environ Chang 16(3):268–281

Ahern J (1999) Spatial concepts, planning strategies, and future scenarios: a framework method for integrating landscape ecology and landscape planning. In: Klopatek JM, Gardner RH (eds) Landscape ecological analysis: issues and applications. Springer, New York, pp 175–201

Allison HE, Hobbs RJ (2004) Resilience, adaptive capacity, and the "lock-in trap" of the Western Australian agricultural region. Ecol Soc 9(1):3. http://www.ecologyandsociety.org/vol9/iss1/art3/

Anderies JM, Ryan P, Walker BH (2006) Loss of resilience, crisis, and institutional change: lessons from an intensive agricultural system in southeastern Australia. Ecosystems 9:865–878

Armson R (2011) Growing wings on the way: systems thinking for messy situations. Triarchy Press Station Offices, Axminster Devon

Benayas JMR, Bullock JM (2012) Restoration of biodiversity and ecosystem services on agricultural land. Ecosystems 15:883–889

Berkes F, Folke C (eds) (1998) Linking social and ecological systems: management practices and social mechanisms for building resilience. Cambridge University Press, Cambridge

[88] Walker et al. (2002), p. 14; Olsson et al. (2004).

[89] Butler and Goldstein (2010), p. 21.

[90] Fischer et al. (2013), pp. 10229–10233.

Black AE, Morgan P, Hessburg PF (2003) Social and biophysical correlates of change in forest landscapes of the interior Columbia Basin, USA. Ecol Appl 13:51–67
Booth P (2003) Planning by consent. The origins and nature of British development control. Routledge, London
Briassoulis H (2003) Mediterranean desertification. Framing the policy context. European Commission. DG for research, sustainable development, global change and ecosystem. EUR 20731, Brussels
Butler WH, Goldstein BE (2010) The US fire learning network: springing a rigidity trap through multiscalar collaborative networks. Ecol Soc 15(3):21. http://www.ecologyandsociety.org/vol15/iss3/art21/
Carpenter SR, Brock WA (2008) Adaptive capacity and traps. Ecol Soc 13(2):40. http://www.ecologyandsociety.org/vol13/iss2/art40/
Cushman SA, McKelvey KS (2010) Data on distribution and abundance: monitoring for research and management. In: Cushman SA, Huettman F (eds) Spatial complexity, informatics and wildlife conservation. Springer, Tokyo, pp 111–130
Dakos V, van Nes EH, Donagelo R, Fort H, Scheffer M (2010) Spatial correlation as leading indicator of catastrophic shifts. Theor Ecol 3:163–174
Diamond DR (1995) Geography and planning in the information age. Trans Inst Br Geogr 20:131–138
Dijst M, Schenkel W (2002) Urban performance in perspective. In: Dijst M, Schenkel W, Thomas I (eds) Governing cities on the move: functional and management perspectives on transformations of European infrastructures. Ashgate, Aldershot, pp 1–18
FAO (2009) The state of food and agriculture. http://www.fao.org/docrep/012/i0680e/i0680e.pdf
Fischer I, Frid A, Goerg SJ, Levin SA, Rubenstein DI, Selten R (2013) Fusing enacted and expected mimicry generates a winning strategy that promotes the evolution of cooperation. Proc Natl Acad Sci U S A 110(25):10229–10233
Foley JA, DeFries R, Asner GP, Barford C, Bonan G, Carpenter SR, Chapin FS, Coe MT, Daily GC, Gibbs HK, Helkowski JH, Holloway T, Howard EA, Kucharik CJ, Monfreda C, Patz JA, Prentice JC, Ramankutty N, Snyder PK (2005) Global consequences of land use. Science 309:570–574
Folke C, Hahn T, Olsson P, Norberg J (2005) Adaptive governance of social–ecological systems. Annu Rev Environ Resour 30:441–473
Gardner RH (1999) RULE: map generation and a spatial analysis program. In: Klopatek JM, Gardner RH (eds) Landscape ecological analysis: issues and applications. Springer, New York, pp 43–62
Gardner RH, Urban DL (2007) Neutral models for testing landscape hypotheses. Landsc Ecol 22:15–29
Gardner RH, Milne BT, Turner MG, O'Neill RV (1987) Neutral models for the analysis of broad-scale landscape pattern. Landsc Ecol 1:19–28
Gualdi S, Somot S, May W, Castellari S, Déqué M, Adani M, Artale V, Bellucci A, Breitgand JS, Carillo A, Cornes R, Dell'Aquilla A, Dubois C, Efthymiadis D, Elizalde A, Gimeno L, Goodess CM, Harzallah A, Krichak SO, Kuglitsch FG, Leckebusch GC, L'Heveder BP, Li L, Lionello P, Luterbacher J, Mariotti A, Nieto R, Nissen KM, Oddo P, Ruti P, Sanna A, Sannino G, Scoccimarro E, Struglia MV, Toreti A, Ulbrich U, Xoplaki E (2011) Future climate projections. In: Navarra A, Tubiana L (eds) Regional assessment of climate change in the Mediterranean. Springer, Dordrecht
Gunderson LH, Holling CS (eds) (2002) Panarchy: understanding transformations in human and natural systems. Island Press, Washington
Halley JM, Hartley S, Kallimanis AS, Kunin WE, Lennon JJ, Sgardelis SP (2004) Uses and abuses of fractal methodology in ecology. Ecol Lett 7:254–271
Holling CS (2001) Understanding the complexity of economic, ecological, and social systems. Ecosystems 4:390–405

Holling CS, Meffe GK (1996) Command and control and the pathology of natural resource management. Conserv Biol 10(2):328–337

Istituto Nazionale di Statistica (ISTAT) (2011) Banche dati. http://www.istat.it/it/prodotti/banche-dati

Istituto Nazionale di Statistica (ISTAT) (2012) Banche dati. http://www.istat.it/it/prodotti/banche-dati

Jones KB, Slonecker ET, Nash MS, Neale AC, Wade TG, Hamann S (2010) Riparian habitat changes across the continental United States (1972–2003) and potential implications for sustaining ecosystem services. Landsc Ecol 25:1261–1275

Jones KB, Zurlini G, Kienast F, Petrosillo I, Edwards T, Wade TG, Li B-L, Zaccarelli N (2013) Informing landscape planning and design for sustaining ecosystem services from existing spatial patterns and knowledge. Landsc Ecol 28(6):1175–1192

Kapur B, Pasquale S, Tekin S, Todorovic M, Sezen SM, Özfidaner M, Gümü Z (2010) Prediction of climatic change for the next 100 years in southern Italy. Sci Res Essays 5(12):1470–1478

Kato S, Ahern J (2008) 'Learning by Doing': adaptive planning as a strategy to address uncertainty in planning. J Environ Plan Manag 51(4):543–559

Kiers ET, Leakey RRB, Izac A, Heinemann JA, Rosenthal E, Nathan D, Jiggins J (2008) Agriculture at a crossroads. Science 320(5874):320–321

Lambin EF, Meyfroidt P (2011) Global land use change, economic globalization, and the looming land scarcity. Proc Natl Acad Sci U S A 108:3465–3472

Lambin EF, Turner BL, Geist HJ, Agbola S, Angelsen A, Bruce JW, Coomes O, Dirzo R, Fischer G, Folke C, George PS, Homewood K, Imbernon J, Leemans R, Li X, Moran EF, Mortimore M, Ramakrishnan PS, Richards JF, Skånes H, Steffen W, Stone GD, Svedin U, Veldkamp T, Vogel C, Xu J (2001) The causes of land-use and land-cover change: moving beyond the myths. Glob Environ Chang 11:261–269

Levin SA (1999) Fragile dominion: complexity and the commons. Perseus Books, Reading

Li B-L (2000) Fractal geometry applications in description and analysis of patch patterns and patch dynamics. Ecol Model 132:33–50

Li H, Reynolds JF (1994) A simulation experiment to quantify spatial heterogeneity in categorical maps. Ecology 75(8):2446–2455

Li X, He HS, Wang X, Bu R, Hu Y, Chang Y (2004) Evaluating the effectiveness of neutral landscape models to represent a real landscape. Landsc Urban Plan 69:137–148

Lindenmayer DB, Likens G (2009) Adaptive monitoring – a new paradigm for long-term studies and monitoring. Trends Ecol Evol 24:482–486

Marsh WM (ed) (2005) Landscape planning: environmental applications, 4th edn. Wiley, New York

Metzger MJ, Rounsevell MDA, Acosta-Michlik L, Leemans R, Schröter D (2006) The vulnerability of ecosystem services to land use change. Agric Ecosyst Environ 114:69–85

Millennium Ecosystem Assessment (MEA) (2005) Ecosystems and human well-being. Island Press, Washington

Milne BT (1991) Lessons from applying fractal models to landscape patterns. In: Turner MG, Gardner RH (eds) Quantitative methods in landscape ecology. Springer, New York, pp 199–235

Musacchio LR (2009) The scientific basis for the design of landscape sustainability: a conceptual framework for translational landscape research and practice of designed landscapes and the six Es of landscape sustainability. Landsc Ecol 24:993–1013

Naidoo R, Balmford A, Costanza R, Fisher B, Green RE, Lehner B, Malcolm TR, Ricketts TH (2008) Global mapping of ecosystem services and conservation priorities. Proc Natl Acad Sci U S A 105(28):9495–9500

Nassauer JI, Corry RC (2004) Using normative scenarios in landscape ecology. Landsc Ecol 19:343–356

Neel MC, McGarigal K, Cushman SA (2004) Behavior of class-level landscape metrics across gradients of class aggregation and area. Landsc Ecol 19:435–455

Olsson P, Folke C, Berkes F (2004) Adaptive comanagement for building resilience in social–ecological systems. Environ Manage 34:75–90
Opdam P, Luque S, Jones KB (2009) Changing landscapes to accommodate for climate change impacts: a call for landscape ecology. Landsc Ecol 24:715–721
Ostrom E (2009) A general framework for analyzing sustainability of social–ecological systems. Science 325:419–422
Owens S, Cowell R (2011) Land and limits: interpreting sustainability in the planning process, 2nd edn. Routledge, New York
Parise M, Pascali V (2003) Surface and subsurface environmental degradation in the karst of Apulia (southern Italy). Environ Geol 44(3):247–256
Perini L, Ceccarelli T, Zitti M, Salvati L (2009) Insight desertification process: bio-physical and socio-economic drivers in Italy. Ital J Agrometeorol 3:45–55
Petrosillo I, Zaccarelli N, Zurlini G (2010) Multi-scale vulnerability of natural capital in a panarchy of social–ecological landscapes. Ecol Complex 7:359–367
Petrosillo I, Costanza R, Aretano R, Zaccarelli N, Zurlini G (2013) The use of subjective indicators to assess how natural and social capital support residents' quality of life in a small volcanic island. Ecol Indicat 24:609–620
Pielke RA (2005) Land use and climate change. Science 310(5754):1625–1626
Pirsig RM (1974) Zen and the art of motorcycle maintenance. An inquiry into values. William Morrow and Co., Inc., New York
Proulx R, Fahrig L (2010) Detecting human-driven deviations from trajectories in landscape composition and configuration. Landsc Ecol 25:1479–1487
Regione Puglia (2012) Piani di sviluppo socioeconomico e di assetto del territorio. http://www.regione.puglia.it/index.php?page=documenti. Accessed 12 Oct 2012
Repetto R, Allen RB (2006) On social traps and lobster traps: choppy waters on the voyage toward fisheries' harvesting rights. In: Repetto R (ed) Punctuated equilibrium and the dynamics of US environmental policy. Yale University Press, New Haven, pp 110–136
Riitters KH, O'Neill RV, Hunsaker CT, Wickham JD, Yankee DH, Timmins SP, Jones KB, Jackson BL (1995) A factor analysis of landscape pattern and structure metrics. Landsc Ecol 10(1):23–39
Riitters KH, Wickham J, O'Neill RV, Jones KB, Smith E (2000) Global-scale patterns of forest fragmentation. Conserv Ecol 4:27–56
Salvati L, Bajocco S, Ceccarelli T, Zitti M, Perini L (2011) Towards a process-based evaluation of land vulnerability to soil degradation in Italy. Ecol Indicat 11:1216–1227
Scheffer M, Westley FR (2007) The evolutionary basis of rigidity: locks in cells, minds, and society. Ecol Soc 12(2):36. http://www.ecologyandsociety.org/vol12/iss2/art36/
Scheffer M, Carpenter S, Foley JA, Folke C, Walker B (2001) Catastrophic shifts in ecosystems. Nature 413:591–596
Scheffer M, Bascompte J, Brock WA, Brovkin V, Carpenter SR, Dakos V, Held H, van Nes EH, Rietkerk M, Sugihara G (2009) Early-warning signals for critical transitions. Nature 461:53–59
Schusler TM, Decker DJ, Pfeffer MJ (2003) Social learning for collaborative natural resource management. Soc Nat Resour 15:309–326
Slezak P (1999) A critique of radical social constructivism. In: Stillwell J, Geertman S, Openshaw S (eds) Developments in geographical information and planning. Springer, Berlin, pp 3–22
Tilman D (1999) Global environmental impacts of agricultural expansion: the need for sustainable and efficient practices. Proc Natl Acad Sci U S A 96:5995–6000
Tilman D, Cassman KG, Matson PA, Naylor R, Polasky S (2002) Agricultural sustainability and intensive production practices. Nature 418:671–677
Turner BL II, Lambin EF, Reenberg A (2007) The emergence of land change science for global environmental change and sustainability. Proc Natl Acad Sci U S A 104(52):20666–20671
Vernier PR, Preston MI, Bunnell FL, Tyrrel A (2009) Adaptive monitoring framework for warblers at risk in northeastern British Columbia: using models and expert opinion to refine monitoring. Wildlife Afield 6:3–14

Walker B, Carpenter S, Anderies J, Abel N, Cumming G, Janssen M, Lebel L, Norberg J, Peterson GD, Pritchard R (2002) Resilience management in social–ecological systems: a working hypothesis for a participatory approach. Conserv Ecol 6(1):14. http://www.consecol.org/vol6/iss1/art14

Walters CJ (1986) Adaptive management of renewable resources. Macmillan, New York

Wu J, Hobbs R (2002) Key issues and research priorities in landscape ecology: an idiosyncratic synthesis. Landsc Ecol 17:355–365

Zaccarelli N et al (2008) Source/sink patterns of disturbance and cross-scale effects in a panarchy of social–ecological landscapes. Ecol Soc 13(1):26. http://www.ecologyandsociety.org/vol13/iss1/art26/

Zurlini G, Riitters KH, Zaccarelli N, Petrosillo I, Jones KB, Rossi L (2006) Disturbance patterns in a social–ecological system at multiple scales. Ecol Complex 3(2):119–128

Zurlini G, Riitters KH, Zaccarelli N, Petrosillo I (2007) Patterns of disturbance at multiple scales in real and simulated landscapes. Landsc Ecol 22:705–721

Zurlini G et al (2013) Highlighting order and disorder in social–ecological landscapes to foster adaptive capacity and sustainability. Landsc Ecol 28(6):1161–1173

Reshaping Agriculture Toward a Transition to a Post-Fossil Bioeconomy

M. Monteleone

Abstract This work tackles the need to overcome the dependence of our societies on fossil energy sources and to establish a productive model focused on the highest harmonization with biological processes based on renewable energy sources. The first step is to start from agriculture. The productive success of industrial agriculture is entirely based on the *energy of darkness*, the fossil resources, while the traditional form of peasant agriculture trusts entirely on the *energy of light*, i.e. the radiative energy coming from the sun. This energy, through photosynthetic carbon assimilation, is stocked into the soil in the form of organic matter; it represents the "second" energy source (after sunlight) of the agro-ecosystem. Protecting the soil, preserving its organic content, and sustaining its biological functioning are the essential elements of a "paradigm shift": from industrial agriculture to agro-ecology. This paper presents and discusses the contents and the forms of this possible transition, its obstacles, and chances. The new approach should be multifunctional and diversified, local based and self-sustained.

Keywords Ecological approach to agriculture • Farming multifunctionality • Knowledge based bioeconomy • Rural development • Sustainable agriculture

1 Introduction

The starting point of this work is to reveal the close connection that currently exists between agricultural practices and energy. Specifically, this work tackles the need to overcome (rapidly and with no hesitations) the dependence of our societies on fossil energy sources to establish a productive model focused on the highest harmonization with biological processes and skilled enough to utilize, with the optimal efficiency, renewable energy sources. Human activities progressively moved away from the laws of "ecology" (with so many negative consequences) and embraced the laws of "economy," the latter wrongly considered a world

M. Monteleone (✉)
STAR*AgroEnergy Group, Department of Agri-Environmental Sciences, Chemistry and Plant Protection, University of Foggia, Foggia, Italy
e-mail: m.monteleone@unifg.it, http://www2.agraria.unifg.it/schedadocente.asp?IDDocente=43&Scheda=PrimaPagina&IDPagina=5

M. Monteduro et al. (eds.), *Law and Agroecology*,
DOI 10.1007/978-3-662-46617-9_18

dimension apart from nature. Although human activities may oppose to nature for a while, they cannot be in contrast with the ultimate thermodynamics laws that affect both artificial and natural life.[1] This contrast generates environmental decline, resource depletion, social poverty, and discontent. Reconciling human societies with natural systems is not an easy task, and a complete reconsideration of the current paradigm we are following in producing and consuming goods, as well as in generating and disposing wastes, should be achieved (a "paradigm shift", we might say). A progressive but comprehensive restructuring of our modern energy system calls to establish a new and creative link with the fundamental rules affecting nature (whether physical or biological). To shape or mimic how nature works and is organized, how its structures have been built in the course of a long evolutionary history, and how its functions can spontaneously proceed without generating environmental burdens should be the aim of our knowledge effort in life science.[2] The (re)discovering of this ancient knowledge can lead to innovation, promoting a new technological attitude in close tune with societal needs.

Probably the first step to take in this process is to start from agriculture. Agriculture, indeed, strongly depends on the native energy coming from the sun; for this reason, it might be the first sector among human activities to be deeply rearranged with respect to nature, basically in its ability to capture and effectively use the native energy from the sun.

2 The Rise of Industrialism

Every technological advancement in human civilization was marked by the ability of men to recognize and properly use new forms of energy, previously unknown.[3] An increasing power characterized these new available energies; "power" is the capacity to obtain more energy in the same unit of time. This greater amount of energy in a shorter time span hugely enhanced the work capacity of the organized societies. We will not discuss, for sure, how the discovery of fire by man can be considered the first "domestic" form of energy (used to cook the food, to heat the place of residence, and to keep out dangerous predators); a fundamental role was also played in the technical performance of metal casting. No comments will be reported about the Neolithic revolution in terms of an increasing amount of net energy obtained by food (cultivated for the first time close to the village and not necessarily hunted very far from it) and animal domestication (to be employed as draft animals). It is worth just to remember the technical ability reached by man in using properly the power of wind and water (fine examples of windmills and water mills may still be found in many European countries).[4] Sailing ships are another

[1] Smil (1991).

[2] Benyus (1997).

[3] Rifkin (2011).

[4] Cipolla (1965).

important form of technical exploitation of the wind power; this technology contributed to promote not only commerce but also the explorations that led to the discovery of new continents on which was based the colonial regime of many European countries.

Moreover, it is impossible to forget the significant historical function exerted by slavery in a vast number of civilizations, from the ancient time until few decades ago, and the contribution it offered in the capital accumulation of private and state organizations.[5]

In the Agricultural Revolution, man came to the control of biological organisms and succeeded to increase significantly the availability of plants and animals for his own consumption. Differently, in the Industrial Revolution, a process started allowing the exploitation and the large-scale use of new forms of energy using fossil fuels. Everything started with the "steam engine," at the turn of the eighteenth century and in the first decades of the nineteenth century.

A greater availability of motive power allowed obtaining a greater amount of coal and transporting it at a higher rate. In turn, a greater amount of coal allowed having a greater motive power. In this way, coal became a strategic component for the rise of industrial civilization and its further dissemination. During the second half of the nineteenth century, both oil and electricity began to be used and progressively characterized the new processing technologies. At this point, all of the essential components of the industrial takeoff were ready.

During the first and second industrial revolution and never before in human history, fossil type of energy sources were used (coal at first, oil some decades after). The convergence of particular conditions has boosted the emergence and the increasing expansion of the "industrial production model": a large availability of fossil energy, a huge accessibility to raw materials, easiness for goods transport, and speediness in communication. Improved transportation methods, such as railroads rapidly built in the most developed countries, favored the exchanges of raw material and finished products, thus stimulating the growth of commerce and further boosting the economic development.[6]

The first phase of the industrial takeoff was characterized by a progressive concentration of capitals, factories, and workers, together with the setting up of large and intensive urban and industrial agglomerations. A drastic change from hand craftsmanship to machine manufacturing took place, firstly in England. Textile industry equipped with mechanical spinning mills was, in general, the first kind of factories powered by steam engines fueled by coal. This kind of coal was no longer the one from wood but the mineral one, obtained from mining.

Corresponding to this socio-technical transformation, a parallel evolution and a gradual adaptation in the forms of political and institutional organization were also observed.[7] The form of the "National State" as institutional organization played a

---

[5] Diamond (1997).

[6] Barraclough (1964).

[7] Mommsen (1969).

central role in this process. It was considered the most appropriate system to ensure an effective development of capitalism in its initial conditions. The State as a complex machinery and a highly centralized system, its organizational gigantism, its strongly hierarchical bureaucracy define a model of society very representative of the modern era, between '800 and '900. Likely, this model revealed to be functional to the industrial capital development until the end of World War II. Then this kind of system went into crisis and evolved in that form that today we all know as integrated, multinational capitalism, overcoming previous territorial boundaries and every other type of barriers. Today we are witnessing an almost complete market integration characterized by a global geographic scale, a monopolistic business concentration, the delocalization of the productive processes in different world areas in relation to the economic affordability of each process with respect to the cost of feedstock, machineries, and workers.

3 Why Fossils Are Different

The evidence that fossil resources like coal, oil, and natural gas are quantitatively limited (nonrenewable) and therefore subjected to a progressive depletion was quite clear already at the beginning of the fossil era. Beyond this, the main feature of these new energy sources relates to their specific nature or quality: while every kind of energy used by man before fossils was clearly a *fund-flow* energy supply, fossils are undoubtedly a *stock-flow* energy supply. It was Georgescu-Roegen to claim firstly this pertinent distinction. This difference is very useful in order to characterize very clearly a renewable from a nonrenewable resource. Georgescu-Roegen wrote:[8] "if the count shows that a box contains twenty candies, we can make twenty youngsters happy now or tomorrow, or some today, and others tomorrow, and so on. But if an engineer tells us that one hotel room will probably last one thousand days more, we cannot make one thousand room-less tourists happy now. We can only make one happy today, a second tomorrow, and so on, until the room collapse. Take also the case of an electric bulb which last five hundred hours. We cannot use it to light five hundred rooms for an hour now. The use of a fund (i.e., its "decumulation") requires a duration. Moreover, this duration is determined within very narrow limits by the physical structure of the fund. We can vary it only little, if at all. If one wishes to "decumulate" a pair of shoes, there is only one way open to him: to walk until they become waste. In contrast with this, the decumulation of a stock may, conceivably, take place in one single instant, if we wish so."

This specific quality revealed by fossils (i.e., to be a stock-flow energy source) had an unprecedented effect because of their extremely high energy "density" and the large amount of deposits actually available at the beginning of the fossil exploitation period. At that time, the only constraint to a large exploitation was

[8] Georgescu-Roegen (1971).

the limited technological extraction capacity. Of course, this capacity rapidly improved over time, reaching an extraordinarily high rate, until the easiest extractable deposits were completely exhausted. Oil fields more difficult to exploit are gradually undermined today.

Until now, the modern industrial development was completely focused on the use of energy from fossils, therefore a finite resource. The story can be summarized as follows: "for millions and millions of years a treasure was accumulated. Then, someone in the family discovered the "nest egg" laboriously saved up and began to dissipate it. Humanity is now living in a period of unprecedented dissipation. It consumes now, in one year, an amount of coal higher than that produced in one hundred centuries of progressive fuel formation." These are the prophetic words of Carlo M. Cipolla in 1962.[9]

At this point, there is the problem of knowing how long this availability will last. Observers, in fact, are considering that the so-called peak oil time had come already or it is very close to be reached. Peak oil, an event based on Hubbert's theory, is the time when the maximum rate of petroleum extraction is attained; soon after, the rate of oil production is expected to enter terminal decline. A second very important issue is the following: whether the harmful effects that such a frenetic consumption is causing to the environment (i.e., global warming) will lead to a forced stop in the use of fossil fuels even before their complete exhaustion.

All this considered, it is quite clear that, within the general scheme of natural and human history, man's dependence on fossil fuels to get the energy he needs can only be a short episode, very short indeed.

4 The Agriculture Paradox

This industrialization process did not fail to involve agriculture too, although it is difficult to specify which of the two productive sectors (industry or agriculture) firstly influenced the other before being in turn influenced by it. A "positive loop" is the mechanism usually identified to explain this kind of process: a systemic change takes place very rapidly, progressively reinforcing the same transformation factors. Anyhow, the starting point is an accumulation of additional capitals to be invested to catalyze the boosting of a new productive process. At the start of the first industrial revolution, agriculture was involved in this process. New forms of cultivations and the general increase in productivity, the restructuring of the land ownership and a greater consolidation of landholdings, a more dynamic farm management, a larger proletarianization of farmers are the blend of reasons that contributed to a radical change in the traditional socio-economic structure of the agricultural sector, thus contributing to the industrial takeoff. A large part of the agricultural workforce was rejected from the countryside and moved in urban areas;

[9] Cipolla (1962).

most took jobs in factories. This population shift and the abundant availability of labor remarkably supplied the new factories and promoted the industrial development of the countries.

It is therefore necessary to mark an apparent "paradox." On the one hand, the increased agricultural productivity made it possible a progressive transfer of resources (food) from the countryside to the city. This fueled further the process of urbanization, allowing an expansion of the industrial base of the workforce at the expense of the agricultural one. The technological progress transferred to agriculture and the consequent agro-industrialization made possible a strong contraction in its workforce and allowed that a limited fraction of the total workforce was dedicated to agriculture. Today, a fraction even less than 5 %, in general, characterizes the employed in agriculture compared to the total. This condition allows to state that a very limited number of workers in the agricultural sector are able to assure an adequate supply of food resources to meet the needs of a complex society, composed by a plurality of productive sectors far away from agriculture. This, undoubtedly, is the sign of a remarkable success. On the other hand, it should be noted that a deep process of social and economic marginalization has involved the agricultural activity, precisely because of the same development model that enhanced its productive capacity. This "paradox" may be reported as follows: its success has been the cause of its failure. The reverse is also true: its failure is the sign of its success. This "paradox" is only apparent, of course, and it can be solved just considering that the development of agriculture was fully functional to that of the industry. Then agriculture traced exactly the same production model and arrived at a total integration with the industrial system. A unique productive model, indistinguishable and undifferentiated from that of the industry, was thus finally achieved. This is exactly what Vandana Shiva defines as "monoculture of the mind."

5 The Energy Dependency of Agriculture

The following considerations by Piero Bevilacqua are fully revelatory about the definite judgment on modern agriculture. According to Bevilacqua,[10] today we are able to perceive very clearly that a *massive energy transfer* from fossils to crop cultivations has almost completely surrogated the traditional good care of soil fertility. The transition from an organic to a chemical form of agriculture was fully accomplished in this last half century and led to a radical change of a long-standing energy paradigm. Indeed, this paradigm was completely reversed. This observation should be considered an outstanding scientific acquisition and suddenly shed light on the fragility, we might almost say the "trick" that explains a so relevant page of the history of modern capitalism in our farmland. Fossil energy

[10] Bevilacqua (2010), pp. 9–10.

sources have radically replaced the energy generated by the soil, a vital form of energy because founded on living processes occurring within the active soil layer. Considering the modern form of a highly intensive agriculture, the natural, living processes are almost completely ignored and bypassed through chemical inputs. Agrochemicals consist in a large amount of auxiliary energy obtained from fossils and supplied to the soil in order to speed up, affect, and interfere with naturally occurring processes strictly related to the "work" of a living system. According to this "modernist" attitude, the soil should be considered simply as the surface layer of the earth, made of aggregated mineral debris obtained from rock fragmentation and weathering, added with organic matter coming from litter deposition of dead plants, animals, and microorganisms, nothing but a simple support to sustain crop growth when sufficient amount of water and fertilizers are supplied to the crops and the best physical and chemical conditions are assured.

Differently, in preindustrial agriculture, farmers ensured soil fertility through countless practices of organic recycling, thus storing the energy captured from the sun and sequestering carbon into the soil. Soil organic fertilization has always been the cornerstone of a good agriculture. Traditional farming practices primarily focused the attention on operations able to preserve soil fertility for a long time. This, indeed, should be considered the original concept of "sustainability": to use native and renewable resources at a rate proportional to their natural regeneration.

Organic recycling is the main form to compensate for the systematic subtraction of agricultural products from farmlands toward the neighboring towns. Today, food processing, conditioning, packaging, storage, and transportation allow the establishing of a very huge and almost worldwide "food-system." As a result, food distribution takes place over long distances and does not take into account traditional barriers (such as geographic and climatic zones or seasonality) that were very constraining just few decades ago. The problem of restoring organic matter, subtracted from highly productive agricultural systems and supplied to urban systems, is progressively worsening and generally neglected.

Differently, minerals (especially nitrogen) are intensively restored into the soils with fertilization supplied by industrial products of chemical synthesis. On this respect, man is significantly affecting the nitrogen cycle through an unprecedented high-rate synthesis (Haber-Bosh process) of nitrogenous compounds that are transferring nitrogen from the atmosphere (N_2) to the soil and, by leaching, to deep or surface water bodies, thus producing environmental impacts.

Considering the need of organic matter recycling, very relevant considerations are reported by Draghetti in his pioneering book "Principles of farming physiology" (1948). According to Draghetti, the farm should be considered a sort of complex "living machine." Similarly to any other organism, the farm is characterized by a proper structure and organization, but it is also provided with a circulatory system. This concept is very close to the idea of the Planet Earth as "super-organism" (Gaia), proposed in 1979 by James Lovelock[11] and consolidated by research on

[11] Lovelock (1979).

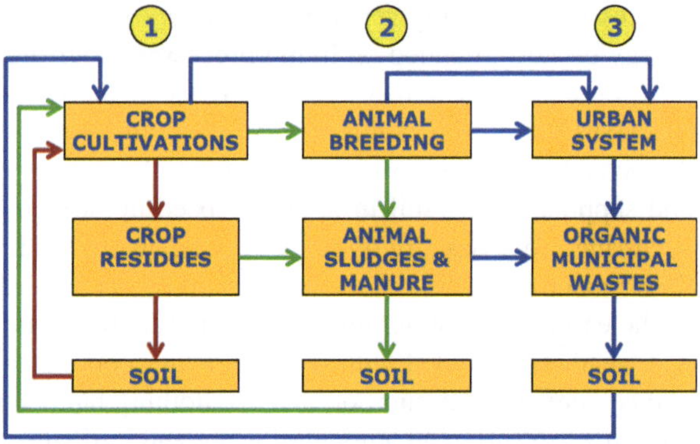

Fig. 1 Small (1—*red arrows*), large (2—*green arrows*), and huge (3—*blue arrows*) energy circulation and material flow across the agro-urban ecosystem. The first two cycles (1 and 2) are from Draghetti (1948); the third one (3) is a personal elaboration

endosymbiosis by Lynn Margulis.[12] While a bloody circulation characterizes animals, the farm circulatory system is made of "organic carbon" strictly coupled with the energy transfers occurring along the trophic chains: the grazing and detritus food chains that constitute every ecosystem. This circulation, in turn, split into a "small" and a "large" circulation. The "small" one supplies the soil with organic matter directly from crop residues that progressively degrade, forming humus, and partially mineralized, releasing nutrients into the soil. The "large" one supplies the soil with organic matter through manure and sludges obtained from animal breeding, where forages cultivated in the farm are used. To complete what Draghetti was considering, it is worth to identify and routinely apply a third type of "circulation." We may define it as a "huge" circulation, the one connecting the farmland with the urban areas, thus allowing the agronomic valorization of organic municipal wastes or agro-industrial by-products that are generated in the urban and industrial districts (Fig. 1).

Currently, the majority of this organic waste is simply disposed in the landfills, thus creating several environmental burdens. Qualified and controlled recycling procedures may address this organic material to anaerobic digestion or composting (or a combination of the two transforming processes) with the ultimate objective to restore soil fertility through organic fertilization.

The direct testimony of J.W. Goethe, who spent a visit in Naples in 1787, is suggesting that, in the preindustrial time, the huge amount of trash, garbage, and organic waste collected from farmers with great care decisively contributed to the fertilization of the Vesuvian lands, especially the farmland close to the city. The following is what Goethe wrote in his diary during the "Italian journey" about the "garbage collectors" observed in Naples.[13] *Naples, 28 May 1787*: "A very large number of people, some middle-aged men, some boys, all very poorly dressed, are occupied in carry the refuse out of the city on donkeys. The immediate area around Naples is simply one huge kitchen garden, and it is a delight to see, first, what

[12] Margulis and Sagan (1987).

[13] Goethe (1813–17), pp. 314–315.

incredible quantities of vegetables are brought into the city every market day, and, second, how human industry immediately returns the useless parts which the cooks reject to the field so as to speed up the crop cycle. Indeed, the Neapolitans consumes so many vegetables that the leaves of cauliflowers, broccoli, artichokes, cabbages, lettuce, and garlic make up the greater part of the city's refuse. Two large, flexible panniers are slung over the back of a donkey: these are not only filled to the brim, but above them towers a huge mound of refuse, piled with peculiar cunning. No garden could exist without a donkey. A boy or a farm hand, sometimes even the farmers himself, hurry as often as possible during the day into the city, which for them is a real gold mine. You can imagine how intent these collectors are on the dropping of mules and horses. They are reluctant to leave the streets at nightfall, and the reach folk who leave the opera after midnight are probably unaware of the existence of the industrious men who, before daybreak, will have been carefully searching for the trail of their horses."

Is still Bevilacqua[14] to remark the strong difference between modern and traditional agriculture: "the productive success of industrial agriculture is entirely based on the "*energy of darkness*", on the fossil heritage received from the early history of the Earth, on the finite resources that are running out very quickly. On the contrary, the traditional knowledge that for millennia ensured the success of peasant agriculture trusts entirely on the "*energy of light*", i.e. the radiative energy coming from the sun able to activate the photosynthetic process of carbon assimilation" (Fig. 2).

The limitless oil consumption appears to be the key lever on which the technological arrogance (*hybris*) of industrial agriculture is based. In fact, if we look at the strong mobilization of fossil energy sources put in place by industrial agriculture over the last half century, then the stunning productivity goals achieved not only should be significantly reduced but should appear in all their impermanence. Even the very successes of genetics, in fact, are inseparable from the looting of fossils subtracted from the bowels of the Earth.

6 Protecting the Soil Is Healing the Man[15]

The soil should be considered an ecosystem in itself, integrated into a larger ecological structure represented by the agricultural system.[16] Prior to complying with the laws of market and economy, an agricultural ecosystem must satisfy the laws of biology and ecology. The "agro-eco-system" is a complex combination of nature-based and man-managed processes; its degree of "naturalness" largely depends on how much the system is *specialized* (reduced biodiversity in order to

[14] Bevilacqua (2010), pp. 9–10 (authors's translation).

[15] Quotation from the title of the book by Aubert (1974).

[16] Caporali et al (2010).

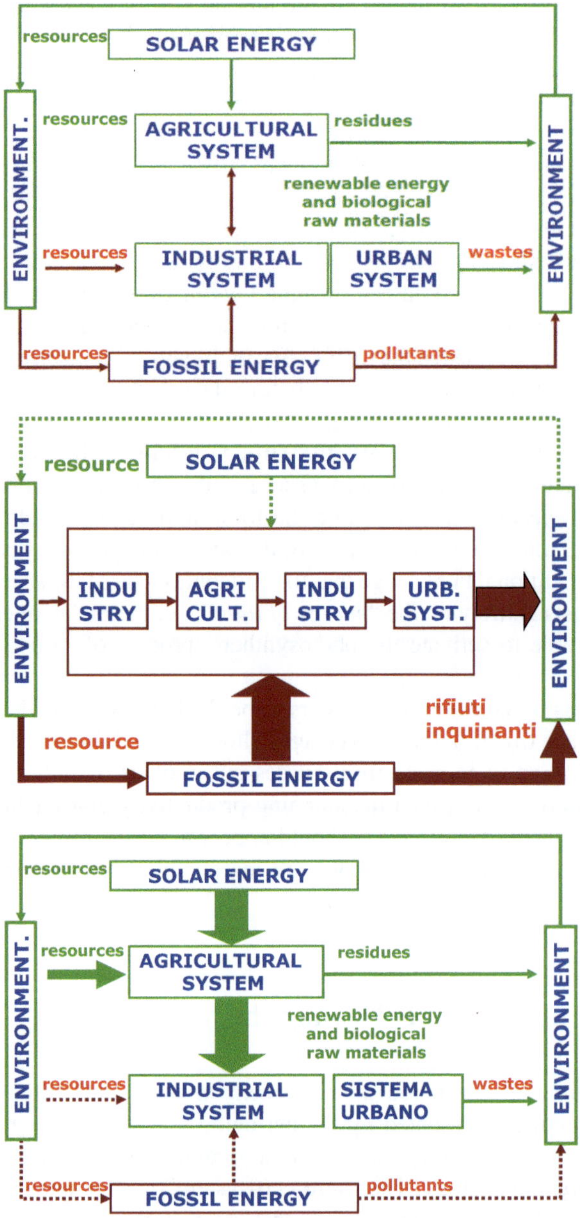

Fig. 2 Energy relationships among the human productive sectors, 1—solar and fossil energy are in a perfect balance, 2—fossil energy is the predominant one, 3—solar energy prevails on fossils

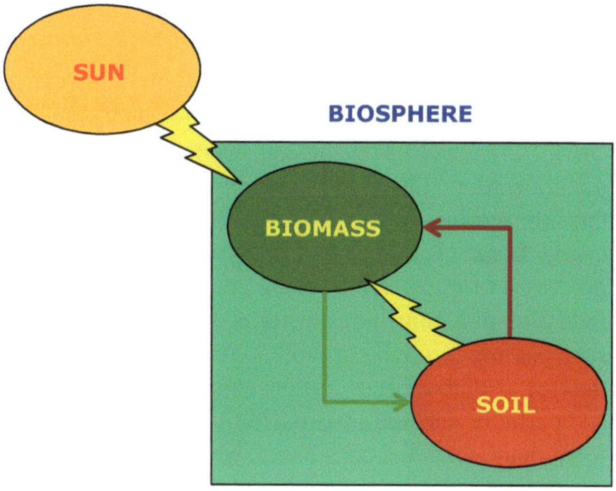

Fig. 3 The soil organic matter may be considered the "second engine" (i.e., energy source) of every natural and agricultural ecosystem (the first being the sun). The mineralization of this organic stock releases energy and minerals used as nutrients for plants

cultivate a limited number of crop species) and *simplified* (limited biological composition in order to address the saved energy to productivity). The level of agro-technical inputs is a general measure of the degree of subrogation the system is adapted: the higher the level is, the lower is the ability of the system to function independently accordingly to self-regulating processes.

Soil quality is the capacity to perform all possible functions generally ascribed to the soil, within specific ecosystem and land use boundaries, to sustain productivity, maintain environmental properties, and promote plant and animal health. *Soil health* relates, in turn, to its functional integrity and the resulting homeostatic adjustments that dynamically control soil conditions.[17] It is the ability to keep the soil stable in its functioning even in the presence of disturbances (*resistance*) or return to the former conditions soon after disturbances (*resiliency*). Some of the main soil functions can be summarized in the following:

- sustaining biological activity, diversity, and productivity;
- regulating and partitioning water and solute flow;
- filtering and buffering, degrading, immobilizing, and detoxifying organic and inorganic materials, including industrial and municipal by-products and atmospheric deposition;
- storing and cycling nutrients and other elements within the earth's biosphere;
- providing a proper physical support of socioeconomic structures and infrastructures.

An interesting analogy is to consider the soil as the "second engine" of every ecosystems, both natural and agricultural, the first being the sun (Fig. 3).

The soil shall keep and maintain stable its precious reserves of organic matter. These reserves are a stock of potential energy. When mineralization takes place,

[17] Hausmann (2005).

then the energy is dissipated and mineral nutrients obtained from the process become available to plant growth. In the meantime, humic substances improve soil quality and stimulate the activities of soil microorganisms (very important in many soil biochemical processes). Soil organic matter plays a function similar to an energy "flywheel." This, indeed, is one of the main features of the "second engine": mineralization slowly and gradually releases the energy incorporated into the soil, thus preventing excess or shortage conditions. The sun originally conveys the energy needed to carry out photosynthesis; this energy activates all the downstream processes of transferring organic matter (together with energy) along the complex food web that constitutes every ecosystem on Earth: from producers to decomposers, passing through consumers. At each ecosystem level, along the trophic chain, the free available energy is consumed: partially stored in the form of organic matter and partially released (dissipated) in the form of entropy. The grazing branch of the food chain concerns consumers, while the detritus branch relates to decomposers, thus allowing the incorporation of organic reserves into the soil.

A homeostatic, complex, and dynamic equilibrium is secretly kept by the soil, and man's interference alters this perfect balance.[18] On this respect, very instructive is the experience reported by Liebig. Justus von Liebig (1803–1873) is the great scientist who made major contributions to agricultural chemistry. He is considered the "father of the fertilizer industry" for his discovery of nitrogen and other minerals as essential plant nutrients. The following text is attributed to Liebig, and, in fact, it turns out to be a collection of excerpts from some of his writings.[19] Liebig wrote: "Unfortunately, the true beauty of agriculture, with its stimulating intellectual principles, is mostly underestimated. The art of agriculture will be lost because of ignorant, unscientific, and short-sighted teachers who will convince farmers to place all their hopes in universal remedies that do not exist in nature. By following their advice, blinded by short-term results, the farmers will forget the soil and lose sight of its intrinsic value and its influence. I willingly admit that the use of chemical fertilizers was based on assumptions that do not exist in the real world. These fertilizers were supposed to lead to a complete agricultural revolution. Manure was to be completely abandoned, and mineral fertilizers were to be used for the replacement of the minerals absorbed by crops. Fertilizers would make farming the same crops possible on the same field, continuously and unfailingly, according to the farmer's whims and needs. I have sinned against the wisdom of the Creator and I have received the proper punishment. I wanted to bring an improvement to his work. In my blindness I believed that in the wonderful succession of laws that connect Life to the Earth's surface, continually renewing it, a link had been forgotten and that I, poor helpless worm, had to provide this missing link."

[18] Vandana (1993).

[19] Liebib (1855). This version is from CNS Ecologia Politica (January–July 2003) 13(54):78. http://www.ecologiapolitica.org/wordpress/web2/200301/articoli/testamento_liebig.pdf. Accessed 10 Sept 2014.

7 A Dilemma: Food or Energy First?

The close interaction between agriculture and industry, discussed in the previous sections, is showing today an additional feature represented by the use of agricultural biomass to provide biofuels as an alternative source of energy. Biofuels, indeed, are generally considered as opportunities to enhance energy security and significantly reduce fossil consumption causing global warming. This issue is very critical and sensitive to diverse considerations, especially with respect to the effective capacity of biofuels to reduce greenhouse gas emissions. Many observers and opinion leaders invoke, not without reasons for alarm, the strategic priority of crops to food destination rather than energy. It is often forgotten, however, that having in mind the current levels of crop intensification, the agri-food production requires a massive input of energy in different forms (fertilization, mechanization, irrigation, pest treatments, etc.), as we explained already. Therefore, we are facing a vicious circle, which offers no solution until it remains anchored to the same loop, the one that has generated the problem. From one side: to meet an increasing food demand (by a growing population consuming larger amount of energy per capita) a higher agricultural productivity is absolutely needed. To achieve this goal, cropping systems need more fossil energy as technical inputs, but this contributes to a more rapid depletion of fossil resources, exacerbating the environmental impact of agriculture and worsening the climatic disorders that affect (among other human activities) also agriculture productivity. Alternatively, if climate change mitigation should be a priority, the process described above is traced in exactly the opposite direction. To secure energy supply and be free from fossils, the risk is sacrificing agricultural production. This could lead to serious food problems, especially to the most vulnerable populations, those who, for reasons of poverty, would be more willing to replace food crops with energy crops. This would limit even more the access to food by the same communities, causing famine and starvation.

How is it possible to get rid of this "double bind" that offers no viable solution? Is it to produce energy at the cost of insufficient food or, conversely, prioritize food production and remain without oil? The only option is to get away from this dilemma, take a "quantum leap" and promote a "paradigm shift", i.e. make a drastic change of reference system and assume a completely new approach with respect to the old system. This means an active step to put an end to the paradigm of "modernization" and target the transition to a "bio-based" economy (and a bio-based society too).

8 Old and New Forms of Agriculture: The Transition Path

The current changes in European agriculture should lead to the gradual emergence of a new paradigm, intended as an alternative one to that of modernization. I will call it "agro-ecological" paradigm. The sociology school lead by van der Ploeg defines it "rural development paradigm." More generally, not only with respect to agriculture but considering society as a whole, it could be framed within the vision of the so-called knowledge-based bioeconomy. These definitions, although from different perspectives, have much in common. First, let us understand the new paradigm with respect to the current one.

At the core of this new paradigm, practices of sustainable agriculture are firmly assumed. Such practices are endogenous (meaning that farmers directly and consciously undertake them without any suggestions or impositions from the outside) and must be considered a kind of response to a business management dominated by objectives in which farmers no longer believe, very briefly: the myth of an increasing productivity, the need to progressively cut the production costs, the strict market dependence, the agro-industrial submission, the trust on the CAP and on the economic strategy of policy makers, the confidence in a technology that does not fit the real farm needs any more.

The theory of agricultural modernization is linked to the concept of economic development measured only in terms of monetary growth. This prevailing approach intended to overcome the traditional forms of agriculture, because they are considered old, immature, undersized, or even useless. The problems related to the application of this modernization paradigm are becoming increasingly evident, and the adverse effects of this "productivist" model are today easily recognized, for example with regard to the health of consumers, the safeguard of the environmental quality, and the protection of the agro-ecological resources.

The massive introduction of external inputs on farms suggests the use of these productive factors in accordance with the rationality of the system that delivered them (a system that is completely foreign to farmers), based on rules of technical application established by an unknown technological apparatus, under requirements dictated by a state bureaucracy far away from farmers.[20]

Under these conditions, the farmer is fully expropriated by his professional knowledge and skills; he becomes a mere executor of externally prescribed instructions. In this way, a complete standardization of the production process is determined, progressively disconnected from the local agricultural milieu. This means a "commodification" of agriculture: the products no longer have any distinctive value, but they are completely homogeneous and indistinguishable on a globalized market.

Another powerful breakup is observed from the side of the environmental management: the restructuring of the farms, according to the modernization paradigm, separates what was previously unified, namely the "production" activities (to obtain goods) and the "reproduction" activities (to conserve the natural factors also exerting an important productive role). In this way, the reproduction of agro-ecological factors (soil, water, plants, and animals) is no longer a direct responsibility of the farmer, but it becomes an externality managed by structures outside and above the farm. Serious ecological problems were created in this way; an exclusive

[20] Cavazzani (2006), pp. 8–11.

focus on the functioning and management of the production unit has determined that the ecological context in which the farm is inserted was almost completely neglected.

The rural development strategy is achieved by practicing "innovative" behaviors ("farming styles" according to van der Ploeg)[21] that appeal to a traditional way of farming, but rediscovering and reinventing appropriate technologies to address the negative features associated to the "productivist" model. Social practices, even before technical practices, define a new space of relationships; independent choices are developed based on the local resources actually available and adapting farms to the connective tissue making up the "rural" milieu. This capacity to adapt to local specific features generates a wide range of highly diversified agricultural practices. This diversification becomes the most characteristic trait of this new farm model, centered on the multifunctionality concept.

The differentiation of agricultural practices is deployed starting from innovations introduced in the farm management that aims at reducing the degree of farm commodification as well as the level of exogenous technological incorporation into the farm. These innovations are, on the one hand, strategies for strengthening noncommercial circuits for the regeneration of productive resources (thus reducing the dependence on the markets) and, on the other hand, the appropriation, reconstruction, and adaptation of "technology packages" consistent with the farm style.

Multifunctional diversification mainly relies on the classification scheme outlined by van der Ploeg. Three different alternatives are identified: *deepening* concerns activities pursuing product innovation and valorization along the supply chain (i.e., organic farming, short chain production, quality innovation, in-farm food processing, regional or typical production). *Broadening* relates to the activation of new functions in the rural area in which farms are located to offer nonfood goods and services aimed at satisfying new needs and provide community services (farm holidays, social agriculture, pet therapy, landscape management, agro-energy production). *Regrounding* concerns managerial or ownership reorganization also turning to activities other than agriculture (for instance off-farm labor) but integrated at the household level.

[21] van der Ploeg (2008).

Table 1 Paired comparison between different traits related to the two contrasting paradigms

Modernization paradigm	Multifunctional paradigm
Increase productivity, maximize the profit	Increase product quality (emphasis on the concepts of product safety and healthiness), maximize well-being
Productive specialization	Productive diversification
Market globalization	Local decentralization, short-chain production, "farmer-market"
Minimization in the productive costs (but generation of environmental "externalities")	Ecological sustainability and environmental compatibility (agro-ecological approach)
Progressive loss of the "added value" in the products	Progressive appropriation of "added value" in the products
Capital-intensive activities	Labor-intensive activities
Social marginalization and regional imbalance	Social integration, participation approach, farm aggregation network

Table 1 schematically lists the main traits related to the two contrasting paradigms. The comparisons are very clear and do not need further explanations.

9 Concluding Remarks

Although characterized by a significant increase in productivity and a drastic reduction in the production costs, industrial agriculture remarks a number of limitations and drawbacks so that, today, a deep and radical rethinking of this paradigm is required and, hopefully, its progressive replacement. The costs ignored by industrial agriculture are schematically summarized as follows: *public health costs* (obesity, diabetes, cardiovascular disease, contamination in the food system), *environmental costs* (soil erosion, fertility losses, threats to biodiversity, water pollution), *social costs* (disruption of rural society, tradition losses, land abandonment, and landscape degradation), *economic costs* (growing dependence on subsidies).

In accordance with Giampietro:[22] "the paradigm of industrial agriculture is rooted in the endorsement of a series of assumptions about the role the agriculture should perform in society. The validity of these assumptions is more and more contested by the public." Now the question that arises is the following: what are the strengths that hamper the exit from this paradigm and the construction of an alternative one? Still following Giampietro's reasoning, it would be a clear example of "Concorde syndrome." The Concorde is the well-known turbojet-powered supersonic passenger airliner; it entered service in 1976 and continued commercial flights for 27 years. In game theory, the "Concorde syndrome" (or "Concorde

[22] Gianpietro and Mayumi (2010), pp. 223–224.

fallacy") is referring to the fact that the British and French governments continued to fund the joint development of the aircraft even after it became apparent that there was no longer an economic reason for it. The project was regarded privately by the governments as a "commercial disaster" that should never have been started, but political and legal issues had ultimately made it impossible for either governments to pull out.

With reference to the European model of agriculture, it is fully demonstrated the need for a transition to a different production model, but the "sunk costs" (i.e., the large amount of money and the political investments already spent in the application of this dominant paradigm) are generating the conditions of a "Concorde syndrome." This constitutes the major obstacle to operate a decisive "paradigm shift."

In conclusion, it is useful to summarize the essential points of the new agricultural paradigm focused on agroecology and rural development. The new model is pursuing:

- the restoration of agricultural activities based on native energy and no longer on fossil predominance, the substitution and replacement of fossil fuels or fossil energy sources with renewables;
- the control and limitation of impacts on biodiversity and on environmental compartments (such as water bodies, soil, atmosphere), revitalizing the ecological functioning of agro-ecosystems;
- the closing of soil nutrient cycles and the stabilization of soil fertility, the reuse and recycling of waste organic materials, residues, and by-products;
- a decisive increase in the use efficiency of the main productive factors and a significant save of energy in terms of agro-technical inputs;
- the adoption of multifunctional and diversified management criteria, the promotion of a low-carbon agriculture, together with conservation techniques;
- the integration of food with the production of renewable energy (small-scale energy farming) and biomaterials (industrial products derived from biomass and not fossils) obtained from integrated biorefinery platforms and cascading processes to reach a zero-waste standard;
- the setting up of interconnected "value chains," according to a local "farm network" that enables the efficient utilization of all products, by-products, waste, and residues.

This, schematically, is the new vision in response to globalization. Innovation, today, means an agricultural model still based on the peasantry way of farming, despite all the attempts to resize this category and transform farmers into businesspersons. The independence and pride of farming should be revalued.

References

Aubert C (1974) Soignons la terre pour guerir les hommes. Le Courrier du Livre, Paris. [Italian edition: Aubert C (1981) Curare la terra per guarire gli uomini (trans: Caizzi A). RED/Studio redazionale, Como]

Barraclough G (1964) An introduction to contemporary history, first edn. Basic Books, New York. [Italian edition: Barraclough G (1975) Guida alla Storia Contemporanea, 4th edn (trans: Andreose M). Laterza, Bari]

Benyus JM (1997) Biomimicry. Innovation Inspired by Nature. Harper & Collins Publisher, New York

Bevilacqua P (2010) Prefazione. In: Tino P (ed) Le radici della Vita. Storia della fertilità della terra nel Mezzogiorno (secoli XIX–XX). XL Edizioni, Roma

Caporali F, Campiglia E, Mancinelli R (2010) Agroecologia. Teoria e Pratica degli Agroecosistemi. Città Studi Edizioni, Torino

Cavazzani A (2006) Introduzione. In: van der Ploeg JD (ed) Oltre la modernizzazione. Processi di sviluppo rurale in Europa. Rubbettino Editore, Soveria Mannelli (CZ)

Cipolla CM (1962) The economic history of world population. Penguin, London. [Italian edition: Cipolla CM (2013) Uomini, Tecniche, Economie. Il Mulino, Bologna]

Cipolla CM (1965) Guns and sails in the early phase of European expansion 1400–1700. Collins Sons & Co. Ltd, London. [Italian edition: Cipolla CM (2003) Vele e Cannoni. Il Mulino, Bologna]

Diamond J (1997) Guns, germs and steel. The fates of human societies. W.W. Norton & Company Inc., New York. [Italian edition: Diamond J (1998) Armi, Acciaio e Malattie. Einaudi, Torino]

Draghetti A (1948) Principi di fisiologia dell'azienda agraria. Istituto Editoriale Agricolo, Bologna (2nd edn by Edagricole, Bologna, 1991)

Georgescu-Roegen N (1971) The entropy law and the economic process. Harvard University Press, Cambridge, MA

Gianpietro M, Mayumi K (2010) The biofuel delusion. The fallacy of large-scale agro-biofuel production. Earthscan, London, pp 223–224

Goethe JW von (1813–17) Italienische Reise. [English edition: Goethe JW von, Italian Journey, 1786–1788 (1962) (trans: Auden WH, Mayer E). Pantheon Books, New York]

Hausmann G (2005) La Terra come Placenta. Libreria Editrice Fiorentina, Firenze

Liebig von J (1855) Die Grundsätze der Agricultur-Chemie mit Rücksicht auf die in England angestellten Untersuchungen. Vieweg, Braunschweig. [Italian edition: Liebib von J (1856) I principi fondamentali della Chimica Agraria in relazione alle ricerche istituite in Inghilterra (trans: Cossa A). Vallardi, Milano]

Lovelock JE (1979) Gaia. A new look at life on earth. Oxford University Press. [Italian edition: Lovelock JE (1981) Gaia. Nuove Idee sull'Ecologia. Boringhieri Editore, Torino]

Margulis L, Sagan D (1987) Microcosmos: four billion years of evolution from our microbial ancestors. HarperCollins, New York

Mommsen WJ (1969) Das Zeitalter des Imperialismus. Fischer, Frankfurt am Main. [Italian edition: Mommsen WJ (1970) L'età dell'imperialismo (trans: Ascheri H). Feltrinelli, Milano]

Rifkin J (2011) The third industrial revolution. Palgrave MacMillan. [Italian edition: Rifkin J (2012) La terza rivoluzione industriale (trans: Canton P). Arnoldo Mondadori, Milano]

Smil V (1991) General Energetics. John Wiley & Sons, New York

van der Ploeg JD (2008) The new peasantries. Struggles for autonomy and sustainability on the era of empire and globalization. Earthscan, London-Sterling. [Italian edition: Van der Ploeg JD (2008) I Nuovi Contadini. Le Campagne e le Risposte alla Globalizzazione. Donzelli editore, Roma]

Vandana S (1993) Monocultures of the mind. Perspectives on biodiversity and biotechnology. Zed Books Ltd, London. [Italian edition: Vandana S (1995) Monocolture della mente (trans: Ricoveri G). Bollati Boringhieri, Torino]

Biofuel Regulation in the EU: A Failure in the Path Towards Environmental Sustainability and Food Security?

M. Alabrese

Abstract This chapter examines the European Union policy on biofuels from a legal point of view in order to discuss whether and to what extent the regulations have been shaped by the social and environmental issues connected to biofuels. Legal and policy documents will be analyzed in order to see how EU policy has addressed the questions arising from the growth of the biofuel sector and whether the efforts to solve the problems of fossil fuel dependence and greenhouse gas emissions have created other issues. The starting point will be the growing global demand for biofuels triggered by their public promotion in a number of countries worldwide. This will help to understand the magnitude of the phenomenon and of the concerns believed to stem from it. In the last part of this chapter, the current biofuel regulations will be discussed in order to see whether the proclamations became commitments, that is, if each social and environmental need was fully integrated into the regulations. Some of the environmental pressures have pushed for an amendment to current EU legislation, which could lead to a new 'indirect land-use change'—'indirect food insecurity' (ILUC-IFI) dilemma. This could be interesting in terms of understanding how the EU is proceeding towards environmental and social sustainability in the production and use of biofuels. The main conclusion of this review is that although the EU policy articulates a range of principles and criteria on sustainability, each of which can be agreed upon, the real challenge has been in implementing the policy, which has not been without its problems.

Keywords EU biofuel policy • Food security • Land use • Sustainability

M. Alabrese (✉)
Institute of Law, Politics and Development (Dirpolis), Sant'Anna School of Advanced Studies, Pisa, Italy
e-mail: mariagrazia.alabrese@sssup.it, http://www.cdg-lab.dirpolis.sssup.it/en/staff/academic/mariagrazia-alabrese/

M. Monteduro et al. (eds.), *Law and Agroecology*,
DOI 10.1007/978-3-662-46617-9_19

1 Establishing the Context

What was the international context with regard to the energy sector while the EU policy on biofuels was emerging? What were the advantages of promoting bioenergy? How were environmental issues and food security addressed worldwide?

This section provides a brief overview of the global context and international legal framework. This will help in understanding the magnitude of the phenomenon and the concerns that are believed to stem from it. At the same time, it will set the scene for a review of the development of the EU's policy on biofuels.

Biofuels have increasingly emerged as a plausible alternative to oil. According to the recent HLPE Report on *Biofuels and Food Security*, in less than just one decade, world biofuel production has increased five times, from less than 20 billion liters/year in 2001 to over 100 billion liters/year in 2011.[1] A number of countries with important agricultural sectors and pressing needs for domestic energy supplies have opted to support the production of biofuels. As a result, such an increase in the world's supply and demand for biofuels depends on the public policies[2] that have thus played a central role in the creation of markets via obligatory or highly stimulated blending targets, coupled with a range of tax exemptions, subsidies and favorable credit.

In the 1970s, in response to the oil shocks, many countries instigated proposals for alternative fuel policies, and in the 1980s Brazil and the USA created a biofuel market and a related productive sector. In fact, the early 1970s signaled a change in the direction of the energy markets. Global demand for energy and oil rose more than had been predicted by oil companies. The United States began to import more oil. Finally, oil-producing countries, which were increasingly aware of the centrality of oil to their own economies and also to the economies of countries that imported it, were strengthened with the formation of OPEC. As a result, there was an increase in oil prices.

In Brazil,[3] the National Alcohol Program (Proalcool) was launched in 1975. It addressed both supply and demand with investment subsidies, the mandatory placement of ethanol pumps, fixed pricing and taxation of gasoline; thus, production increased rapidly. Currently, the growing use of ethanol in Brazil is linked to the development of the flex-fuel vehicle industry, which introduced vehicles that can switch between ethanol and conventional gasoline. Additional demand for Brazilian biofuels came from the United States to fill the advanced biofuel mandate.[4] In Brazil, the blending of ethanol with gasoline fuels is regulated, with a

[1] High Level Panel of Experts (HLPE) on Food Security and Nutrition of the Committee on World Food Security (2013).

[2] Sorda et al. (2010), pp. 6977–6988.

[3] Hira and de Oliveira (2009), pp. 2450–2456.

[4] OECD/FAO (2012).

required ethanol content of between 20 and 25 %, depending on governmental decisions (which in turn depend on market conditions).

In the USA, ethanol production started to rise in the 1980s with the Energy Tax Act of 1978. This legislation introduced a subsidy aimed at blending ethanol with fossil gasoline. Later, many other policy tools were established, which provided insured loans for small biofuel producers, benefits to automobile makers for flex-fuel vehicles, and a tariff on imported ethanol. With a global production share of about 50 %, the USA is today the largest ethanol producer, especially since the enactment of the Energy Independence and Security Act of 2007.[5]

As far as the global biodiesel production is concerned, it has been estimated to increase to above 42 billion by 2021. The European Union is expected to be by far the largest producer and user of biodiesel. Other significant players are Argentina, the USA, Brazil, as well as Thailand and Indonesia.[6] In fact, in the last 10 years, public support for bioenergy has increased in both developed and developing countries, even though the three major markets continue to be Brazil, the EU and the US.

After an initial phase in which the stated objectives of biofuel policies focused on boosting domestic energy production and reaching self-sufficiency, in the 1990s the range of targets widened. For many countries, the list of objectives for the implementation of biofuel policies included not only the security of the energy supply but also regional development, the creation of new outlets or demand for agricultural products in order to boost farm incomes and, overall, environmental improvement (including mitigation of climate change).[7] In the US, for example, the amendments to the Clean Air Act in 1990 saw the beginning of a shift in renewable fuel policies towards environmental concerns.[8] Likewise, in the EU, among the bioenergy policy drivers was the goal of combating climate change, arising from the Kyoto commitments.

International action on environment and climate change helped to give importance to the question of renewable energy in the 1990s. The UN Framework Convention on Climate Change was signed in 1992 at the UN Conference on Environment and Development in Rio de Janeiro. In 1997, the Kyoto Protocol was adopted, which legally bound developed countries to emission reduction targets.[9] Besides the objective to improve the security of energy supplies, developed countries have been implementing biofuel policies in response to growing environmental concerns associated with climate change and global warming. The development of renewable fuels is seen by a number of countries as one way to reduce their greenhouse gas emissions, as part of established Kyoto commitments arising from the Convention on Climate Change. That development at this point

[5] OECD/FAO (2012).

[6] OECD/FAO (2012).

[7] OECD (2008).

[8] Mayer et al. (2013).

[9] Danish (2007), pp. 31–56.

was not expressly connected to the opportunity to improve food security, though rural and social developments were taken into consideration.

2 The Main Social and Environmental Issues Linked to Biofuels

There are many reasons for the provision of more support for renewable energy: environmental, economic, and political. However, some think that these objectives have not always been fulfilled. In recent years, the contribution of biofuel use to reducing GHG emissions has been strongly contested and a number of unintended impacts of biofuel policies have been reported.[10] In the wake of such criticism against biofuel policies, some even argue about the term "biofuel," "saying that the prefix 'bio' masks harmful social and environmental effects. Using 'agrofuel' instead, they stress the threat it poses 'because of the intensive, industrial way it is produced, generally as monocultures, often covering thousands of hectares, most often in the global South'. For them, biofuels development implies changes in land use and/or land property relations, in ways undermining ecosystems and/or poor people's access."[11]

In 2007, Jean Ziegler, the UN Special Rapporteur on the right to food, called for a 5-year moratorium on biofuels on the basis that converting crops such as maize, wheat and sugar into fuels was driving up the prices of food, land and water. As a consequence, the poorest countries would not have been able to import enough food for their people: "It is a crime against humanity to convert agricultural productive soil into soil which produces food stuff that will be burned into biofuel," he said using very harsh words.[12]

The biofuel and food price debate is controversial in scientific the literature. Since the sharp rise in food prices in 2007/2008, many studies and official reports have investigated whether the increase in biofuels production has had a significant impact on the development of agricultural commodity prices.[13] The evidence is not conclusive, and it seems that the impact of biofuels on the price of food also depends on the crop, on the local conditions and whether the discussion is on a global or local level.

These contradictions were recognized by the aforementioned HLPE Report on *Biofuels and Food Security*, which stated that "biofuel production and the policies used to support its development can relate both positively and negatively with each of the four dimensions of food security—availability, access, utilization (nutrition) and stability."[14] This statement referred to the widely accepted FAO definition of

---

[10] Doornbosch and Steenblik (2007) and Pimentel (2012).

[11] Franco et al. (2010), pp. 661–698.

[12] UN News Centre (2007).

[13] World Bank (2008); Ajanovic (2010); Zilberman et al. (2013), pp. 275–281.

[14] High Level Panel of Experts (HLPE) on Food Security and Nutrition of the Committee on World Food Security (2013).

food security, according to which "Food security exists when all people, at all times, have physical and economic access to sufficient, safe and nutritious food that meets their dietary needs and food preferences for an active and healthy life" (World Food Summit, 1996).

Following on from the food security concerns, another issue is natural resources, especially land and water. Land use for energy production has been focused on by all strata of stakeholders and policy makers at a European and international level. The growing demand for bioenergy crops has led to great competition for land and water among food-oriented agricultural activities, energy production and the use of agricultural land for nature conservation and urbanization. In addition, biofuel policies have been considered as some of the key drivers behind the recent wave of large-scale land acquisition in developing countries, as many companies have been oriented by the increasing demand for biofuels in their land investment strategies.[15]

Besides the risk of the reallocation of land from the production of food crops to biofuel crops, international large-scale land acquisitions have a number of socio-economic impacts. Food insecurity is often the principal effect of biofuel land deals, but the displacement of local and indigenous communities should not be underestimated from a social point of view.

Competition for land also involves environmental questions when biofuel production requires land devoted to the protection of biodiversity and carbon sequestration, which is when direct and indirect land-use changes can occur. The former happens when the expansion of crops for biofuels takes place at the cost of forests or grassland. The latter results when the feedstocks for biofuels are not directly established on valuable environmental land but on arable soil, triggering the displacement of agriculture elsewhere to forested land or other natural areas.[16]

This phenomenon, which will also be addressed in the last part of this chapter, not only carries the risk of significant environmental damage, nullifying the GHG emission reductions attributed to biofuels and exacerbating anthropogenic climate change, but is also relevant to food security "since what is an 'indirect' land-use change relative to biofuels, is a 'direct' land-use change relative to food (food crops expanding directly on other lands), and vice versa. In other words, minimizing 'ILUC' effects could be at the expense of food security and create 'indirect food insecurity' (IFI). In turn, minimizing IFI could lead to ILUC effects."[17]

[15] Cotula et al. (2008). High Level Panel of Experts (HLPE) on Food Security and Nutrition of the Committee on World Food Security (2013).

[16] Commission (EC) Report on indirect land-use change related to biofuels and bioliquids (Communication) COM(2010) 811 final, 22 December 2010.

[17] High Level Panel of Experts (HLPE) on Food Security and Nutrition of the Committee on World Food Security (2013).

The complexity of these questions confirms that biofuels should be placed within the framework of the "food-energy-environment trilemma,"[18] which still appears as the right approach to the subject.

3 The EU Biofuel Policy Evolution: Intersection (and Possible Integration) with Social and Environmental Concerns

Bearing in mind the major issues emerging from the development of biofuels, it is worth considering whether and how the European legislation on bioenergy takes into account these unintended impacts of biofuel policies and if it can be said to have been shaped by them.

Europe is at the forefront of energy policy and renewable energy development.[19] The European Union has acted both on energy and the environment for many years with the awareness that greenhouse gas emissions come largely from energy use and production. The importance of the environment in developing the energy sector was recognized in the Treaty of Lisbon, which came into force on 1 December 2009. It gives a new legal basis to energy that was lacking in the previous treaties: article 194 of the Treaty on the Functioning of the European Union declares that "In the context of the establishment and functioning of the internal market and with regard for the need to preserve and improve the environment, Union policy on energy shall aim, in a spirit of solidarity between Member States, to: (a) ensure the functioning of the energy market; (b) ensure security of energy supply in the Union; (c) promote energy efficiency and energy saving and the development of new and renewable forms of energy; and (d) promote the interconnection of energy networks."

As is the case with other countries, in the 1980s the EU paid increasing attention to renewable energy, which was considered a good means for tackling energy security problems. Thus, at this time, the environmental advantages related to renewable energy were not yet at the center of the European Community's considerations. The scenario slowly started to change from 1987, when a title devoted to the environment was introduced by the Single European Act into the Treaty of Rome of 1957. In addition, the Fourth Environment Action Programme of the European Community (1987–1992), published in the same year, specifically referred to the relationship between environmental policy and energy production. Thus, the EU's energy policy took on an environmental dimension. No longer was the security of energy supply the only issue.[20] Integrating the environment into

[18] Tilman et al. (2009), pp. 270–271.

[19] Alabrese (2012), pp. 38–45.

[20] Commission (EC) Energy for the Future: Renewable Sources of Energy—Green paper for a Community Strategy (Communication) COM(1996) 576 final; Commission (EC) Energy for the

Community energy policy was the specific subject of the Communication from the Commission in 1998 entitled *Strengthening environmental integration within Community energy policy*,[21] where several measures were proposed to incorporate the environmental dimension into energy policy objectives and actions. However, the first part of twenty-first century has been characterized by an increase in dependency on energy imports combined with an increase in oil prices. Yet again, the issue of energy security became central, and the EU laid down the *Green Paper—Towards a European strategy for the security of energy supply*,[22] in which not only dependence on external energy sources but also environmental concerns were confronted. Indeed, the document gave prominence to climate change and promoted the development of new and renewable energies. This policy was put into action in Directive 2001/77 on the promotion of electricity produced from renewable energy sources.

This was followed by Directive 2003/30 on the promotion of the use of biofuels or other renewable fuels for transport. The promulgation of the Biofuels Directive represents the first significant milestone in the development of a coherent EU policy on biofuels.[23] It aimed to introduce a nonbinding target of a 2 % market share for biofuels in 2005 and a 5.75 % share in 2010. A proper review of the Directive, which would have given an exact assessment of the measures set down, was not delivered, but in 2006 the Commission's suggestion of a new strategy on biofuels was published.[24] The detailed *EU Strategy for Biofuels* was considered by legal scholars as a document that paved "the way for the development of a more mature EU policy on biofuels."[25] It ratified that the 2005 target share of 2 % had not been achieved and stimulated the production and use of biofuels.

It is interesting to note that the promotion was done in consideration of the sustainability of biofuel production in terms of environmental, economic and social issues. While exploring the opportunities of biofuel production in developing countries, the Commission warned that the exploitation of their resources would have to be conducted with respect to good agricultural practice, namely that sustainable biomass production would have to be compatible with environmental requirements and that it could not impact negatively on the production of foodstuffs. The document highlighted the opportunities that would come from developing countries, considering that biomass productivity is higher in tropical areas and the production costs are comparatively low. Even the benefits for these countries

future: renewable sources of energy—White Paper for a Community strategy and action plan (Communication) COM(1997) 599 final.

[21] Commission (EC) Strengthening environmental integration within Community energy policy (Communication) COM (1998) 571 final.

[22] Commission (EC) Green Paper—Towards a European strategy for the security of energy supply (Communication) COM(2000) 769 final.

[23] Switzer and McMahon (2011).

[24] Commission (EC) An EU Strategy for Biofuels (Communication) COM(2006) 34 final, 8 February 2006.

[25] Switzer and McMahon (2011), p. 714.

were addressed: "the production of biofuels from suitable feedstocks could also generate economic and environmental benefits in a number of developing countries, create additional employment, reduce energy import bills and open up potential export markets. In particular, the production of bioethanol could offer a feasible alternative for some sugar-producing countries affected by the reform of the EU sugar regime."[26]

At the same time, social and environmental issues are taken into account as it is underlined that "in countries where a large-scale expansion of feedstock production is likely to take place, environmental concerns relate to pressures on eco-sensitive areas, like rainforests. There are also concerns regarding the effect on soil fertility, water availability and quality, and pesticide use. Social effects concern potential dislocation of communities and competition between biofuel and food production. These concerns need specific investigation and quantification and, if necessary, should be addressed through strong regulatory frameworks."[27] In addition, the 2006 European Parliament Resolution on the promotion of crops for nonfood purposes,[28] which was completely devoted to the creation of new income opportunities for farming, recognized—although in passing—the problem of competition between food and fuel.

To increase the use of renewable energy sources, the European Commission drew up the *Renewable Energy Road Map. Renewable energies in the 21st century: building a more sustainable future*,[29] where its long-term strategy was set out. In the Road Map, the Commission proposed a mandatory target of 20 % for the renewable energy share of energy consumption in the EU by 2020 and a mandatory minimum target of 10 % for biofuels. The proposal was accepted by the European Council,[30] and for the first time a legally binding target was introduced.

The introduction of a binding biofuel mandate was subject, among other things, to production being sustainable. Even the European Parliament stressed the importance of sustainability criteria for biofuels and requested that the Commission should undertake action towards a mandatory certification system for biofuels in its Resolution of 2007 on the Road Map for Renewable Energy in Europe.[31] A stronger position of the European Parliament can be found in the Resolution on sustainable agriculture and biogas.[32] This document clearly declared that biogas

[26] Commission (EC) An EU Strategy for Biofuels, 4.

[27] Commission (EC) An EU Strategy for Biofuels, 7.

[28] European Parliament Resolution on the promotion of crops for non-food purposes (2004/2259 (INI)).

[29] Commission (EC) Renewable Energy Road Map: Renewable energies in the twenty-first century: building a more sustainable future (Communication) COM(2006) 848 final, 10 January 2007.

[30] Council of the European Union (EU), Presidency Conclusions—Brussels 8/9 March 2007, 2 May 2007, 21.

[31] European Parliament Resolution on the Road Map for Renewable Energy in Europe, 25 September 2007.

[32] European Parliament Resolution on sustainable agriculture and biogas: a need for review of EU legislation (2007/2107(INI)).

production should be preferred that used organic waste and crop by-products unsuitable for food and feed production. In addition, the possible connection between bioenergy production (primarily bioethanol and biodiesel) and rising grain and food prices on the world market was recognized.

In 2007, a set of policy measures was introduced.[33] In the objectives put forward by the "Energy Package," reducing greenhouse gas emissions was central. The Commission also proposed creating a new legislative framework to enhance the promotion and use of renewable energy. Discussions on the draft Directive stressed the need to develop effective sustainability requirements for biofuels, as well as giving prominence to the link between biofuels and food security. Consequently, Directive 2009/28[34] was issued.

This brief review of the main steps that created the EU biofuel policy shows that, together with the great opportunities for sustainable development and climate change mitigation, bioenergy growth has also met many challenges over the years. In the last part of this chapter, the current biofuel regulations will be examined in order to see if the proclamations became commitments, that is, if each social and environmental need turned into an obligatory rule.

4 The Current Scenario and the Future Horizon in the EU Biofuel Legislation

Directive 2009/28/EC on the promotion of the use of energy from renewable sources established mandatory targets to be achieved by 2020 for a 20 % overall share of renewable energy in the EU and a 10 % share for renewable energy in the transport sector. At the same time, through Directive 2009/30/EC,[35] the EU adopted a mandatory target to achieve a 6 % reduction in the greenhouse gas intensity of fuels used in transport by 2020. Each Member State would decide its own renewable energy mix and the means by which this target would be reached, but the contribution of biofuel towards these goals is significant.

According to the Directive,[36] different renewable energy sources are allowed within the framework of the mandatory target in the transport sector, such as renewable electricity, hydrogen and biofuels, though the actual implementation of

---

[33] Commission (EC) An energy policy for Europe (Communication) COM(2007) 1 final.

[34] European Parliament and Council Directive (EC) 2009/28 on the promotion of the use of energy from renewable sources and amending and subsequently repealing Directives 2001/77/EC and 2003/30/EC.

[35] European Parliament and Council Directive (EC) 2009/30 amending Directive 98/70/EC as regards the specification of petrol, diesel and gas-oil and introducing a mechanism to monitor and reduce greenhouse gas emissions and amending Council Directive 1999/32/EC as regards the specification of fuel used by inland waterway vessels and repealing Directive 93/12/EEC.

[36] EC Renewable Energy Directive 2009/28; Art 3 (4).

the regulation focused mainly on biofuels. "The latest projections by member countries for their National Renewable Energy Action Plans show that 88 % of the target will be covered by traditional biofuel, and within biofuels, three-quarters will come from biodiesel."[37] Biofuel development was promoted not only through a quota obligation scheme but also through direct subsidies. These targets cannot be fully met using only EU domestic biomass, which is basically why "the EU biofuel policy has triggered the creation of an increasingly globalized biofuels and biofuels feedstock market, involving a key role for developing countries agriculture."[38]

The European need to address the growing challenge of climate change made the Renewable Energy Directive (2009/28/EC) and Fuel Quality Directive (2009/30/EC) set out a sustainability scheme for biofuels and bioliquids, which, however, only focused on environmental concerns.

According to the Directives, the increasing worldwide demand for biofuels and bioliquids should not encourage the destruction of biodiverse lands. Thus, sustainability criteria are defined ensuring that biofuels and bioliquids can qualify for the incentive rewards only when there is a guarantee that they do not originate from biodiverse areas or that they do not endanger ecosystems. Although these sustainability criteria are not mandatory in general, their fulfilment permits Member States to include energy from biofuels and bioliquids in measuring their achievement of national energy targets and their compliance with renewable energy obligations and, overall, for their eligibility for financial support. Thus, biofuels and bioliquids can be counted as renewable energy for the purposes of the Directive, and financial support can be obtained only when it can be guaranteed that they meet these criteria.

The requirements, set forth in article 17,[39] relate to greenhouse gas reductions, land with a high biodiversity value and land with a high carbon stock and agri-environmental practices. In brief, to qualify for financial support, biofuels and bioliquids have to reduce greenhouse gas emissions and must not be produced using raw materials from the land with a high carbon stock or high biodiversity value. In addition, where biofuels and bioliquids are made from raw materials produced within the European Union, they should also comply with European environmental requirements for agriculture.

However, and this is a crucial point, according to the Directive, the Commission will report every two years to the European Parliament and the Council on the impact on social sustainability of the increased demand for biofuel in the EU and in third countries and on the impact of the EU biofuel policy on the availability of foodstuffs at affordable prices, in particular for people living in developing countries. The reports will also address to what extent land-use rights have been respected. They will state, both for third-world countries and Member States, whether the country has ratified and implemented some Conventions of the

[37] Mayer et al. (2013).

[38] High Level Panel of Experts (HLPE) on Food Security and Nutrition of the Committee on World Food Security (2013).

[39] EC Renewable Energy Directive 2009/28; EC Renewable Energy Directive 2009/30, Art. 7.

International Labour Organisation mentioned in the Directive, such as the Convention concerning Forced or Compulsory Labour, the Convention concerning Equal Remuneration of Men and Women Workers for Work of Equal Value, the Convention concerning the Abolition of Forced Labour, the Convention concerning Minimum Age for Admission to Employment, etc. In addition, the Commission will, if appropriate, propose corrective actions, in particular if evidence shows that biofuel production has a significant impact on food prices.

The Commission is also invited to submit a further report concerning the impact of Indirect Land Use Change (ILUC) on greenhouse gas emissions and address ways to minimize that impact. The report can be accompanied by a proposal containing a concrete methodology for emissions from carbon stock changes caused by indirect land-use changes.[40]

In the wake of these provisions, a Report from the Commission on indirect land-use change related to biofuels and bioliquids was issued in 2010[41] and a Proposal for a directive amending the Renewable Energy Directive and the Fuel Quality Directive was published on October 2012.[42] According to the Report, the increased use of biofuels in the EU raises the existing demand for agricultural commodities; thus, the biofuel feedstock needed may be produced on land *directly* converted from another status (such as forest or grassland, etc.) to agricultural land. However, in order to comply with the sustainability criteria introduced by the Directives, raw material could instead be cultivated on existing agricultural land. In this case, as the nonfuel demand (for food, feed or fiber) will still need to be satisfied, it could mean that nonagricultural land will be brought into production. Through this route, the extra biofuel demand could lead indirectly to land-use change. Moreover, the land-use change may not be induced locally but could take place even on different continents.

Although land-use change can have a wide range of impacts (greenhouse gases, biodiversity, social issues, etc.), the amendment proposal of the current European legislation only focuses on the consequences of the greenhouse gas emissions of biofuels, as required by the Directives. This phenomenon takes place when ILUC involves the conversion of high carbon stock land, which can lead to significant CO_2 emissions, thus reducing the GHG emissions savings that biofuels would provide by the displacement of fossil fuels.

The main aims of the proposal should be to start the transition to advanced biofuels, such as those made from wastes and algae, and to address ILUC. For these purposes, the proposal introduced a cap of 5 % for conventional biofuels (those produced from food crops) to count towards the targets of the Renewable Energy

[40] EC Renewable Energy Directive 2009/28, Art. 19.6.

[41] Commission (EC) Report on indirect land-use change related to biofuels and bioliquids (Communication) COM(2010) 811 final, 22 December 2010.

[42] Commission (EC) Proposal for a Directive of the European Parliament and of the Council amending Directive 98/70/EC relating to the quality of petrol and diesel fuels and amending Directive 2009/28/EC on the promotion of the use of energy from renewable sources COM(2012) 595 final, 17 October 2012.

Directive and a new incentive scheme to further promote sustainable and advanced biofuels from feedstocks that do not create an additional demand for land.

As far as the indirect land-use change is concerned, the Commission's proposal (assuming that a number of uncertainties were associated with the available models used to quantify indirect land-use change) did not include the ILUC factors among the criteria required for biofuels. Thus, although the proposal could act under the framework of the precautionary principle that might have allowed the introduction of such measures, it only established the introduction of reporting estimated emissions from carbon stock changes caused by indirect land-use change based on the best available scientific evidence. The following step within the legislative procedure, which is likely to proceed very slowly, took place in September 2013 when the European Parliament approved the draft with amendments without introducing ILUC factors into the sustainability criteria.

It is not worth going into more detail on the proposal and its weaknesses, given that it may never be passed; instead, it seems more appropriate to focus on some questions related to the sustainability criteria currently in force.

As already mentioned, EU policy only takes into account environmental sustainability issues which, however, just looking at the EC's Report on ILUC, do not seem to have been effectively tackled. Moreover, in order to deal with negative impacts on social sustainability and the availability of foodstuffs at affordable prices, and to guarantee the respect of land rights and the well-being of employees, the Directives set out the need for a biannual report from the Commission instead of a mandatory system of rules. However, the current amending proposal still does not address the social impacts of biofuels.

Is the proposal a missed opportunity for giving prominence to the social issues connected to biofuel production? It is worth noting that under the framework of the Policy Coherence[43] for Development, the EU is carrying out studies on the impact on developing countries of the European biofuel policy and of the increased demand for biofuels. In addition, international working groups, such as the Global Bioenergy Partnership,[44] and a number of NGOs advocate a comprehensive certification of biofuels to ensure the production of biomass in a socially and environmentally sustainable way.

So why does the EU not consider expanding the sustainability criteria to include social standards, food security, access to natural resources? Maybe one answer lies in international trade regulation. Given that the Directive affects global trade flows, the EU has to ensure that the sustainability criteria issued are compatible with the applicable World Trade Organization law. In other words, the implementation of certain trade measures in terms of biomass sustainability criteria need to be compliant with the rules of the GATT Agreement.

---

[43] Commission (EC) Policy Coherence for Development Work Programme 2010–2013 (Commission Staff working document) SEC(2010) 421 final, 21 April 2010.

[44] Global Bioenergy Partnership [GBEP] (2011).

The complexity of the question can be simplified by underlining, firstly, that according to the principles of nondiscrimination, Member States cannot discriminate between *like products*, though an equal treatment of 'like' products is mandatory. Secondly, sustainable and nonsustainable biomass and biofuels can be regarded as 'like' products. Thirdly, GATT article XX lists a number of exceptions that could, however, justify trade restrictive measures that would otherwise be incompatible with WTO obligations. Therefore, the main question is whether the enforcement of socially and environmentally oriented sustainability criteria can be considered as part of the exceptions presented in article XX. As the exceptions only refer to environment-related measures, the provision of criteria to avoid competition with food products and social standards are most probably not compliant with WTO rules.

The proposed ILUC factors could also be highly sensitive because the available numerical assessment models are not robust enough to be defended within an agreed market regime, such as the WTO. This topic[45] is truly vast, and a thorough analysis is beyond the scope of the present paper. However, the matter was already being taken into serious account by the EU when Directive 2009/28/EC was still at the proposal stage.[46] Indeed, if the EU wants to introduce social sustainability criteria and ILUC factors, these international challenges need to be considered among the other issues at stake.

5 Concluding Remarks

The initial surge in biofuel production in many countries worldwide was driven by energy security goals and rising fossil fuel prices. It was soon realized that market forces alone were not sufficient to drive the process, which required heavy policy support (subsidies, quota obligation schemes and tariffs for imports). Yet the gap between the demand for biofuels and the potential domestic supply in developed countries entailed expanding biofuel production in developing countries, which had the land and the climate suitable to produce raw feedstocks on a large scale.[47] This condition triggered a number of social and environmental issues related to biofuel production and trade.

Since the emergence of its biofuel policy, the European Union has attempted to give appropriate attention to such concerns, as revealed from the communications, resolutions, proposals, drafts and legislation issued by its Institutions over the years. Looking at the EU biofuel policy from this perspective may make it seem that policy is being totally shaped by the worries stemming from the growth of the biofuel sector.

[45] Lendle and Malorie (2010); Switzer and McMahon (2011) at pp. 713–736.

[46] Btg Report prepared for DG TREN [EC] (2008).

[47] FAO (2013).

However, in an attempt to further check the level of coherence and effectiveness of the measures set down, one could argue that the objectives have not been completely met and even some external issues have arisen: the efforts to solve the problems of fossil fuel dependence and greenhouse gas emissions have actually created several other problems. Indeed, the same climate change mitigation goal seems to have failed, considering the last proposal for the amendment of the Renewable Energy Directive. The initial aim to introduce the ILUC factors into the current regulation proves that biofuel may not be sufficiently efficient to reduce GHG emissions.

Moreover, current sustainability criteria do not properly address social sustainability: social dimensions tend to revolve around a few documents and reports that are supposed to be issued by the Commission.

In this connection, the EU biofuels policy could face a complex dilemma in the next few years. In fact, its development may lead to a situation in which the choice to introduce indirect land-use change factors into the regulations could result in an increase in "indirect food insecurity" (IFI).

ILUC emissions are extremely important if we want to guarantee the real effectiveness of biofuel in reducing GHG emissions. However, introducing the ILUC factors, without seriously improving the whole system, could lead to a worse effect in many countries, encouraging producers to shift from producing food crops to biofuel feedstocks. This may result in the effect that providing environmental protection—by preventing biofuel companies from the conversion of certain areas such as high-value forests or peatlands to grow feedstocks—could increase food insecurity. In the same way, drawing attention to food availability could affect the ability of biofuels to reduce GHG emissions and their impact on climate change and environmental protection.

Apart from this 'ILUC-IFI dilemma,' and, perhaps making the situation worse, it has also been noted that "another limitation of the existing biofuel certification schemes is the concern that small-scale farmers are left out of these biofuel developments because of the dominant governance structure of these certifications led by large-scale agro-industry and the cost structure of certification, which is out of reach for most smallholders."[48] Although this assumption actually refers to all kinds of certification schemes, both public and private, the EU biofuel sustainability criteria themselves lack a structure that would enhance full participation by smallholders. This once again highlights the weak link between the EU biofuel legislation and its social implications.

To conclude, it is worth wondering whether the EU biofuel legislation, defined by the Commission as the "the most comprehensive and advanced sustainability scheme anywhere in the world,"[49] has been failing in the path towards environmental sustainability and food security objectives. While the policy structure articulates a range of principles and criteria on sustainability, each of which can

[48] FAO (2013).

[49] Commission (EC) Report on indirect land-use change related to biofuels and bioliquids.

be agreed on, the real challenge has been its implementation on the ground—which has revealed several flaws.

References

Ajanovic A (2010) Biofuels versus food production: does biofuels production increase food prices? Energy. doi:10.1016/j.energy.2010.05.019

Alabrese M (2012) Land use, bioenergy and sustainability. EU Agrarian Law 1(2):38–45

Cotula L, Dyer N, Vermeulen S (2008) Fuelling exclusion? The biofuel boom and poor people's access to land. IIED/FAO, London

Danish KW (2007) The international regime. In: Gerrard MB (ed) Global climate change and U.S. law. ABA Publishing, Chicago, pp 31–56

DG TREN [EC] (2008) Sustainability criteria and certification systems for biomass production. http://ec.europa.eu/energy/renewables/bioenergy/doc/sustainability_criteria_and_certification_systems.pdf. Accessed 24 Mar 2014

Doornbosch R, Steenblik R (2007) Biofuels: is the cure worse than the disease? OECD Publishing, Paris. http://www.oecd.org/sd-roundtable/39411732.pdf. Accessed 24 Mar 2014

FAO (2013) Biofuels and the sustainability challenge: a global assessment of sustainability issues, trends and policies for biofuels and related feedstocks. FAO, Rome. http://www.fao.org/docrep/017/i3126e/i3126e.pdf. Accessed 24 Mar 2014

Franco J, Les Levidow DF, Goldfarb L, Hönicke M, Mendonça ML (2010) Assumptions in the European Union biofuels policy: frictions with experiences in Germany, Brazil and Mozambique. J Peasant Stud 37(4):661–698

Global Bioenergy Partnership [GBEP] (2011) The global bioenergy partnership sustainability indicators for bioenergy. FAO/GBEP, Rome. http://www.globalbioenergy.org/fileadmin/user_upload/gbep/docs/Indicators/The_GBEP_Sustainability_Indicators_for_Bioenergy_FINAL.pdf. Accessed 24 Mar 2014

High Level Panel of Experts (HLPE) on Food Security and Nutrition of the Committee on World Food Security (2013) Biofuels and food security. HLPE, Rome. http://www.fao.org/fileadmin/user_upload/hlpe/hlpe_documents/HLPE_Reports/HLPEReport5_Biofuels_and_food_security.pdf. Accessed 24 Mar 2014

Hira A, de Oliveira LG (2009) No substitute for oil? How Brazil developed its ethanol industry. Energy Policy 37:2450–2456

Lendle A, Malorie S (2010) Sustainability criteria in the EU renewable energy directive: consistent with WTO rules? ICTSD, Geneva

Mayer S, Schmidhuber J, Barreiro-Hurlé J (2013). Global biofuel trade: how uncoordinated biofuel policy fuels resource use and GHG emissions. ICTSD, Geneva. http://www.ictsd.org/downloads/2013/05/global-biofuel-trade-how-uncoordinated-biofuel-policy-fuels-resource-use-and-ghg-emissions.pdf. Accessed 24 Mar 2014

OECD (2008) Biofuel support policies an economic assessment. OECD Publishing, Paris

OECD/FAO (2012) Biofuels. In OECD-FAO agricultural outlook 2012. OECD Publishing, Paris. http://dx.doi.org/10.1787/agr_outlook-2012-6-en. Accessed 24 Mar 2014

Pimentel D (ed) (2012) Global economic and environmental aspects of biofuels. Taylor and Francis Group, Boca Raton

Sorda G, Banse M, Kemfert C (2010) An overview of biofuel policies across the world. Energy Policy 38(11):6977–6988

Switzer S, McMahon JA (2011) EU biofuels policy—raising the question of WTO compatibility. Int Comp Law Q 60(3):713–736. doi:10.1017/S0020589311000285

Tilman D, Socolow R, Foley JA, Hill J, Larson E, Lynd L, Pacala S, Reilly J, Searchinger T, Somerville C, Williams R (2009) Beneficial biofuels—the food, energy, and environment trilemma. Science 325(5938):270–271

UN News Centre (2007) UN independent rights expert calls for five-year freeze on biofuel production (October 26, 2007). http://www.un.org/apps/news/story.asp?NewsID=24434&#.U1ud4cY47K9. Accessed 24 Mar 2014

World Bank (2008) Rising food prices: policy options and World Bank response. World Bank, Washington, DC. http://siteresources.worldbank.org/NEWS/Resources/risingfoodprices_backgroundnote_apr08.pdf. Accessed 24 Mar 2014

Zilberman D, Hochman G, Rajagopal D, Sexton S, Timilsina G (2013) The impact of biofuels on commodity food prices: assessment of findings. Am J Agric Econ 95(2):275–281

Part III
Rural Sustainable Development and Rural Civilization

Agroecology and Politics: On the Importance of Public Policies in Europe

M. González de Molina Navarro

Abstract In this paper, we discuss about the most appropriate strategy for spreading agroecology and getting agricultural sustainability in Europe, where the agriculture is highly industrialised, very dependent of the public subsidies, and the peasant has disappeared practically. We claimed for a sustainable food system, that they are necessary not only for the health of European agroecosystems and the well-being of the farmers but also for the food autonomy of countries whose production is overturned too much to meet the European demands of animal feeding and to maintain an unsustainable diet. From an agroecological perspective, the most coherent solution is promoting the sustainable degrowth of the European food system. The organic agriculture and the fair consumption could be the most suitable way to achieve it, two proposals that should go indissolubly united. But this will not be possible without a change in public policies and institutional framework.

Keywords Agroecology • Organic agriculture • Public policy • Sustainable degrowth

1 Introduction

Agroecology arose as a response to the ecological crisis in the countryside, promoting the sustainable management of natural resources and equitable access to those resources.[1] Under its standard, there have been many experiences in production, distribution and consumption, which, being innovative, are the avant-garde of an alternative food system. These experiences in social innovation are the basis on which a more sustainable future will be built, though, in themselves, they are not sufficient to produce changes at a higher scale of social organisation or even for their own survival as successful experiences. The simple sum of these experiences does not guarantee change, given that local experiences depend on the limiting

[1] Guzmán Casado et al. (2000).

M. González de Molina Navarro (✉)
Department of History, Geography and Philosophy, University Pablo de Olavide, Seville, Spain
e-mail: mgonnav@upo.es, http://www.historiambiental.org/en/miembros

M. Monteduro et al. (eds.), *Law and Agroecology*,
DOI 10.1007/978-3-662-46617-9_20

capacity of the institutional arrangements that exist at higher scales of social organisation, especially at the national scale.

This text asserts the relevance of an agroecological strategy to achieve agrarian sustainability in Europe, which overcomes the limited framework of agricultural activity and concentrates on how the feeding of the people of Europe is organised. This strategy must be based on the sustainable degrowth of the food system as a whole. Individual or collective actions with regard to production or of citizens in the marketplace are not enough to achieve this; it also requires the introduction of public policies that make degrowth possible and, therefore, active participation in the political struggle. Agroecology and politics are, therefore, two terms that should be intimately linked together with law, an essential instrument for the construction of new institutions and regulations that encourage agroecological transition.

2 The Food System and Its Environmental Impact

The way in which the people of Europe feed themselves has changed very significantly, and these changes are some of the main causes of unsustainability, not only as regards human health but also as regards the health of the ecosystems and the stock of natural resources[2] and not only for Europeans but also for third countries.[3] New and increasingly costly processes have appeared between production and consumption. The feeding of the people now involves the use of new and more sophisticated "artefacts" powered by gas or electricity, which have increased the energy cost of food. Food processing and distribution has taken on an importance never before seen. The food market is now global and involves foodstuffs that incorporate high energy and material inputs (transport, processing, logistics, etc.). All of the foodstuffs that we find today on our table have a long story behind them, with high consumption of energy and materials, emissions and imbalanced economic trading models that turn the food supply into a process with heavy environmental loads. A recently published United Nations report recognised that agriculture and the consumption of fossil fuels are the two main sources of the planet's unsustainability.[4]

Meeting the food needs of the people of Europe requires huge tracts of productive land in third countries to be "subordinated" to the food system and the companies that control it. For example, for the people of Spain to be able to consume 3,000 calories per day (3,405 kcal), 109 million tonnes of animal and plant biomass is needed or, to put it another way, 2.43 tm/person/year or 6.65 kg/

[2] Alonso and Guzmán Casado (2004), pp. 471–541; González de Molina et al. (2005), pp. 119–144; González de Molina and Guzmán Casado (2006).

[3] International Assessment of Agricultural Knowledge, Science and Technology for Development (IAASTD) (2009).

[4] United Nations Environment Programme (2011), p. 3.

person/day.[5] Spain has 42.16 million hectares of farmland suitable for the production of biomass, of which only 41 % is cultivated land.[6] However, although there has been a significant increase in the productivity of the land, the cultivated crop area has paradoxically fallen and domestic production is unable to meet domestic demand. After depopulating the countryside, turning agriculture into a subsidised sector and seeing rural life lose its prestige, the food demands of the Spanish people cannot be met by the country's own agroecosystems. Only by resorting to the international market is it possible to maintain food habits as opulent as those seen in Spain. The basis of the traditional diet, carbohydrates, has lost weighting to fats, which now represent over 40 % of all calories consumed.[7] Meat, milk and other dairy products are directly responsible for this increase. This has a high territorial cost: to produce 1 kg of vegetables requires 1.7 m^2 of crop area, whereas 1 kg of meat requires[8] 7 m^2.

Over the last decade, Spain has exported 20 million tonnes of foodstuffs, more than half of which were horticultural products, this being the main speciality of Spanish agriculture. This specialisation has a high social and environmental impact.[9] On the other hand, the country has imported almost 31 million tonnes, giving a deficit of over 10 million tonnes. Just the cereal, seed and animal feed requirements alone equal the total of all exports. Most of these imports are used to feed livestock or are processed by the food industry. Spanish eating habits, like those of rich and developed countries, require large crop areas to be devoted to the production of grain and fodder in peripheral countries in order to breed sufficient livestock to meet the high demand for meat and dairy products. It can, therefore, be understood how ideas such as "unequal ecological exchange"[10] and "ecological debt"[11] have proliferated in the political and academic debate. Although Europe has not resorted excessively to "land grabbing", the subordination of the production of large areas of land in developing countries to the production of food to meet the unsustainable Western diet may be considered, paraphrasing the words of the former Director General of the FAO, Jaques Diouf, to be a new form of colonialism. Witzke and Noleppa[12] estimated the amount of "virtual agricultural land" that Europe imports. The figures are unequivocal: UE-27 countries export around 14.1 million hectares, while soya on its own represents imports of 19.2 million. In total, the deficit is 35 million hectares, approximately the area of Germany.

In recent works, we have estimated the energy cost of the Spanish food system from six activities of the food chain: the consumption generated by the national and

[5] González de Molina and Infante (2010), pp. 113–137.

[6] Ministerio de Medio Ambiente y Medio Rural y Marino (2010).

[7] Schmidhuber (2006).

[8] Carpintero (2006), pp. 31–44, 41.

[9] Delgado and Aragón (2006), pp. 423–474.

[10] Hornborg (1998), pp. 127–136.

[11] Martínez Alier and Oliveres (2003).

[12] Von Witzke and Noleppa (2010).

international transport of food and agricultural products, their processing, their packaging, their packing, the energy cost of the sale in food outlets and the cost of conservation and preparation in the home. The high food mileage and the duration of the distribution and marketing process require that foodstuffs be kept in a good state of conservation during this period. This need, together with the need to take care of the appearance of the product, which is even more important in our culture than the natural properties of the foodstuff, requires the massive use of packaging and packing.

The food eaten in Spain, then, requires a very significant input of energy, the great majority of which comes from fossil fuels and which is used outside the agricultural sector. If we incorporate the rest of the activities necessary to put food on the table in each home, we see that the agricultural sector is responsible for a little over a third of the total consumption of primary energy in the Spanish food system. The transport, industrial processing, packing, sale, conservation and consumption of food account for the remaining 66 %. In total, over 1,400 Pj is needed to meet the endosomatic metabolism requirements of the Spanish people, while the energy contained in the foodstuffs consumed only amounts to 235 Pj.[13] That is to say, for each unit of energy consumed as food, six have been used in its production, distribution, transport and preparation. The inefficiency of the human feeding process is a faithful reflection of its unsustainability.

3 Reducing the Metabolic Profile of Developed Societies

The metabolic profile of developed societies, including European societies, is impossible to maintain indefinitely, and its environmental and social impacts are extremely serious. Any future economic strategy must aim to reduce this profile to levels compatible with the conservation of the ecosystems, thereby ensuring their long-term survival. In view of the data given in the previous section, such a strategy must, then, pay special attention to how human food requirements are met while achieving two main objectives: (a) the promotion of sustainable ways of managing agroecosystems and (b) the promotion of a food consumption pattern that is less costly in social, energetic and territorial terms. All of this must be achieved without reducing the quality of life of all of the players involved in the process (producers, distributors, consumers, etc.) while avoiding situations in which improvements in, for example, energy efficiency facilitate a new rise in consumption.

As regards the first objective, organic agriculture is the starting point for the promotion of sustainable methods of managing European agroecosystems. It is, a priori, the production method that comes closest to agrarian sustainability in Europe, despite being a sector that is not without its problems.[14] The territorial

[13] Infante and González de Molina (2013), pp. 27–35.

[14] González de Molina et al. (2007), pp. 47–73.

development of organic agriculture in Europe, the agricultural management methods that it promotes, its association with local markets and the consumption of fresh, seasonal products make it particularly suited to deliver significant degrowth of the metabolic profile of the European economy while also improving the quality of our diet.

It has seen spectacular growth in recent years and has become a real alternative to the conventional production model. It has gone from little more than 6,000 farms with only 100,000 ha to 197,000 farms and over 7.6 million hectares under cultivation in 2008.[15] In relative terms, this is 4.3 % of the farmland used in the European Union as a whole, and the producers number around 1.9 % of the total number of farmers, an apparently high percentage, but this is explained by the fact that most organic farmers work full time, which occurs to a lesser extent in conventional agriculture. The member states with the largest area devoted to organic agriculture were Spain, Italy, Germany, the United Kingdom and France.

The evolution of organic agriculture in the EU as a whole has been strongly influenced by institutional support that has been offered since, in the early 1990s, a regulatory basis was established to govern the sector, Regulation (EEC) 2092/91, and the subsequent introduction of economic support measures (especially agri-environmental measures). Other factors, such as the expectations of new markets and food scandals, have also significantly influenced its development.[16] In 2005, the agri-environmental measures came to €3,830 million in the EU-25, of which €660 million were spent on organic agriculture (17.2 %). Over the period 2004–2006, 46 % of the total organic crop area in the UE-25 received agri-environmental aid.[17]

Meanwhile, consumption is growing at an annual rate of almost 10 % in the main member countries.[18] The sale of organic products in 2007 represented 1.9 % of family food consumption in the Union, with a turnover of €14,381 million, almost €36 per capita. However, 80 % of the market is concentrated in four member states: Germany, the United Kingdom, France and Italy. The organic agriculture market is relevant in Austria (almost 5 % of the total food market), Germany (3.7 %), Denmark and Luxembourg (3.8 %). However, in the more recent members to join the European Union (the UE-12 countries), consumption is below 0.2 %. In all events, new consumers concerned about their health and also about the environment are joining the traditional consumers.

Furthermore, the environmental and health benefits afforded by organic agriculture increase the quality of life of the citizens, especially as regards their diet, while reducing the energy cost. Available studies speak of organic production reducing

[15] European Commission. Directorate-General for Agriculture and Rural Development (2010), p. 1.

[16] Lampkin and Padel (1994).

[17] European Commission. Directorate-General for Agriculture and Rural Development (2010), p. 3.

[18] European Commission. Directorate-General for Agriculture and Rural Development (2010), p. 41.

carbon dioxide emissions by between 40 and 60 % in the transition from conventional to organic agriculture, depending on the orientation of production, since nitrogen-based fertilisers and chemical pesticides are not used and potassium and phosphorus-based fertilisers and concentrated foodstuffs are used very little.[19] To this must be added the savings that can be achieved through the on-farm production of biofuels (bioethanol, for example, which is compatible with most mechanical technologies) and the introduction of photovoltaic solar energy to pump irrigation water. We shall address this point below. Studies of organic agriculture agree that this method of production, if used correctly, avoids contamination of agricultural origin (by eliminating the use of fertilisers and synthetic pesticides and by better water management). It also avoids illness related to the use and handling of pesticides, which affect the population in general but, more specifically, the farmers. Organic agriculture, moreover, maintains the genetic biodiversity of the agrarian system and its surrounding area, including the protection of the habitats of wild flora and fauna.

The impressive development of organic agriculture in Europe is due in large part to the crisis suffered in the agricultural sector, especially in those agroecosystems that have difficulty in competing with intensive production, production under plastic or intensive housed livestock farming. Organic agriculture has become a profitable alternative for farmers whose land is found in these territories and who, if it were not for the market opportunities and greater subsidies associated with it, would probably have abandoned farming. This is particularly evident in extensive livestock farming and in many traditional crops, both herbaceous and ligneous. According to the recently published European Union report on organic agriculture, organic production is strongly present in regions with extensive livestock farming systems based on permanent pastureland. The importance of organic agriculture is generally lower in flat regions where conventional intensive production predominates.[20]

Organic agriculture is also producing a rejuvenation of the agricultural sector since the age of organic producers is lower than the average. Fifty-six percent of conventional farmers are over 55 years of age, while in organic agriculture, the percentage is only around 36 %. Farmers under the age of 55 represent 64.3 % of the organic sector.[21] In the same way, the incorporation of women as full-time farmers was greater than the average in the sector as a whole. There are no studies on the impact that organic agriculture is having on rural development other than the increase in agricultural income that it appears to generate. In other countries, such as Italy and in some districts of Andalusia, organic agriculture appears to be a very good complement to and a stimulus for rural tourism and, therefore, for the

[19] Alonso and Guzmán Casado (2004), pp. 471–541; Stolze et al. (2000); Aguilera et al. (2010).

[20] European Commission. Directorate-General for Agriculture and Rural Development (2010), p. 13.

[21] European Commission. Directorate-General for Agriculture and Rural Development (2010), p. 22.

diversification of economic activities in the rural world. A recent study maintains that organic agriculture is allowing the generation of positive socioeconomic impacts within the framework of European rural development,[22] adding to the generation of income and additional employment with respect to conventional agriculture.[23] According to a study by the Sustainability Observatory in Spain and the Biodiversity Foundation, the organic agriculture sector generated 49,867 jobs in 2008, which represents 0.25 % of the working population in the whole Spanish economy.[24]

Organic production, furthermore, is at the centre of some strategies that are structured around short marketing circuits or channels offering traditional varieties that are better adapted to local taste and that are leading to the resurgence of seasonal consumption.[25] Part of the increase seen in the consumption of organic products in Spain that is still difficult to quantify is that part due to the rise of short marketing channels, that is, the increase in sales that involve direct contact between the producer and consumer and the growing presence of organic products in local markets. In recent years, there has been an increase in the number and membership of associations of producers and consumers, consumer cooperatives organised around groups of producers, retail outlets, home delivery of fresh and even processed foods, and the supply of local foodstuffs to health and educational institutions.[26] It would be useful to evaluate the positive impact that short channels are having on the configuration of an alternative food system with much lower energy costs and that is healthier from the environmental and human health point of view. The benefits that this type of channel bring to the farmer, in terms of income, and the consumer, in terms of final price, should also be assessed, but it appears clear that experiments with direct consumption lead to lower final prices and higher, surer profits for the farmer.

However, the reduction of the metabolic profile of the European food system requires a drastic reduction in intensive livestock farming (which is, incidentally, facing increasing problems of falling profitability), which will only be possible through a change in the regulations governing the food markets and in the public policies that encourage the consumption of meat and dairy products. Extensive livestock farming, especially ecological livestock farming, can meet only part of the demand for foodstuffs of animal origin, and so a change in consumption patterns towards a more vegetarian diet is essential in this respect.[27] This change is advisable not only because of the incapacity of European agroecosystems to feed a much smaller livestock pool sustainably or to reduce the energy consumption of the food

[22] Van der Ploeg et al. (2002).

[23] Offermann and Nieberg (2000).

[24] Observatorio de la Sostenibilidad en España, Fundación Biodiversidad (2010), p. 87.

[25] González de Molina (2009).

[26] Consejería de Agricultura y Pesca de la Junta de Andalucía (2007).

[27] Erb et al. (2009); Dutilh and Kramer (2000), pp. 98–101; Jones and Crane (2009), p. 18; Kramer (1996), pp. 289–293.

system as a whole but also because of the criteria of social equitability and the redistribution of wealth on a world scale, reducing the enormous amounts of grain that Europe imports in order to feed its livestock and which represents the removal of a very significant amount of land from feeding human beings, thereby prejudicing countries that suffer serious food security problems.

4 The Risk of "Conventionalisation"

However, a very wide range of situations are to be found under the umbrella of European Regulations that govern ecological production (ECR 834/2007): from agroecosystems that are effectively managed in a sustainable way to situations in which there is merely a substitution of inputs. In these cases, the environmental benefits of organic production tend to become diluted, and the optimum provision of environmental services suffers.

The greater profitability of organic farms has encouraged the entry into the sector of a type of producer who is more concerned with subsidies and the price premium than with the way in which the food is produced. As they do not seek a substantial change in the way their land is farmed, they have become or are becoming a captive market for large commercial suppliers of inputs that have already created a specific "bio" sector. The regulations allow the use of natural pesticides and authorised fertilisers that, in certain circumstances and with certain crops, allow the soil to be worked more intensively, with shorter rotations, etc. And so, for example, in organic woody crops grown on sloping ground, the soil can be over-tilled, causing soil erosion problems that can be just as serious as those caused by conventional tilling and the use of herbicides. This is no different from the essence of the conventional agricultural model, the cause of evident unsustainability: reduction of the energy efficiency of farms, external dependence and loss of farming profitability as a result of significant costs outside the sector, maintenance of the opening up of the energy and nutrient cycles, etc.[28]

The result is usually an increase in "conventionalisation". This is understood to mean the process by which organic agriculture becomes a version that mirrors conventional agriculture, reproducing the same path and sharing the same social, technical and economic characteristics.[29] Conventional food market logic pressures organic producers towards intensification if the pressures are not counteracted by the action of the administrations.[30] By means of the control of food marketing and processing and through the introduction of industrial inputs, farmers are obliged to adopt conventional farming techniques if they are to survive.

[28] Guzmán Casado and Alonso (2008), pp. 167–176.

[29] Buck et al. (1997), pp. 3–20; Hall and Mogyorody (2001), pp. 399–422; Darnhofer et al. (2010), pp. 67–81.

[30] Guthman (2008).

Organic agriculture as it is practised in Europe is, to a certain extent, "decoupled" from its corresponding agroecosystems. Farmers with greater awareness face serious difficulties in closing the cycles, given the lack of organic matter, with livestock farmers suffering a shortage of organic feed and raw materials for its production. The divide between crop farming and livestock farming is a phenomenon that strongly affects organic agriculture and reduces its level of sustainability. In the same way, the lack of machinery adapted to ecological farming practice that maximises energy efficiency in the use of fossil fuels and the lack of incentives to use biofuels (on the farm) mean that organic agriculture is today contributing less than it could to sustainable degrowth.

Furthermore, a quantitatively relevant part of organic agriculture also contributes to maintaining unsustainable marketing channels, with very high energy costs and a considerable loss of added value and of farmers' autonomy. A significant portion of the growth in demand comes from the generic supermarket sector (non-specialised, including *discounters*), which sells products that have travelled long distances. The fact that the countries that consume most are not those that produce most gives an idea of the active "internal" trade that takes place within the Union. The case of Andalusia is paradigmatic: the greater part of organic production, over half, is destined for export. Horticultural and citrus products have the highest export levels, with percentages of 73 % and 78 %, respectively, of the products put on the market.[31]

These channels unnecessarily raise the price of the product, and decision-making is very far removed from the producer. These channels also tend towards the homogenisation of varieties of plants and breeds of livestock, encouraging the loss of genetic diversity. The preferences expressed by these markets differ little from those of conventional markets and attract buyers with a high disposable income, with the result that a substantial part of the population is unable to access this type of healthy food and, in turn, the price tends to prevent its popularisation. But perhaps the most worrying aspect is that, in general and in practical terms, the absence of social initiatives in the field of distribution and the lack of associations in the sector could undo the efforts that are being made to encourage local consumption. The imbalance between growing demand and insufficient, poorly organised supply encourages the entry of large-scale distributors and reproduces the same conventional model in which a ridiculously low percentage of the final price is earned by the farmer.[32] The risk that distribution ends up in the same hands as in the conventional system, with the same unsustainable mechanisms of operation, exists and cannot be ignored. This is a field in which a choice between two food supply models is still possible: the conventional model and another alternative model, based on short channels and different patterns of consumption.

[31] Soler et al. (2009), pp. 135–148.

[32] European Commission. Directorate-General for Agriculture and Rural Development (2010), p. 42.

Organic production is the strongest bastion of an alternative to the current configuration of the European food system. But, as we have seen, some important aspects of the way it currently works, which are heading in the wrong direction, need to be corrected. Above all, its farming area and territorial impact on agroecosystems must be enhanced so that it can be a real alternative to conventional agriculture. It must be as sustainable as possible. Only then will it be able efficiently to provide the environmental services that society demands. Organic production, though, will not be an efficient alternative bringing degrowth if it is not accompanied by a significant change in food consumption habits and the values that drive them. If these do not change, bringing about a reduction in the consumption of meat, eggs and dairy products, even if they were organic, the pressure to import foodstuffs from countries with food security problems and shortages will intensify and the progress achieved will be insufficient. Solidarity with the poorest, then, requires a change in the way in which Europeans meet their endosomatic needs. Organic production and responsible consumption are, therefore, the two fundamental pillars on which a more sustainable food system must be based.

5 How Can This Be Made Possible?

Firstly, a change is essential in our individual and family food consumption patterns. This change should favour local, seasonal products and should tend towards a more vegetarian, less carnivorous diet, which takes health and quality as the main criteria in food purchases. The preferences of European consumers are already being expressed through the existence of green markets, among them the market for organic products. But without political and social intervention (above all, by the State and also by the political parties, movements and social networks), it will not be possible to guide the growth of the market and of green markets (so that these appear and develop) along the path of sustainability.

The food market, where organic products must compete, is a good example of this. Food market forces, among which the strength of the concentrated, large-scale distributors prevails against a fragmented agricultural sector, generate a trend towards "conventionalisation" in organic agriculture. The pressure for prices that are perceived as being lower stimulates a response among organic farmers towards greater externalisation of territorial costs (less rotation, less crops, high-response seeds, more plant health products, etc.) and, therefore, greater dependence on external inputs and higher energy costs. In this way, organic products are encouraged to take a shortcut in order to generate more profits at the cost of sustainability. This trend is encouraged by a regulatory structure (European regulations) that allows and even encourages the use of external resources.

An individual change in production and consumption patterns is not, therefore, sufficient. The market is a reflection of the balance of power, and this can only be confronted by presenting an organised front in order to compete within its limits or to resist outside it. In this regard, it is essential to undertake and multiply collective experiences in ecological production and responsible consumption through the creation and strengthening of production and consumption groups, producers and consumer associations, etc. Many of these experiences, which are fortunately already under way, demonstrate that another food system is possible without losing quality of life. Throughout Europe, numerous agroecological experiences have been seen, in both rural and urban areas, in production and consumption, and these are the vanguard of this new food system.

It should, though, be asked whether it is possible for sustainable food consumption *per se* to become a relevant percentage. The two action plans, both individual and collective, are certainly essential, but they are not sufficient. Sustainable food experiences, created by social networks and movements, will not be able to develop, expand or even survive in more favourable conditions without an appropriate institutional framework. In the same way, the food consumption patterns of the first world may change voluntarily but too slowly, and it is also possible that in a fairly large segment of the population, they do not change at all. In this regard, the role of the State and of political agroecology, as an inspiration for public policies, is essential.[33] In a society such as European society, it is to be expected that degrowth will not enjoy wide social support, especially among the European middle class, which is the majority social class and which has benefited more than any other from the post-war model of economic growth and from the Welfare State. Degrowth appears to be a threat to their lifestyle. The processes of individualisation described by Beck and consumer selfishness will make it difficult.[34] The role of the State and of social movements in the field of ecology and responsible consumption are, then, vital for the introduction of institutional changes that encourage change in consumption patterns, whether this be through new regulations or tax burdens and stimuli or other instruments.

But, moreover, on a wider social scale, sustainability problems arise that can only be addressed by the State. For example, territorial planning, which directly affects agroecosystems, falls outside the scope of individual decisions or of social movements. The design of public policies on this scale is the exclusive competence of the administrations involved (state, regional, local, etc.). This brings up the problem of how to achieve presence in those administrations, either alone or through alliances with other social and political forces, in order to promote public policies for degrowth. The debate about how to make this possible is, perhaps, one of the most important debates pending, and it could take place not just within the field of political ecology but also in the field of agroecology. Until this debate takes place, we can propose some criteria for the design of such policies with an agroecological focus that will facilitate sustainable degrowth.

With regard to production, policies of this type should try to close the nutrient cycles and reduce direct energy consumption. It is no coincidence that these are the main consumers of primary energy in the agricultural sector. The encouragement of

[33] González de Molina (2013), pp. 45–59.

[34] Beck (1998).

composting, with the creation of networks of local facilities, promoting the self-sufficiency of farms in the replenishment soil fertility, is an essential policy. Such a policy has already been applied successfully in Andalusia.[35] The creation of these networks promotes the integration of producers, encouraging them also to group together for other purposes such as integrated pest treatment, joint marketing, the exchange of seeds, etc. In all events, greater and better integration between crop and livestock farming can be encouraged through relatively simple measures. For example, the establishment of priorities based on ecological livestock farming on public pastureland and forests, favouring the production of organic matter (natural parks, common land for livestock, etc.) through networks of manure storage facilities and local organic matter banks.

Public agroecological policies should pay special attention to energy. There has, to date, been little development of mechanical technologies adapted to the needs of organic agriculture. Nevertheless, there is a wide margin for improvement, for example, in the use of solar energy to extract and transport irrigation water or in the local production of biofuels (ethanol).[36] Another important consumer of energy is livestock farming. Here, greater efforts are required to achieve degrowth. The feeding of livestock with animal feed is responsible for a very high percentage of the consumption of primary energy within the food system as a whole. The only type of livestock farming that can be sustainable is extensive livestock farming, but this can only support relatively limited livestock loads. European livestock farming is maintained mainly by its own pastureland and fodder, limiting its growth. Obviously, it is not possible to maintain the number of animals currently farmed in Europe since they are fed mainly through massive imports of feed and fodder and are kept in a stabled regime.

Although they may not have an obvious direct impact on organic production, some institutional regulations are fundamental. The right of organic producers to continue to be organic producers must be guaranteed. To do so, regulations must be introduced and actions implemented to combat direct and diffuse contamination of organic farms by chemical products and, of course, to prevent contamination by GM crops, today the most direct threat to organic production.

Public degrowth policies must also pay special attention to distribution. Transport, processing, packing and sale in shops, that is to say, the distribution chain, is responsible for 47.5 % of the primary energy used in the Spanish food system. The administrations must become actively involved in the expansion and consolidation of other shorter and more sustainable distribution and marketing channels. There are many possible measures. Agroindustry can be encouraged to locate in areas close to farms, the use of renewable energies can be encouraged, legislation can be changed to favour artisan industries, the use of recyclable materials and, above all, the minimisation of the amount of materials used can be promoted, etc. But the main battleground for degrowth is transport. Transport is responsible for almost

[35] Consejería de Agricultura y Pesca de la Junta de Andalucía (2007).

[36] Guzmán Casado et al. (2011), pp. 825– 835.

18 % of the direct consumption of primary energy of the Spanish food system. This figure does not include the cost of manufacture and the maintenance of the vehicles and the necessary logistics. Public policies affecting transport should be oriented towards encouraging short distribution channels that require less transport. This has been called the "kilometre zero" strategy. Many measures may be taken to favour shorter circuits (reduction in VAT and other tax exemptions for activities such as home-delivered organic produce parcels, bio-fairs, consumer cooperatives, municipal markets, supply to restaurants and public catering establishments in the area, etc.) or to penalise high food mileage (a tax on every mile travelled by a product or simply the labelling of the product with that information).

As regards consumption, public policies should favour a change in food habits: less meat and livestock products that have received animal feed, less out-of-season products, local foodstuffs, etc. Here, incentives are required to encourage this type of behaviour, especially incentives that affect the final purchase price. The measures described in the previous paragraph, which tend to promote local products and local consumption, will have a positive effect on the final price and, almost certainly, on the diversity and security of supplies. But, by means of publicity and public information campaigns, public policies can also do a lot to change consumers' habits and values. The public administrations are, moreover, the largest consumers in a country. In this regard, they can make a very relevant contribution by implementing responsible procurement policies. The introduction of organic food in public institutions (hospitals, primary and secondary schools, universities, armed forces, etc.) has an important bandwagon effect. As well as providing a healthy, residue-free diet to users of these services, it is also a powerful instrument for dietary education and for the dissemination of the virtues of organic foodstuffs among patients and their families, schoolchildren, parents, etc. But it can also serve as a valuable instrument to stimulate production and shorter channels if priority is given to procurement from small and medium-sized organic producers located close to the centres of consumption. This has been demonstrated by the Andalusian experience.

It is supposed that organic agriculture undertaken with agroecological criteria will reduce out-of-sector expenditure and raise the net added value. At the same time, organic production does not necessarily mean a reduction in productivity per hectare at farm level, and there are even some crops with a greater yield than with conventional production. However, on a larger scale, this new sustainable coupling of crop and livestock farming in the territory requires a diversification of land use that is clearly contrary to monoculture, to over-specialisation and, in general, to the orientation of sowing decisions in line with market price and expected profits. The application of a number of practices that make agriculture, and especially organic agriculture, more sustainable could bring an increase in costs for the farmer or, in other words, a reduction of income. Such costs must be appropriately compensated by the beneficiaries of the services provided, in this case, society. The way in which they could be compensated may be through payment for environmental services (PES). These payments also represent a necessary paradigm change with respect to agricultural activity: the consumer remunerates the farmer through the markets for agricultural products but does not pay anything for the provision of environmental services.[37] It is also a question of equitability in the damaged relationship that exists in exchanges between the agricultural sector and other economic sectors. In short, this and other instruments that we have mentioned demonstrate that it is possible to achieve degrowth in the food system without destroying jobs, losing income or diminishing the quality of life and without harming the expectations of development and food security in third countries.

References

Aguilera E, Lassaletta L, Gimeno B, Porcuna JL (eds) (2010) GHG emissions and C sequestration in Mediterranean croplands: available information and gaps (preliminary results). In: Paper presented at the round table on organic agriculture and climate change first workshop FiBl; Frick, 10–11 May 2010. http://www.organicandclimate.org/fileadmin/documents_orca/orca-centres/orca-centre-6/meetings-events/2010-05-10/Input3_LasalettaRTOACCWorkshop_10_5_2010.pdf. Accessed 28 April 2014

Allen P, Kovach M (2000) Capitalist composition of organic: the potential of markets in fulfilling the promise of organic agriculture. Agric Hum Values 17:221–232

Alonso A, Guzmán Casado GI (2004) Sostenibilidad y Agroecología: Oportunidades para el sector agrario andaluz. Informe Anual del Sector Agrario en Andalucía, 2003. Unicaja, Málaga, pp 471–541

Beck U (1998) La sociedad del riesgo: hacia una nueva modernidad. Editorial Paidós, Barcelona

Buck D, Getz C, Guthman J (1997) From farm to table: the organic vegetable commodity chain of Northern California. Sociologia Ruralis 37:3–20

Carpintero O (2006) La huella ecológica de la agricultura y la alimentación en España, 1955–2000. Áreas. Revista Internacional de Ciencias Sociales 25:31–44

Consejería de Agricultura y Pesca de la Junta de Andalucía (2007) II Plan Andaluz de Agricultura orgánica (2007–2013). Consejería de Agricultura y Pesca de la Junta de Andalucía, Sevilla

Darnhofer I, Lindenthal T, Bartel-Kratochvil R, Zollitsch W (2010) Conventionalisation of organic farming practices: from structural criteria towards an assessment based on organic principles. A review. Agron Sustain Dev 30:67–81

Delgado M, Aragón MÁ (2006) Los campos andaluces en la globalización. Almería y Huelva, fábricas de hortalizas. In: Etxezarreta M (ed) La agricultura española en la era de la globalización. Ministerio de Agricultura, Pesca y Alimentación, Madrid, pp 423–474

Dutilh CE, Kramer KJ (2000) Energy consumption in the food chain. Comparing alternative options in food production and consumption. AMBIO 29(2):98–101

Engel S, Pagiola S, Wunder S (2008) Designing payments for environmental services in theory and practice: an overview of the issues. Ecol Econ 65:663–674

Erb KH et al (2009) Eating the planet: feeding and fueling the world sustainability, fairly and humanely – a scoping study. Social ecology working paper 116, Wien

European Commission. Directorate-General for Agriculture and Rural Development (2010) An analysis of the EU organic sector. European Commission, Brussels

Food and Agriculture Organization (2007) Annual report. FAO, Rome

González de Molina M (ed) (2009) El desarrollo de la agricultura orgánica en Andalucía. Crónica de una experiencia agroecológica. Editorial Icaria, Barcelona

[37] Allen and Kovach (2000), pp. 221–232; Lomas et al. (2005); Wunder (2005); Food and Agriculture Organization (2007); Engel et al. (2008), pp. 663–674.

González de Molina M (2013) Agroecology and politics. How to get sustainability? About the necessity for a political agroecology. Agroecol Sustain Food Syst 37:45–59
González de Molina M, Guzmán Casado GI (2006) Tras los pasos de la insustentabilidad. Agricultura y medio ambiente en perspectiva histórica. Icaria, Barcelona
González de Molina M, Infante J (2010) Agroecología y decrecimiento. Una alternativa sostenible a la configuración del actual sistema agroalimentario español. Revista de Economía Crítica 10:113–137
González de Molina M et al (2005) Sobre la sostenibilidad de la agricultura andaluza. In: Moreno L, Calvo M (eds) Introducción a la sostenibilidad en Andalucía, Consejería de Medio Ambiente. Junta de Andalucía, Sevilla, pp 119–144
González de Molina M, Alonso A, Guzmán Casado GI (2007) La agricultura orgánica en España desde una perspectiva agroecológica. Revista Española de Estudios Agrosociales y Pesqueros 214:47–73
Guthman J (2008) Agrarian dreams. The paradox of organic farming in California. University of California Press, Berkeley
Guzmán Casado GI, Alonso A (2008) A comparison of energy use in conventional and organic olive oil production in Spain. Agric Syst 98:167–176
Guzmán Casado GI et al (2000) Introducción a la Agroecología como desarrollo rural sostenible. Ediciones Mundi-Prensa, Madrid
Guzmán Casado GI et al (2011) The land cost of agrarian sustainability: an assessment. Land Use Policy 28:825–835
Hall A, Mogyorody V (2001) Organic farmers in Ontario: an examination of the conventionalization argument. Sociol Rural 41:399–422
Hornborg A (1998) Towards an ecological theory of unequal exchange: articulating world system theory and ecological economics. Ecol Econ 25(1):127–136
Infante J, González de Molina M (2013) Sustainable de-growth in agriculture and food: an agro-ecological perspective on Spain's agri-food system (year 2000). J Clean Prod 38:27–35
International Assessment of Agricultural Knowledge, Science and Technology for Development (IAASTD) (2009) Agriculture at a crossroads. Global report of the International Assessment of Agriculture Knowledge, Science and Technology for Development. Island Press, Washington DC
Jones P, Crane R (2009) England and Wales under organic agriculture: how much food could be produced? CSA report, p 18
Kramer KJ (1996) Energy consumption in food products life cycles. In: Proceedings international conference of life cycle assessment in agriculture, food, non-food agro-industry and forestry: achievements and prospects. Ceuterick D, Flemish Institute for Technology Research [VITO], Mol, Belgium, pp 289–293
Lampkin NH, Padel S (eds) (1994) The economics of organic farming. An international perspective. CAB International, Wallingford
Lomas PL, Martín B, Louis C, Montoya D, Montes C, Álvarez S (2005) Guía práctica para la valoración económica de los bienes y servicios ambientales de los ecosistemas. Publicaciones de la Fundación Interuniversitaria Fernando González Bernáldez, Madrid
Ministerio de Medio Ambiente y Medio Rural y Marino (2010) Estadísticas 2009. Agricultura orgánica. España. Ministerio de Medio Ambiente y Medio Rural y Marino, Madrid 2010
Martínez Alier J, Oliveres A (2003) ¿Quién debe a quién? Deuda ecológica y deuda externa. Icaria, Barcelona
Observatorio de la Sostenibilidad en España, Fundación Biodiversidad (2010) Informe Empleo verde en una economía sostenible. Fundación Biodiversidad, Madrid
Offermann F, Nieberg H (eds) (2000) Economic performance of organic farming in Europe. University of Hohenheim, Stuttgart
Schmidhuber J (2006) The EU diet – evolution, evaluation and impacts of the CAP.: FAO working paper, Rome. http://www.fao.org/fileadmin/templates/esa/Global_persepctives/Presentations/Montreal-JS.pdf. Accessed 31 March 2014

Soler M, Pérez D, Molero J (2009) Cuentas económicas de la agricultura y ganadería ecológicas en Andalucía 2005. In: González de Molina M (ed) El desarrollo de la agricultura orgánica en Andalucía. Crónica de una experiencia agroecológica. Editorial Icaria, Barcelona, pp 135–148

Stolze M, Piorr A, Härin A, Dabbert S (2000) Environmental impacts of organic farming in Europe. In: Offermann F, Nieberg H (eds) Economic performance of organic farming in Europe. University of Hohenheim, Stuttgart

United Nations Environment Programme (2011) Decoupling natural resource use and environmental impacts from economic growth. A report of the working group on decoupling to the international resource panel [Fischer-Kowalski M, Swilling M, von Weizsäcker EU, Ren Y, Moriguchi Y, Crane W, Krausmann F, Eisenmenger N, Giljum S, Hennicke P, Romero Lankao P, Siriban Manalang A, Sewerin S], United Nations Environment Programme, Le Mont-sur-Lausanne

Van der Ploeg DJ, Long A, Banks J (eds) (2002) Living countrysides. Rural development processes in Europe: the state of the art. Elsevier, Doetinchem

Von Witzke H, Noleppa S (2010) EU agricultural production and trade: can more efficiency prevent increasing "Land Grabbing" outside of Europe? OPERA Research Center, University of Piacenza, p 14. http://www.appgagscience.org.uk/linkedfiles/Final_Report_Opera.pdf

Wunder S (2005) Payments for environmental services: some nuts and bolts. Bogor: CIFOR, Occasional paper no. 42

Sustainability Science: A Possible Strategy to Enhance Resilience to Climate and Rural Ecosystem Changes

S. Grego and V. Naso

Abstract Sustainability science is an increasing internationally recognized field that combines together social, natural, technical, and health scientists as well experts in the humanities, decision makers, and private enterprises to find solutions and strategic planning to deal with the huge challenges facing human societies.

Keywords Agriculture • Biodiversity • Climate change • Sustainability science • Transdisciplinary

1 Introduction

Improving the sustainability of humanity's relationship with Planet Earth is firmly established as a societal goal for the twenty-first century. To achieve it, we need a better understanding of how to govern "the process of moving towards greater sustainability," with a new style of governing, called *governance*, which is more pluralist and decentralized than the conventional state-centered government style.

With climate change, biodiversity loss, global water and energy crises, the growing problem of desertification, the phenomenon of massive urbanization, and many other manifestations of global environmental change becoming more and more evident, there is a widespread and increasing feeling in the society at large that the concept of sustainability is not sufficient to counteract the complex and problematical situations.

As it is known, *sustainability* is a concept dealing with the way humans should act towards the environment and how they are responsible to each other and towards future generations. The use of the word *sustainability* exploded in

S. Grego (✉)
Department of Agriculture, Forestry, Nature and Energy, University of Tuscia, Viterbo, Italy
e-mail: grego@unitus.it, http://www.dafne.unitus.it/web/interna.asp?idPag=1250

V. Naso
Department of Mechanical and Aeronautical Engineering, University "La Sapienza" of Rome, Rome, Italy
e-mail: vincenzo.naso@uniroma1.it, http://www.cirps.it/didattica-prof-naso

M. Monteduro et al. (eds.), *Law and Agroecology*,
DOI 10.1007/978-3-662-46617-9_21

academic, professional, and public communication in the last decade of the twentieth century.

Sustainable development has been defined in many ways, but the most frequently quoted definition is from *Our Common Future*, also known as the *Brundtland Report*[1]: "Sustainable development is development that meets the needs of the present without compromising the ability of future generations to meet their own needs."

It contains within it two key concepts:

- the concept of *needs*, in particular the essential needs of the world's poorest population, to which main priority should be given;
- the idea of *limitations*, imposed by the state of technology and social organization on the environment's ability to meet present and future needs.

All definitions of *sustainable development* intrinsically require that we see the world as a system, a system that connects *space* (all continents), a system that connects *time* (what we do now has an effect in the future), a system that is strongly influencing the *quality of life* of all humans.

It is evident that *sustainability* is based on a simple principle: everything that we need for our survival and well-being depends, either directly or indirectly, on our natural environment. Sustainability creates and maintains the conditions, under which humans and nature can coexist in productive harmony, that permit satisfying the social, economic, and other requirements of present and future generations. A *sustainable approach* is a system-based methodology that requests to understand the interactions among the *three pillars* (*environmental*, *social*, and *economic*) in an effort to better understand the consequences of our actions. Ideally, research that seeks sustainable solutions to protect the environment also strengthens our communities and promotes prosperity.

Sustainable development was fundamental to the conventions on climate change and biological diversity agreed at the United Nations Conference on Environment and Development (UNCED), held in Rio de Janeiro in 1992; the Desertification Convention (UNCCD) was agreed afterwards, including sustainable development in the crucial way to fight desertification.

However, emerging recognition of *two fundamental errors* characterizing past polices for natural resource issues is an indication of understanding of the need for a worldwide fundamental change in thinking and in the practice of environmental management.

The *first error* has been an implicit assumption that ecosystem responses to human use are linear, predictable, and controllable. The *second one* has been an assumption that human and natural systems can be treated independently. In fact, evidence that has been accumulating in diverse regions all over the world suggests that natural, economical and social systems behave in nonlinear ways, exhibit

[1] World Commission on Environment and Development [WCED] (1987).

marked limits in their dynamics, and that ecological-social-economical systems act as strongly coupled, complex, and evolving integrated systems.[2]

At the beginning of the twenty-first century, we face significant new challenges, including rapid climate change, the degradation of fresh water resources, the massive use of nonrenewable primary energy sources (that is, fossil and nuclear fuels), the globalization of diseases, the economical crises, the massive immigration with evident social consequences, and the more complicated question of long-term environmental security. The foot print of human activity continues to expand to the point that is having a significant impact on nearly all of Earth's environmental systems. We are participating in and increasingly becoming designers and managers of the complex relationships among people, their necessities, ecosystems, and biosphere. Human and environmental health are highly complex, and human well-being is inextricably linked to the integrity of local, regional, and global ecosystems. Environmental research and education are therefore key elements of local, national, regional and global security, health, and prosperity.

We will analyze some very complex systems such as agriculture, climate change, and desertification as examples of nonlinear relationships between *development* (which is not the same as *economic growth*), *sustainability*, *environmental policy making*, *ecosystem quality*, and *human well-being*.

2 Agriculture

Agriculture as developed in the recent decades is not a sustainable activity, but it is becoming an area of problems. Agriculture is the first, most intensive user of nonrenewable land, water, primary energy sources, ecosystems, and biodiversity. It is generally recognized that in 2050 we will need more food, more water, and more energy, and that is a real challenge.

In fact, we don't have any more new land to use for food production. The suitable land is reduced. Only 10 % of land is utilized for human activity, but the competition among industrialization, energy uses (biofuel cultivations), civil activity (roads, houses, resorts), and many others strongly reduces the amount of land suitable for food production. It has been estimated that out of 13×10^9 available ha, only 1.5×10^9 ha are utilized for crops, while 3.30×10^9 are dedicated to pasture. The rest of the land is marginal, rocky, too dry, too wet or too cold. So only the 12 % of valuable land is utilized for food production. The amount of cultivated land per person decreased significantly from 1960, when each person had 0.44 ha, to the actual value of 0.22 ha. It has been estimated that in 2050 it will drop to 0.15 ha per person. The competition between food and biofuel production certainly will not alleviate the trend even in the near future.

[2] Folke et al. (2002), pp. 437–440.

2.1 Agriculture and Water

One-third of water used in Europe goes to the agricultural sector. As a matter of fact, agriculture affects both the quantity and the quality of water available for other uses. In some parts of Europe, pollution from pesticides and fertilizers, used only in agriculture, remains a major cause of poor water quality.[3] Our industries and lifestyles, together with the personal needs of our growing populations, are also nature's rivals for the use of clean water. Climate change adds an additional element of uncertainty to the availability of water resources. In southern European countries, such as Greece, Italy, Portugal, Cyprus, Spain, and southern France, the arid or semiarid conditions necessitate the use of irrigation. In these areas, nearly 80 % of water used in agriculture currently goes to irrigation.[4]

Agricultural practices may also have negative impacts on water quality: improper agricultural methods may elevate concentrations of nutrients, fecal coliforms, and sediment loads. Increased nutrient loading from animal waste can lead to eutrophication of water bodies, which may eventually damage aquatic ecosystems. Animal waste may also introduce toxic fecal coliforms, which threaten public health. Grazing and other agriculture practices may intensify erosion processes by raising sediment input to nearby water sources.[5] The increased incidence and severity of flooding could mobilize sediment loads and associated contaminants and exacerbate impacts on water systems, while more severe droughts may reduce pollutant dilution, thereby increasing toxicity problems. But whatever the impacts on water systems are, the task of achieving water quality objectives in agriculture will become more difficult in the coming years as a result of climate change, although this is a poorly understood and researched aspect of climate change science to date.[6]

2.2 Agriculture and Energy

In recent years, the overall world consumption of primary energy sources ranges around 13.5–14 Gtoe (billions of *t*ons of *o*il *e*quivalent), including biomass and other noncommercial energies.[7] Up to 85 % of such a total amount comes from *nonrenewable sources* (fossil and nuclear ones), while 80 % are also *nonsustainable*, in terms of climate change effects, due to the massive greenhouse gas production caused by the combustion.

[3] Moss (2008), pp. 659–666.

[4] Folke et al. (2002).

[5] Thornes (2007), pp. 13–26.

[6] Organisation for Economic Co-operation and Development [OECD] (2012).

[7] International Energy Agency [IEA] (2013).

Direct final energy consumption for agricultural and agro-industrial uses reflects such a critical data: both for direct energy uses and for indirect ones (mainly water and heat), rural areas are in appropriate energy starvation, and critical aspects are both surviving and the quality of life. 2,500 million people (94 % in rural areas) depend on biomass fuel, while death for pulmonary problems deriving from the unsafe burning of biomass for cooking and/or heating is the second most diffused social cause of death, after AIDS.[8] Finally, 1.7 million people have no access to electricity, 92 % of them living in rural areas.

2.3 *Agriculture and Biodiversity*

Agricultural land use affects large parts of terrestrial area, so its contribution to biodiversity is critical for successful conservation in the future.[9] A landscape vision is needed to understand why agricultural land use has the well-known negative and less-known positive effects on biodiversity and related ecosystem services. Agricultural land use and biodiversity conservation have been traditionally viewed as incompatible. Ecologists and conservationists often focus on unspoiled or little intervened habitats to save the remains of wild nature.

Only recently has there been an increasing recognition that such a conservation focus is of limited value. Intensive land use in agriculture and forestry is irrefutably the main cause of global change and biodiversity loss, but low-intensity land use systems may be important elements of large-scale programs on biodiversity conservation. In fact, the importance of biodiversity in multifunctional agriculture and for ecosystem services, such as pollination and biological control, is well known. During the last decades, worldwide losses of biodiversity have occurred at an exceptionally increasing rate, and agricultural intensification has been a major driver of this global change.[10] The modernization of traditional agroecosystems, which in tropical regions are still much under traditional management, strongly affected biodiversity.[11]

The dramatic land use changes include the conversion of complex natural ecosystems to simple-managed ones and the intensification of resource use, including the application of more agrochemicals and a generally higher input and output, which is typical for agroecosystems as relatively open systems. Not only the biodiversity of unspoiled habitats and traditional, low-intensity agroecosystems but also the biodiversity of intensively used agroecosystems that has been greatly reduced during the last decades.

---

[8] World Health Organization [WHO] (2013).

[9] Thrupp (2000), pp. 265–281.

[10] Tilman et al. (2001), pp. 281–284.

[11] Perfecto et al. (1997), pp. 935–945.

The main biodiversity losses are due to the postwar transformation of traditional to modern, high-intensity land use systems in simple landscapes. The decline of biodiversity may affect ecosystem functioning and yield, although the functional role of biodiversity is not yet completely understood.[12] More recently, the focus on biodiversity in undisturbed habitats has also been challenged. Attention has been called to the fact that 95 % of contemporary terrestrial ecosystems are managed ones, including agricultural systems.[13] The conventional view is that agroecosystems are at best insignificant with respect to biological diversity, and at worst they reduce diversity to negligible levels.

Although no one would affirm that a modern agroecosystem may have as much biological diversity as a rainforest, it is doubtful that certain agroecosystems indeed have a very high diversity of plants.[14]

2.4 Agriculture and Soil

Unlike air, water, and biota, which are mobile systems, soil is site specific, and although it is more stable than the other three systems, it shows great variability in space and time.

Soil is the Earth's living skin, essential for life on our planet. Nevertheless, increasing areas are being covered with impervious materials as a result of urban development and the construction of new infrastructure. This "soil sealing" causes an irreversible loss of the soil's natural functions and can lead to floods as water can no longer seep and drain away. Soil sealing can also affect human health, as well as medium- and long-term economic development and food security. Soil is a nonrenewable resource: its health is important for the world's sustainable development and therefore needs to be preserved and managed carefully.

Soil can be affected by physical, chemical, and biological degradation. Soil health, biodiversity, and soil resilience are severely limited in extreme environments and are more sensitive to anthropogenic disturbance.[15] Agricultural activities contribute to these negative effects. Land use practices such as deforestation, overgrazing, some agricultural cultivation practices, removal of vegetative cover or hedgerows can exacerbate physical degradation of the soil due to agriculture. The increasing demand for water, the sometimes excessive mechanization and ploughing are further causes of such degradation.

However, it must be borne in mind that industrialization, urbanization, road construction, fire, other human activities, and, more generally, anthropic and demographic pressure and climate changes are also major factors. Dramatic is the

[12] Daily (1997).

[13] Western and Pearl (1989).

[14] Paoletti and Pimentel (1992).

[15] Doran and Zeiss (2000), pp. 3–11.

current rate of soil loss by sealing through urban expansion and infrastructure in Europe, like in Germany (120 ha/day), Italy (100 ha/day), Austria (35 ha/day), and Switzerland (10 ha/day). This urban expansion increases the costs of urban infrastructure, traffic in urban areas, and energy consumption and has negative effects on the quality of the countryside and the environment. This development is in direct competition with agricultural land uses and is threatening valuable agricultural soils all over Europe.[16]

3 Climate Change

The changing climate impacts society and ecosystems in a broad variety of ways.

For example, climate change can increase or decrease rainfall, influence agricultural crop yields, affect human health, cause changes to forests and other ecosystems, or even impact our energy supply.

Climate-related impacts are occurring across regions of the country and across many sectors of our economy.[17] Many state and local governments are already preparing for the impacts of climate change through "adaptation," which is planning for the changes that are expected to occur.

Starting from the beginning of the nineteenth century, people worldwide began burning more coal and, later, oil for homes, factories, and transportation. Massively burning these fossil fuels releases carbon dioxide and other greenhouse gasses into the atmosphere. These added greenhouse gasses caused Earth to warm more quickly than it has in the past. The rising of temperature is having many side effects at the planet level, such as changes in rainfall pattern, melting of glaciers and sea ice, sea level rise, and increased intensity and/or frequency of extreme events. These changes in physical processes have impacts on biological and socioeconomic factors such as shifts in crop growing seasons, changes in disease vectors, increased rates of extinction for many species, severe water shortage, and heavy deluges and flooding. Moreover, climate change plays a significant role on people's health, causing more frequent, more severe, and longer heat waves in summer time.[18]

How much warming has happened? Scientists from around the world within the Intergovernmental Panel on Climate Change[19] showed us that during the past 100 years, the world's surface air temperature increased an average of 0.89 °C, while the CO_2 percentage in the atmosphere rose up to the value of 395 ppm (never reached in the past 400,000 years).[20] This may not sound like a great change, but even such increase can meaningfully affect the Earth's equilibrium. The IPCC

[16] Scalenghea and Ajmone Marsanb (2009), pp. 1–10.
[17] Parmesan and Yohe (2003), pp. 37–42.
[18] Karl et al. (2009).
[19] Intergovernmental Panel on Climate Change [IPCC] (2013).
[20] Rohde (2013).

Report is supposing that the number of cold days and nights has decreased at the global scale between 1951 and 2010. It is probable that, since 1950, the number of heavy precipitation events over land has increased in more regions than it has decreased. Regional trends vary, but confidence is high for North America with trends towards heavier precipitation events. On the other hand, there is solid conviction for the occurrence, during the last millennium, of droughts of greater magnitude and longer duration than observed since 1900 in many regions.

4 Desertification

Desertification is often triggered by initial conditions of environmental fragility. Causes are linked to several underlying factors (of both natural and anthropic nature) that act as a complex system of interactions.

People cause desertification by cutting trees, logging, and diverting river water to cities for human consumption.

Another consequence of desertification at local and global levels is the reduction in biodiversity since it contributes to the destruction of the habitats of animal and vegetable species and microorganisms. It encourages the genetic erosion of local livestock and plant varieties and species living in fragile ecosystems. It is extremely difficult to put a figure on this loss because of our inadequate familiarity with the features, the role, and the economic importance of the biodiversity of the dry zones. The destruction of the natural grass and woody vegetation cover in dry areas affects the topsoil temperature and the air humidity; consequently, it influences the movements of atmospheric masses and rainfall. Furthermore, the drying of the soils and the destruction of soil cover encourage air erosion.

The primary reasons for desertification are overgrazing, overcultivation, increased fire frequency, water impoundment, deforestation, overuse of groundwater, increased soil salinity, and global climate change. Climatic changes are both a consequence and a cause of desertification. In this context, climate change makes ecosystems even more sensitive and fragile because it increases the preexisting climate aggressiveness. Instead, the socioeconomic causes are generated from the impacts of anthropic pressure linked to urban expansion and economic activities, especially when the above factors involve an unsustainable exploitation of natural resources.

Each of these environmental hazards, even if not producing immediately observable desertification effects, can create instability in the ecosystem equilibrium.[21]

[21] Perini et al. (2009), pp. 45–55.

5 The Sustainability Science Approach

How can the difficult global social-economical-environmental changes be managed? It is readily evident that we are facing many urgent sustainability challenges, including poverty, epidemics, violent conflicts, economic crisis, besides climate change, aggressive agriculture, and biodiversity.

These problems range in scale from global to local and are expected to affect future generations.[22] We considered all the changes that are currently occurring in our environment and in our society in a sectorial way. Promoting sustainable development requires research on a wide range of social, economic, institutional, and environmental issues. The aim to understand the dynamics of coupled social-ecological systems stimulated an innovative, problem-driven research, which has been called *Sustainable Science*.

Sustainability science has emerged over the last two decades as a vibrant field of research and innovation.[23] Today, this field carried out a basic research agenda, an increasing production of results, and a growing number of researchers committed to teaching its methods and findings. Like *agricultural science* and *health science*, Sustainability science is a field defined by the problems rather than by the disciplines it employs. From its core focus on advancing the understanding of coupled human–environment systems, sustainability science has reached out with focused problem-solving efforts based on urgent human needs. As most recently delineated by the World Summit on Sustainable Development (Rio +20), these efforts include improving access to water supplies of adequate quality and quantity, enhancing agricultural production and food security, developing cleaner energy and manufacturing systems, mitigating the human health impact of pollution and environmentally mediated disease, encouraging governance of rapid urbanization, and, more generally, making more effective use of environmental and natural resources to promote poverty alleviation. Similarly, sustainability science is being applied to devise practical protection for the Earth's key life-support systems. Special attention in recent years has been paid to mitigating pressures on the global climate, conserving ecosystem services, and protecting biodiversity. Finally, and most ambitiously, sustainability science is seeking to support the integrative task of managing particular places where multiple efforts to meet multiple human needs interact with multiple life-support systems in highly complex and often unexpected ways.[24]

Generally, research relevant to the goals of sustainable development has been carried out from diverse disciplines as geography and geochemistry, agriculture and heath, ecology and economics, or physics and political science. But its research programs should transcend the knowledge of its foundational disciplines and focus instead on understanding the complex dynamics that arise from interactions

[22] Van der Leeuw et al. (2012), pp. 115–120.

[23] Clark and Dickson (2003), pp. 8059–8061.

[24] Dodds et al. (2012).

between human and environmental systems.[25] The participatory, interdisciplinary research can contribute to the solution of complex persistent problems. This is the transdisciplinary research that is, essentially, team science. In a transdisciplinary research attempt, scientists contribute their unique expertise but work entirely outside their own discipline. They make every effort to understand the complexities of the whole project rather than one part of it. Transdisciplinary research allows investigators to transcend their own disciplines to inform one another's work, capture complexity, and create new intellectual spaces.[26]

Brandt (with other scholars)[27] identified five key themes to adopt transdisciplinary approaches to sustainability science.

The first theme is the *lack of coherent structure* between scientists and stakeholders due to different prospective and the *lack of interaction*. Sustainability science could be the right attempt to increase the exchange and integration of different disciplinary and nonacademic knowledge, allowing mutual learning between scientists and stakeholders.

The second is the *integration of methods to use in the transdisciplinary research* as a crucial point in the establishment of efficient and coherent research frameworks.

The third theme is *research and knowledge production*, due to the fact that transdisciplinary projects need a collaborative identification of the problem, its analysis with the cocreation of solution-oriented and transferable knowledge, and the implementation of the results into practice.

The fourth theme is the crucial element of *link between stakeholders and scientists*. Is this information, a consultation, a collaboration, or empowerment? It is evident that the involvement of stakeholders is a vital goal for transdisciplinary projects, but it is not clear to what extent this goal could be reached within published transdisciplinary research.

The fifth theme is related to the *impact of transdisciplinary research at global level*. In fact, much of transdisciplinary research originates from developed countries, but the sustainability problems are global and not local or regional.

6 Conclusion

It is readily evident that there is a need to develop a systematic approach that allows the integration of knowledge across disciplines, ecosystem health, economic development, and social needs. We should understand how to integrate the dynamic interactions among the Earth system, social and economic development, and sustainability and how long-term trends can remodel the interactions between

[25] Clark (2007), pp. 1737–1738.

[26] Hirsch et al. (2008), pp. 433–441.

[27] Brandt et al. (2013), pp. 1–15.

nature and society. Science and technology could be more effectively bound to reach sustainability, and research should be addressed to the factors that limit the resilience, enhancing the vulnerability, of the nature–society interactions. The implementation of sustainability requires a considerable scientific and technical knowledge, coupled with a constructive political will. Sustainability without the scientific-technological and political ability to complete it is meaningless.

However, it is not clear how transdisciplinary approaches will integrate in the future scientific research and how the boundary among scientific research, political decision making and societal organizations will further evolve. Active interaction and effective debates are taking place on the necessity, adequacy, and capability to bridge science and political decision to make the link nature–society sustainable.

References

Brandt P et al (2013) A review of transdisciplinary research in sustainability science. Ecol Econ 92:1–15

Clark WC (2007) Sustainability science: a room of its own. Proc Natl Acad Sci U S A 104:1737–1738

Clark WC, Dickson NM (2003) Sustainability science: the emerging research program. Proc Natl Acad Sci U S A 100:8059–8061

Daily GC (1997) Nature's services: social dependence on natural ecosystems. Island Press, Washington, DC

Dodds F, Strauss M, Strong MF (2012) Only one Earth: the long road via Rio to sustainable development. Routledge, London/New York

Doran JW, Zeiss MR (2000) Soil health and sustainability: managing the biotic component of soil quality. Appl Soil Ecol 15:3–11

Folke C et al (2002) Resilience and sustainable development: building adaptive capacity in a world of transformations. AMBIO 31:437–440

Hirsch HG et al (eds) (2008) Enhancing transdisciplinary research: a synthesis in fifteen propositions. In: Handbook of transdisciplinary research. Springer, Heidelberg, pp 433–441

Intergovernmental Panel on Climate Change [IPCC] (2013) Working group I contribution to the IPCC fifth assessment report climate change 2013: the physical science basis. https://www.ipcc.ch/report/ar5/wg1/. Accessed 11 Mar 2014

International Energy Agency [IEA] (2013) World energy outlook. http://www.worldenergyoutlook.org/publications/weo-2013/. Accessed 11 Mar 2014

Karl TR, Melillo JM, Peterson TC (eds) (2009) Global climate change impacts in the United States. United States Global Change Research Program. Cambridge University Press, Cambridge/New York

Moss B (2008) Water pollution by agriculture. Philos Trans R Soc B 363(1491):659–666

Organisation for Economic Co-operation and Development [OECD] (2012) Water quality and agriculture: meeting the policy challenge. Organization for Economic Co-Operation and Development – Directorate for Trade and Agriculture. http://www.oecd.org/tad/sustainable-agriculture/49849932.pdf. Accessed 11 Mar 2014

Paoletti MG, Pimentel D (eds) (1992) Biotic diversity in agroecosystems. Elsevier, Amsterdam

Parmesan C, Yohe G (2003) A globally coherent fingerprint of climate change impacts across natural systems. Nature 421:37–42

Perfecto I, Vandermeer J, Hanson P, Cartín V (1997) Arthropod biodiversity loss and the transformation of a tropical agro-ecosystem. Biodivers Conserv 6(7):935–945

Perini L, Ceccarelli T, Zitti M, Salvati L (2009) Insight desertification process: bio-physical and socio-economic drivers in Italy. Ital J Agrometeorology 3:45–55
Rohde DL (2013) Environmental values and behaviors – GLOBAL WARMING ACT, Heinonline. http://home.heinonline.org/. Accessed 11 Mar 2014
Scalenghea R, Ajmone Marsanb F (2009) The anthropogenic sealing of soils in urban areas. Landsc Urban Plan 90:1–10
Thornes JB (2007) Modelling soil erosion by grazing: recent developments and new approaches. Geogr Res 45(1):13–26
Thrupp LA (2000) Linking agricultural biodiversity and food security: the valuable role of agrobiodiversity for sustainable agriculture. Int Aff 76(2):265–281
Tilman D et al (2001) Forecasting agriculturally driven global environmental change. Science 292:281–284
Van der Leeuw S, Wiek A, Harlow J, Buizer J (2012) How much time do we have? Urgency and rhetoric in sustainability science. Sustain Sci 7:115–120
Western D, Pearl MC (eds) (1989) Conservation for the twenty-first century. Oxford University Press, Oxford/New York
World Commission on Environment and Development [WCED] (1987) Our common future. Oxford University Press, Oxford. http://www.un-documents.net/our-common-future.pdf. Accessed 11 Mar 2014
World Health Organization [WHO] (2013) Data and statistics. http://www.who.int/research/en/. Accessed 11 Mar 2014

Agriculture, Climate Change and Law

S. Vernile

Abstract The climate change phenomenon can significantly affect the agricultural sector. Agriculture, indeed, is at the same time one of the causes of climate change, a sector among the most harmed by it and an opportunity to mitigate climate change. Even though scientists should care about climate change, a very important role is played by the law through the implementation of different legal tools. In particular, those tools range from command and control instruments, characterized by their mandatory nature, to market based tools. The last type, indeed, is very frequently used in the agricultural sector, in which it is important to promote autonomous sustainable production and environmental-friendly activity in order to guarantee for future generations the same resources that we have at our disposal.

Keywords Agriculture • Climate change • Command and control instruments • Intergenerational equity • Legal reaction • Market based tools • Precautionary principle

1 Introduction

Climate change represents nowadays one of the most threatening issues for environmental safety and stability. Climate is a very important element for human activities and life since it has a significant impact on natural resources and has serious economic and social consequences.

Many of the direct and immediate effects of climate change are on agriculture.[1] Reduced water resources, freezing crops and periods of drought are just a few examples of climate's influence on the agricultural system. Because of the undoubted effects of climate change on agricultural production, it is necessary to modify or adapt current practices in the face of new climate events, such as regional

[1] Hamilton (2011), pp. 341–360.

S. Vernile (✉)
Department of Law, Bocconi University of Milan, Milan, Italy
e-mail: scilla.vernile@unibocconi.it, http://didattica.unibocconi.eu/docenti/cv.php?rif=115184&cognome=VERNILE&nome=SCILLA

M. Monteduro et al. (eds.), *Law and Agroecology*,
DOI 10.1007/978-3-662-46617-9_22

temperature shifts, increased frequency of heavy precipitations, extended dry periods and extreme weather events.

Like other environmental fields, climate change has been addressed by both natural and social sciences. Climate change indeed, more than other environmental issues, transcends the boundaries between legal and nonlegal disciplines, even if the legal instruments appear to play a very significant role to guarantee sustainable development and a safe environment.

For these reasons, the question we are trying to answer in this paper aims at assessing the role played by law in climate policy implementation, specifically with regard to the agricultural sector.

The awareness of the growing nature of climate change, the degradation it causes to agriculture and natural resources and the risks involved, both to human health and to the quality of life, have led scholars from around the world to address the problem. Because of the importance of the climate problem and the difficulty of implementing appropriate solutions and tools to limit it, regulation cannot be based on a purely technical-scientific perspective.

This means that a response should address several fronts, trying to guarantee immediate and effective results. In this regard, law represents the best way to impose a faster reduction of polluting emissions.

In order to analyze the different legal tools to react to climate change, we should consider both the legislative and administrative levels, and we cannot ignore that the need for a governmental response has been felt at international, European and local levels.

Considering every measure adopted to reduce the climate change phenomenon, in fact, we should care also about the relationship among different levels of regulation since environmental problems, and climate change as well, are usually cross-border issues, and accordingly the solution should be cross-border too.[2]

Indeed, the first answer to climate change was derived from the United Nations Framework Convention on Climate Change, also known as the Earth Summit, held in Rio de Janeiro from 3 to 14 June 1992, in which climate change was defined as a common concern of humankind. The same convention declared the need for international action, characterized by individual responsibility for every State, proportioned to its own capacities and economic and social conditions.[3]

Prior to examining the relationship between climate change and agriculture and how the legal system can react to eliminate or reduce the negative impact of sudden and significant climate events, we should briefly recall what we mean by climate change.

Using a common and nontechnical language, we could say that climate change represents the effect produced by the so-called phenomenon of global warming,

[2] Carlarne (2011), pp. 255–257.

[3] Montini (1999), pp. 133–148.

which consists in the rising of Earth's temperature due to the excessive amounts of greenhouse gases.[4]

The increasing presence of greenhouse gases creates, in fact, a kind of film around the planet, which reduces the dispersion of heat from Earth's surface. In this way, global temperatures increase.

The climate, indeed, represents the point of balance between the thermal energy entering in the biosphere (mostly coming from the sun) and the dispersion of that energy into space. If dispersion is reduced by greenhouse gases, global temperatures increase, causing significant changes in global climate such as higher rainfall variability, flooding, extreme heat or persistent drought.

Even if global climate has always been characterized by instability and unpredictability, over the last years climate variability has increased. Indeed, one should also keep in mind that climate change is a long-term phenomenon and not a forecast of what the weather will be like. It means that it is quite hard to identify climate change phenomenon in a very short time or in a few years, but we should consider the effects that global warming may produce in the long term.

Another aspect one should take into account is the fact that climate change is connected with industrialization. Analysis of climate change has focused on the alleged connection between human activities and global warming.

More and more human activities are in fact based on the use of fossil fuels that produce CO_2 emissions, one of the main greenhouse gases.[5]

According to the Fourth Assessment Report of the Intergovernmental Panel on Climate Change (IPPC 2007), indeed, the increase in global temperatures in recent decades is mostly due to human activities such as fossil fuel burning and land use.[6]

However, although the natural sciences can demonstrate the production of CO_2 emissions by some human activities, nowadays it is not possible to verify with complete certainty the relationship between all human activities, global warming and climate change. The climate change challenge, indeed, presents one of the typical limits of environmental law: the lack of scientific certainty.[7]

Even if a number of studies show the close connection between emissions, rising temperatures and climate change or at least the existence of a serious and imminent danger in the impact of human activities on climate, we cannot forget the so-called Climategate. In 2009, some emails of researchers of the University of West Anglia showed that, despite the official reports, the existing data had not demonstrated a particular worsening climatic condition.

This case does not eliminate the need to seek legal solutions to reduce polluting emissions but brings to light an issue very much connected with climate change,

---

[4] If on one hand the greenhouse gases are needed, since in their absence the solar radiation could not be restrained in any way and temperatures would be so low as to prevent the survival of the planet, on the other hand an excessive increase could lead to the opposite condition, preventing life on Earth.

[5] Onzivu (2009), pp. 1311–1336.

[6] Jaykus et al. (2008).

[7] Abler et al. (2004), pp. 117–126.

uncertainty.[8] We don't know how and how much climate will change and what the effects will be on the environment as well as on human activities. For this reason, managing uncertainty is a very important aspect of climate change policy.[9]

Thus, the importance of the precautionary principle in developing policies aimed to reduce climate change is evident.[10] The precautionary principle indeed allows public institutions to act at an earlier stage where there are serious risks of irreversible damages and the lack of full scientific certainty could not be used as a reason for postponing measures to contrast climate change consequences.

The risk of harming the environment and human health—because of extreme increases in temperatures, unpredictable weather events, coastal inundations, spread of disease, destruction of biodiversity and threats to water or food availability—modifies environmental strategies, in a way so that, despite the lack of certainty, society should adopt protective measures.[11]

But respecting the precautionary principle does not only mean anticipating the action notwithstanding the lack of scientific certainty. It also means that all the instruments and policies we might adopt should be analyzed in the light of the mentioned principle.

Then since there is still uncertainty about the level and degree of climate change, every measure considered by the law to address the climate change challenge must be proportionate to the specific goal.

2 The Effects of Climate Change on Agriculture

The problem of climate change can be viewed from three different perspectives with respect to the agricultural sector.

Foremost, as we already said at the beginning of this work, agriculture is one of the areas most affected by frequent extreme weather events, such as floods or heat waves and droughts.

Although climate change can sometimes be positive for agriculture (just think about northern areas that have experienced an increase in productivity due to milder climatic conditions), most of the climate change's effects are negative. We can just recall some examples like the reduction of water availability, the spread of parasites caused by higher temperatures and humidity, the increased frequency of heavy precipitation events and extended dry periods.

The final consequences of climate change vary greatly, and they impact both farmers, because of the significant reduction of incomes, and citizens/consumers. The last group indeed suffers the impact on food supply, resulting in the increase of

[8] Fracchia (2013), pp. 4–42.

[9] Mearns (2010), pp. 998–1101.

[10] Manfredi (2011), pp. 28–39.

[11] Tedsen and Homann (2013), pp. 90–100.

imports of certain products, higher price instability, the reduction of production in certain regions and changes in seasonal patterns.[12]

Actually, referring to the impact of climate change on food, we must remember that global warming can produce both positive and negative effects. According to some studies, indeed, the overall impact on food production will not be so huge, but it will impact most notably developing countries that usually do not have the capacity to adapt to climate change because of the lack of technical and economic resources.[13]

It means that the consequences of this phenomenon are inequitably distributed since the majority of the actions causing climate change stems from the developed countries, but the effects, for the most, impact the less-developed regions, which have lesser ability to adapt.

Without going through the analysis of equity between states, which will be examined at the end of this work, it is worth highlighting that the negative effects of climate change on agriculture also include soil consumption.

More specifically, climate change impacts soil consumption in two ways: first of all, greenhouse gas emission significantly threatens both natural and rural environments.[14] Second, quite often efforts to combat climate change can have negative consequences for the soil.[15]

An example of this consists of the use of biofuels as a replacement for conventional motor fuels in order to comply with European provisions regarding the use of renewable energies. According to Directive 2009/28/EC, each Member State shall increase the use of renewable energies in order to reach a 20 % target (10 % in transport) in 2020.

So in the last decade, governments encouraged the production and use of biofuels (combustible materials deriving form biomass produced by plants, animals and organic wastes) as a partial solution to reduce climate change.[16] However, reacting to concerns that biofuel production might increase emissions, reducing food availability and exacerbating degradation of land, forest and water resources,[17] the European Union has introduced a sustainable scheme of biofuel production.[18]

It means that, in order to obtain a certification for biofuels, it is necessary to comply with specific requirements set out at the EU level aimed, among other targets, to ensure the conservation of soil and biodiversity. According to Directive 2009/30/EC, indeed, biofuel production shall be sustainable and cannot encourage the destruction of lands that support high levels of biodiversity.[19]

---

[12] Coderoni (2013), pp. 84 et seqq.

[13] Charles (2011), pp. 44–47.

[14] Carraro (2011), pp. 231 et seqq.

[15] Lin (2011), pp. 34–43.

[16] Pelsy (2008), pp. 119–136.

[17] Ottinger (2009), pp. 253–264.

[18] Cafagno (2007).

[19] As you can read in Directive 30/2009/EC, "Consumers in the Community would, in addition, find it morally unacceptable that their increased use of biofuels could have the effect of destroying

But the use of agriculture to produce biofuels is only one of the causes of soil erosion related to climate change. Another example, indeed, consists of the use of agricultural lands to produce renewable energy, such as solar or wind. For this reason, in fact, the relevant domestic laws require taking into account the soil and biodiversity protection when determining the location of renewable energy facilities.[20]

However, if on one hand, as said above, the soil consumption may be partly considered as a consequence of mitigating climate change,[21] on the other one soil erosion harms the efforts to combating climate change.[22]

The loss of land, in fact, not only due to the climate change mitigation efforts, but even more due to increased urbanization, reduces, just to pick an example, the ability of the soil to absorb water, when instead climate change is causing a significant increase of rainfall events with high levels of precipitation in very short periods.[23] This brings to light that soil consumption can be at the same time the cause and consequence of climate change, as for agriculture.

But if so far we have drawn attention to the negative effects of climate change on food production and biodiversity, now it is worth focusing on the second perspective regarding agriculture as a partial cause of climate change. Agricultural practices, such as concentrated production of livestock, deforestation or use of microbial fertilizers, are important contributors to the release of greenhouse gases into the atmosphere.

If, on one hand, agriculture is one of the areas most affected by climate change, on the other one it is also a source of greenhouse gases, and therefore it must contribute to global mitigation efforts by reducing their own emissions. In particular, the main sources of greenhouse gases from agriculture are N_2O emissions from agricultural soils, N_2O and CO_2 emissions from cultivated organic soils, CH_4 emissions from enteric fermentation and CH_4 and N_2O emissions from manure management.[24]

From this point of view, it is necessary to analyze whether the use of the most damaging practices can be prohibited by the law and more sustainable ones can be promoted instead.

biodiverse lands. For these reasons, it is necessary to provide sustainability criteria ensuring that biofuels can qualify for the incentives only when it can be guaranteed that they do not originate in biodiverse areas or, in the case of areas designated for nature protection purposes or for the protection or rare, threatened or endangered ecosystems or species, the relevant competent authority demonstrates that the production of the raw material does not interfere with those purposes."

[20] Barelli (2014), pp. 1–27, who recalls Legislative Decree n. 387/2007, according to which the renewable energy facilities can be even located in agricultural soils, but it is necessary to respect the regulation about environment and biodiversity protection.

[21] Boscolo (2014), pp. 129–146.

[22] Cartei (2014), pp. 45 et seqq.

[23] Di Gennaro (2014), pp. 3 et seqq.

[24] ECCP (2001).

The last perspective is to consider agriculture as an opportunity, a resource to tackle the problem of climate change, not only through the reduction of greenhouse gases (from a perspective that could be named "passive") but also in an active manner, in the sense of increasing and improving the capacity of carbon sequestration of the soil and contributing to the production of alternative energy sources.

Enhancing agricultural practices, indeed, can strengthen the sector's capacity to contribute to the conservation of biodiversity and water, at the same time preserving and sequestering carbon in the soils. This goal can be achieved by measures such as the maintenance of grassland, restoration of wetlands and peatlands and the absence of treatment or minimum tillage of the soil.

From this point of view, on one hand, for instance, it could be appropriate to encourage the installation of anaerobic digestion plants, able to produce biogas that can be converted into heat and electricity, as well as the reduction of fossil fuel consumption. On the other hand, instead, a partial solution can be represented by the use of organic farming practices in order to increase the function of absorption of carbon dioxide in the soil.

In this perspective, agriculture may be an efficient and hopeful tool to resolve the problem of climate change by promoting, at the same time, the development of new techniques and mechanisms able to not only produce food but even reduce GHGs emissions.

3 Law's Reaction to Climate Change

The problem of climate change, as well as many environmental issues, requires a multisectorial and global intervention since the effects of carbon dioxide emissions are not geographically limited. Those effects, indeed, could occur in different places from those where polluting emissions are more concentrated.

The global nature of the climate change problem needs major cooperation between different governments because only joint action can be effective to reduce the emissions that are causing global warming.

It is not a coincidence, in fact, that the most important response is represented by the emission trading mechanism, developed at the Kyoto Conference in 1997, imposed on all the signatory States and transposed at the European level too since 2003.

However, despite the need for joint action, efforts to combat climate change must also be made at the local level, according to the subsidiarity principle and based on administrative decentralization. It means that not only the State but also the other governmental levels should implement policies and actions to reduce emissions and to contrast climate change.[25]

[25] In this sense, in Italy, for instance, Regions have signed, in 2001, a Protocol in order to coordinate every intervention to reduce polluting emissions. Compiani (2002), pp. 365–387.

The need for cooperation between national governments, on one hand, and regional and local authorities, on the other, has been highlighted also by the European Commission's 2009 White Paper on adaptation, which stated that policies at the local level are lagging behind due to lack of resources and tools to support local authorities in the assessment of risk and vulnerability and adaptation in the preparation of plans.

Nevertheless, the joint consideration of mitigating and adapting policies in the context of climate plans requires both local and national measures. The two levels (national and local), indeed, differ from each other in the timescale of actions, costs and benefits; in the geographical dimension of benefits; and in the sectors in which the measures apply.[26]

Once clarified that climate change should be addressed in different ways, it is now important to focus on the various issues brought up by the legal system to reduce greenhouse gas emissions, even if, still now, the measures adopted have been more words than action in a way that it makes it possible to see a kind of gap between the demanding targets established at international levels—by entering into agreements or protocols[27]—and an appropriate local strategy to contrast climate change.[28]

By the way, law can react to the climate change challenge in many ways: some measures focus directly on greenhouse gas emissions; others instead refer to related aspects, such as energy, transportation, agriculture, land use, waste management, construction and so on.[29]

Within the various legal instruments that are generally used in environmental policies, we should distinguish between "command and control" and "market based" tools. Even in agricultural matters, we have to keep in mind the above classification and that we can use both those kinds of measures in referring to climate change too.

The problem of climate change requires both regulatory and incentive instruments, arranged to ensure at the same time legal certainty and the necessary degree of flexibility.

First of all, as to the authoritative tools, also known as "command and control" instruments, we must recall, for example, the establishment of standards (just think, for example, about emission limits for electric utilities: the law often requires increased energy performance of products and new buildings and the implementation of renewable sources, such as solar and wind),[30] the powers of planning and, even more important, emissions' permits. We cannot even forget to refer to the IPPC, Integrated Prevention Pollution and Control, an integrated environmental permit that takes into account all the emissions deriving from an industrial activity.

---

[26] Croci (2010), pp. 147–164.

[27] Liakopoulos (2005).

[28] Rodgers et al. (2011), pp. 245–250.

[29] Meltz (2013).

[30] Rosso Grossman (2010), pp. 223–255.

While it is immediately clear that the IPPC takes into account climate change, since it considers any type of polluting emission deriving from industrial activities, the importance of other authorizations should be pointed out as well. Chiefly, we can refer to the Strategic Environmental Assessment, which focuses on the environmental compatibility of public plans and programs. As known, indeed, the SEA is an environmental report where the probable significant effects on the environment and the reasonable alternatives of a proposed plan or program are identified.

As to this particular instrument to protect the environment, the European Commission highlights the need of integrating the climate problem in the SEA procedure. Even if the original regulation of the SEA regarded the climate as one of the environmental aspects that should have been taken into account, actually no Member State has paid the appropriate attention to this element.

For this reason, in 2009, the Commission found it worthwhile to clarify the need to consider the relation between plans and programs and climate in a double sense: for the effects that the relevant plan or program can produce on climate and, on the opposite, for the impact of climate change on the implementation of plans or programs.

Otherwise, even referring to the other well-known environmental permit, the Environmental Impact Assessment, it would be important to foster attention to the climate problem. Indeed, it should also take into account climate change's adaptation criteria.

All the mentioned instruments can be included among the command and control tools. They represent the best way to ensure the respect of limits, by virtue of their authoritative nature, necessary to provide mandatory land use controls when autonomous measures and voluntarism are not efficient enough to deliver climate policy.

But speaking about all the different methods used to contrast climate change, we should also take into account the importance of market-based tools, especially in a sector in which, as said above, there is not a scientific certainty about the extent of the problem of climate change. So, in this case, a good way to address the problem could be to improve the private sector's sensibility or to provide market-based mechanisms, able to influence the strategies of consumers and enterprises.

In this respect, indeed, the most important instrument adopted to contrast the problem of polluting emissions at global level is the emission trading system, provided in 1997, which is nothing more than the creation of an artificial market.

It is based on a double approach, called "cap and trade." At the first level, indeed, the Kyoto Protocol establishes the maximum of polluting emissions that can be produced, whereas, at the second one, contracting parties are allowed to negotiate emission permits.

As part of this second system, the Kyoto Protocol creates both a joint implementation mechanism and a clean development one. The three instruments share a common feature in the fact that they are all based on a balance between environmental protection goals and economic sustainability.

In particular, according to the joint implementation system, each party listed in Annex 1 may transfer or purchase to any other party of the same Annex emission reduction units, resulting from projects undertaken in any sector of the economy, in order to reduce emissions or increase removals of greenhouse gases by absorption, afforestation and reforestation.

The clean developing mechanism, instead, provides that the industrialized states may acquire emissions' permits, by implementing in the developing countries clean development projects, based on new technologies, high efficiency and low emissions of greenhouse gases. Specifically, two basic conditions are established to obtain the emission permits: the reduction of emissions in developing countries must be certified, and the projects must contribute to their economic growth according to the criteria of sustainable development.

So both of the mechanisms are project based, in the sense that they require the ideation of a specific project by a state included in Annex I that should be realized in a different country, counted in Annex I in the case of joint implementation or in a third state in the case of the clean development mechanism.[31]

In conclusion, climate change must be addressed through a mixed approach, based on both theoretical and pragmatic points, involving different governmental levels and integrating simultaneously regulatory or authoritative instruments and market or voluntary tools. The last ones, indeed, are not sufficient to significantly reduce the risk of anthropogenic interference with the climate system for at least two reasons. First, they focus on industrial emissions, but industrial emissions, although they are the most significant causes of air pollution, are not the only source of concentrations of greenhouse gases; thus other tools are needed for agriculture. Second, market mechanisms are based on the dynamics of free trade; they work only as long as they are convenient for business.

Accordingly, regulations and mandatory measures are necessary to guarantee a high level of environmental protection, even in the context of climate change, in order to achieve the far-reaching targets established at the international level to reduce the emissions of GHGs.

4 Legal Instruments to Control Climate Change in the Agricultural Sector

As pointed out above, the problem of the relationship between climate change and agriculture should be addressed in various manners, including the use of new scientific-technical tools and the implementation of effective public policies and legal instruments.

Because agriculture relates to climate change in very different ways, various legal instruments must be designed to limit air pollutant emissions in order to ensure sustainable agriculture.

Foremost, indeed, a distinction should be drawn between mitigation strategy, which aims to reduce polluting emissions, and adaptation one, i.e. coping with the

[31] D'Auria (2010), pp. 21 et seqq.

climate change that cannot be avoided.[32] The second strategy seems to be more a technical-scientific matter, in the sense that it requires farmers adapting their management to the new climate conditions.

Since it is not possible to avoid all the climate change consequences (it also derives, as noted above, from natural factors), from a practical point of view it is necessary to use farming techniques that may adapt to climate change, for example by altering the mix of crops grown, seed varieties, planting and harvesting dates, crop rotations and generally adjusting agricultural infrastructures to shifting crop zones.

Even adaptation, indeed, can be of two types: autonomous or planned. The first one involves farmers' response to climate change based on their expertise on available technologies and resources. But autonomous adaptation may not be sufficient.

In this case, private strategies against climate change may be improved by institutional or policy actions. Planned adaptation, indeed, includes education efforts, specific research studies and governmental policies to address climate change, such as providing appropriate transport and storage infrastructures, promoting efficient management of markets and offering financial aid.

Going back to the first strategy, instead, since it moves from the need of reducing emissions, the legal reaction acquires much more importance in order to achieve this goal. Only by implementing legal instruments is it possible to impose real emission limits and to ensure that human activities will be carried out in a sustainable way.

For this reason, it is worth recalling the role of regulation and standards as a command and control approach to reduce GHGs emissions. For example, one path is represented by imposing the use of certain fertilizers and/or spreading methods or concentration of herds or even cycles of crop rotation.

The regulatory approach also guarantees the necessary updating, given the difficulty to establish restrictions to respond to a long-term phenomenon, characterized by unforeseeable changes that are strongly connected to economic conditions and technology.[33]

Even if many different legal instruments are available to contrast climate change, it is worth focusing on the ones most used by agricultural policies and questioning which nature a further intervention should have. The EU agricultural policies began to address the problem of climate change only in 2003. The CAP (Common Agricultural Policy) of 2003 indeed, for the first time, untied financial aid to farmers from production and connected aid to a sustainable land management and organic production respecting the natural systems and cycles. Prior to this date, indeed, at the European level and at the national one, the rural development policies were initially developed on the basis of very different objectives with respect to the

[32] Roggero et al. (2010), pp. 359–377.

[33] Sigman (2007), pp. 289–306.

mitigation of climate change. The same, in fact, were based on competitiveness, environmental protection and the development of rural areas as a priority.

This does not mean, of course, that some of the measures and actions planned as part of the Rural Development Programs (RDP) were not characterized by goals related to the mitigation of climate change or, in some cases, the adaptation of agricultural ecosystems and forestry to climate change, but a clear and explicit consideration of the goals of "climate" policies in rural development has taken place only in 2003.

Nevertheless, on January 2009, by approving Regulation no. 74/2009, the European Union has introduced significant changes to rural development policies, which enable the competent authority to amend the previous strategies, providing new measures of farming in order to meet the following priorities: climate change, renewable energy, water management, biodiversity, measures accompanying restructuring of the dairy sector.[34]

The European approach has been indeed very important, not just because of the supranational extent of many climate events but also because the European intervention can favor potential economies of scale for research, information, collection and sharing of data and knowledge. Moreover, climate change is affecting the European market, and common policies to react to climate change can give rise to a new context of solidarity between the Member States.

Another very important and recent step concerning the European strategies to improve climate sustainability in rural policies is represented by the provision of specific instructions in the document "Principles and recommendations for integrating climate change adaptation considerations into rural development programs 2014–2020," accompanying the European strategy of adaptation.

But general agricultural policies considering climate change should be implemented to reduce effectively the problem in question.

Looking at the various instruments provided by the legislation of states, one of the most recurrent tools lies with the mechanism of virtuous farmers' incentives, not depending on the quantities produced but on the techniques used in order to reduce the amount of greenhouse gases released into the atmosphere. Still now, the sector of agriculture enjoys a regime of voluntary actions to reduce GHG emissions: governments, for example, grants to encourage environmentally friendly farming.[35]

Even if grants do not constitute the sole instruments used, given the sensitivity of this sector to consumers, probably the most appropriate tools are in fact the ones based on market rules, encouraging the adoption of agricultural techniques less polluting due to consumer preferences.

A useful measure that perhaps could be further enhanced is represented by labeling. A rapidly rising new concept is "biological certification," related to the "climate sustainability" of products, as a result of agricultural techniques based on a reduced use of fertilizers.

[34] Cesaro (2010), pp. 31–34.

[35] Scannell (2010), pp. 437–466.

Therefore, the use of grants or voluntary systems, such as labeling, can help overcome the problem of the impossibility of directly applying environmental principles to private people, in any case in which there is not a specific regulation. Indeed, while public bodies are always required to take into account the environmental issues in their activities (just think of the integration principle or the sustainable development one), applying all the environmental principles to private activities requires a specific provision of law.[36]

This appears to be the reason why in agriculture, a sector where just a few provisions establish specific limits, climate change is addressed for the most part by using incentives and grants or spontaneous measures.

5 Conclusion

The challenge of climate change could inspire a new evaluation of the agricultural sector. The adoption of new agricultural policies, biologically and environmentally sustainable, can confirm the valuable role that agriculture plays in the environmental equilibrium and, in general, for social welfare.

In this sense, we also should consider the impact of climate change on future generations and agricultural resources in the future. Global climate change, caused partly by human activities, raises serious problems even in relation to intergenerational justice.

Every measure against climate change must be adopted taking an intergenerational prospective, in order to slow the rate of changes, to reduce the direct effects and to ensure that future generations have the necessary resources and tools to adapt to climate change.[37]

Because climate change is a long-term phenomenon, steps to address it require long-term measures. In this sense, every generation should take care of this problem.

Climate change indeed is an intergenerational problem, but it involves relationship between present generations as well. The impact of global warming in fact is usually more negative for developing countries, which have less resilience and capacity to adapt. At the same time, since we are dealing with a long-term phenomenon, the well-being of future generations depends on the actions that present generations take today.

It means that every measure aimed to address climate change must consider both these aspects: equity between generations and between countries.

But the sustainable development principle can be useful even when there are no specific legal instruments to contrast the climate change. The principle, indeed,

[36] Fracchia (2010).

[37] Brown Weiss (2008), pp. 615–628.

allows to limit some human activities in order to avoid climate change even if there are no particular policies to prevent global warming.

References

Abler D, Shortle J, Fischer A (2004) Climate change and Pennsylvania agriculture. Penn State Environ Law Rev 12(1):117–126

Barelli U (2014) I limiti alle energie rinnovabili con particolare riferimento alla tutela della biodiversità. Rivista giuridica dell'ambiente 29(1):1–27

Boscolo E (2014) Oltre il territorio: il suolo quale matrice ambientale e bene comune. Urbanistica e appalti 18(2):129–146

Brown Weiss E (2008) Climate change, intergenerational equity and international law. Vt J Environ Law 9(3):615–628

Cafagno M (2007) Principi e strumenti di tutela dell'ambiente come sistema complesso, adattativo, comune. Giappichelli, Torino

Carlarne C (2011) Exploring methodological challenges within the context of climate change law and policy. Am Soc Int Law 105:255–257

Carraro F (2011) Consumo di suolo e federalismo fiscale ambientale. Rivista di diritto finanziario e scienza delle finanze 70(2):231 et seqq

Cartei GF (2014) Consumo di suolo: la prospettiva dell'Unione europea. In: Cartei GF, De Lucia L (eds) Contenere il consumo di suolo. Saperi ed esperienze a confronto. Editoriale Scientifica, Napoli, pp 45 et seqq

Cesaro L (2010) I cambiamenti in itinere dei programmi di sviluppo rurale per contrastare i cambiamenti climatici. Agriregionieuropa 6(21):31–34

Charles S (2011) Climate change: impacts of food safety. Nat Resour Environ 26(1):44–47

Coderoni S (2013) Agricoltura e cambiamenti climatici: dalle politiche comunitarie ai Psr. Agriregionieuropa 9(35):84 et seqq

Compiani V (2002) L'adempimento dei trattati internazionali sui cambiamenti climatici da parte dell'Italia e della regione Lombardia. Rivista giuridica dell'ambiente 17(2):365–387

Croci E (2010) L'azione dei governi locali per il clima. In: Fracchia F, Occhiena M (eds) Climate change: la risposta del diritto. Editoriale Scientifica, Napoli, pp 147–164

D'Auria M (2010) Il protocollo di Kyoto. In: Fracchia F, Occhiena M (eds) Climate change: la risposta del diritto. Editoriale Scientifica, Napoli, pp 21 et seqq

Di Gennaro A (2014) Urbanizzazione e territorio rurale in Italia: lo spazio, il tempo, le parole. In: Cartei GF, De Lucia L (eds) Contenere il consumo di suolo. Saperi ed esperienze a confronto. Editoriale Scientifica, Napoli, pp 3 et seqq

ECCP (2001) Agriculture. Mitigation Potential of Greenhouse Gases in the Agricultural Sector. Working Group 7, Final report of European Climate Change Programme, COMM (2000)88, European Commission, Brussels

Fracchia F (2010) Lo sviluppo sostenibile. La voce flebile dell'altro tra protezione dell'ambiente e tutela della specie umana. Editoriale Scientifica, Napoli

Fracchia F (2013) The science of administrative law, juridical method and epistemology: the roles of paradigms in the era of the crisis of modernity. Ital J Public Law 5(1):4–42

Hamilton ND (2011) Farming an uncertain climate future: what COP 15 means for agriculture. Univ Ill Law Rev 2011(2):341–360

IPPC (2007) Climate change 2007, mitigation of climate change. Contribution of Working Group III to the Fourth Assessment Report of the IPPC. In: www.ipcc.ch

Jaykus L-A, Woolridge M, Frank JM, Miraglia M, McQuatters-Gollop A, Tirado C, Clarke R, Friel M (2008) Climate change: implications for food safety. Food and Agriculture Organization of the United Nations. http://www.fao.org/docrep/010/i0195e/i0195e00.HTM

Liakopoulos D (2005) Il regime giuridico dei cambiamenti climatici nell'ambito internazionale e comunitario. http://www.ambientediritto.it/dottrina/Dottrina_2005/cambiamenti_climatici_Liakopoulos.htm
Lin J (2011) The environmental regulation of biofuels: limits of the meta-standard approach. Carbon Clim Law Rev 5(1):34–43
Manfredi G (2011) Cambiamenti climatici e principio di precauzione. Rivista quadrimestrale di diritto dell'ambiente 2(1):28–39
Mearns LO (2010) Quantification of uncertainties of future climate change: challenges and applications. Philos Sci 77(5):998–1101
Meltz R (2013) Climate change and existing law: a survey of legal issues past, present and future. Congressional Research Service. http://digital.library.unt.edu/ark:/67531/metadc98036/
Montini M (1999) Le politiche climatiche dopo Kyoto: interventi a livello nazionale e ricorso ai meccanismi di flessibilità. Rivista giuridica dell'ambiente 14(1):133–148
Onzivu W (2009) Tackling the public health impact of climate change: The role of domestic environmental health governance in developing countries. Int Lawyer (ABA) 43(3):1311–1336
Ottinger RL (2009) Biofuels – potential, problems and solutions. Fordham Environ Law Rev 19(2):253–264
Pelsy F (2008) The European Commission 2008 directive proposal on biofuels: a critique. Law Environ Dev J 4(2):119–136
Rodgers C, Franks J, Lowe P (2011) Implementing climate change policy for agriculture – what role for the law? Environ Law Rev 13(4):245–250
Roggero PP, De Sanctis G, Seddaiu G (2010) Cambiamenti climatici e sistemi agrari: da "Comando e controllo" a "Azioni concertate". In: Fracchia F, Occhiena M (eds) Climate change: la risposta del diritto. Editoriale Scientifica, Napoli, pp 359–377
Rosso Grossman M (2010) Climate change and the law. Am J Comp Law 58:223–255
Scannell Y (2010) Innovations and adaptations to climate change in Irish law. In: Fracchia F, Occhiena M (eds) Climate change: la risposta del diritto. Editoriale Scientifica, Napoli, pp 437–466
Sigman H (2007) Legal liability as climate change policy. Univ Pa Law Rev 155(6):289–306
Tedsen E, Homann G (2013) Implementing the precautionary principle for climate engineering. Carbon Clim Law Rev 7(2):90–100

Knowledge-Based Greening as a New Bioeconomy Strategy for Development: Agroecological Utopia or Revolution?

E.G. Koukios

Abstract The paper starts with definitions of the concepts of "green economy" and "greening" and an introduction into their complexity, leading to the formulation of a "greening alphabet." Then, based on a historical review and the experiences of the speaker, examples of pathways to greening are presented and mapped with the help of a simple model. The paper advocates the focus on green, knowledge-based bioeconomy as a new development model for the crisis-plagued Southern European and Mediterranean economies and societies. To assess the feasibility of this shift, several strategic factors have been identified as critical, along with driving forces, barriers, and scenarios for growth. These factors include research modes and contents, the broader policy environment, required skills, fragmentation risks, and societal factors, especially related to stakeholders' involvement and consensus building. Some key points on the revolutionary character of adopting the green bioeconomy based strategy are raised as concluding remarks, with particular reference to Southern European regions.

Keywords Bioeconomy • Bio-greening • Development • Green economy • Southern Europe

1 Introduction

According to UNEP's 2008 Green Economy Initiative (2008), as used by Wikipedia,[1] "Green Economy" is an economy that results in improved human well-being and social equity while significantly reducing environmental risks and ecological scarcities. Also, according to the same source, "Greening" is the process of transforming artifacts, such as a space, a lifestyle, or a brand image into a more environmentally friendly version.

[1] http://en.wikipedia.org/wiki/Green Economy; http://en.wikipedia.org/wiki/Greening.

E.G. Koukios (✉)
National Technical University of Athens, Bioresource Technology Unit, Athens, Greece
e-mail: koukios@chemeng.ntua.gr, http://www.star-agroenergy.eu/emmanuel-koukios.html

M. Monteduro et al. (eds.), *Law and Agroecology*,
DOI 10.1007/978-3-662-46617-9_23

On the other hand, bioeconomy or biobased economy is a term used recently to express the broad spectrum of potential applications of biological sciences and their associated technologies for improved performance and quality of products and services in various fields of the economy, both established and emerging.[2]

A particular feature of the value and strategic attraction of bioeconomy, especially of its sustainable version, to policy and decision makers is its potential "greening" effect on processes, products, and systems. Such a bio-greening strategy could follow two main pathways:[3]

- That of "green" chemistry and engineering, i.e., of the employment of biobased solutions to specific environmental and energy problems, e.g. treatment of biowastes, development of "cleaner" bioindustries, recycling of nutrients, production of bioenergy and biomaterial vectors;
- The one of integrated "green" management, i.e., the formulation and use of biobased tools and methods for the rational management of communities, ecosystems, and other natural and anthropogenic biosystems, i.e., soils, aquatic systems, forests, and cities.

The increasing demand for "greener" processes, products, and production systems is usually associated with a postindustrial (or metaindustrial) phase of world economies.[4] This is due to the fact that the industrial mode of production, which dominated the world for three centuries, was particularly aggressive to natural resources and ecosystems.[5] The symbolic start of this greening-friendly phase is provided by the publication of the Club of Rome report in 1972,[6] and its 40-year-long history is paved with follow-ups of the Club of Rome work,[7] with new approaches to model the world economy,[8] and significantly with the UN 1987 publication that first defined "sustainable development."[9]

Developing biobased technological solutions to the growing greening demand has been enhanced by the significant progress in biological sciences and technologies, initiated by the discovery of the DNA double helix molecular structure.[10] The parallel history of the "bio" and the "greening" components is shown in Table 1.

---

[2] European Commission (2012).

[3] Koukios (2013a, b, c).

[4] Strategic Assessment of Science and Technology (1994).

[5] Cipolla (1978) and Ashton (1975).

[6] Meadows et al (1972).

[7] Laszlo et al. (1978).

[8] Leontief et al (1977).

[9] United Nations (1987).

[10] Watson (1968).

Table 1 Brief comparative history of green economy and bioeconomy paths (Koukios 2013a, b, c)

Greening path	Bio-path
1972: "Limits to Growth," Club of Rome Report 1973: E.F. Schumacher's "Small is Beautiful" 1970s: Oil Crises, Renewable Energies R&D 1980: "The Third Wave" by Alvin Toffler 1987: "Our Common Future," UN/Brundtland Report 1988: Intergov. Panel on Climate Change (IPCC) 1990s: Climate Change Debate—UN Earth Summits Rio Agreement, 1992—Kyoto Protocol, 1998 2000s: Greening Strategies and Policies	1962: Nobel Prize for DNA 1960/1970s: Molecular Biology 1970s: Biochemical Engineering 1980s: Genetic Engineering, Genome Mapping 1990s: Crises Phenomena (e.g., "Mad Cows"), Public Concerns on GMOs 2000s: Biobased Development, Bioeconomy, Bio-Info-Nano Hybrids

2 Methodology

Bio-greening processes, products, and systems will be analyzed according to feasibility and sustainability benchmarks available in the literature and the author's relevant R&D expertise.[11] To limit the fuzzy character of the sustainability issues, we have developed the following "alphabet" of 26 criteria proposed for the sustainability assessment of any particular technology, project, process, or system, which we can call "the ABC of Sustainability":[12]

A. Substitution of fossil-based organic industrial feedstocks by biobased ones;
B. Complete utilization of raw materials—"zero-waste" target;
C. Recycling of materials;
D. Reduction of the amount and volume of any waste generated;
E. Resource saving by making long-life products;
F. Use of renewable or recyclable feedstocks, including renewable energy sources;
G. Rational use of energy;
H. Systematic saving of energy;
I. Design of low energy-consumption systems;
J. Emphasis on water economy, as well as other critical resources;
K. Application of soft production/conversion energy systems;
L. Systematic utilization of agricultural and agro-industrial residues;
M. Enrichment of the agricultural/productive soils in organic matter;
N. Rational use of chemical fertilizers and other chemical inputs in agriculture;
O. Land use: priority of food vs. fuel production;
P. Priority to cover the energy needs of agricultural and food production;
Q. Optimal utilization of wastes;
R. Increasing of the biodegradability of wastes;

[11] Koukios et al. (2010), pp. 147–151.

[12] Koukios (2013a, b, c).

S. Reduction of the industrial scale of operations to meet the ecosystem limits;
T. Design of "closed," circular, or integrated production systems;
U. Application of biosciences and biotechnologies whenever possible;
V. Decentralization of production activities according to space-planning criteria;
W. Early participation of key stakeholders in decision and policy making;
X. Securing of social acceptance;
Y. Improving of health and quality-of-life impacts;
Z. Reduction of GHGs (greenhouse gases) as a key element of the bio-greening strategy.

2.1 Pathways of Greening

Bio-greening is one of the several pathways towards a green economy, which have been emerged in various parts of the world in the last 50 years. The list includes almost a dozen strategies, some already forgotten, others with mediocre results, a few still in progress—to be integrated into the bio-greening efforts:

- appropriate, alternative technologies;
- end-of-pipe solutions;
- clean(er) technologies, ecotechnologies;
- clean, smart products;
- low-carbon, nonfossil resources;
- water, energy, resource, and waste management;
- environmental engineering & management;
- organic systems;
- grassroots societal experiments;
- sustainable development;
- bio-greening, biobased economy, bioeconomy.

2.2 Mapping Bio-Greening Pathways

Managing the combined complexity of greening and the economy is an extremely challenging task, which requires the development and use of novel tools, as well as conceptual, administrational, monitoring, and other tools. Responding to this challenge, we have introduced such a simple tool, which is based on the idea of "mapping" and has the form of a 2-dimensional representation of the interactions between the two components of the greening actions, as shown in Fig. 1[13].

[13] Koukios (2015).

Greening Sciences & Technologies	ECONOMY: Areas of Application							
	A1	A2	A3	A4	...	...	...	Am
B1								
B2								
B3								
B4			B4/A3*					
...								
Bn								

Fig. 1 Mapping the knowledge-based greening landscape of the economy. *Example: B4/A3 means use of waste treatment technology B4 to treat waste of industry A3

3 A New Bio-Greening Strategy

3.1 Scope for Southern European Economies

The following is a list of the 12+1 strategic reasons behind a bioeconomy-oriented development model, as identified by the Advisory Group of the Knowledge-Based BioEconomy (KBBE) thematic area of the now completed 7th Framework Programme for Research of the European Commission. All these arguments are particularly relevant to the crisis-stricken Southern European economies, also in support of the bio-greening development strategy advocated in the present work:[14]

(1) high stakes: food, farming, land use; (2) dynamics of change; (3) innovation potential; (4) environmental aspects; (5) climate change; (6) substitution of fossil fuels; (7) socioeconomic aspects, e.g., unemployment; (8) quality and security—needs for upgrading; (9) business opportunities, especially at local and regional levels; (10) policy coordination (see below); (11) European value; (12) global development—needs for networking and cooperation; and (13) a smart move, i.e., a good basis for smart specialization of Mediterranean and other regions.

3.2 Driving Forces and Growth Scenarios

To evaluate the success dynamics of the bio-greening strategy, we will use the analysis of the Greek Technology Foresight on bioeconomy in the year 2021, where 2 key drivers were identified: (1) excellence in biosciences and biotechnologies and (2) social acceptance and partnership of the key stakeholders. As shown in Fig. 2, the future could follow one of four scenarios: (1) "Fertile Valleys," (2) "Islands of Survival," (3) "Hospitable Plateaus," and (4) "Crystallisation Poles".[15]

[14] Koukios and Lange (2013).

[15] General Secretariat for Research and Technology (2005).

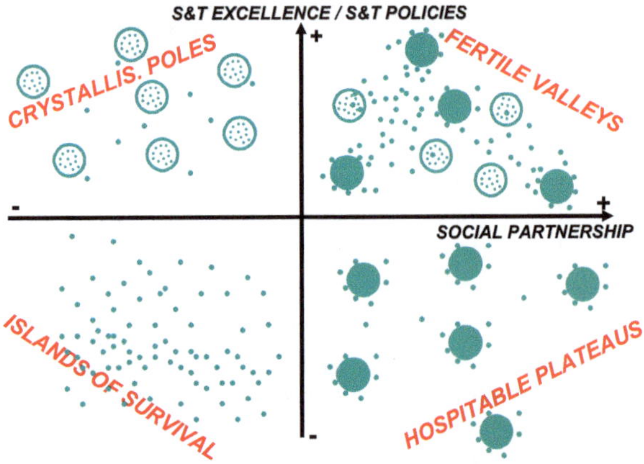

Fig. 2 Scenarios for growth of bio-greening applications in a Southern EU country

4 Critical Factors

In the following paragraphs, we will summarize the major barriers and other key factors determining the success of a knowledge-based bio-greening strategy, with special reference to Southern European economies and societies.[16]

4.1 Multipolicy Environment

Targeting green bioeconomy as a Southern European national and/or regional development strategy is not the subject of a single policy but rather the object of a multipolicy landscape. We can easily identify more than a dozen different policy areas, such as those typically falling under the jurisdiction of a Ministry, which affect the bioeconomy and the greening stories: research, innovation; agriculture, forestry; rural areas; maritime; public health; energy; environment, water, climate; new functional materials; industrial; market, competition; financial; regional; European affairs; international cooperation; crisis/recovery management. The question now is how the treasure-loaded green bioeconomy "vehicle" will get safely through this paper-power battlefield, where each one of the policy areas concerned could affect, delay, alter, distort, block, or even kill the new strategy. Learning to navigate within such a complex policy landscape has to be the top priority of the new strategy, and will be addressed in future works.

[16] Koukios (2013a, b, c).

4.2 Research and Innovation

Promoting greening by research will require a radical shift in the form and contents of research and innovation efforts, according to the following seven "Golden Rules," based on the view of research as the main key to unlock the reservoir of vital and valuable innovation the bio-world constitutes:

1. better understanding of complex phenomena involved;
2. planning and implementing knowledge-based actions;
3. environmental biotechnologies as a potential research flagship;
4. design of environmentally compatible solutions, drawing upon other novel RTD areas and approaches;
5. significant role in social and economic development as a key opportunity field for international cooperation;
6. responding to societal concerns and assessing risks;
7. research to be accompanied by appropriate information, communication, dissemination, and crisis-management components.

4.3 Skills and Dexterities

The quest for a green bioeconomy as a new strategic issue will to a great extent depend upon the efficient action of multiplayer actors, which in turn leads to the identification of the new skills and dexterities needed, presented here in the form of another list of seven "Golden Rules":

- *radical change*: shift in socioeconomic structures, cultures and lifestyles, knowledge modes, and organization patterns;
- *mutually transforming processes*: by learning and cognition;
- *knowledge*: including cognitive and affective elements;
- *consideration of all critical flow systems*: molecular, energy, material, information, financial, and human;
- *to do (1)*: introduce greening skills through problem-oriented university curricula;
- *to do (2)*: add an extra layer to the existing professional education systems;
- *to do (3)*: use the KIC (Knowledge & Innovation Communities) concept as new instrument for change.

4.4 Fragmentation Risks

This is one of the major threats of biobased development models, which could also affect bio-greening by weakening it. It could lead to spreading resources too thin, and even creating internal competition for funds and markets. The coordinating power of the bioeconomy "umbrella" is rather weak and prone to get challenged by some of its better established components, i.e., agriculture and medicine. What could mitigate this risk and glue together the loose bioeconomic sectors, fields, and areas is the momentum of a Great Vision that could effectively mobilize people, resources, and ideas towards national and regional targets. In addition to vision, other potential remedies include proper risk management, bridging and integration of efforts, and a healthy catalytic role of regulation.

4.5 Societal Factors

This is a very important group of factors, both for their direct impact on the feasibility and sustainability of any new knowledge-based strategy, as well as for their indirect impacts through synergies with the other types of factors presented above, e.g., lobbying for policies, interest in research careers, demand for education and training, adoption of a vision on green bioeconomy. Our analysis of this field reveals an urgent need for action according to another seven "Golden Rules," which are summarized as follows:

1. *Promote awareness*: appropriately informing the public especially on complex sociotechnical subjects, with the use of special tools and methods now in short supply.
2. *Seek and secure acceptance*: a minimum of social acceptance is necessary for takeoff and deployment, preferably by educated decisions based on transparent agendas.
3. *Perform early "soft" studies*: identification and assessment of critical societal issues are instrumental for optimal orientation and "tuning" of the technical work.
4. *Map social dynamics*: stakeholders and other social actors can only be defined by a comprehensive and case-specific effort to map the particular social dynamics.
5. *Organize actors' involvement*: timely and structured involvement of stakeholders and other critical actors constitutes the "heart" of the sociotechnical decision process.
6. *Build consensus upon strategic alliances*: stability of decisions, based on converging interests and expectations of actors, is a prerequisite for sustainability.
7. *Resolve conflict*: serious dissonance of views/opinions among actors can jeopardize consensus unless properly resolved, e.g., by a culturally compatible mediation.

The successful application of most of the above guidelines requires a new generation of tools and methods for group consultation and decision making; an example of this vital element taken from the author's experience is given below in a special Annex.[17]

[17] Karaoglanoglou et al. (2013).

5 Concluding Remarks

Greening as a strategy is fully symbiotic with sustainable economic and societal innovative solutions, including those of the bioeconomy. In particular, bio-greening aims at tapping the huge reservoir of the biological world in order to promote greening, along with achieving other important targets, with respect to food, feed, fertilizer, energy, materials, chemicals, health, quality, security of supply, etc.

The most promising path of bio-greening as an innovation strategy is through product and function/service innovation; this could revitalize sectors and regions, including enhanced employment, by smart interdisciplinary applications. On the other hand, the process innovation-based bio-greening actions should focus on novel technologies for clean solutions supporting the new products and the efficiency of their end-uses and life cycles. As for the third type, i.e., of system innovations, such as the biorefinery, they do offer valuable possibilities to combine product with process innovations and harvest their synergies.

Therefore, our proposal of this strategy as a priority development target for the crisis-plagued Southern European economies is well justified and matches their relevant needs and challenges. Of course, as shown in several parts of this paper, the shift from the present model to the new one will not be easy, as it will depend upon major and "deep" changes in several key sectors and fields. However, this feature could enlarge the scope for broader changes regarding management of human, biological, natural, and other resources of these countries. So, in this sense, we are not facing another Utopia but a true "Neo-technical" Revolution.

Acknowledgments This paper was prepared in the frame of the research cooperation of its author with the STAR Team of the University of Foggia, in Italy. The work reported here is also based on the long relevant experiences of the author's research team, the Bioresource Technology Unit, at the National Technical University of Athens, in Greece. The author gladly acknowledges useful discussions with the STAR Coordinator, Prof. Massimo Monteleone. Preliminary forms and/or parts of this work were presented, prior to the Lecce 2013 Workshop, at conferences and other meetings in Copenhagen, Vienna, Foggia, Sorrento, and Athens.

Annex: A New Group Decision Method for Agro-Bio-Greening

A novel decision-making method, which follows a structured and quantitative group consultation, will be demonstrated by a simulation game. It was performed in the class of the elective Bioenergy course of the M.Sc. program on Energy Production and Management at the author's university. The object of the exercise was to assess the prospects of a green bioenergy project as affected by its stakeholders, in a typical "Rural European South" case.

THE PROJECT: In a rural region of Greece, an agrobiological hydrogen generation plant of 10 MW capacity is under discussion; its feedstock will be 5 dry t/h (or 40 dry kt/year) of a carbohydrate-rich energy crop to replace other crops and raise income and employment.

STAKEHOLDERS: After consultations with the class, representing the various societal and economic audiences, the following key stakeholders were defined and invited to the decision process, as six groups of persons were asked to play the corresponding roles:

(a) farmers (biomass producers);
(b) bioenergy producers (biomass-to-biohydrogen conversion plant);
(c) biohydrogen distribution company;
(d) local community;
(e) national government;
(f) other players (financial players, the EU, scientists, eco-NGOs, etc.): scientists were selected.

DECISION CRITERIA: Stakeholders' groups were asked to rank their priorities in the following decision-making criteria on a scale of 1–9 (with 1 and 9 being the most and the least important criteria for each stakeholder group):

1) return on investment;
2) annual revenues;
3) national economy;
4) regional/local economy;
5) application of new/innovative technologies;
6) employment;
7) environment;
8) political and institutional benefits;
9) other criteria (to be specified by the stakeholders).

STAKEHOLDERS' RESPONSE: The preference vectors of all stakeholders are presented in Table 2, along with a first synthesis of the findings leading to a final, overall ranking of the 9 decision criteria according to the dynamics of the situation.

Table 2 Map/monitor stakeholders' response

	Rank of each interest group/stakeholder for the decision criteria							
Decision criteria	(a)	(b)	(c)	(d)	(e)	(f)	Total	Overall ranking of importance
1	4	1	2	8	9	4	28	III_A
2	1	2	1	4	5	8	21	I
3	6	6	6	5	2	5	30	IV
4	2	8	5	1	3	6	25	II_B
5	7	4	3	6	8	1	29	III_B
6	3	7	4	2	4	2	24	II_A
7	5	5	7	3	7	3	30	IV
8	8	3	8	7	1	7	34	V
9	9	9	9	9	6	9	51	VI

Quantification by preference vectors

Fig. 3 Use of preference vectors to identify strategic alliances between actors

	(a)	(b)	(c)	(d)	(e)	(f)
(a)		22.0	12.0	14.0	28.0	22.0
(b)			16.0	30.0	34.0	26.0
(c)				24.0	32.0	20.0
(d)					24.0	18.0
(e)						36.0
(f)						

In the top position, we find economics in the form of annual revenues, followed by an interesting tie for the second place of importance between employment and local economy.

STRATEGIC ALLIANCES BETWEEN STAKEHOLDERS: The next Fig. 3 presents the sums of the differences (in absolute values) between the preference vectors of each pair of actors. The smaller the number, the closer are the views of a pair of stakeholders. The closest link found was of the (a)–(c) pair, followed by the (a)–(d) and the (b)–(c) ones. In the other end of the spectrum, we find the (e)–(f) and (b)–(e) pairs, which show potential conflicts.

References

Ashton TS (1975) The industrial revolution 1760–1830. Oxford University Press, London

Cipolla CM (1978) The economic history of world population. Penguin, London

European Commission (2012) Innovating for sustainable growth: a bioeconomy for Europe. COM (2012) 60 final, Brussels: EC

General Secretariat for Research and Technology (2005) Technology foresight in Greece (2001–2021). Synthesis Report, Ministry of Development, GSRT, Athens, in Greek

Karaoglanoglou L, Koullas D, Monteleone M, Koukios EG (2013) Best practices for building sustainable biomass-to-biofuel chains in Southern EU regions: The cases of Capitanata (IT) and Thessaly (GR). In: Proceedings of the 21st European Biomass Conference, Copenhagen, June 2013

Koukios EG (2013a) Biological raw materials for energy and chemicals, course notes. NTUA, Athens (in Greek). http://www.chemeng.ntua.gr/courses/bpy/files/check_list.pdf; http://www.chemeng.ntua.gr/courses/bpy/files/sygxrones%20taseis%20antirrypansis.pdf

Koukios EG (2013b) Environmental Management Potential of Developing Sustainable Bioeconomy Solutions – An Overview. Keynote Paper presented at the 7th International Conference of Environmental Engineering and Management, Vienna, 18–21 September 2013

Koukios EG (2013c) Panel presentation on a new biobased development strategy for Southern Europe. In: Proceedings of international Sorrento 2013 meeting (Sorrento, Italy: OBI – Observatory of Banks and Enterprises, November 2013). http://www.youtube.com/watch?v=KIE0QnIBy2Q

Koukios EG (2015) Technology management for the biobased economy: mapping, dynamics and policies – The case of Greece. In: Trzmielak D (ed) Internationalisation, innovation, commercialisation and technology transfer. University Press Company, Lodz (in press)

Koukios EG, Lange L (2013) Strategy- and policy-related reflections on biomass within a bioeconomy frame by the FP7 KBBE advisory group. In: Proceedings of the 21st European biomass conference, Copenhagen, June 2013

Koukios EG, Daouti-Koukios I, Koullas DP, Avgerinos E (2010) Critical parameters for optimal biomass refineries: the case of biohydrogen. Clean Technol Environ Policies J 12(2):147–151

Laszlo E et al (1978) Goals for mankind. Signet, New York

Leontief W et al (1977) The future of the world economy. Oxford University Press, New York

Meadows DH et al (1972) The limits to growth. Earth Island Ltd., London

Strategic Assessment of Science and Technology (1994) Greening of industry. Report of SAST activity (Brussels: European Commission, 1994)

United Nations (1987) Our common future – The Brundtland report. United Nations, New York

Watson JD (1968) The double helix – A personal account of the discovery of the structure of DNA. Penguin Books, London

Agrobiodiversity, Intellectual Property Rights and Right to Food: The Case of Andean Countries

M. Pierri

Abstract This essay provides a knowledge contribution in order to understand the relationship between agrobiodiversity, intellectual property rights and the right to food in the context of Andean "Nuevo Constitucionalismo." The distinctive feature of this regulatory experience is that Andean constitutions are ecologically oriented and they ensure some tools to guarantee respect for the right to food not only in terms of quantity but also in terms of quality. To achieve this goal, these countries have introduced in constitutional texts the principle of "food sovereignty," which holds back the danger of monopoly patents over seeds and promotes the preservation of local agrobiodiversity, which is particularly concentrated in the area.

Keywords Agrobiodiversity • Intellectual property rights • Nuevo Constitucionalismo • Right to food

1 Introduction: What Is the Relationship Between Agrobiodiversity, Intellectual Property Rights and the Right to Food?

The title of this essay initially seems a little eccentric. What do agrobiodiversity, intellectual property rights and the right to food have to do with each other?

1.1 The Concept of Agrobiodiversity

The first term refers, as defined by the Food and Agriculture Organization (FAO), to the variety and variability of animals, plants and microorganisms that are important for food and agriculture and that are the result of interactions between the

M. Pierri (✉)
Department of Legal Sciences, University of Salento, Lecce, Italy
e-mail: maurizia.pierri@unisalento.it, https://www.unisalento.it/web/guest/scheda_personale/-/people/maurizia.pierri

M. Monteduro et al. (eds.), *Law and Agroecology*,
DOI 10.1007/978-3-662-46617-9_24

environment, genetic resources and management systems and practices used by men.[1] In other words, agrobiodiversity can be described as the diversity of the species cultivated in agriculture, in which the cultural factor makes a crucial contribution, so that domestication has been called the foundations of the biodiversity of crops.[2]

Agrobiodiversity is now more than ever a fundamental aspect of agricultural activity and an important source of raw materials for technological innovation, and not just for the benefit of agriculture but also for that of medicine, industry, pharmaceuticals and cosmetics. The range of action of agrobiodiversity is very wide. It includes plant genetic resources for food and agriculture; farm animal genetic resources, aquaculture and other animal species, such as insects; abiotic factors that have a decisive effect on various aspects of agrobiodiversity; finally, also the economic, cultural and social factors affecting agricultural activities, such as the traditional knowledge of local communities, cultural factors and participatory processes, tourism, agriculture and other socioeconomic elements related to agricultural activity.[3]

Domesticated biodiversity includes genetic variation between species, breeds, crops and types of animals, plants and microorganisms manipulated for agricultural production systems. Because of its complexity, we distinguish three levels of conceptualization of agrobiodiversity, which are made up of communities, plants, animals and microorganisms, in addition to the wildlife living in natural conditions related to domestic species.

It's not easy to locate the exact moment in which the concept of agrobiodiversity assumed a central role in the scientific debate on agriculture and its controversial effects on biodiversity,[4] but it happened in relatively recent times. Evidence of the delayed interest in agrobiodiversity is the data on the reduction of existing biological species: the first evidence of the use of plants and animals for cultivation and animal husbandry dates back to between 10,000 and 14,000 years ago, but only a small fraction of the existing biological diversity has been domesticated and actually contributes to food and to the world's agriculture. In fact, just 15 varieties of plants and seven animals provide 90 % of the food consumed. So, while in theory domestication should not have been detrimental to the conservation of biological diversity, in fact it has caused an impoverishment.

[1] FAO (1999), September 1999, p. 4.

[2] Frankel et al. (1995), p. 40.

[3] Agrobiodiversity can be defined as an interdisciplinarity science, such as agroecology: see Dalgaard et al. (2003), pp. 39–51.

[4] Probably at the beginning of the 1990s, with the stipulation of the Convention on Biological Diversity (CBD) signed Rio de Janeiro on 5 June 1992 at the United Nations Conference on Environment and Development (the Rio "Earth Summit"). The need for an ecological conversion of agriculture is well described by Gliessman and Rosemeyer (2010).

1.2 Intellectual Property Rights over Seeds

Intellectual property rights, seen as a bond between a subject and an object, which is the result of the subject's research or creation or invention, is protected both at international and national levels through legal instruments that guarantee the inventor's exclusive use of the work of the intellect, i.e. its financial exploitation. In particular, patents confer on the holder the legal capacity to prevent third parties from commercially exploiting the object patented or the procedure.[5]

The issue of intellectual property is related to both biodiversity and agrobiodiversity.[6] The first represents a limitless source of compounds and processes that men discover and use, thanks to the development of science and technology and the evolution of technological innovation processes that use these elements and biological processes in industrial applications, creating new goods and services. All this generates and assigns value to the elements of biodiversity as it allows them to convert their commercial potential into economic benefits of a different kind. Nevertheless, intellectual property and biodiversity management operate in different legal contexts and in tension with each other. The biogenetic resources fall within the sovereignty of the respective countries, and the access to them has to be regulated by national legislation. Paradoxically, when these resources are used for an invention, the intellectual property systems of the most developed countries, which recognize national sovereignty over biological and genetic resources, allow their intellectual appropriation, without requiring disclosure of the origin of that resource or that access was done legally. The weakness or absence of legal limitations on access and the permissiveness of patent law means that genetic resources are in fact granted to the private intellectual property of their users, turning them into owners.

With regard to the relationship with agrobiodiversity, the thorniest issue is related to the intellectual property rights over seeds. It is clear that for companies (especially multinationals) the transformation of seeds into "merchandises" is a desirable goal, attainable through the exercise of intellectual property rights over seeds. The risk is represented by the disappearance of local seeds and the proliferation of patents and laws that promote monocultures, monopolies and privatizations.

Patentability, sometimes radically challenged,[7] has been used by large corporations to obtain the absolute control of the seed, through genetic engineering. In fact, the industrial patentability of the seed, regarded as an invention, has become possible, thanks to the artificial insertion of a gene that modifies the original structure. Once this is achieved and once the monopoly on the product is secured, the multinationals act privileging uniformity instead of the expense of diversity and

[5] United Nation Conference on Trade and Development (1996).

[6] On the relation between agrobiodiversity, the right to food and farmers' seed systems, see De Schutter (2009).

[7] Shiva (2005), pp. 3 et seqq; see also Shiva (2001), p. 22.

operate a selection that is independent of taste, quality and nutritional value but is designed for industrial processing and transport over long distance. The patent system and the idea of the seed as the creation of man that may be the subject of exclusive rights are also in contrast with some principles rooted in the culture of the indigenous peoples,[8] guardians of ancestral agricultural practices, based on seed selection by farmers (who focus on diversity), seed preservation but, especially, free trade.

Patent protection in this area, however, is internationally recognized by the clauses contained in the TRIPS Agreement (Trade-Related Aspects of Intellectual Property Rights), which is included in Annex 1C of the Marrakesh Agreement Establishing the World Trade Organization, signed in Marrakesh, Morocco on 15 April 1994. In particular, it challenged Art. 27.3 (b), which gives Member States the possibility of excluding from patentability plants and animals, *other than microorganisms* and essentially biological processes for the production of plants and animals, *other than non-biological and microbiological processes* (such as genetic modification of the seed). In addition, the agreement envisages that countries take steps to protect plant varieties either by patents or by other ad hoc systems or with both things: this type of protection is an obstacle to the free exchange of seeds among farmers, which as mentioned is a fundamental element in the traditional farming culture of the indigenous peoples.

At the same international level, a different orientation emerges from the clauses contained in the so-called International Treaty on seeds (International Treaty on Plant genetic Resources for Food and Agriculture), a multilateral agreement, adopted in the FAO Conference (Resolution 3/2001) on November 2001, coming into force 3 years later. The multilateral international agreement has as its objectives the conservation and the sustainable use of plant genetic resources for sustainable agriculture and food security as well as the equitable sharing of the benefits that may arise from the use of plant genetic resources, in accordance with the Convention on biological Diversity (CBD). In recognition of the role of farmers, especially those who work in the areas of greatest biodiversity, development and conservation of genetic resources, the agreement commits Member States to preserve agricultural crops, ensure their free exchange, and recognize farmers as the custodians and users of genetic diversity.

The Treaty on seeds is part of a trend (of which a perfect example is the Via Campesina movement[9]), calling upon States, in drawing up rules on seeds, to restore the official status of the ecological and biological laws of nature such as diversity and adaptation, which are in contrast with the tendency to shape global laws on intellectual property and patents. The latter, as we have mentioned, have the drawback of referring to seeds as a creation or invention of the businesses that

[8] Glenn (2010), pp. 140 et seqq, refers to the chthonic legal tradition such as an ecological tradition. See also Goldsmith (1992), p. XVIII.

[9] The "Via Campesina" is the international movement, born in 1993, that brings together millions of peasants, small and medium-size farmers, landless people, women farmers, indigenous people, migrants and agricultural workers from 73 countries in Africa, Asia, Europe and the Americas. It defends small-scale sustainable agriculture as a way to promote social justice and dignity.

own them, at the expense of the possibility for farmers in indigenous communities to self-seed, store and exchange them freely, contributing to the enrichment of the agrobiological heritage.

1.3 The Right to Food

Finally, the right to food, ranked by international law (the Universal Declaration of Human Rights)[10] among the fundamental human rights instrumental for a dignified life, entails the possibility of every person to have access to healthy, nutritious, safe food, adequate in both quantity and *quality*. In fact, the right to food is composed of two complementary elements: the *availability* of food and the *accessibility* of the same. Availability, according to the definition given by the Committee on Economic, Social and Cultural Rights, must be seen as the opportunity "to obtain its food directly from the land or from other natural resources, or to dispose of efficient distribution processing and market systems, capable of transporting food from the place of production to where it is needed according to the demand." As regards accessibility, the same committee stated that it refers to each person's physical and economic access to a *sufficient* amount of food. Sufficiency is not simply and reductively considered in the quantitative sense (minimum ration of calories) but in the qualitative one, with reference to the adequacy of the food for their needs and their culture, the need to ensure a combination of elements that guarantee physical growth, mental development and physical activity. This must all reflect the physiological needs of the human being at all stages of life, according to sex and occupation. In this context, the issue of food security becomes crucial, considered not only as certainty of long-lasting access to food but also as the need of every individual to have access to safe and healthy nourishment. In the International Covenant on Economic, Social and Cultural Rights, adopted by the United Nations on 16 December 1966 and coming into force 10 years later, the nations made a commitment to take all measures to improve methods of production, conservation and distribution of food by making full use of technical and scientific knowledge, by disseminating knowledge of the principles of nutrition and by developing or reforming agricultural organization in order to achieve the most efficient development and utilization of natural resources. Later, at the United Nations World Food Summit, the Rome Declaration on World Food Security and Plan of Action was signed, which emphasized the need for a commitment by States to implement policies aimed at ensuring a supply of *nutritionally adequate* food. The Committee on Economic, Social and Cultural Rights,[11] in its Observation Générale of 12 May

[10] Article 25, paragraph I of the Universal Declaration of Human Rights adopted by the General Assembly of the United Nations, December 10, 1948.

[11] It is a commission established under Resolution 1985/17 of the Economic and Social Council (ECOSOC) of 28 May 1985 and composed of independent experts to oversee the implementation of the International Covenant on Economic, Social and Cultural Rights.

1999, interpreted Article 11 of the International Covenant on Economic, Social and Cultural Rights, giving the expression "adequate supply" a cultural connotation: "availability of food in sufficient quantity and quality to meet the physiological needs, free from harmful substances, and *acceptable within each culture*" (at para. 6).

The cultural element, relative to the right to food, is further reinforced in the debate within the United Nations Commission on Human Rights, in the years since Resolution 2000/10 of April 2000, by which the right to food gets its definitive consecration as the "right to have regular, permanent and free access, either directly or by purchase, to a qualitatively and quantitatively adequate and sufficient food supply, *corresponding to the cultural traditions of the population* to which the consumer belongs, and able to ensure a physical and mental, individual and collective life which is fulfilling, dignified and free from anxiety."

On the other hand, the implementation of the right to food involves the implementation of policies to enhance the national communities' ability "to cover their food needs," improving not only the methods of production, conservation and distribution but also the prospects of work through agrarian reforms for the benefit of those who do not own land.

The connection between the international recognition of the right to food and the protection of agrobiodiversity is clearly shown in the Declaration of Cordoba, presented on December 10, 2008, on the 60th anniversary of the Universal Declaration of Human Rights, which reflects the results of the international assessment on science, knowledge and technology for agricultural development adopted by the World Bank and FAO and signed by 60 countries on April 2008. This document recognizes the need for a complementary, diversified approach to agriculture for its sustainability and in order to fight the world food crisis. Here it is also emphasized that agricultural models based on small plantations may be the right alternative for food security based on human rights. Among the factors calling for greater attention, the declaration cites the lack of protection of communities of small landholders and of indigenous peoples against agro-industrial plantations; land reforms that are inadequate to guarantee land for poor rural communities, indigenous peoples, agricultural workers, especially women; the lack of support for small-scale food production in relation to access and control of seeds, water, infrastructure, information, credit and marketing; the overemphasis on the international sale of agricultural products, to the detriment of local markets for local crops responding to local needs and food habits; the lack of safeguards to prevent the abuse and the negative consequences of excessive intellectual property rights over seeds; excessive emphasis on the forms of agricultural production based on a high level of external inputs to the detriment of local crops; the lack of recognition and appreciation of traditional eating habits and crops, leading to marginalization and underutilization of the same; the lack of adequate protection against the loss of biodiversity caused by the expansion of monoculture plantations in food production; the lack of recognition of the need for differentiated interventions/ solutions for different agroecological conditions, distinct cultural traditions and different types and levels of national development. The issuing of patents that concern seeds is therefore recognized as a cause of impoverishment of agrobiodiversity and furthermore as a contributory cause of the food crisis.

These last observations justify the tendency to a reframing of international policies on the right to food and, similarly, of the fight against hunger, which is one of the eight millennium goals: the real problem is not the lack of food but its unequal distribution and inadequate quality, associated with the phenomenon of *malnutrition*, which cannot be countered simply by the introduction of large-scale cropping to the detriment of the local agricultural systems.

From the above, we can see the clear relationship between agrobiodiversity, intellectual property rights over seeds and the right to food, which though seemingly unrelated issues are actually connected. The enhancement of agrobiodiversity serves for the protection of the right to food, not in terms of quantity but in terms of quality. If the right to food is not confined to demanding the enforceability of a varied diet, adequate in both quantity and quality, then the protection of agrobiodiversity becomes a target instrumental in the enjoyment of the right to food and can enter into contrast with the protection of intellectual property rights over seeds by multinationals.

What regulatory instruments can be used to break up the harmful relationship binding the monopoly on seeds to the erosion of agrobiodiversity and, therefore, to the full enjoyment of the right to food? A very interesting hotbed of juridical experimentation is to be found in the constitutional and infraconstitutional experience that the countries of the Andes have been developing in recent years.

2 The Characteristics of the Andean Region

The Andean region, in particular Venezuela, Bolivia and Ecuador, is an invaluable laboratory for testing the relationship between the right to food, in the sense of a constitutionally guaranteed fundamental claim of the individual, the conservation and the enhancement of agrobiodiversity and the protection of intellectual property rights over seeds.

The reasons that make this context particularly significant are of both a geobiological and constitutional kind:

a) In terms of geobiology, the Andean area is renowned for its high concentration of natural and agricultural biodiversity, and it is therefore a reservoir for experimentation of practices and regulations designed to safeguard a situation that needs to be protected since it may be threatened by the professionalization of the seed trade to the detriment of the right to food. In fact, populations like those in the Andes, who can rely on a rich heritage of biological diversity, along with the corresponding agricultural cultures, and could therefore meet their own food needs, have to cope with the fact that large companies or multinational corporations control the intellectual property rights over seeds. The effect of depriving the original community of the rights related to the agrobiological resources and to the related knowledge results in a depletion of agrobiodiversity (for example, but not exclusively, due to the widespread sale restricted only to certain seeds, related to the monopolistic control of biodiversity) and in the impoverishment of farmers with limited economic resources.

b) From a constitutional point of view, this region is well suited to a comparative analysis because of its homogeneity, going beyond the geographical dimension and disregarding the social-political differences of the countries that are part of it, and considering the Bolivarian basis of their constitutionalism and the constant call for mutual integration. The roots of this tendency towards political and economic union are to be sought in the project of Simon Bolivar, shortly after independence from Spanish colonialism. His plan to link together the territories that formed the Viceroyalty of New Granada (Cundinamarca) and the Captaincy of Venezuela was the first embodiment, though of a federalist kind, of the Tratado de Alianza y Federación signed at Santa Fe de Bogota on May 28, 1811.[12] The treaty, signed by José Cortés de Madariaga, one of the leaders of the 1810 revolution for the emancipation of Venezuela, and President Jorge Tadeo Lozano, proclaimed the friendship, alliance and federation of the two nations and the integrity of their territories. The most significant element is the idea of American solidarity and openness to all States wishing to join a federal agreement on equal representation and rights.[13]

The idea of unity is expressed by Bolivar in the famous Carta de Jamaica[14] and confirmed at the Congress of Angostura in 1819, in the final part of his speech, when he expresses his hope in the common desire of the peoples and governments for the union of the sister nations New Granada and Venezuela into one great nation.

The common destiny of the Andean populations therefore has its roots in the thought of the "Liberator," who inspired the Constitution (Ley fundamental) of Angostura (1819) for the newly formed Great Colombia (which included Panama, Venezuela, Ecuador and Colombia), and also the subsequent constitutions of Peru (1823) and Bolivia (1826). The same year saw the closure of the "anfictiónico" Congress of Panama,[15] which was preceded by an intensive network of contacts initiated by Bolivar, who sent emissaries to Peru, Chile, Buenos Aires and Mexico with a mission to negotiate and sign treaties of "unión, liga y Confederación perpetual." These agreements represented the cornerstone of the Congress of Panama, convened on December 7, 1824, by the governments of Colombia, Mexico, the United Provinces of Rio de la Plata, Chile and, months later, Central

---

[12] Barbas (1985), pp. 75–78.

[13] Acuäna (1983), p. 13. http://www.bdigital.unal.edu.co. Accessed 14 Sept 2014.

[14] Bolivar (2000). See also Salcedo (1982), pp. 154–193.

[15] See De la Reza (2010).

America, with plans to establish a confederation capable of Hispanic–American dialogue at the international level.

The legacy of these experiences, which bind the Bolivarian countries, can be found in the Acuerdo de Integracion Subregional Andean (Acuerdo de Cartagena), signed in 1969 between Bolivia, Colombia, Chile, Ecuador and Peru (Venezuela, which had joined in 1973, left in 2006, and Chile in 1976), which gave birth to the Andean Community of Nations (CAN). As confirmation of the homogeneity of the region, it should be noted that in the last decade, Venezuela, Bolivia and Ecuador have been engulfed by a wave of political and constitutional renewal that has led these countries to try out an "inclusive" economic model, characterized by strong state control, the protection of individual and social rights as an alternative to the neoliberal approach, challenged as exclusionary also because of the particularly aggressive mode employed by Latin American capitalism.[16]

2.1 Regulations at Regional Level

At the regional level, we find some sources of *hard law*, relating to the management of the relationship between agrobiodiversity and intellectual property rights.

The most important is Decision 391, adopted by the Andean Community of Nations, on February 28, 1998, in accordance with the content of Article 15, paragraph 2, of the Convention on Biological Diversity (CBD), on the common regulation of access to genetic resources. Andean countries, by nature multiethnic and culturally pluralistic, recognize the value in the Decision of the ancestral knowledge of indigenous peoples in farming cultures (i.e., with reference to genetic resources, biodiversity and related know-how), pledging to protect them "in situ" and "ex situ." The Convention, in fact, committed the States to deal with the analysis of aspects related mainly to the systems of intellectual property, including the issue of the knowledge, innovations and practices of traditional indigenous techniques (Art. 8, paragraph j) and the customary use of resources through sustainable traditional cultural practices (Article 10, paragraph c).

Article 7 of Decision 391 states that the Member States, in accordance with the Decision and with the complementary domestic legislation, recognize and enhance the rights and decision making of the indigenous communities, concerning knowledge, innovation and practices traditionally associated with genetic resources and their derivatives.

It is important to note that the CBD and Decision 391, while enshrining the principle of national sovereignty over genetic resources and derived products (Art. 3 CBD, Art. 5 Decision 391), do not prevent countries with high technological development from accessing such resources but govern the terms and establish the goals.

[16] See Valadés (2003), pp. 471 et seqq.

It is clear that the practical implementation of the Convention and the Decision will depend on the development of complementary methods of compromise with the receiving countries, in order to ensure respect of the principle of fair and equitable participation in the benefits deriving from the use of genetic resources, which also requires the technological transfer of these to the countries of origin.

At the level of regional *soft law*, the issue of biodiversity, linked to that of the indigenous peoples and their knowledge, had been addressed even before the Rio Summit of 1992, with reference to the legal basis of the Amazon Cooperation Treaty (TCA). In the declaration of the second meeting of the Presidents of Amazon countries (10–11 February 1992), the Presidents met to examine the themes of the United Nations Conference on Environment and Development and adopted a declaration pledging to adopt some tools to strengthen indigenous peoples' land rights. In the Joint Declaration of Amazon Countries in view of the United Nations Conference on Environment and Development, the countries affirmed (Chapter II, paragraph 8) the need to exploit and protect the traditional methods and knowledge of indigenous peoples and local communities, their participation in the economic and commercial benefits of exploitation of biological diversity, as a necessary tool to ensure economic and social development. The same document, in Chapter IX, paragraph 1, states that it is necessary to recognize the value of local knowledge and to create mechanisms to protect traditional knowledge and compensate for the appropriation and commercial use of this knowledge. For this reason, it is necessary to ensure the right conditions for its development within the communities of origin. Finally, in the Draft Inter-American Declaration on the Rights of Indigenous Peoples, there are several references to the environment from the point of view of indigenous cultures (Preamble, paragraph 3), the right to restitution of property confiscated or their compensation, the right of indigenous peoples to attain official recognition of ownership, control and protection of intellectual property rights over their cultural heritage, which (in agreement with paragraph 2 of the same article) includes genetic resources, seeds and knowledge of plant and animal life.

2.2 Constitutional Regulations

Although it has not abandoned the original Iberian-Roman-pre-Columbian matrix, the Venezuelan Constitution (1999) and, especially, the Bolivian (2007) and the Ecuadorean ones (2008) represent an innovative reality for what has been defined as their "bio-centric and ecological shift," witnessed by the central role accorded to the rights of nature.[17] The preamble of the Constitution of Ecuador of 2008 celebrates Nature, the "Pacha Mama," of which humans are a part and which is vital to their

[17] Fundación Pachamama, Recogniting Rights for Nature in the Ecuadorian Constitution, p 3. http://www.therightsofnature.org/wp-content/uploads/pdfs/Recogniting-Rights-for-Nature-in-the-Ecuadorian-Constitution-Fundacion-Pachamama.pdf.

existence. The hallmark of the new Ecuadorian citizenship is a harmonious relationship with nature, an essential condition to achieve *buen vivir*,[18] i.e. *sumak kawsay* (which is very different from the "well-being" or "welfare" formula of utilitarian origin that is the ultimate goal of European social policies).[19] Nature becomes the subject of rights, the first of which is the respect for its existence, the maintenance and regeneration of its cycles of life, its structure, its functions and its evolutionary processes. The Constitution also preempts the immediate objection that Western constitutionalism, traditionally tied to a subjectivist approach of public rights, might make about the impossibility of Nature to demand respect for its rights. Article 71 of the Constitution of Ecuador recognizes that everyone, individuals, communities, populations, nationalities, is entitled to denounce infringements of the rights of nature to the public authorities. It might be objected that this action is the equivalent of the collective actions recognized by European jurisprudence, but there is a substantial difference: what is demanded is respect for the rights of others (that of nature to be safeguarded) and not their own. Man as citizen becomes a necessary tool to vindicate the rights of the Earth. And this is the real breech with the Western legal tradition, the element that overturns and almost cancels the categories of subjective rights and legitimate interests underlying the distinction between those who are authorized to act judicially and those who are not.

All this leads the doctrine to assert that Andean constitutionalism makes a leap, from environmentalism to a deeper level of ecological thinking.[20] The space accorded to the so-called new rights, among which there is the health of the environment, in the European context is rooted mostly in the evolutionary, adaptive interpretation of constitutional clauses by the courts, while here it is the text of the Constitution that recognizes the dignity of the new demands. There is a reversal of the hermeneutic horizon: the fulcrum of the Constitution ceases to be the individual or, at best, the "person" and becomes the cosmos, nature, of which man is only a part. It is a vision dear to chthonic legal traditions, which finds recognition in the Andean's "nuevo constitucionalismo."

New constitutions grant significant protection to the right to food, not separated from the preservation of traditional food cultures. On that basis, there have been adopted (or there are in the pipeline) some legal instruments relating to food sovereignty,[21] agrobiodiversity and access to land. There are some clauses in the constitution and infraconstitutional texts that propose a balance of the interests and

---

[18] See Acosta and Martínez (2005).

[19] See Zaffaroni (2012), pp. 422–434; Bagni (2013). See also Weber (2011), pp. 13 et seqq; Lanni (2011), p. 80, note 146.

[20] See Carducci (2012), pp. 323–324.

[21] The term "food sovereignty" was introduced in 1996 by the international peasant movement Via Campesina, at its meeting in Tlaxcala (Mexico), and reaffirmed in the forum parallel to the World Food Summit in Rome. It finds a definitive elaboration in the Declaration of Nyéléni 2007 during the International Forum on Food Sovereignty in Mali. See Corrado (2010), pp. 23–26; Patel (2009), pp. 663–706.

rights at stake that is quite different from the classical approach: for this reason they raise some eyebrows. However, the doctrine of comparative law is freeing itself of its Eurocentric skepticism, taking an interest in the experiences of these countries, for too long relegated to the category of states undergoing modernization, or in transition to democracy. The presumption of Western constitutionalism and its liberal matrix, of being the natural cradle of law and democracy and of the protection of rights and freedoms is gradually giving the way to an awareness of its limitations: excessive individualism and unbridled capitalism.

The other fundamental principle used by the Andean constitutions and by the rules of application is the proclamation of "food sovereignty,"[22] in the sense of a policy option that is intended to safeguard identity and national security. In a broad sense, food sovereignty is the right of people to have access to nutritious and culturally appropriate food[23] that is accessible and produced sustainably as well as the right to decide on their own food and agriculture systems. Clearly, this involves the protection and regulation of domestic agricultural production and domestic markets with the goal of food self-sufficiency in a regime of endogenous sustainable development.

2.2.1 Ecuador's Experience

The constitution that best embodies the ideals of the *nuevo constitucionalismo* is that of Ecuador, adopted by the Constituent Assembly on January 15, 2007.

From the very first articles, it is evident that the new ecological rights are central, inspired by the principle of "sumak kawsay": the population's right to live in a healthy and ecologically balanced environment is recognized, and the public interest in the protection of the environment, ecosystems, biodiversity and the genetic integrity of the common heritage is declared (Art. 14).

The most interesting aspect, which can simply be mentioned here because it only marginally affects the theme of this essay, is the original perspective in dealing with the *rights* of the environment, where the environment is, on the one hand, the *object* of the population's claims (rights related to the environment and related to the right to health) and, on the other, the *subject* having rights (however indirectly claimed).

On the first point, the environmental rights are closely related to "buen vivir" (the right to enjoy a healthy environment), which in turn is an instrumental practice to exercise the right to health (Art. 32), which is divided, by including them, also into other rights (water, food, etc.) set forth in the Constitution. Consequently, there is the repudiation of all polluting practices (the development, production, possession, sale, import, transport, storage and use of chemical, biological or nuclear agents and highly toxic persistent organic contaminants, agrochemicals banned internationally, technologies and harmful experimental biological agents and

[22] See also Rubio (2010).

[23] See Cavazzani (2008), pp. 43–47.

genetically modified organisms harmful to human health or threatening food sovereignty or ecosystems, as well as the bringing of nuclear waste or toxic waste into the country) and the espousal of the strengthening of clean technologies and renewable energy sources (Art. 15).

On the second point, there is a whole chapter devoted to the rights of nature (Art. 71 et seq.) or "Pacha Mama," which includes the safeguarding of its existence and the maintenance and regeneration of its life cycles, structures, functions and evolutionary processes. Nature has the right to be repaired, *independent* of state obligations to compensate the individuals and communities that depend on the natural systems damaged (Art. 72)

As for the theme of this essay, namely the relationship between agrobiodiversity, the right to food and intellectual property rights over seeds, the system of checks and balances can be reconstructed on the basis of the rules already mentioned and others found in Title VI of the Constitution, which regards the regime of development.

The conservation of biodiversity and agrobiodiversity is a priority of state policy: for collective knowledge related to the agricultural heritage, of ancestral knowledge, of genetic resources containing biological diversity and agricultural biodiversity, any form of appropriation of the relevant knowledge, innovations and practices is prohibited (Art. 57, paragraph 12). The protection of biodiversity goes so far as to limit the activities that may lead to the extinction of species, the destruction of ecosystems or the permanent alteration of the natural cycles; it prohibits the introduction of organisms and organic and inorganic material that can alter the final national genetic heritage (Art. 72). Having adopted a model of sustainable development,[24] respectful of cultural diversity, biodiversity and regenerative capacity of natural ecosystems, so as to guarantee the satisfaction of the needs of present and future generations, the Ecuadorian State explicitly states the concept of *sovereignty of biodiversity* (Article 400). The administration of biodiversity and its management, with particular attention to agriculture, forestry and genetics, is the duty of the State, which must adopt a responsible attitude towards future generations and avoid, except in cases of national interest, the cultivation and use of transgenic seeds.

With regard to the right to food, Article 13 assures people and communities the right to safe and permanent access to healthy, sufficient and nutritious food, which should locally be produced in accordance with their different identities and cultural traditions. But it is in the enforcing of the principle of food sovereignty (Art. 282) that the political weighting of the balance between the needs represented by the rights at stake (the right to food and the protection of agricultural biodiversity, on the one hand, and intellectual property rights, on the other) is settled, clearly in favor of the first of the two terms. By closely linking the protection of agrobiodiversity and the right to food (implicit in the commitment to ensure that individuals, communities, peoples and nations can permanently achieve healthy

[24] See Carrión and Herrera (2012).

and culturally appropriate food self-sufficiency), the State accepts the responsibility of implementing the domestic agricultural production by encouraging small producers. To this end, it undertakes to *promote the conservation and recovery of ancient knowledge and agrobiodiversity related to this, as well as the use, conservation and the free exchange of seeds*, preventing monopolistic practices and all kinds of speculation with food. Finally, Art. 402 prohibits the granting of rights, *including intellectual property rights*, to derived products, or synthesized ones, obtained from collective knowledge linked to national biodiversity.

At a subconstitutional level, let us point out the Ley Orgánica de la Soberanía Alimentaria (LORSA), signed on May 5, 2009, which applies and further explicitly states the constitutional principle of food sovereignty. It is significant that the law treats them together, evidently considering closely related the aspects of the quality of the food supply, the right to health (Articles 24–30) and those relating to the protection of agricultural biodiversity (Arts. 7 and 8), also through the principle of free movement of native seeds. The germplasm, seeds, native plants and ancestral knowledge associated with them constitute the assets of the Ecuadorian people and consequently may not be subject to appropriation in the form of patents or other modes of intellectual property, in accordance with Art. 402 of the Constitution.

2.2.2 Bolivia's Experience

The Constitution of Bolivia, adopted on October 2008 and subsequently ratified by referendum on 25 January 2009, manifested from the very first articles its ecological vocation and respect for the holistic chthonic cultures,[25] having some ethical and moral principles (Article 8 I.), including that of *ñandereko* (harmonious life) and *suma qamaña* (well living). Harmony with the environment is then underlined in other articles (Articles 33 and 34) that recognize the right of individuals to a healthy, protected and balanced environment, which is projected to future generations, and the right to take legal action to defend the environment. For this purpose, it envisages the sustainable use of natural resources (Art. 380).

With regard to the protection of biodiversity and agrobiodiversity, it envisages the state's commitment to promote conservation (Art. 354) and enshrines the principle of sovereignty over agrobiodiversity, which emerges from a combined reading of Articles 381, 382 and 383, which require the state to protect all its generic resources and microorganisms found in the ecosystems of the area, as well as knowledge about their use and exploitation, establishing restrictive measures, partial or total, temporary and permanent, on extracting biodiversity resources, as well as punishing criminal cases of possession, management and illegal trafficking of biodiversity species.

Of particular significance, also in view of what has been mentioned in section 1, paragraph 2 of this essay, are the instruments indicated in the Constitution to

[25] Cfr. Vargas Lima (2012), pp. 251–267.

protect agrobiodiversity: on the one hand, Art. 381, section II, envisages a "patent system that will safeguard the existence, as well as the intellectual property of the State or local social subjects that request it"; on the other hand, Article 382 establishes the right and duty of the State to defend, recover, safeguard the defense, recovery and security and bring back to the nation the biological material from natural resources, ancestral and other knowledge originating in the territory (on this point, Art. 407 indicates as one objective of rural policies the control on entry and exit of the country's genetic resources). The first measure is in line with what was agreed in the TRIPs but not with the trends in the development of free trade envisaged in the International Treaty on seeds. The second, however, seems to want to put a stop to the loss of biogenetic resources fueled by multinational companies, which take out patents for seed in countries different from the supplier and then exploit it. This thus follows in the steps of the most recent agroecological trends in enhancing the role of local farmers, especially those belonging to indigenous populations.

Finally, the rights to food and food security are recognized as human rights, with the concomitant obligation of the State to ensure food security not only in quantitative terms but also in qualitative terms (Art. 16): the principle of security exists alongside that of food sovereignty (Art. 405 and 407) and is the goal at which rural development policies must aim.

The Constitution of President Morales of 2009 is not the first document in which we can find the political tendency to balance interests in favor of the conservation of agrobiodiversity in Bolivia. Previously, a rural, agricultural and forestry revolution was triggered, which was expressed in the Plan of the Ministerio de Desarrollo Rural, Agropecuario y Forestall, presented on November 2007 and oriented to an environmentally sustainable rural development, pursuing the principle of food sovereignty.[26] It was based on a communitarian, associative and individual system of agriculture and forestry production, more efficient in terms of the use of natural resources. The starting point of the agrarian reform of Morales'government was the rationalization of the land distribution, through the approval of seven decrees, nicknamed "siete surcos" (seven furrows), and the Ley de Reconducción Communal de la Reforma Agraria (Ley 3545/2006). It legalized the land ownership from 5,000 up to 10,000 ha, redistributing millions of hectares through the "Instituto Nacional de Reforma Agraria (INRA)" to effectively rationalize the rural situation. The agrarian revolution also formed the basis for the new production model, which is focused on food sovereignty ("productivo Programa para la Seguridad Alimentaria, 2008"), whose beginning coincided with the dawn of the global food crisis and the consequent rise in prices, which the program aimed at stabilizing. In 2008, Morales also created the Fundo para la Reconstrucción, Seguridad Alimentaria y Apoyo productive, aimed at addressing situations of increasing prices and lack of food security. The three pillars of food sovereignty policies of the government were the redistribution of land, which favored the indigenous

[26] See Ormachea (2008).

peoples of the Amazon; the public support to increase basic food production by small producers and also for medium- and large-scale production for the domestic market; finally, the objective of overcoming the agricultural export model of development that was causing social inequality and a development that eroded natural resources. To achieve this, the government introduced a policy designed to dismantle soybean monocropping, thus destabilizing the big farms that had an interest and converting a system based on monoculture into one involving a plurality of farmers producing different foods, primarily in favor of the internal market and then in favor of exporting.

2.2.3 Venezuela's Experience

The Venezuelan Constitution is less recent than those just dealt with, and its content reflects this temporal *hiatus*. Nevertheless, it contains some references to the protection of traditional knowledge of indigenous peoples that put it in a leading position in the Andean regional space. In fact, with reference to the situation of indigenous peoples, it recognizes the right to preserve and promote the traditional productive activities and economic practices, based on relations of exchange, reciprocity, solidarity (Article 123). But next, Article 124 effectively provides the maximum protection to traditional knowledge: it safeguards the intellectual property of the *collective* knowledge of indigenous peoples and the social vocation of the benefits arising from activities related to the exploitation of genetic resources. Over these and over traditional knowledge, it is prohibited to take out patents. Consistent with the approach of intergenerational environmental (Article 127) and biodiversity protection, it envisages a ban on patenting the genetic code of living beings.

This framework of Constitutional discipline saw the enactment of the Ley de Tierras y Desarrollo Agrícola authorized by the "Ley Habilitante" on November 2001. Article 305 envisages that the State should promote sustainable agriculture as a strategic foundation of rural development in order to guarantee *food security* for the population, achieved by developing and prioritizing domestic food production. To this end, in 2000 the Plan Nacional de Desarrollo Agrícola y de la Alimentación was introduced. The concept of food security accepted by the law of agricultural development is actually still unbalanced in favor of "quantity" because it is interpreted, according to Art. 305, as granting a certain and sufficient availability of food. Subsequently, the Ley Orgánica de Seguridad y Soberanía Agroalimentaria, passed in 2008, determined that food sovereignty and security must be guaranteed in accordance with the features, principles and purposes of constitutional and legal security and of overall defense of the nation.[27] Therefore, the principle of food security is absorbed into the broader area of security and

[27] Parker (2008), pp. 121–143.

national defense, so much so that exemplary punishments are laid down for the destruction and deterioration of strategic resources.

3 Conclusions

This essay raises the question of *whether* there is a connection and *what* that connection is between agrobiodiversity, intellectual property rights and the right to food. The implicit corollary of this question is *whether* there is a point of equilibrium and *what* that point of equilibrium is between the interests underlying the assets at stake and, finally, *whether* there exist constitutional experiences effectively achieving an optimal balance.

On the one hand, there is the right to food, whose connection with the protection of agrobiodiversity is indisputable, given the emergence of the qualitative component of food security, which is constructed by the most recent international conventions as being related to the protection and implementation of agrobiodiversity policies. On the other hand, there is the intellectual property right, also protected at the international level, which extends to the patenting of seeds, conceived as a human creation, being the result of processes of genetic modification. The patenting of seeds entails their exclusive ownership and the exploitation of their economic potential. At the same time, it places a limit on the free exchange of seeds, a very common practice among indigenous farmers, who are the guardians of ancestral knowledge related to agriculture and whose practices contribute to enhancing the agrobiodiversity of the areas where they live.

The Andean region is a cradle of biodiversity and place of numerous indigenous communities. In recent years, in the context of regional agreements aimed at controlling the common system of genetic resources, it has experienced a constitutional evolution directed towards some development principles related to the protection of nature and introducing some provisions protecting biodiversity, sustainable development, indigenous communities and their traditional knowledge related to agriculture. A cornerstone of the *nuevo constitucionalismo*[28] is "food sovereignty," alongside the principle of "sovereignty over biodiversity": both of these justify the policies to safeguard the genetic resources of the country, the ancestral knowledge of indigenous communities and their relationship with the territories in which they are located and work. In this perspective, the (individual) guarantee of intellectual property rights recedes with respect to the exploitation of (collective) indigenous community rights: the patenting of seeds is instrumental to the protection of the latter, and the regime of free trade is recognized as a good praxis, serving to maintain agrobiodiversity.

[28] Andean "nuevo constitucionalismo" is different from the "neoconstitucionalismo," which has affected the post-Second War constitutions, resulting in some changes in constitutional theory and the functioning of the rule of law. On this topic, see Carbonnel (2003).

References

Acosta A, Martínez E (eds) (2005) El buen vivir. Una vía para el desarollo. Flacso, Ildil, Quito
Acuäna ER (1983) Bolívar: pensamiento constitucional. Universidad Externa de Colombia, Bogotà
Bagni S (2013) Dallo Stato del *benestare* allo Stato del *buen vivir*. Innovazione e tradizione nel costituzionalismo latino-americano. Filodiritto, Bologna
Barbas DU (1985) Las Constituciones de Colombia, vol I. Ediciones Cultura Hispánica. Instituto de Cooperación Iberoamericana, Madrid
Bolivar S (2000) Carta de Jamaica. Fundación Editorial Epigrafe, Bogotá D.C
Carbonnel M (ed) (2003) Neoconstitucionalismo(s). Trotta, Madrid
Carducci M (2012) Epistemologia del Sud e costituzionalismo dell'alterità. Diritto pubblico comparato ed europeo 14:323–324
Carrión D, Herrera S (2012) Ecuador rural del siglo XXI. Soberanía alimentaria, inversión publica y política agraria. Instituto de Estudios Ecuatorianos, Quito
Cavazzani A (2008) Tra sicurezza e sovranità alimentare. Sociologia Urbana e Rurale 87:43–47
Corrado A (2010) Sovranità alimentare: la proposta alternative della Via Campesina. Agriregionieuropa 6(22):23–26
Dalgaard T, Hutchings NJ, Porter JR (2003) Agroecology, scaling and interdisciplinarity. Agric Ecosyst Environ 100:39–51
De la Reza G (2010) Documentos sobre el Congreso anfictiónico de Panamá. Fundación Biblioteca Ayachcho y Banco Central de. Venezuela, Caracas
De Schutter O (2009) Seed policies and right to food. enhancing agrobiodiversity, encouraging innovation. Background document to the report (A/64/170) presented at the 64th session of the UN General Assembly
FAO (1999) Agricultural biodiversity, multifunctional character of agriculture and land conference, background paper 1. Maastricht, Netherlands
Frankel OH, Brown AHD, Burdon JJ (1995) The conservation of plant biodiversity. Cambridge University Press, Cambridge
Glenn HP (2010) Legal traditions of the world – sustainable diversity in law. Oxford University Press, Oxford-New York
Gliessman SR, Rosemeyer M (eds) (2010) The conversion to sustainable agriculture – principles, processes and practices. CRC Press, Boca Raton – London – New York
Goldsmith E (1992) The way: an ecological world-view. Rider, London
Lanni S (2011) Sistema giuridico latinoamericano e diritti dei popoli indigeni. In: Lanni S (ed) I diritti dei popoli indigeni in America Latina. ESI, Napoli
Ormachea E (2008) Soberania y seguridad alimentaria en Bolivia: politicas y estado de la situación. Centro de Estudios para el desarollo laboral y agrícola de. Bolivia, La Paz
Parker D (2008) Chávez y la búsqueda de una seguridad y soberania alimentarias. Revista venezolana de Economia y Ciencias Sociales 14(3):121–143
Patel R (2009) What does food sovereignty look like? J Peasant Stud 36(3):663–706
Rubio B (2010) El nuevo modelo de soberanía alimentaria en América latina. VIII Congreso de la Asociación latinoamericana de Sociología Rural, Porto del Galinhas
Salcedo JA (1982) Bolívar. Un continente e un destino. Ed. C.E.B.B, Caracas
Shiva V (2001) Protect or Plunder? Understanding Intellectual Property Rights. Zed Books, London
Shiva V (2005) Earth democracy. South End Press, Cambridge
United Nation Conference on Trade and Development (1996) The TRIPS agreement and developing countries. United Nation Publications, New York – Geneva
Valadés D (2003) El nuevo constitucionalismo iberoamericano. In: Fernandes Segado F (ed) La Constitución de 1978 y el constitucionalismo iberoamericano. CEC, Madrid

Vargas Lima AE (2012) El derecho al medio ambiente en la Nueva Constitución Política del Estado Plurinacional de Bolivia. Anuario de Derecho Constitucional Latinoamericano 18:251–267

Weber G (2011) Introducción. In: Weber G (ed) Debates sobre cooperación y modelos de desarrollo. Perspectivas desde la sociedad civil en el Ecuador. Centro de Investigaciones CIUDAD, Quito

Zaffaroni ER (2012) Pachamama, Sumak Kawsay y Constituciones. Diritto pubblico comparato ed europeo 14:422–434

The Constitutional Dimension of Traditional Rural Skills: Protection and Promotion

A. Denuzzo

Abstract The identity of a territory is expressed in terms of local knowledge in the formula "traditional rural skills," which contain the condensed and concretely tangible local biography of a community. The law is called on to preserve such knowledge from the risk of disappearing and to promote it as an instrument of responsible cultural tourism and new employment opportunities. We need to identify the best instruments to protect this immaterial patrimony since the norms typical of the Code of Cultural Heritage and Landscape that generally imply a *res* cannot be used. In order to promote these time-honored rural skills and intangible cultural goods in general, we need to involve all levels of government and organized actors in civil society.

Keywords Promotion • Protection • Subsidiarity • Territory • Traditional rural skills

1 Intangible Cultural Goods: The State of the Art

At the entry to the Museum of Kabul, a simple and poignant plaque reads: "A Nation stays alive, when its Culture stays alive." This extraordinary declaration had withstood the fiery inferno of Afghanistan and brings to mind Winston Churchill's response during World War II when he was asked to block public subsidies to the arts in order to concentrate revenue on the army: "Then what are we fighting for?"[1]

Yet for at least 30 years, the *leitmotiv* of Italian cultural policy has been the scarcity of available financial resources. Yet the lack of funds to safeguard cultural goods is not a natural catastrophe; it is a political choice – a regressive and irresponsible choice. Literally, cultural heritage is our collective inheritance, the legacy of the generations that went before us. It is what defines us as a family and as a community.

---

[1] Montanari (2014), p. 42.

A. Denuzzo (✉)
Faculty of Educational, Political and Social Studies, University of Salento, Lecce, Italy
e-mail: antonellodenuzzo@gmail.com

M. Monteduro et al. (eds.), *Law and Agroecology*,
DOI 10.1007/978-3-662-46617-9_25

Moving from the well-known definition of the Commissione Franceschini (1967) that defined a cultural good as something that testified to the value of civilization,[2] Massimo Severo Giannini has shown how such testimony constitutes an intangible value that is inherent in a material entity yet distinct from it.[3] While the *res* itself can have no material value and hence is not definable as a heritage asset, it is the cultural immanence intrinsic to a ruined building that makes it a cultural good.

Giannini's thesis tends to stress that a cultural good is the product of an intangible value (for example, the technique used to build Puglia's *trulli*) that overcomes the *res*, which may instead be of insignificant value. The superiority of the cultural component is indisputable for so-called identity goods or goods whose cultural content depends on the fact that they have been the site of an important historical event.[4]

Yet even these hypotheses have a basis represented by a heritage good. In any case, a *res* becomes a cultural good by virtue of the value of its inherent testimony or for its characteristics (*qua* document, artistic work, etc.). These characteristics express its historical, artistic or ethnoanthropological worth. In conclusion, Giannini's argument highlights the *quid* of "intangible" present in any cultural good.

Marco Dugato has defined an intangible cultural good as a common sentiment generated by a network of intangible goods identifiable in a city, territory, ancient *borgo* or artistic expression that evoke such places. Furthermore, this perspective is based on a substratum of material goods and also intersects with cultural goods and goods that are not cultural, as well as with memories and sensations from our own experience.[5]

The underlying thesis of this study is that intangible goods originate in traditional activities typical of a particular *locus*. These can be the products of an artisan or craft tradition dating back in time, the testimony of distant and deeply felt practices, from the feast of the patron saint to the proverb, from a commemorative horse race to a religious procession. In particular, the identity of a territory is expressed by its local knowledge contained in the expression "traditional rural skills." The spiritual biography of a community is condensed and concretely tangible in such practices. It is as if the lives, the aspirations and the collective and individual histories of our forebears are at least partially enshrined in the artifacts and traditions that we guard so jealously today.

The law is called on to preserve such knowledge from disappearance and to promote mass instruments to promote responsible cultural tourism (Art. 117(3),

---

[2] Luther (1999), p. 2.

[3] Giannini (1976), pp. 3 et seqq.

[4] Art. 10, Para. 3, Lett. d), of the Code. On the difference between cultural goods "of historical reference" or "of indirect historical interest"—according to Cantucci (1953), p. 111—and goods "di testimonianza identitaria," see Morbidelli (2011), pp. 125–134, exp. at p. 27.

[5] Dugato (2014), pp. 139–146.

Italian Constitution). This should be done with a view to developing culture (Art. 9) and to create new work opportunities. With the rediscovery of traditional rural skills, we can propose a model of civil economy able to create widespread income without inhibiting the key constitutional function of cultural heritage, that is, generating citizenship values through knowledge. Experience of the past may be a vital antidote in times such as ours, nailed as they are to the immanent horizons of breaking news.

The aim of the UNESCO Conventions of Paris (2003, 2005) was to safeguard intangible cultural heritage,[6] in the shape of oral traditions, languages, artisan skills, the performing arts, social practices and rituals (songs, folklore, games), food, knowledge of nature and the universe. The Conventions promote displays of cultural identity ontologically without a material or bodily substratum, where the ideal extrinsic value is transcendent.

More recently, on 14 January 2014, European Parliament Resolution 2013/2098 (INI) to promote competitiveness in rural economies, entitled "On regional branding: towards best practice in rural economies," linked the theme of territorial quality branding to rural development understood in the broad sense, that is, which also takes material and immaterial rural cultural heritage into account. In fact, the existence of intangible cultural goods has long been recognized in Italian legal theory.[7] Yet, to date, Italian law on cultural goods treats them as material goods.[8] Art. 7 *bis* of the Italian Code of Cultural Heritage and Landscape stipulates that "the expressions of collective cultural identity" outlined by the UNESCO Conventions "are adaptable to the provisions of the current Code if they are represented by material testimonies," thus confirming the symbiotic relationship between the cultural good and *res*. Moreover, the Italian Constitutional Court has observed that "culture does not assume an autonomous importance, separate and distinct from goods of historical, artistic, archaeological and ethnographic interest, but co-penetrates things that constitute its material form and, consequently, which cannot be protected separately from the good itself" (Judgement 118/1990).

[6] Art. 2 of the Convention to safeguard intangible cultural patrimony of 17 October 2003 came into force on 20 April 2006 and was ratified by Italy by virtue of Law 167/2007. It includes as cultural heritage "the practices, representations, expressions, knowledge, skills that communities, groups and, in some cases individuals recognize as part of their cultural heritage," then specified in terms of oral traditions and expressions, including language performing arts; social practices, rituals and festive events; knowledge and practices concerning nature and the universe; traditional craftsmanship. Law 19/2007, of ratification and execution of the Convention on the protection and promotion of the diversity of cultural expression signed in Paris on 20 October 2005, sets out to protect expressions cultural in danger of disappearing.

[7] Cassese (1976), pp. 160–188, exp. at p. 177; Ainis and Fiorillo (2003), p. 106.

[8] Severini (2014), pp. 119–128.

2 Intangible Cultural Goods and Territory

In the judgment of the constitutionality of Regional Law (Lazio) 31/2001, on the "Protection and promotion of historical locations," the Italian Constitutional Court suggested an open notion of cultural goods, whose "particular historical or cultural value can be recognized by a regional or local community, without this involving their being designated as cultural goods as per Decree 490/1999 (now repealed) and a consequent special conformation of their legal regime."[9]

Continuing this orientation in a more recent declaration, the Court asserted that "Art. 40 of the Veneto Regional Law 11/2004 does not establish new criteria to identify cultural goods in order to achieve the goals of its own regime in the sphere of regulations. On the contrary it stipulates that the regulation of the government of the territory – and hence its particular nature – take into account not only cultural goods identified according to State law, but also other, provided that they constitute part of a territory with its own form and its own history."[10] Indeed, there is no lack of regional laws with their own lists of intangible goods. Lombardy Regional Law 27/2008 planned an Intangible Inheritance Register (*Registro di eredità intangibile*) in four sections: the *Libro dei Saperi*, *Libro delle Celebrazioni*, *Libro delle Espressioni* and *Libro dei Tesori umani viventi*. Decree 77/2005 of the Regional authority for cultural goods and the Region Sicilia took similar steps to protect Sicilian identity.

Normally, the recognition of this type comes with forms of financing or measures of organizational stability (events) or "memorization" (cataloguing, collections, inventories). In short, all the forms of protection and promotion cited by the repealed Art. 153, Para. 3 of Decree 112/1998, summarized as "support intervention for cultural activity by means of financial aid, the arrangement of structures or management," the "organization of initiatives to increase knowledge of cultural activities and to promote their diffusion" or the "development of new cultural and artistic expressions and of less known expressions, also with the use of developing technologies."

The work of collecting and documenting intangible goods and defining their content (for example, transcribing the words and music of a popular chant) is not reserved to public bodies but has traditionally been carried out by cultural institutions or single individuals. We only need to think of Italo Calvino's collection of *Fiabe italiane* or manuals of good manners starting with Monsignor Giovanni Della Casa's *Galateo*. From another perspective, through the collections of trade customs, the Chambers of Commerce have canonized those consolidated mercantile, maritime, agricultural and forestry traditions, which also testify to civilization.[11]

---

[9] Judgement 94/2003 (author's translation). Carpentieri (2003), pp. 1017 et seqq.

[10] Not by chance is the Court referring to its earlier Judgement 94/2003.

[11] For example, the distances for planting trees set by local usage (see Arts. 892 and 893 of the Italian Civil Code) are inherited from a plantation culture that, among other things, helped shape the form of the landscape.

The range of intangible cultural goods is such that their identification, as well as measures to promote them, compete "naturally" with the territorial community in question and constitute values and identity of that same community. In the Judgement 194/2013, the Constitutional Court declared this with reference to material goods, but the reasoning can also be extended to intangible goods. In particular, it can apply to traditional rural skills: "the fact that a specific thing is not classified by the State as of artistic, historical, archeological or ethno-anthropological interest and hence not considered as a cultural good, does not mean that it is excluded and it can, instead, have, even residually, some cultural interest for a particular territorial community: where this interest is hypothetically anchored in an inalienable heritage identity, ideals and experience and even symbols, of that single and specific community."

3 Legal Instruments for the Protection and Promotion of Traditional Rural Skills

In the wake of the Paris Conventions on the Protection and Promotion of the Diversity of Cultural Expressions (UNESCO 2005), Italy included Sardinian pastoral songs (*canto a tenore*), the Sicilian puppet theater (*Opera dei pupi*) and the Mediterranean diet to its list of intangible cultural goods.[12]

In order to avoid any sort of panculturalism,[13] with the attendant risk of depreciation related to the difficulty of identifying an open-ended and unitary legal notion of a cultural good, we need to ask what criteria to use to identify what counts as part of such a vast and heterogeneous category and also which are the best instruments to safeguard these intangible cultural goods. It is clear that we cannot extend to intangible goods the norms typical of the Code that are based on the ban on making changes or the control of the movement of cultural goods, which generally presuppose a *res* (e.g., coercive guardianship, commodatum, deposit, access, use). Neither can we hypothesize the attribution of economic or exclusive rights, given that there are no owners claiming such intangible goods, if not in very rare cases.

Protective measures must certainly include the cataloguing and documentation in special records. Naturally, intangible goods require "dynamic" protection in harmony with their continuing state of transition, which is the essence of the phenomenon to be preserved, hence, the care taken to recognize traditional phenomena that have fallen into disuse and the continual updating of any related documentation.

[12] UNESCO distinguishes between lists dealing with the cultural heritage of humanity, prepared by an Intergovernmental Committee, and national lists of intangible cultural heritage in urgent need of protection by the national States; see Bartolini (2013), pp. 110–111.

[13] The warning is made by Severini (2000), pp. 12 et seqq.

The State and the Regions can introduce measures that provide protection for traditional rural skills. The cultural "recognition" of a manufactured article is a sort of controlled denomination of origin, able to stimulate public interest. After having surveyed the active subjects able to carry out traditional workmanship, we need to set down operational rules for those working in the sector. On the other hand, events focusing on the display of traditional productive techniques can use protective instruments to safeguard their image; the entire expressive context of such events, represented by custom, rituality and by a brand name itself, can be protected insofar as these are expressions of the personality of the good in question.[14]

The subject competent to exercise this form of protection is the Ministry for Cultural Assets and Activities. Before inserting an agri-food product in the sphere of intangible cultural goods, the Ministry must not stop at its organoleptic characteristics (if anything, the competence of the Ministry of Agriculture) but should consider the identification and stratification of the history of a population and its territory inherent in that product. The underlying problem is that, to date, the Ministry treats territory where cultural goods are found as an opportunity for bureaucratic decentralization and sets up peripheral bodies such as the Superintendences (*Soprintendenze*), in the form of a separate apparatus according to the typology of the goods in question: archaeological, historical-artistic, architectural landscape or archival. This organizational fact has legitimated a rather reductive interpretation of the "Republic" as a subject called on to guarantee protection as per Art. 9 of the Constitution. The expression, which takes in all levels of government covered by the law, has not prevented the territorial autonomies—precisely because they lack the competences allocated to the central State—from being excluded from the exercise of the functions of protection and relegated to a merely collaborative function. In any case, the work of identifying intangible goods cannot be carried out by the Ministry alone but must necessarily take place through forms of center-periphery linkages because it is precisely in the local context that we find those phenomena that are the living expression of tradition and cultural identity.

The range of intangible cultural goods and their contiguity with territorial communities mean that such goods[15] lend themselves to regulation or at least recognition, particularly at the local level. Traditional rural skills allow us to configure forms of *traditio* of the memory and values generated by such skills and forms of promotion of knowledge through legal sources at various levels, from State legislation to regional legislation, from provisions by local bodies to deliberations in cultural institutions and customs canonized in the documentation of public and private actors, whose legal importance is not limited to particular measures or to the attribution of a marketable "credit."

---

[14] Comporti (1997), pp. 540 et seqq.

[15] These are defined by some as "light" cultural goods or separate from cultural goods *sensu stricto* and according to law *ex* Art. 117, Para. 2, lett. s), of the Constitution. Vitale (2010), pp. 171 et seqq., exp. at p. 176.

Thanks to what can be defined as the positive externalities of cultural heritage, the survival of intangible goods is primarily guaranteed by promotional activities.[16] For example, the "profit" created by restoration activity and giving the public access to material cultural goods also helps safeguard the artisan tradition. Its promotion is the history of a functional sphere open to the intervention of all levels of government, public and private (profit and nonprofit), with their different aims and different forms, depending on the cultural goods and the institutional and territorial contexts involved. In contrast to protection, promotion consists of a series of coordinated positive actions and is not an activity automatically assigned to the exclusive competence of one or other institutional subject involved.

In this perspective, the participation of organized actors in civil society is essential. Currently, the instruments and the production of traditional rural skills are safeguarded in small folklore, ethnological or anthropological museums, with some incursions into history. The number of these museums has grown exponentially from the late postwar period to the present day,[17] thanks to the intervention of voluntary cultural initiatives. Today, the promotion of the knowledge of traditional rural skills needs to be advocated beyond formal museums because it is precisely this knowledge, and its growth, that allows us to measure the incessant barrage of initiatives of the presumed promotion of heritage. In conclusion, the theme of traditional rural skills engages two basic elements: the anthropological-territorial datum and the importance of the principle of subsidiarity in its double, vertical and horizontal, sense.

On the one hand, it is the task of the national legislator—in their protective function—to identify intangible goods legally. On the other hand, it is in the territory that generates them, that conserves their memory and where they appear and take form that such goods and memory can be preserved with the support of the community of reference (association, third sector, guardians of tradition). In this horizontal sense, subsidiarity can help the legislator not only in the initial phase of identifying what represents the origin or tradition of a population, and hence that counts as a cultural good, but also in defining the objectives of the promotion of intangible heritage, including forms of center-periphery linkages. We need to protect our immaterial heritage. If we do not defend its ability to generate knowledge, it will be lost, even if it is perfectly conserved. Only with this particular awareness can we promote the objectives of the Paris Convention of 20 October 2005, that is, diversity of cultural expressions and their transmission from one generation to the next.

[16] Ainis and Fiorillo (2003), p. 106.

[17] Baldin (2012), p. 13.

References

Ainis M, Fiorillo M (2003) L'ordinamento della cultura. Giuffrè, Milano

Baldin L (2012) Musei, volontariato e sussidiarietà orizzontale. Museo in-forma. Rivista Quadrimestrale della Provincia di Ravenna. Notiziario del sistema museale provinciale 16 (44):13

Bartolini A (2013) Beni culturali. In: Enciclopedia del diritto. Annali, vol VI. Giuffrè, Milano, pp 110–111

Cantucci M (1953) La tutela giuridica delle cose di interesse artistico o storico. Cedam, Padova

Carpentieri P (2003) Tutela e valorizzazione dei beni culturali. Urbanistica e Appalti 7(9):1017 et seqq

Cassese S (1976) I beni culturali da Bottai a Spadolini. In: Cassese S (ed) L'Amministrazione dello Stato. Giuffrè, Milano, pp 160–188

Comporti M (1997) Sfruttamento abusivo dell'immagine altrui e dei segni distintivi delle persone giuridiche. AIDA 7(6):540 et seqq

Dugato M (2014) Strumenti giuridici per la valorizzazione dei beni culturali immateriali. In: Bartolini A, Brunelli D, Caforio G (eds) I beni immateriali tra regole privatistiche e pubblicistiche. Jovene, Napoli, pp 139–146

Giannini MS (1976) I beni culturali. Rivista trimestrale di diritto pubblico 26(1):3 et seqq

Luther J (1999) Principi per il diritto dei musei pubblici in Italia. Fondazione G. Agnelli, Torino

Montanari M (2014) Istruzioni per l'uso del futuro. Il patrimonio culturale e la democrazia che verrà. Minimum fax, Roma

Morbidelli G (2011) Commento all'art. 10. In: Sandulli MA (ed) Codice dei beni culturali e del paesaggio. Giuffrè, Milano, pp 125–133

Severini G (2000) La nozione di bene culturale e le tipologie di beni culturali. In: Caia G (ed) Il Testo unico sui beni culturali ambientali. Analisi sistematica e lezioni. Giuffrè, Milano, pp 12 et seqq

Severini G (2014) Immaterialità di beni culturali. In: Bartolini A, Brunelli D, Caforio G (eds) I beni immateriali tra regole privatistiche e pubblicistiche. Jovene, Napoli, pp 119–128

Vitale C (2010) La fruizione dei beni culturali tra ordinamento internazionale ed europeo. In: Casini L (ed) La globalizzazione dei beni culturali. Il Mulino, Bologna, pp 171 et seqq

On the Humanity of Land

A. de Nitto

Abstract The land, which men inhabit and cultivate nonexclusively, is the place of history: *homo* comes from *humus*. A "rural humanism" seems naturally "whole" or "integrated": it joins or includes fragments of experience according to criteria of practical intelligence. The "land question" regards, in conclusion, above all the qualities of coexistence between persons, and it expresses the abilities to cultivate that experience.

Keywords Coexistence • Community • Culture • Experience • Rural humanism

1 Life and Life

Umanità della terra [Humanity of land] is, etymologically speaking, almost a tautology: that is, a proposition in which the elements that make it up repeat themselves or refer to each other or, even, rather, in which they seem reciprocally implied, as if through a kind of reflecting back and forth of meanings.

In this sense, the expression may be the mirror image of the "terrestriality of man": one can equally say, generically, that land is "human," at least in the sense that it constitutes the natural *habitat* of men, and that men are "terrestrial" creatures, in the sense that their life is intimately tied to land (*pulvis es*: "dust thou art," Gen. 3, 19).

We are speaking—beyond, possibly, anthropocentric emphases or more or less romantic or deterministic idealizations and in a mental perspective oriented, as far as possible, toward the "multiplicity"—of a "union" of life with life:[1] on one hand, the "world of human actions" or the human world of "history" (*prágmata*, the "contents" of the *práxeis*, in the sense of "things done"), individual men and men together, reunited, generation upon generation, in the community of all times; on the other hand, the "secret life of the land," that mysterious unfolding, in certain places and times, of energies and vital thrusts and counterthrusts, of all those

[1] Capograssi (1959), pp. 269 et seqq. (original work published 1952).

A. de Nitto (✉)
Department of Legal Sciences, University of Salento, Lecce, Italy
e-mail: achille.denitto@unisalento.it

M. Monteduro et al. (eds.), *Law and Agroecology*,
DOI 10.1007/978-3-662-46617-9_26

silently living forces and species.[2] Inside the "material" (the physicality, biodiversity, etc.), inside their processes and dynamics of change,[3] but, if one may say, even more: lives, in some way, intersected, interfering, and reciprocally, in some way, belong to each other, if the criterion of belonging or appropriation made sense.[4]

"Nature" and "history," moreover, are used, also here tautologically, as huge "containers" or "reservoirs" of "life," though communicating variously, notwithstanding the long time that we have tended, in the European experience, to keep them separated in awareness (such as neglecting the ancient tradition of "natural history"): humanistic (or, *lato sensu*, literary) culture on one hand and scientific on the other.[5]

The biological experience of living beings is carried out through continual "communications": "systems" result, as a rule, structurally "open" and—as we know from classical mechanics—in continual transformation. Each internal element necessarily draws from a source, to get sap or nutrition. The "rooting" of *ambiente* [environment] (space that surrounds a thing or person and in which these move or live[6]), always necessarily specific, however great the dimensions, regards, therefore, not only stability (e.g., of a plant in the ground) but the entire metabolism, therefore the life itself.

Human "actions," on the other hand, or the "human world of experience" make use of inexhaustible decanting, of experience and of knowledge, or of successive transplanting or repeated grafting or of unforeseen percolating: despite the extraordinary originality of someone, no one does everything by oneself. Events are connected even in unknowable ways and are "rooted" within tissues of humanity that—with an abstract word generally used to express the concrete—we call, without qualification, "history": thus they exude, in every way, their own insuppressible "historicities."

I would like to continue speaking in the spirit of an explorer in the woods rather than that of a gardener: of one who, without worrying too much about keeping trees and shrubs in order, ventures forth, for passion and with the perceptiveness of which he is capable, into the undergrowth and along paths he does not know and that he only, perhaps, let's say, intuits. The effort will still produce some fruit, even if only that of the pleasure of the journey.

It goes without saying that the experience of the aspiring explorer is, like all others, variably conditioned, and it bears the weight, even when unaware, of its limits and partiality, at least as far as the available tools and the ability to handle them go. But this, in its unavoidable obviousness, might not constitute its main defect.

[2] Capograssi (1959), pp. 273 et seqq.

[3] Careri (1982).

[4] Graziani (2013), pp. 1–35, 3 et seqq. In analogous terms, by the same author: Graziani (2011), pp. 11 et seqq.; Graziani (2006a), pp. 172 et seqq.; Graziani (2006b), pp. 71 et seqq.; For some affinities, Demetrio (2013).

[5] Preti (1968).

[6] Vocabolario della lingua italiana (1986), vol. I. Treccani, Roma (see also www.treccani.it).

2 *Homo* Comes from *Humus*

"Humanity," in the sense of specific characteristic of the "humans" (or of "men," despite the "independence" of the adjective from the noun), is connected to *homo*, which would derive from the Indo-European name of the "land," which, in Latin, would be preserved in *humus*.[7] By reason of this derivation (*homo* from *humus*), *homo* (male and female) indicates, precisely, the "human being" as such or, properly, the "being born from the land" and thus the "terrestrial" creature (in opposition to "celestial," the quality of the gods, of those who live above or beyond the land[8]).

As a generic term, *homo* is distinguished, moreover, from *vir* (as, in ancient Greek, *ántropos* is distinguished from *anér*), which refers specially—in opposition to *mulier* and *femina*—to the precisely virile qualities of *uomini maschi* [male men] (whence also the meaning of *marito* [husband] or of *eroe* [hero], in the sense of "man worthy of this name," or also of *soldato* [soldier]; from *vir*, moreover, *virtus*, to designate "strength pure and simple," but later any "type of quality or merit").[9]

"Man," therefore, in a generic sense, is a "terrestrial creature" but, also, in a particular sense, is a "rational" creature (in opposition to *fera*, *bestia*) and, however "fallacious" or "fallible" (in opposition, once more, to the infallible gods), as well as "living" creature though "mortal" (in opposition, yet again, to the gods or immortals or, precisely, to *morti* [the dead].

From *humus*, furthermore, *humilis* (humble), he who remains at the level of the soil, almost flattened to the land, without lifting himself above it, and, in a moral sense, he who therefore does not become haughty, and, in Italian, *umiliare* (both in the sense of *abbassarsi* [lowering oneself] or *inchinarsi* [bowing oneself down] in a sign of respect and also *mortificare* [to mortify oneself] or *avvilire* [to debase oneself], or again, in a material sense, "inhume" or also "exhume," in reference to practices, arising from *pietas*, of depositing in the land (or of restoring to it) or, vice versa, *trarre i resti* [to draw the remains], as it is said, of a human person.

3 Land and Soil

Land, on the other hand, in Latin, besides *humus* is also *terra*[10] or, again, *tellus* (in a synonym of poetic type directed at personification or divinization): in this sense, it is opposed not only to "heaven" but also to "sea" or "water" and designates the planet or the globe or an inhabited part of it (like *regione* [region], from *regio*, a term used in the augural language to designate a space delimited by straight lines, or

---

[7] Ernout and Meillet (2001), p. 297 and p. 688 (1st edn 1932).

[8] Benveniste (1969), p. 420.

[9] Ernout and Meillet (2001), p. 739.

[10] Ernout and Meillet (2001), p. 688.

paese [village], from *pagus*, in the sense of rural territory delimited by cippus or boundary stones, whence also *pagano* [pagan]).

4 Around the Land

I would like to enumerate, in no particular order nor exhaustively, other words that, in Italian, we consider, in various ways, associable with the Italian *terra* [land], with their relative variations of meaning.

It goes without saying that the translation—an operation itself refractory to exactness—must take into account in a general way the unavoidably controversial character of human communication,[11] with the fact that this, most of the time, happens by "rustle"[12] and, precisely, with the weight of the mentalities, the ways of living and of conceiving and of representing the world, all subject to "perennial mutation."[13]

Besides complex terms—some already mentioned—like "globe" (or terrestrial "sphere"), "planet," *mondo* [world][14] (or even "continent," in the proper sense of continual land not interrupted by sea), or derivations like *terraferma* [mainland],[15] *retroterra* [hinterland], *entroterra* [inland], *sottoterra* [underground] and besides the generic *immobile* [property], I would point out, in reference to "surfaces," "terrain" (noun), "soil," "zone," "area" (or, also, "areal"), *pianale* ("floor" [plain], *spianata* [clearing], *pianura* [lowland]) and then *fondo* [acreage], *campo* [field], *campagna* [countryside], *podere* [estate], *agro* [countryside in the vicinity of a city], *appezzamento* [plot], *tenuta* [holding], *landa* [moor] (usually uncultivated), *radura* [glade] (usually without trees), *sodaglia* [fallow] or also, by extension, *gleba* [glebe] (from *zolla* [clod]) or *contado* [county], *contrada* [district] and, in relation to "territory"—which, in certain more recent usages (e.g., in the expression "government of the territory," see Art. 117 third paragraph, Italian Constitution), appears also promiscuously assimilated to *habitat*, *ambiente* [environment], or *paesaggio* [landscape][16]—or, rather, to its inhabited localities, obviously, *città* [city] and then *paese* [town] (already mentioned), *borgo* [borough] (or *borgata* [township]), *villa* [manor] (echoes in names like Francavilla, Villafranca, etc.),

---

[11] Giuliani (1965), pp. 281–287.

[12] Barthes (1984).

[13] Steiner (1975).

[14] Regarding geographic comprehension of the world, based that is on the reduction of the latter to a geographic map and with regard to the problem of the relations between "map" and "territory," Farinelli (2003), p. 3, and Farinelli (2009).

[15] In relation to which one can have, like a protrusion into the sea or like the extremity of a peninsula or continent, a *capo* [cape] (like that at Leuca or that at Buona Speranza).

[16] For the idea, instead, of geographic space as a social product or of the social production of space (with reference to H. Lefebvre), in the diachrony of the sovereignty–territoriality relations, Di Martino (2010), especially pp. 12 et seqq.

villaggio [village], *casale* [farmhouse], or, also, *pieve* [parish]. And, in reference to ways of communication, for men and livestock, there are the words *sentiero* [path], *mulattiera* [muletrack] (from *mulo* [mule]), *tratturo* [sheep trail], *trazzera* [dirt road]. Some derived adjectives, besides *terrestre* [terrestrial], are *terreno* [earthly] (e.g., *la vita terrena* [earthly life]), *terraneo* [terranean] (relative to the lands and thus, e.g., *medi-terraneo* [mediterranean]), *conterraneo* [fellow countryman] (one who comes from the same land), as well as *estraneo* [foreign] or *straniero* [foreigner] (one who comes from another land), *terrier* [landed], *terroso* [earthy], *terreo* [ashen] or, with Greek etymologies, of a different sound, autochthonous, indigenous, aboriginal.

5 Land as Mother Nature

The *terra* [land] (also *Gea*, *Gaia*), in a representation, moreover, close to myth, is, one would say by vocation, "mother" (as in "Mother Nature," from *mater* also *materia*): inasmuch as *fertile* (from *fero*), she generates and bears fruit and therefore, by reason of her fecundity, "nurtures" and "feeds"[17] and then also "raises" and preserves (or also, perhaps, "protects," in the sense that she welcomes and she makes to *abitare* [inhabit] (frequentative of *habeo*, originally understood in the sense of *tenere* [holding], then of *possedere* [possessing] or *occupare* [occupying], and finally of *avere* [having][18]).

She, therefore, on the part of men, is cultivated (for food) and is inhabited (for rest), and so also, in her turn, she makes them multiply and replenish (as, again, in the first book of the *Genesis*): "in the sweat of thy face," through work and effort, as well as those necessary to counter dangers, pitfalls, calamities, and devastating forces.

6 "Ager" and "Rus"

Terra recalls, at the same time, in Latin, two other words, probably used for a long time, that carry an uncoincidental expressive freight: (1) *ager*, in the sense of cultivated field (closed) (in opposition both to *urbs*, the city, and to *domus*, the home[19]), and (2) *rus*, in the sense of *campagna* [countryside] (open or free field, not

[17] As in the *Cantico delle creature*: "Laudato si', mi' Signore, per sora nostra matre terra, la quale ne sustenta et governa, et produce diversi fructi con coloriti flori et herba" (Praise be, my Lord, for our sister Mother Earth, who sustains and governs and produces different fruits with colorful flowers and grass).

[18] Ernout and Meillet (2001), p. 287.

[19] Benveniste (2001), pp. 226 et seqq.

necessarily cultivated) (even here, however, in opposition both to *urbs* and to *domus*).[20]

Ideally linked to *ager*, we have, in Italian, words like *orto* [vegetable garden] or *giardino* [garden] and to *rus*, words like *parco* [park], *riserva* [reserve], or *oasi* [oasis] (natural), while, in specialist languages, we have, *mutatis mutandis*, *unità fondiaria* [property unit] or *poderale* [farm unit] or even *colturale* [crop unit] or *aziendale* [business unit].

Summing up, with some stretching, we would say that, in the dimension of *ager*, the land, properly, is cultivated or, following Art. 44 Italian Constitution, the soil is exploited (making, typically, *agricoltura* [agriculture] and, in lands destined as *silvicoltura* [forestry, woods], in the alternative, naturally, from ancient times, to *pastorizia* [sheep farming][21]): the agronomic techniques aspire to optimize, in prevalently quantitative terms, the cultivation toward the goal of production,[22] and in the dimension of *rus*, the land, properly, is lived (on) (and one makes, if anything, *ruralità* [rurality] or *sviluppo rurale* [rural development][23]): diverse disciplines are integrated to improve the overall qualities of *vita in/di campagna* [life in or of the countryside].[24]

7 Agrarian and Rural

Again, on the subject of words and of the usages that can seem variably approximative or promiscuous[25]—besides the variations *diritto agrario/droit rural*, to indicate, in two different languages, a discipline that should be the same—it

[20] On the theme of the relation between "open fields" and "closed fields" (or walled or fenced fields, *en-* or *inclosures*), it is always well to start over in reference to experiences that are not archaic, by Bloch (1931); in Italy, by Sereni (1974) (1st edn 1961). For some profiles, relative to a specific geographical area, D'Elia (1959) [from the Annuario of Liceo-ginnasio statale "G. Palmieri" of Lecce, (1958–1959)], p. 41 on, inter alia, *chesure* [enclosures] and, p. 54, on *ortali* [little vegetable garden adjacent to the house]; and D'Elia (1982).

[21] Bloch (1931), pp. 28–29.

[22] Instead, on agriculture without production (already *set aside*: different, of course, the events of the uncultivated lands [*terre incolte*]), in relation to the maintenance of the lands "no longer used for production purposes" in "good agricultural and environmental condition," see Regulation (EC) 1782/2003 (Regulation of the Council that establishes common norms regarding systems of direct support in the area of common agricultural policy and institutes certain support systems in favor of farmers) and that modifies Regulations (EEC) 2019/93, (EC) 1452/2001, (EC) 1453/2001, (EC) 1454/2001, (EC) 1868/94, (EC) 1251/1999, (EC) 1254/1999, (EC) 1673/2000, (EEC) 2358/71, and (EC) 2529/2001), especially Art. 5, and Regulation (EC) 73/2009, especially Art. 6 (in Art. 146, Regulation (EC) 1782/2003 is repealed). Differently, now, Regulation (EC) 1307/2013 (Regulation of the European Parliament and of the Council establishing norms on direct payments to farmers within support systems provided for by the common agricultural policy and which repeals Regulation (EC) 637/2008 of the Council and Regulation (EC) 73/2009 of the Council).

[23] In this perspective, Albisinni (2000), pp. 421 et seqq.

[24] Graziani (2006b), pp. 71 et seqq.

[25] Fiorelli (2008) L'italiano giuridico dal latinismo al tecnicismo. In: Fiorelli P, pp. 71 et seqq. (original work published 1998).

seems significant that, in Italian language, we speak, for example, of "agrarian economy and policy" but, mostly, of "rural legislation"; of "agrarian mechanics (or also "hydraulics")" but of "rural valuation"; of "rural property" (or "rates" or also "tillage") but of "agricultural (or "agrarian") business (or undertaking)"; of "rural savings (or "chest") but of "agrarian credit""; of "agrarian landscape" but of "rural building (or "construction")"; of "agrarian reform" (or "rates") but of "rural classes"; of "agrarian universities" but of "rural communities"; of "rural sociology" but of "agrarian usages (or "customs")"; of "agricultural tax rates" or of "agricultural middle class" but of "rural family."

And it is also significant that we use the adjectives *agreste* [agrarian] and *rustico* [rustic] also with evaluative connotations (absent, nevertheless, from expressions like *fondo rustico* [rural fund] or *servitù rustiche* [rural easements], which can both approach *campestre* [country] or *selvatico* [wild], and of which remain, instead, prevalently immune *agrario*, *agricolo*, or *rurale*.[26]

On the other hand, the two words *agrario* and *rurale* seem then habitually reunderstood in one semantic space, which, referring equally to persons or things, includes many other terms:[27] together with *contadino* [farmer or peasant, "who stands in the "*contado*"" [county] or "*agricoltore*" [farmer], "*coltivatore*" [farmer], "*bracciante*" [labourer], "*zappaterra*" [land digger], "*zappatore*" [digger], "*mezzadro*" [sharecropper], "*colono*" [settler], "*massaro*" [steward], "*ortolano*" [gardener] or, less common, "*manente*" [tenant], "*oprante*" [day labourer], "*gualano*" [teamster], "*mesarolo*" [labourer hired monthly], "*pastinatore*" [seed drill operator], or names derived from agrarian contracts no longer in use, like "*livellario*" [recipient of a *livello* contract to farm land on certain conditions]; or other words, connotative, and sometimes derogatory, like "*contadinesco*" [peasant], "*campagnolo*" [yokel], "*villico*" [villager] o "*villano*" [villein], "*cafone*" [boor] (etymology uncertain), "*terrone*" [southern Italian], "*bifolco*" [bumpkin], "*zotico*" [hick], "*paesano*" [rustic] (also used in the sense of typical, characteristic, and, however, also, homemade, related to French *paysan*), almost synonyms of "*rozzo*" [rough], "*rude*" [coarse], "*grossolano*" [crass] "*scostumato*" [dissolute], "*incolto*" [uneducated, uncultivated], "*volgare*" [vulgar], "*caprone*" [billygoat] or, in some regional variations, "*burino*" [redneck], "*bùttero*" [cowboy], "*buzzurro*" [hillbilly], "*boattiere*" [oxherder], "*boaro*" [crude, rude] (linked to "*boarìa*" [agrarian contract] between the *boaro* and the landowner), "*forese*" [labourer], "*pacchiano*" [tawdry], "*tanghero*" [lout], "*tamarro*" [of low esteem]; or, especially in reference to the work of he who deals with livestock, "*bovaro*" [cowherd], "*vaccaro*" [herdsman], "*pecoraio*" [shepherd], "*capraio*" [goatherd], "*porcaro*"

[26] The adjective *rurale* [rural], in particular, is combined with nouns like "life," "civilization," "world," "environment," "landscape" and again with "community," "society," "population," "village" and with "economy," "policy," "development" and then with "area," "zone," "building," "street," "viability," and so forth.

[27] For many insights, Fiorelli et al. (1962).

[swineherd], "*cavallaro*" [horseman], "*mandriano*"[drover] "*stalliere*" [groom] "*barrocciaio*" (*birocciaio*) [carter], "*carrettiere*" [carter], "*carradore*" [carter] or, to other livelihoods, "*fattore di campagna*" [director of an agricultural business], "*campaio*" [watchman].

All these terms are used in opposition to who, living in the city in the condition of *cittadino* [citizen] (*civitas*/*urbs*),[28] uses, instead, *civile* [civil] or *urbano* [urban] (or also, in reference to the fact of living in a *borgo* [town], *borghese* [bourgeois], in the sense also of *ammodo* [nicely], *garbato* [genteel], *gentile* [kind], *cortese* [polite], *fine* [fine], *educato* [trained], etc., on the characteristics of *civiltà* [civilization] or *urbanità* [urbanity] brought about by the circumstance, precisely, of living (together) in the city.[29]

From this collection, one would gather, in a generic first and almost paradoxical approximation, not only that *coltivare* [cultivating] the land appears, as a rule, incompatible with becoming *colto* [cultured] but that, further, *cultura* [culture] (in the sense of "intellectual and moral awareness of oneself, of one's place, and of one's rights and duties with respect to society and humanity, of one's historical condition, based on a personal process of direct and indirect experiences"[30]) implies, for reasons connected above all with the use of time of life (and, therefore, of the relation *otium*/*negotium* = *quod non sit otium*), distancing oneself from the land, to dedicate oneself to nurturing the spirit or the mind. As has been said, oversimplifying or forcing the contrast, by which, in this perspective, either one works, and therefore does not think (in the sense that one has no time for thinking) or one thinks, and therefore does not work (in the sense that one has no time for working).[31]

In a successive approximation, one could also derive, with more forcing, that "the construction of social reality"[32]—inasmuch as it requires reflexive or introspective activity—is undertaken, by choice, in places, as far as possible demarcated, of "inside" (of one's own interior, of the house, of the city) and not in spaces, by definition uncertain, of "outside": as if the latter were unavoidably "dissipating" or of heightened entropy, or anyway required, to achieve comparable results (for example, the in "closed" fields), an employment and a waste of energy vastly superior.

[28] Crifò (2000), pp. 26–27. Still on the relations between *pólis* and *urbs*, Benveniste (2001), vol. 1, p. 281. For the events of the medieval and modern world, Costa (1999–2001).

[29] Silone (1970), p. 20 (original work published 1930). Among the many studies, Braudel (1979), p. 450.

[30] De Felice and Duro (1993), p. 523.

[31] Emblematic, still, Dewey (1929), p. 4.

[32] Searle (1995).

8 Country Life

Life in the countryside "joins" or "includes," in the sense that it puts into relations, on a substantive plane, events or activities or persons, whereas the city "separates" or "excludes," in the sense that it organizes the factors of existence on a formal plane above all, parceling or breaking up or specializing (in terms of competence) the lives of the persons.

In the countryside, the gaze, while firmly fixed on the land on which one's feet are placed, pushes naturally to the distance: the imminence of rain determines decisions and behaviors that would not even be imagined in the city. If, on the other hand, a misfortune occurs in the city, one turns to the competent expert; if the misfortune happens in the countryside, one attempts, first of all, to repair it oneself. It is not merely wishful thinking or stinginess but, *lato sensu*, the mentality or the economy of the organization.

Moreover, on communicative events, from the distances between the plants of an olive orchard or from their bearing or from their foliage, we will be able to extract information, besides the techniques of planting and pruning, in general, of crops, also, so to say, of possible "sociologies" of the harvest: distant rows, or nonrows, recall images of premodern types; persons instead of machines; long times and slow and cadenced rhythms in the periodicity of operations, labor, and furthermore mere subsistence; oil, foreseeably, with high acidity.[33]

9 Peasantry and Town People

The hypothesis of a "rural humanism" (rather than a "rural anthropology")—derived, with provisional and conventional emphasis, for willing interlocutors, from that on the man–land relations—is linked to the perception that men of any kind involved in regular relations with the countryside (agritourism seems, rather, a metropolitan type of variation) are identifiable—completely independently of their "social position," as that identified—according to specific qualities or capabilities, not reducible to those of "citizens."[34] Quality and capability mean, especially, regardless of evaluations, mentality and sensitivity, modes, therefore, of conceiving and feeling life and the world, besides themselves: different ways of perceiving time and space, as well as "using them" or "occupying them" with the permissible stability.[35]

Naturally, it would be senseless to envisage the physiognomy of an abstract "man of the countryside" or even the model of an ideal "life in the countryside." We can only try to sketch, almost facetiously, some traits that, on the basis of

[33] For a witness, Bacile di Castiglione (1873).

[34] Silone (1970), p. 40.

[35] So also taught in conversation, Eugenio Cannada-Bartoli.

experience, also literary, appear constant or characteristic, for as much as we can succeed in knowing, of propensities, tendencies, or attitudes.[36]

10 For a Rural Humanism

A "rural humanism" seems, above all, intrinsically "whole": "whole" not in the sense of Maritain (for whom one cannot "propose to the man only the human," with the risk that he becomes "inhuman")[37] but rather in an elementary sense or, so to speak, domestic sense, that of unrefined flour, which conserves all its own components.

In this perspective, every single fragment of experience (which is always, properly, an ensemble of experiences) contains, to scale, or in proportion, the same structure or conformation of the relative "complex" (of experience):[38] in the countryside, for example, a work day, independently of contractual or union listings, does not appear, at most, divisible in "occupied" time (precisely, by labor) and "free" time (from that same labor). In the countryside, time is, qualitatively, always "continuous" (and therefore generic, that of the duration) and, furthermore, also "punctual" (and therefore specific, that which is "correct" or "appropriate") so that, in its fractions, there is, contemporaneously, space for a labor while one rests and for rest while one labors. There is always something to do; it doesn't matter, or it's not clear whether for necessity or for fun or for both.

In this experience, emphasizing and generalizing, it is as if someone carried everything back or as if, ultimately, everything regarded everything, or almost, without solutions of continuity or without possibility of contrived mutilations, in senses most diverse and also contradictory: and so, for example, it is always immediately clear, and even implicit, that "in life nothing is free, and nothing is sure"[39] and

---

[36] Among the many authors, including classics, de Balzac H (1855, unfinished) Les paysans. Scènes de la vie de campagne or, in Italian literature (southern or *meridionalistica*) of the twentieth century, Levi C, Jovine F, Scotellaro R, and also, in his genre, Calamandrei (1989) (original work published 1939–1941). Not completely outside of this context appears, moreover, that scattered literature, often biographical or autobiographical, between the nineteenth and twentieth centuries, about the experience of farmers lived by men like Cavour, Jacini, the Fortunato brothers, De Viti de Marco, Einaudi (among many less well known: some trace in the lists of the universal Expositions, e.g., Torino 1884). And neither seem negligible, in this regard, that various intellectual tradition or those sensitivities and competencies, which, more or less directly, appear attributable, in southern Italy, to the lesson of the abbot Genovesi (already a professor of metaphysics and then, from 1754, of "civil economy") and, in northern Italy, to those of Romagnosi and Cattaneo (which, among others, permit connections between statistics and history, for a nonintellectualist study of social phenomena). Humanisme intégral: problèmes temporels et spirituels d'une.

[37] Maritain (1936), p. 15.

[38] On this expression, used also in the plural, Orestano (1987), p. 370.

[39] Capograssi (1959), p. 307.

that "he who doesn't work doesn't eat";[40] that you take things as you find them, without throwing away (or being able to throw away) anything; that any "thing" can suddenly become useful or relevant; that you don't congratulate yourself and there's no time to waste; that you are not, ultimately, the master of your "own" time, not being able, sometimes, to anticipate or hurry and, other times, to postpone or slow down; that you are not even the master of your choices, having to consider—to the ends, for example, of planting a plant or rotating crops—climate, exposure to sunlight and winds, humidity, chemical-physical properties of the land, available water, and so many other things still: that, besides producing, you need above all to think about earmarking the products, in relation to the characteristics or dimensions of the markets, the alimentary habits and tastes of the persons, novelties, or traditions; that indifference or inattention can constitute a luxury as much as scruples and fussiness; that isolating yourself can be as mistaken as crowding; and so on.

Here we get to the point *festina lente*:[41] concreteness, good sense, urging not to scatter, express, precise, a "natural" propensity to inclusion and integration. Cuts (de-cisions, from the Latin *de-caedo*), being traumatic events, are adopted as last resort. Here, moreover, generally, you doubt, but then you trust: wary but willing, keeping back but risking.

Causations appear necessarily complex and itineraries multiple or not necessarily linear nor progressive, entrusted to the exclusiveness of oral language or that of symbols or to the effectiveness of concluding behaviors: the information is valid for those present and participating, not for those absent or alien, though they might seek it.

Reasoning (*cervello fino* [pure brain]), which is an expression of "practical intelligence" rather than "calculable" rationality, becomes, therefore, spontaneously "topical," anchored in the "here and now": but the empirical foundation, or pragmatic setting, instead of an abstract choice of "method," rather appears as a further extension of experience of one's own or another, taking place in the wake of know-how passed down by experience and knowledge acquired by imitation or emulation and consolidated by custom: that is, again, by experience.

11 Land and People: "Constitutional" Features of Living Together

On these bases, the "issue of the land" ends up regarding, under a juridical view, not only, nor even mainly, *in rebus* relations, the relations, that is, between men and things,[42] whatever their legal arrangements, but also, properly, the "personal" relations, namely the multiple relations that, relatively to the things (especially

---

[40] Saint Paul, *Second letter to the Thessalonians*, 3, 10.

[41] For another "classic," Ridolfi (1868), especially vol. I, p. 21 (original work published 1857–1858).

[42] For a reference almost of "juridical archeology," Guiraud (1893) (Guiraud was the pupil and biographer of Fustel de Coulanges ND).

concerning immovable), are established and developed among the members of the communities, determining, as it were, their disciplines.[43]

In other words, beyond the merely "manorial" or "assets ownership," individual or collective[44] or "obligatory" profiles in agrarian relations—traditionally entrusted, by us, the one and the other, to the doctrine of "private" and "agrarian" law[45] (mostly in reference to codified norms or so-called special laws, albeit progressively revised in the constitutional or "communitarian" context)—the issue assumes a specific and important relevance on the "constitutional" plane: not already, of course, only in the sense of written dispositions in the "constitutional papers" (or in the "fundamental laws")[46] or of "constitutional law" conceived as the subject of specialists, versed in the study above all of "organizational" dynamics, but also of "constitutions" in the proper sense, meaning the many and various modes in which communities constitute themselves and live together; or of the various processes through which, besides the institutional engineering, the living together continually takes and retakes form; or of the identities that they variably express in relation to their "histories."

As a manifestation of the *status salutis*, of tendencies, inclinations, sensitivities, and even feelings, or, in a word, of qualities, slowly and, mostly, silently sedimented in the life of the communities, the constitutions express, as the languages—a phenomenon, *par excellence*, anonymous of the speakers—not only the fundamental rules of living together but also the living together itself and its deepest characteristics: that which is beyond the limits of writing, which could be a simple, though meaningful, transcription. "Deepest characteristics" means connotations impressed, like genetic traits, into the very folds of behaviors, or of nonbehaviors, or in the most stored or least attended corners of the "common conscience" or of the "collective memory," in an even prelogical dimension of experience.[47]

It is obvious that these characteristics can be considered subsistent, inasmuch as they become perceived, recognized, named, narrated. The problem is how.

---

[43] Spantigati (2007), pp. 119–136.

[44] For everyone—within the framework of a long and uninterrupted critical remeditation of many "mythologies" (it is in the 1977 "Un altro modo di possedere" and 20 years earlier the study on the Abbazie benedettine)—Grossi (1992) La proprietà e le proprietà nell'officina dello storico. In: Grossi P, pp. 603 et seqq. (original work published 1988 and republished 2006, with a prologue entitled *Venti anni dopo*, by ESI, Napoli).

[45] In the fervor of the doctrinal debate in Italy in the 1930s—and of the many concrete initiatives for "reform" in the sector (especially, "land reclamation")—these are reported, in particular, in the contributions of Bolla G, Finzi E, Maiorca C, Pugliatti S, Vassalli F, as well as of Arcangeli A and of Maroi F: on this subject, Jannarelli (2006), pp. 39 et seqq.

[46] Among the first comments, yet on the project of the Constitution, Mortati (1944–1947), pp. 3–13 (response to an inquiry on land ownership in the constitutional reform of the Italian State), and then Esposito (1949), pp. 157 et seqq. (republished in Esposito (1954), pp. 181 et seqq.), and also Mortati (1954), pp. 262 et seqq.

[47] Giuliani (1984), pp. 101 et seqq.

Now, excluding that these operations can be performed through deliberations adopted by a majority, whatever criterion of calculation be used, and excluding, above all, that the "fundamental things" can be put to a vote, since they are destined to remain, forever the fundamentals, invisible and therefore, most of the time, undecidable and ineffable, or at least not exactly, it will be necessary to identify forms and modalities, and not simply procedures, that allow us, within the limits of the possible, to decipher and bring publicly to light this sort of unavailable heritage of identity and to interpret it, evaluate it, discuss it with the necessary pondering and with all possible wisdom.

It will be necessary, naturally, in the first place, to contrast the constant danger and the deception of fraud and falsehood, bullying, arrogance, violence in the thousand possible forms. It will never be easy nor vested.

But the impression is that, in the final analysis, after having adopted and tried even the most sophisticated formal remedies, we cannot but return, in one way or another, to trust in "culture," in the capacity to see and to understand experience and therefore to cultivate it as if it were land, to insistently work it, and then to work it again so that it does not stop being fruitful.

For this, we will not be able, in conclusion, to do more than trust: men more than simple procedures—those who, in the various, most appropriate venues, seem to us to know how to really see and understand and who we ourselves know how to recognize and choose as capable, for their qualities, not their competences.

12 About Community

It is almost a postscript, to conclude. Speaking of *convivenza* [living together], it has been almost automatic to use the word "community," which in common parlance—and notwithstanding the endless reflection of all times—is, mostly, considered equivalent to "collectivity" or even to "society,"[48] in reference to images abstractly representative of a "we," no matter how extensive.

"Community"[49] would, without theoretical pretense, recall the idea of a discrete and not occasional ensemble of persons who share something without having

[48] In the text of the Italian Constitution—as it is easy to see—the nouns "community" and "collectivity" appear each only once (respectively, in Art. 43, in the "community of laborers or users," and in Art. 32, in health also as "interest of the collectivity"); "society" appears twice (in Art. 4, in the duty of the citizen to undertake activities or functions that contribute to the "material or spiritual progress of society," and in Art. 29, in the family as "natural society based on marriage"). As far as the subjective perspective and reference nouns, the most often used is "citizen" (Arts. 3, 4, 16, 17, 18, 26, 38, 48, 49, 50, 51, 52, 54, 59, 75, 84, 102, 118, and 135), the least is "individual" (Art. 32). Also found are "man" (Arts. 2, 48, 51, and 117), "single" (Arts. 2, 18, and 118), and, a little more frequently, "person" (Arts. 3, 32, 111, 119, and 120).

[49] Community comes from *munus*. *Munus* expresses the idea of duty, charge, or office or of official charge, but it implies, at the same time, that of exchange, in the sense of give in change: according to a secondary but very frequent meaning, the gift from the side of he who does (Ernout and

stipulated to it, as perhaps happens in aesthetic experience. "Share" in the sense that they have or "feel" or "discover" that they have something in common, above all "immaterial," "have" not in an ownership or appropriative sense and "in common" not each one *pro quota* but, as far as possible, for the whole. Persons are those who "know" each other or, rather, "recognize" each other as being born, perhaps, or as living in the same place (*patria*, the land of the fathers; "nation," from *natio*, derived from *nasci*, *nascere* (to be born)) or in the same period of time or as speaking the same language or attributing meaning to the same symbol (e.g., a flag or a hymn) or for having lived together, even without being aware of it, through apparently insignificant experiences (studying the same books, hearing the same songs, eating or dressing in the same way, etc.). "Together" implies not by chance and neither artificially aggregated or organized group of "individuals" (*átomoi*, that is, indivisible) brought together, in representation, in a dimension that is authentically "public": "public" (which would recall *populus*) in the sense of "ultra-individual" or "inter-individual" and, therefore, "open," "communicating," "plural," "multiform," that is to say, the opposite of "closed," "detached," "reserved," "exclusive," or, even concisely, "private" (from *privare* [deprive] or, precisely, *render privo*).[50]

The "places," in their uniqueness, the lands, appear as a formidable "glue," so also the times, the impermanence. But all these have outlined that are completely approximate: they change, at least by reason of the contemporary relevance with respect to different planes of reference and of the plural belongings of each. The community of a town, autonomous and exclusive of itself, is nevertheless, in some way, also part of a broader community of a region, as in a progression of concentric circles; the community of the living in a given moment includes, under many and various forms, also those who, in the meantime, are no longer living, besides those who are not living yet.

These are ensembles that are unavoidably open, communicating, and in movement. The watchword, if we remain in a dialogic perspective, becomes "reciprocity": which, according to its etymology, expresses precisely the idea of going backwards and then forwards,[51] like the sea on the shore. Dynamic relations are conceived, or discovered, as mirrors: capable, that is, of linking the protagonists through continual rebounds in one direction and in the other, motions interfering even when not contemporary or of the same intensity, so what happens in one part provokes appreciable effects also in another part.

Meillet (2001), p. 422), which nevertheless obligates the recipient to an exchange (Ernout and Meillet (2001), p. 422; Benveniste (2001), vol. I, p. 71). It is developed right from the etymology—as starting point or hermeneutic point to escape from the dialectic of the debate on the community in modern political philosophy—in the philosophical study of Esposito (2006).

[50] Dewey (1927).

[51] Ernout and Meillet (2001), p. 566.

References

Albisinni F (2000) Ruralità come regola di diritto per uno sviluppo sostenibile. In: Various Authors, Agricoltura e diritto. Scritti in onore di Emilio Romagnoli, vol I. Giuffrè, Milano, p 421

Bacile di Castiglione F (1873) Intorno all'usanza di estrarre l'olio da olive riscaldate. Poche considerazioni proposte ai signori proprietari della provincia di Lecce. Tipografia Editrice Salentina, Lecce

Barthes R (1984) Essais critiques, vol 4: Le bruissement de la langue. Paris, Editions du Seuil. Italian edition: Barthes R (1988) Il brusio della lingua. Saggi Critici, IV (trans: Bellotto B). Einaudi, Torino

Benveniste E (1969) Le Vocabulaire des institutions indo-européennes. Paris, Les editions de Minuit. Italian edition: Benveniste E (2001) Il vocabolario delle istituzioni indoeuropee, vol 2 (trans: Liborio M). Einaudi, Torino

Bloch M (1931) Les caractères originaux de l'histoire rurale française. Les Belles-Lettres, Paris. Italian edition: Bloch M (1973) I caratteri originali della storia rurale francese (trans: Ginzburg C). Einaudi, Torino

Braudel F (1979) Civilisation matérielle, économie et capitalisme (XV-XVIII siècles). Vol 1: Les structures du quotidien: le possible et l'impossible. A. Colin, Paris. Italian edition: Braudel F (1982) Civiltà materiale, economia e capitalismo (secoli XV-XVIII). Vol 1: Le strutture del quotidiano (trans: Vivanti C), 2nd edn. Einaudi, Torino

Calamandrei P (1989) Inventario della casa di campagna. Vallecchi, Firenze

Capograssi G (1959) Agricoltura, diritto, proprietà. In: Capograssi G (ed) Opere, vol V. Giuffré, Milano, p 269

Careri G (1982) Ordine e disordine nella materia. Laterza, Roma-Bari

Costa P (1999–2001) Civitas. Storia della cittadinanza in Europa, vol I-IV. Laterza, Roma-Bari

Crifò G (2000) Civis. La cittadinanza tra antico e moderno. Laterza, Roma-Bari

D'Elia M (1959) Aspetti della tecnica delle confinazioni della proprietà terriera dell'agro galatinese (Terra d'Otranto) nel Quattrocento, alla luce dei riflessi linguistici. "Terreni magnoligiati". Adriatica Editrice Salentina, Lecce

D'Elia M (1982) Una denominazione del 'maggese' nei dialetti dell'Italia meridionale. Adriatica Editrice Salentina, Lecce

De Felice E, Duro A (1993) Vocabolario italiano. S.E.I.-Palumbo, Torino-Palermo

Demetrio D (2013) La religiosità della terra. Una fede civile per la cura del mondo. Raffaello Cortina, Milano

Dewey J (1927) The public and its problems. H. Holt and Company, New York

Dewey J (1929) The quest for certainty: a study of the relation of knowledge and action. Minton-Balch, New York. Italian edition: Dewey J (1966) La ricerca della certezza. Studio del rapporto fra conoscenza e azione (trans: Becchi E, Rizzardi A). La Nuova Italia, Firenze

Di Martino A (2010) Il territorio: dallo Stato-Nazione alla globalizzazione. Sfide e prospettive dello Stato costituzionale aperto. Giuffré, Milano

Ernout A, Meillet A (2001) Dictionnaire étymologique de la langue latine, 4th edn. Klincksieck, Paris

Esposito C (1949) Note esegetiche sull'art. 44 della Costituzione, in Rivista di diritto agrario 28 (1):157 et seqq

Esposito C (1954) La Costituzione italiana. Saggi. Giuffré, Milano

Esposito R (2006) Communitas. Origine e destino della comunità, 2nd edn. Einaudi, Torino

Farinelli F (2003) Geografia. Un'introduzione ai modelli del mondo. Einaudi, Torino

Farinelli F (2009) La crisi della ragione cartografica. Einaudi, Torino

Fiorelli P (2008) Intorno alle parole del diritto. Giuffré, Milano

Fiorelli P, Bandini M, Grossi P (eds) (1962) Bibliografia del diritto agrario intermedio, vol I: Gli studi. Giuffré, Milano

Giuliani A (1965) Stampa e opinione pubblica. Rivista trimestrale di diritto e procedura civile 29 (1):281–287
Giuliani A (1984) Droit, mouvement et réminiscence. Archives de philosophie du droit 29:101 et seqq
Graziani CA (2006a) Conserviamo la nostra terra (per una nuova iniziativa). Ritorno al diritto 2 (3):172 et seqq
Graziani CA (2006b) Proprietà della terra e sviluppo rurale. Diritto romano attuale 8(16):71 et seqq
Graziani CA (2011) Terra bene comune. Studi parlamentari e di politica costituzionale 44 (173–174):11 et seqq
Graziani CA (2013) La terra: oggetto di possesso o bene comune? Archivio Scialoja-Bolla 11:1–35
Grossi P (1977) 'Un altro modo di possedere'. L'emersione di forme alternative di proprietà alla coscienza giuridica postunitaria. Giuffré, Milano
Grossi P (1992) Il dominio e le cose. Percezioni medievali e moderne dei diritti reali. Giuffré, Milano
Guiraud P (1893) La propriété foncière en Grèce jusqu'à la conquête romaine. Hachette, Paris
Jannarelli A (2006) Il dibattito sulla proprietà privata negli anni Trenta del Novecento. Diritto romano attuale 8(16):39 et seqq
Maritain J (1936) Humanisme integral: problèmes temporels et spirituels d'une nouvelle chrétienté. Aubier, Paris. Italian edition: Maritain J (1946) Umanesimo integrale (trans: Dore G). Studium, Roma
Mortati C (1944–1947) Indirizzi costituzionali nella disciplina della proprietà fondiaria. Rivista di diritto agrario 23–26(1):3–13
Mortati C (1954) La Costituzione e la proprietà terriera. In: Orlando Cascio S (ed) Atti del terzo convegno di diritto agrario, Palermo 19–23 ottobre 1952. Giuffré, Milano, p 262
Orestano R (1987) Introduzione allo studio del diritto romano. Il Mulino, Bologna
Preti G (1968) Retorica e logica. Le due culture. Einaudi, Torino
Ridolfi C (1868) Lezioni orali di agraria, vol I–II, 3rd edn. Cellini & C., Firenze
Searle JR (1995) The construction of social reality. Free Press, New York. Italian edition: Searle JR (2006) La costruzione della realtà sociale (trans: Bosco A). Einaudi, Torino
Sereni E (1974) Storia del paesaggio agrario italiano, 2nd edn. Laterza, Bari
Silone I (1970) Fontamara. Mondadori, Milano, 1970
Spantigati F (2007) La terra e il suo regime giuridico. Agricoltura istituzioni mercati 4(1):119–136
Steiner G (1975) After Babel: aspects of language and translation. Oxford University Press, Oxford. Italian edition: Steiner G (1994) Dopo Babele. Aspetti del linguaggio e della traduzione, 2nd edn (trans: Bianchi R, Béguin C). Garzanti, Milano

MIX
Papier aus verantwortungsvollen Quellen
Paper from responsible sources
FSC® C105338

If you have any concerns about our products,
you can contact us on
ProductSafety@springernature.com

In case Publisher is established outside the EU,
the EU authorized representative is:
Springer Nature Customer Service Center GmbH
Europaplatz 3, 69115 Heidelberg, Germany

Printed by Libri Plureos GmbH
in Hamburg, Germany